Springer Proceedings in Mathematics & Statistics

Volume 92

For further volumes:
http://www.springer.com/series/10533

Springer Proceedings in Mathematics & Statistics

This book series features volumes composed of select contributions from workshops and conferences in all areas of current research in mathematics and statistics, including OR and optimization. In addition to an overall evaluation of the interest, scientific quality, and timeliness of each proposal at the hands of the publisher, individual contributions are all refereed to the high quality standards of leading journals in the field. Thus, this series provides the research community with well-edited, authoritative reports on developments in the most exciting areas of mathematical and statistical research today.

Christoph Bandt • Michael Barnsley
Robert Devaney • Kenneth J. Falconer
V. Kannan • Vinod Kumar P.B.
Editors

Fractals, Wavelets, and their Applications

Contributions from the International Conference and Workshop on Fractals and Wavelets

Springer

Editors
Christoph Bandt
Institut für Mathematik und Informatik
Universität Greifswald
Greifswald, Mecklenburg-Vorpomm.
Germany

Michael Barnsley
Mathematical Sciences Institute
Australian National University
Canberra, Australia

Robert Devaney
Math Department
Boston University
Boston, MA, USA

Kenneth J. Falconer
Mathematical Institute
University of St Andrews
St Andrews, Fife, UK

V. Kannan
Department of Mathematics
and Statistics
University of Hyderabad
Hyderabad, India

Vinod Kumar P.B.
Department of Basic Sciences
and Humanities
Rajagiri School of Engineering
and Technology
Kerala, India

ISSN 2194-1009 ISSN 2194-1017 (electronic)
ISBN 978-3-319-08104-5 ISBN 978-3-319-08105-2 (eBook)
DOI 10.1007/978-3-319-08105-2
Springer Cham Heidelberg New York Dordrecht London

Library of Congress Control Number: 2014949263

Mathematics Subject Classification (2010): 28A80, 81Q35, 42C40, 65T60

Printed on acid-free paper

Springer is part of Springer Science+Business Media (www.springer.com)

Preface

Fractal geometry is a young field. It was initially developed in the 1980s, driven by the motivation to model rough phenomena in nature, and by new opportunities of computer visualization. Towards the end of the 1980s, wavelets were introduced for the needs of signal and image processing. Today, the field of fractals and wavelets has grown into a respected mathematical discipline with specific concepts and techniques, and with plenty of applications inside and outside mathematics.

In November 2013 a workshop and the first International Conference on Fractals and Wavelets in India took place at Rajagiri School of Engineering and Technology, Kochi, Kerala.

In the workshop, from November 9 to 12, leading experts from all over the world gave comprehensive survey lectures on the state of the art in their areas. In the International Conference from November 13 to 16, new research results were presented by mathematicians from ten countries. There were more than 100 participants from India, revealing that research in fractals and wavelets has taken root at many Indian universities, with an emphasis on applications to engineering, medicine, Internet traffic, hydrology, and other fields.

This volume contains all invited lectures of the workshop as well as selected contributions to the conference. Providing readable surveys, it can be used as a reference book for those who want to start work in the field. It documents the present state of research in the area, both in India and abroad, and can help to develop cooperation among widely scattered groups.

The organizers of the conference would like to thank the management of Rajagiri School of Engineering and Technology, Cochin, Kerala, India for the inspiration and support provided to conduct the conference.

The organizers acknowledge the financial support given by the International Centre for Theoretical Physics, the International Mathematical Union, the International Council for Industrial and Applied Mathematics, the National Board for Higher Mathematics India, the Department of Science & Technology India, the Defence Research & Development Organisation India, the Indian National Science Academy, the Kerala State Council for Science Technology & Environment, and The South Indian Bank Limited.

Contents

Part I Fractal Theory

Introduction

This book is divided into three parts: Fractal Theory, Wavelet Theory, and Applications. Each part begins with survey papers written for a general audience, followed by surveys on more advanced topics and by contributed papers presenting recent results.

In the first part, C. Bandt gives an introduction to basic fractal concepts and methods, followed by an introduction to self-similar sets. Self-similar sets are generated by similitudes and form the most simple class of fractals. Another important class, known from appealing computer visualizations, is generated by iteration of polynomials, rational functions, and entire functions of a complex variable. S. Sutherland presents the theory of Julia and Fatou sets, and R.L. Devaney discusses their topological intricacies.

Four other invited lectures provide new concepts and techniques which were developed by the authors. The concept of fractal homeomorphism is introduced by M. Barnsley, B. Harding, and M. Rypka. Dimension results on self-affine sets and measures are simplified by K. Simon by introducing the concept of almost self-affine set. M. Urbański treats the more complicated class of self-affine sets over an infinite alphabet, and A. Tetenov studies projection and rigidity properties of fractal curves in n-dimensional space.

The contributed lectures of Part I deal with new three-dimensional fractals, projections of Mandelbrot percolation sets, and approaches to fractals in more general topological spaces.

Wavelet Theory and fractal functions are studied in Part II. Roughly speaking, wavelets are basis functions with self-similarity properties which ensure an efficient coding of signals and images. General bases in Hilbert spaces called frames are the fundamental concept here. O. Christensen gives an introduction to frame theory. In a second lecture with Hong Oh Kim and Rae Young Kim he presents recent trends and open problems in the field. P.R. Massopust introduces a new class of fractal functions, using the new concept of a local iterated function system.

The contributed talks of Part II concern a variety of different constructions of fractal functions and wavelets with good approximation properties, such as preservation of convexity.

Part III starts with an invited lecture of N. Cohen, the inventor of fractal antennas. Taking examples from his field, he discusses the problems and difficulties which arise on the way before new inventions can be implemented into practice.

The contributed lectures in this part deal with application to cancer detection and brain signal analysis, chemical engineering and hydrology, Internet traffic, image processing, and tomography. They illustrate the rapid development and wide range of applied fractal research in India.

Part I
Fractal Theory

Introduction to Fractals

Christoph Bandt

Abstract This non-technical introduction tries to place fractal geometry into the development of contemporary mathematics. Fractals were introduced by Mandelbrot to model irregular phenomena in nature. Many of them were known before as mathematical counterexamples. The essential model assumption is self-similarity which makes it possible to describe fractals by parameters which are called dimensions or exponents. Most fractals are constructed from dynamical systems. Measures and probability theory play an important part in the study of fractals.

Keywords Fractal • Self-similarity • Box dimension

1 Mandelbrot's Vision of Fractals

1.1 Potential Applications

Benoit B. Mandelbrot coined the term "fractal" and created fractal geometry with his groundbreaking monography "The Fractal Geometry of Nature" [5] in 1982. He begins this work with some words which have become famous: "Clouds are not spheres, mountains are not cones, coastlines are not circles, and bark is not smooth, nor does lightning travel in a straight line."

Workshop on Fractals and Wavelets at Rajagiri School, Kochi, India, 9 Nov 2013.

C. Bandt (✉)
Institut für Mathematik und Informatik, Universität Greifswald, 17487 Greifswald, Germany
e-mail: bandt@uni-greifswald.de

C. Bandt et al. (eds.), *Fractals, Wavelets, and their Applications*, Springer Proceedings in Mathematics & Statistics 92, DOI 10.1007/978-3-319-08105-2_1,

Mandelbrot clearly saw the need of modelling irregular phenomena in science and economy, since he had worked on this field for many years. He knew that traditional methods do not work—a view which the majority of his colleagues did not share. But his main message was positive: there are mathematical concepts which can be applied to all of these phenomena.

Meanwhile fractal geometry is established as a mathematical area with deep theorems and exciting intrinsic problems. Nevertheless, we must not forget that *it is the diversity of potential applications which makes our field so attractive.*

1.2 Mandelbrot's Way

As a rule, new ideas and their inventors do not get accepted right away. Actually, Mandelbrot got his first tenure professorship at the age of 75. In his autobiography [6] which appeared after his death in 2010, he characterizes himself as a "scientific maverick." He was born in 1924 in the Jewish quarter of the Polish capital which became known as Warsaw ghetto during the Nazi occupation in World War II. Many relatives, all neighbors and friends of his childhood were killed by the Germans. Fortunately, his family emigrated to France before the war. When Nazi occupation came to France, Mandelbrot had to cover his identity and live under continuous threat for several years.

After the war, Mandelbrot became a mathematics student, proved his exceptional geometrical talent, and gained scholarships at elite universities in France and the USA. But then, instead of joining mainstream research, he became interested in various obscure phenomena and strange applied problems. Mandelbrot had to struggle 25 years to get recognized. In 1975, he wrote the initial French version of his book which collected his views and results. Physicists started to accept and apply his ideas. And then, within few years, fractals became very popular, and Mandelbrot got a lot of honors.

2 Fractals in Contemporary Mathematics

Before we come to details, let me point out some personal views. In my opinion, fractals form one important facet in the development of twenty-first century mathematics.

2.1 Classical Mathematics

Classical mathematics, centered around analysis, was triggered by applications in astronomy, physics, and engineering, by problems with a moderate amount of data. Some ingenious ideas of Gauss are difficult to comprehend even today, and the

Riemann hypothesis is still not solved—but most questions with impact outside mathematics had a relatively simple structure. Up to the middle of the twentieth century we had no calculators. All numerical calculations were done by hand, with the help of tables. Nevertheless, classical mathematics was a driving force for the development of all our achievements in science and technology: mechanical watches, cars, railways, the atom bomb, the first computers, and the first studies on global change. Mathematics also played a leading role in education. It was considered necessary to understand the modern world.

2.2 New Challenges

Now we live in a world which is extremely complex and difficult to understand. Nobody can oversee all structures he/she is involved in. Computers and huge amounts of data are virtually everywhere. Life sciences, economy, and climate research investigate processes of incredible complexity. They cannot do anything without advanced mathematical methods. At the same time, public reputation of mathematics has shrunk, and mathematical communication and education are in a worldwide crisis.

In my mind, the basic challenge of today's mathematical education is to give people an orientation in a complex world. People must find strategies to comprehend and influence their environment. We must remain masters of computers and not become their slaves. We have to decide the essential things and leave routine work to the machines.

Mathematics has the chance to shape the future, in research as well as in education. Tremendous efforts are required to meet this challenge. Communication practice and curricula must be thoroughly revised and changed. We need classical mathematics as well as new concepts and techniques.

Fractals and networks are among the new concepts. It is no surprise that both are strongly connected with computers. Complex networks became a research field around 2000 in connection with the fast development of the world wide web. The study of fractals was greatly enhanced by the development of computer graphics facilities in the 1980s. Mandelbrot was a long-term fellow at IBM, the leading computer enterprise in the time of mainframe computers.

3 Self-similarity

3.1 Fractal Symmetry

There is no precise mathematical definition of the word "fractal". Most experts agree that negative properties like "something very strange" or "very irregular," or Mandelbrot's first definition "sets with non-integer dimension" are not helpful. The essential property is self-similarity: *small parts and big parts of the figure look similar.* When we see a piece of the figure, we cannot conclude where we are, nor can we say something on the size of the piece. This property is also called *scale-invariance.* The structure of a fractal is nearly the same at every corner, on large scale and on small scale.

Self-similarity is a kind of symmetry which simplifies the analysis of fractals. Every kind of symmetry simplifies problems. To determine the volume of a solid, for instance, we need a triple integral, but for a solid of revolution, we need only a single integral. The real line, one of the fundamental sets in mathematics, is symmetric with respect to translations and reflection. It is also self-similar in the above sense, so it can be called a fractal. The same holds for $\mathbb{R}^n$.

3.2 The Benefit of Self-similarity

In order to study the whole structure of a fractal, it is enough to study small pieces, because the structure is everywhere similar. An everyday example is the process of distribution of public money. When you know how this works in small scale, in a town or university, then you can become a minister—since the mechanisms and problems in the government are similar, only at larger scale.

When we consider a cloud, a mountain scenery, a satellite image of a coastline, a tree or lightning, we can agree that self-similarity is present, at least to some extent. So the definition should apply to reality. However, self-similarity is not a property of nature. It is a *model assumption,* like the concept of a straight line or a circle. In practice, lines are never infinite and they are never straight, but calculations with lines have been successful. When we find sufficiently simple theoretical classes of self-similar figures, we can try to use them as models of reality.

In the sequel we shall consider different relations between small and big pieces of certain sets which lead to different classes of fractals with a rigorous definition. For introductory reading, we recommend the classical treatments which convey fascination in fractals: Mandelbrot's original work [5], Barnsley's well-illustrated textbook [1], and Falconer's mathematical treatment [3]. One may also consult the introduction by Peitgen et al. [7], Schroeder's view of a physicist [9], and Edgar's collection of seminal papers [2]. The web also contains a lot of stimulating material.

4 The Cantor Set

4.1 The Topological Viewpoint

The Cantor set is the basic example of a fractal. It comes in different disguise, see Fig. 1. The topology of $\mathbb{R}^n$ characterizes it as an uncountable compact (closed and bounded) set without isolated points which is totally disconnected—there is

Fig. 1 Self-similar Cantor sets

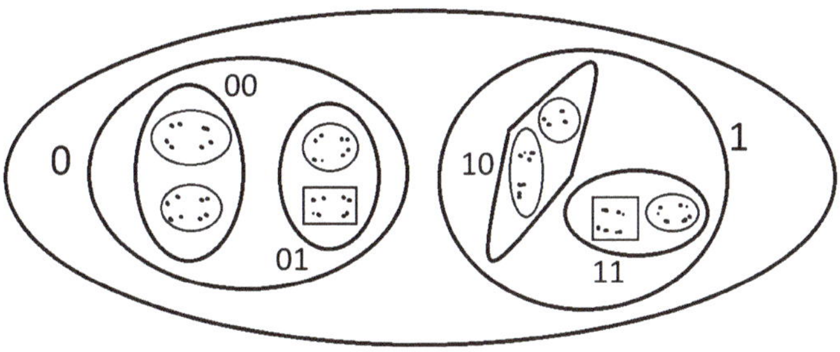

Fig. 2 The mathematical picture of the Cantor set

no continuous curve which connects points within the set. The last property is emphasized by Mandelbrot's name *Cantor dust.*

In some of the examples of Fig. 1, self-similarity is obvious to the eye. In a more abstract way, self-similarity can be introduced to the Cantor set C in many ways. We divide C into two closed and disjoint subsets C_0 and C_1. In the plane, this is done by enclosing the pieces into domains D_0, D_1 bounded by curves which do not intersect C, as indicated in Fig. 2. Next, we divide the set C_0 into two closed disjoint sets C_{00} and C_{01}, and C_1 into two closed disjoint subsets C_{10}, C_{11}. Then we do the same with the new sets $C_w = C_{w_1w_2}$ where $w_1, w_2 \in \{0, 1\}$, and so on. It is clear that all these subsets are Cantor sets again. From the viewpoint of topology, the pieces C_w are equal to C.

In mathematical constructions of a Cantor set C, one starts with a surrounding set D, which may be an interval, rectangle, or ball, and continues with surrounding sets D_w, so that D_{w0} and D_{w1} are disjoint for every word w. This abstract construction of a Cantor set is shown in Fig. 2.

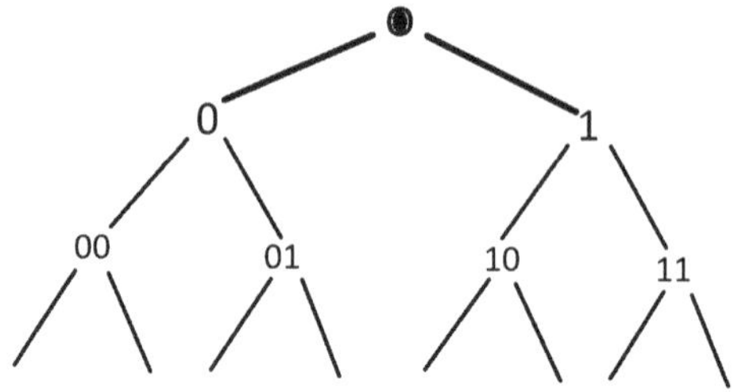

Fig. 3 The binary tree—another picture of the Cantor set

4.2 Algebraic Description of the Cantor Set

We consider the alphabet $A = \{0, 1\}$ with two letters. Each piece C_w and its surrounding set D_w is given by a word $w = w_1 w_2 \ldots w_n \in A^n$ of some length n. The set of all words of the alphabet addresses the pieces of C which we constructed, and their surrounding sets.

Next consider a point $x \in C$. It is contained in some D_{w_1}, then in some $D_{w_1 w_2}$, in some $D_{w_1 w_2 w_3}$ and so on. These sets are nested: $D \supset D_{w_1} \supset D_{w_1 w_2} \supset \ldots$ If we construct the D_w in such a way that their diameter tends to zero with $n \to \infty$ (this is not difficult to arrange), then x will be the intersection of the nested sequence:

$$\{x\} = \bigcap_{n=1}^{\infty} D_{w_1 w_2 \ldots w_n} \qquad \text{for some sequence } s = w_1 w_2 \ldots$$

Thus each point x of the Cantor set corresponds to a unique sequence s in the alphabet A. This is the abstract concept of a Cantor set:

C is the set of all sequences $s = w_1 w_2 \ldots$ over some finite alphabet A.

4.3 The Binary Tree

The binary tree is another strong picture of the Cantor set (Fig. 3). Each node of the tree is denoted by a word $w = w_1 \ldots w_n$, and is connected to its parent $w_1 \ldots w_{n-1}$ as well as to its children $w0$ and $w1$. The root of the tree is denoted by the empty word $\bullet$. The points of C correspond to infinite non-intersecting paths starting in $\bullet$, which can be written in the form $s = w_1 w_2 \ldots$ Such trees appear in the analysis of algorithms and in programming. Of course this tree is usually modified by assigning varying numbers of children to the nodes or by identifying certain words, as indicated in Fig. 4. The structure of many algorithms, as well as the structure of languages, the structure of human thinking and society, show some self-similarity—certainly not as regular as our figures.

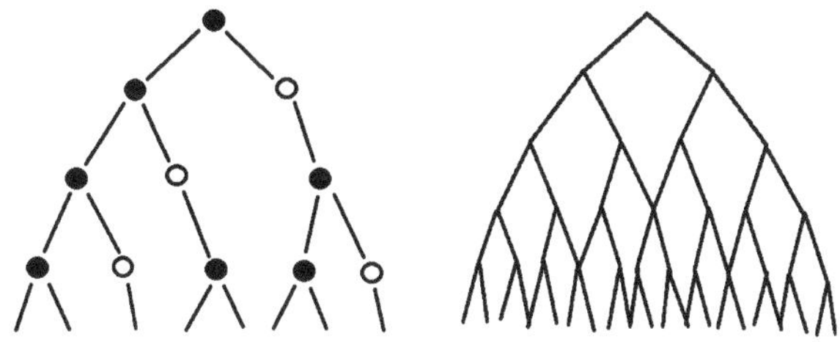

Fig. 4 Two regular modifications of the binary tree. How many nodes does level n contain?

4.4 Description of Self-similarity

How is the whole set C related to C_0, or any other piece C_w? The sequences of points in C_0 are those which start with 0. When we assign to each sequence $s = s_1s_2\dots$ the sequence $0s = 0s_1s_2\dots$ then we get a one-to-one correspondence between C and C_0. More generally, for every word $w = w_1\dots w_n$ we have a function $f_w(s) = ws = w_1\dots w_ns_1s_2\dots$ which maps C onto C_w in a one-to-one way. It is not difficult to show that this function is a homeomorphism with respect to the product topology, or to the usual topology of sets in Fig. 1. *Adding letters in front of a sequence will lead us from the whole set to subsets.*

4.5 The Number System

When you feel uncomfortable with the use of an alphabet, think of our decimal number system. Each real number between 0 and 1 has a decimal expansion $x = 0.a_1a_2\dots$ Here the alphabet is $A = \{0, 1, \dots 9\}$. One instance of the Cantor set is defined by those decimal numbers which involve only digits 0 and 9. It is too tiny to draw, try it! Cantor's original middle-third set from 1888 takes all numbers with digits 0 and 2 in the ternary expansion:

$$C = \left\{x \in [0, 1] \;\middle|\; x = \textstyle\sum_{k=1}^{\infty} a_k 3^{-k} \text{ with } a_k \in \{0, 2\}\right\}. \tag{1}$$

4.6 The Interval

If we do not exclude digits, we get the unit interval, with ten basic pieces for the decimal system and two pieces for the binary system. We can call it C, but it is not a Cantor set, it is connected. The reason is that some points have two addresses, for instance $0.1000\dots = 0.0999\dots$ in the decimal system and $\frac{1}{2} = 0.1000\dots = 0.0111\dots$ in the binary system. Thus the two pieces C_0 and C_1 have a common point. Moreover, the pieces C_{w0} and C_{w1} are also connected since in the binary system $0.w_1 \dots w_n 0111 \dots = 0.w_1 \dots w_n 1000 \dots$ So the self-similarity is preserved, and the unit interval $[0, 1]$ *is a fractal.*

5 Some Fractal Curves

5.1 A Nowhere Differentiable Curve

As modifications of $[0, 1]$, we obtain some fractals which were known as mathematical counterexamples around 1900. Von Koch suggested in 1904 to lift the middle third of an interval $[a, b]$ instead of deleting it. He replaces the middle third of the interval by two intervals of the same length with a common endpoint c, and repeats this procedure with all small intervals again and again. The result is a *continuous curve K which does not possess a tangent in any of its points.*

5.2 A Proof with Self-similarity

We assume there is a tangent in some point $x \in K$ and derive a contradiction. For each $\varepsilon > 0$ there must be a little piece K_w which is inside the double cone around the tangent line with vertex x and angle $\pm\varepsilon$. But since the piece K_w is geometrically similar to K (see Sect. 6), the same must hold for K and a line through some point $y \in K$. Since y is inside the triangle $T = \Delta abc$, every side of the triangle is seen from y under an angle of at least 30°. If a double cone through y contains K, then at least two of the vertices a, b, c are on one side of the cone. Thus the opening angle of the cone is at least 30° which contradicts the assumption for $\varepsilon < 15°$. We proved that K has no tangent.

Mandelbrot did not consider the non-differentiability of K as a bad property. On the contrary, he recommended Koch's curve as a model for coastlines. They have (almost) infinite length when we measure them precisely enough.

The construction of K can also be rephrased as a "decreasing set construction." We delete from the triangle T a maximal equilateral triangle, delete from each of the remaining triangles again an equilateral triangle, etc. Now we can also delete isosceles triangles with a smaller base b and will get wilder Koch curves, cf. Fig. 5.

5.3 A Plane-Filling Curve

One can ask what happens in the limit $b \to 0$. We still have a curve, but with a lot of double points. This curve will cover the whole triangle T which is now right-angled. Such plane-filling continuous curves were constructed by Peano and by Hilbert around 1890, and they were quite disturbing for the mathematical concept of dimension. Mandelbrot turned the property into the positive: look here, this is a

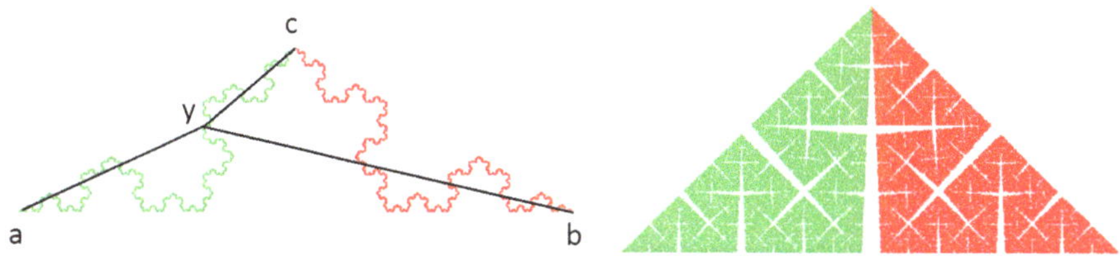

Fig. 5 The Koch curve and an almost plane-filling modification

good model for the system of human blood circulation. This must consist of vessels, but must be space-filling. Because whenever you hurt you anywhere, blood will come out.

5.4 A Simple Curve with Positive Area

There is another variation due to Knopp 1915: we can consider triangle bases b_n which decrease with the level n of construction, in such a way that the sum of the triangle areas which we cut out at level n is half as large as at level $n-1$. In that case we are left with a proper continuous curve K with no double points. But since the sum of deleted areas is less than the area of T, the curve has positive area! A more complicated proof for the existence of such curves was given by Osgood in 1905.

5.5 Different Types of Self-similarity

The pieces of Koch's curve are all geometrically similar to each other. So we call it a self-similar set. The right-angled triangle which comes as the limit Peano curve is also a self-similar set. Knopp's curve with positive area is not self-similar, but there are continuous bijective maps between the pieces K_w and K, and we also consider it a fractal. The Koch and Knopp curves are homeomorphic to the unit interval—but not the Peano curve, due to double points. So the topological relation between pieces and the whole is the same, but the metric properties differ. As a matter of fact, the fractal dimension of the curve with deleted base b is ... (see Sect. 7).

5.6 The Graph of Brownian Motion

To conclude this section, we mention a very important curve construction which introduces *random self-similarity*. We assume that we have a device which yields independent random numbers with standard normal distribution. Any mathematical software on your computer will do, even Excel. Those numbers are between -5

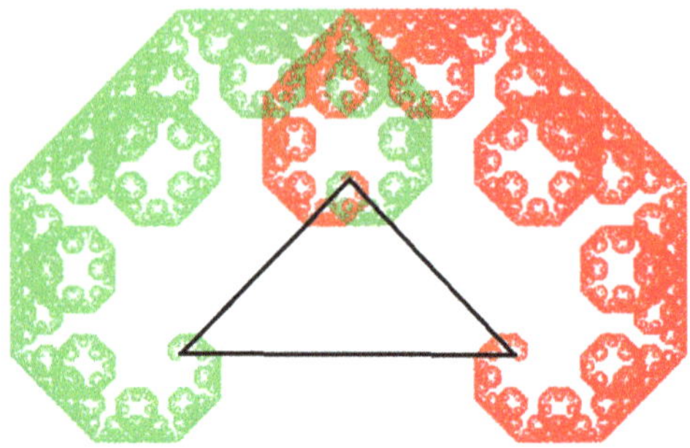

Fig. 6 The Lévy curve and its basic triangle

and 5. Our basic interval I will start at $(0,0)$ and end at $(1,z)$ where z is the first random number. The midpoint of the interval is $(\frac{1}{2},\frac{z}{2})$. Now the midpoint is shifted by $z_\bullet/\sqrt{2}$ in vertical direction, up or down, depending on the random number $z_\bullet$. Next, we shift the midpoints of the resulting intervals I_0 and I_1 by $z_0/2$ and $z_1/2$ in vertical direction, getting four intervals $I_{w_1w_2}$. We proceed by induction: on level n the midpoints intervals I_w are moved up or down by $z_w/\sqrt{2}^{n+1}$. This random construction will converge, with probability one, to the graph of a continuous function f. Of course we get different functions for different random numbers, as shown in Fig. 7 below. The construction is self-affine, not self-similar, since in each step, horizontal direction shrinks by the factor $\frac{1}{2}$ and vertical direction by $1/\sqrt{2}$.

5.7 *Lévy and His Curve*

This is the midpoint displacement algorithm for Brownian motion, the most fundamental stochastic process which was suggested as a model for the financial market by Bachelier in 1905 [5]. The construction was known to the great probabilist Paul Lévy in the 1940s. Mandelbrot considered himself as a student of Lévy: he "came closest to being my mentor" [5, p. 398]. Incidentally, Lévy also discovered a plane-filling curve which is obtained when we repeatedly replace an interval I_w by two intervals I_{w0}, I_{w1} which form an isosceles triangle with right angle over I_w (Fig. 6). This curve is self-similar and it is plane-filling, which is rather difficult to prove. As a young man, Lévy reported this result to a meeting of the French academy in 1912. Probably he was discouraged by the reaction of the audience since he published his work only in 1937 [2]. Thus the dimension of the Lévy curve is 2. The dimension of its *boundary* was found to be 1.955 by several authors only around 2000, which confirms the fact that the interior of Fig. 6 is very fragmented.

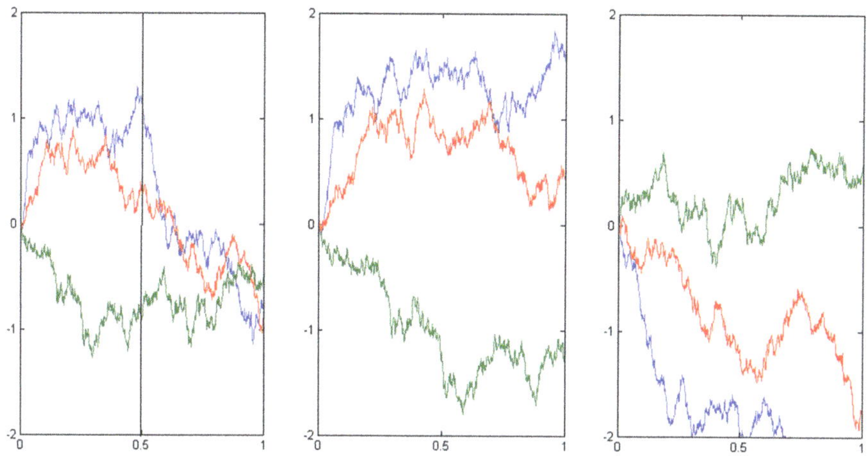

Fig. 7 Illustration for random self-similarity: three graphs of Brownian motion on [0, 1] and their rescaled *left* and *right* parts

5.8 *Random Self-similarity*

For the graphs of Brownian motion constructed in Sect. 5.6 and illustrated in Fig. 7, self-similarity is more complicated. *All pieces are realizations of the same random process, after rescaling*. Rescaling here means that the graph of $f(x)$ with $x \in [a, b]$ is replaced by the graph of $\frac{1}{\sqrt{b-a}} \cdot [f(a+x(b-a)) - f(a)]$ for $x \in [0, 1]$. The rescaled piece is a realization of Brownian motion on [0, 1], probably not the one with which we started. This holds for the graphs of the function over arbitrary intervals $[a, b]$, not only for the dyadic construction intervals. It is this random kind of relation what we observe in clouds and mountains. Random constructions are much more realistic models of nature than the Koch curve, but their study is also much more difficult. Two-dimensional midpoint displacement constructions were used to model mountain scenery in [5] and in various computer games.

6 Fractal Constructions by Mappings

6.1 *Hutchinson's Equation*

Self-similarity can be understood best when it is defined by mappings. The middle-third set 1 of Cantor is transformed into its left piece C_0 by the map $f_0(x) = \frac{x}{3}$, and into its right piece C_1 by $f_1(x) = \frac{x+2}{3}$. Both maps are similarity maps, so C is self-similar, and can be characterized as solution of the equation

$$C = f_0(C) \cup f_1(C) . \tag{2}$$

Since f_0 and f_1 map the whole set onto a subset, it is natural to assume that they are contractive maps. That means, the distance of the images of two points x, y is strictly smaller than the distance of the points themselves. A bit more rigorously, f is a contraction if there is a number $r < 1$ such that

$$|f(x) - f(y)| \leq r \cdot |x - y| \,.$$

Hutchinson [4] proved that for any two contractions f_0, f_1 on $\mathbb{R}^n$ there is a unique compact non-empty set C which fulfils Eq. (2). Thus we have a lot of self-similar sets, one for any choice of two contractions. We can also take three or more, but we stick to the simplest case.

6.2 The IFS Algorithm

Barnsley [1] came up with a computer construction for C. Take one point c of C, for instance the fixed point of f_0. Then the images $c_i = f_i(c)$ must also be in C, because of Eq. (2). The same holds for $f_w = f_{w_1} \cdots f_{w_n}(c)$ for each 0-1-word $w = w_1 \ldots_n$. Since the pieces C_w have diameter smaller r^n times the diameter of C, the collection of these 2^n points for some moderate n, say 15, will be a perfect computer approximation of the fractal C. This requires five lines of code.

A random algorithm for approximating C starts with $c_0 = c$, and takes independent random numbers $z_n \in \{0, 1\}$ for $n = 1, 2, \ldots$ (this is like coin-tossing: head is 1 and tail is 0). Then define

$$c_n = f_0(c_{n-1}) \text{if } z_n = 0 \text{ and } c_n = f_1(c_{n-1}) \text{ if } z_n = 1 \,.$$

This requires only two lines of code, and with high probability, it will soon generate points in all C_w. As a rule, already 30,000 points give a reasonable picture for the eye. Barnsley suggested the name "iterated function system," or IFS, for such algorithms.

6.3 An Exercise in Complex Numbers

Let us model a tree T in the complex plane, with a stem S from 0 to $2i$ and two branches from $2i$ to $3i + 1$ and $3i - 1$. The mappings from the stem to the branches will be $f_0(z) = \frac{z}{2} \cdot (1 - i) + 2i$ and $f_1(z) = \frac{z}{2} \cdot (1 + i) + 2i$. When we turn on the IFS algorithm, we get the Lévy curve! That is too much, so we shall later decrease the ratios of the f_i, replacing $\frac{1}{2}$ by 0.4, say. First we have to care for the branches which are not drawn by the IFS algorithm. Only the leaves of the tree form the self-similar set. One has to add a third map, $f_2(z) = \frac{1}{3} \cdot \operatorname{Im} z$ to obtain stem and

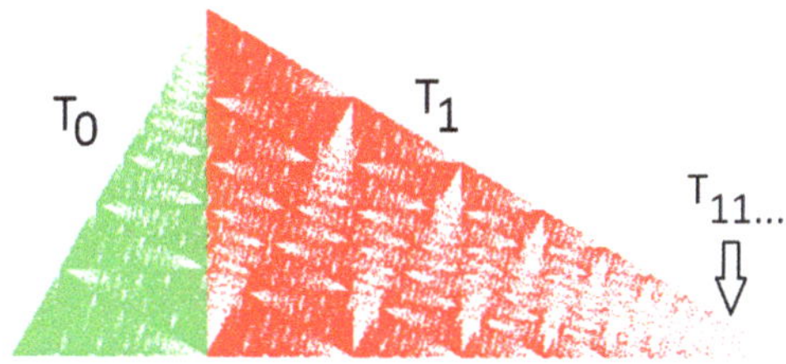

Fig. 8 A fractal measure

branches. The tree is not strictly self-similar, since f_2 is not a similarity map, but it fulfils the equation $T = S \cup f_0(T) \cup f_1(T)$ (cf. [1]). Much more natural trees were constructed by Prusinkiewicz [8] with the related concept of L-systems.

6.4 Fractal Measures

In our first experiment with random IFS in 1987, we tried to draw a triangle with vertices 0, 1, and $(1+\sqrt{3}i)/4$. Any right-angled triangle is self-similar, as you know from high-school. The mappings have the form $f_i(z) = a_i\bar{z} + b_i$. The random IFS with 500,000 points yields the surprising Fig. 8.

After a while, we found that the picture is correct. The IFS will indeed fill the whole triangle, if it runs long enough. But the picture shows what today is called a *fractal measure or multifractal*, with an uneven distribution of the points. The areas of the right piece T_1 of the triangle is three times larger than the area of the left piece T_0, but the number of IFS points in both pieces is the same. What is worse, the area of n-th level pieces $T_{11..1}$ and $T_{00..0}$ are related by the factor 3^n and still both get the same number of IFS points. With nonlinear maps, we get even more impressive examples of fractal measures. Fractal measures have become a separate area of fractal geometry. Actually, measure theory is the mathematical toolbox which has most often been used by mathematicians in the field. Hausdorff defined measures of fractional dimension already in 1918 (cf. [2]), which can be considered as the starting point of fractal geometry. It took more than 10 years before colleagues started to understand Hausdorff's ingenious idea.

6.5 Dynamical Systems

This is another key concept connected with fractals. A dynamical system consists of a set X with some mathematical structure and a mapping $g : X \to X$ which preserves that structure. For a Cantor set C fulfilling Eq. (2), the mapping $g : C \to C$ can be defined as inverse of f_0 and f_1,

$$g(c) = f_0^{-1}(c) \text{ for } c \in f_0(C) \text{ and } g(c) = f_1^{-1}(c) \text{ for } c \in f_1(C)\,.$$

If the f_i are contracting, then g is expanding. The fractals best-known to the public are the Julia sets of complex quadratic maps $g(z) = z^2 + b$ with various constants $b \in \mathbb{C}$. Here f_0, f_1 can be considered as the quadratic roots $f_i(z) = \pm\sqrt{z-b}$, the two inverse branches of g, and the Julia set as the solution of (2). *The Mandelbrot set, that well-known logo of fractal geometry, is the set of all constants b for which the Julia set is not a Cantor set.* The situation is somewhat involved, however. Among others, the f_i are not contractive.

6.6 Attractors

There are important fractals which are generated by a single map g which provides the self-similarity, but in such a way that there is no obvious structure of pieces. One example is the Hénon map on $X = \mathbb{R}^2$ given by the simple formula $h(x, y) = (y+1-ax^2, bx)$ [7, Sect. 12.1]. The fractal which is obtained by repeated application of such a map is called an *attractor of the dynamical system* (X, g). The structure of the Hénon attractors, even for the standard values $a = 1.4, b = 0.3$, is not yet mathematically understood although famous mathematicians have tried their best. Also some properties of the Mandelbrot set are not yet resolved. Even in the unit interval, it is not exactly known for which parameters r between 3.5 and 4 the quadratic mappings $g(x) = rx(1-x)$ have Cantor set or interval attractors. There are lots of open mathematical problems in this field.

7 Dimensions and Exponents

7.1 The Concept of an Exponent

When we adopt the topological viewpoint, there is only one Cantor set, up to topological equivalence. This viewpoint is now too general. We want to study metric properties, we want to distinguish thick and thin Cantor sets in Fig. 1.

How can we describe, measure and classify fractals? There is one important principle: size does not matter. Geometrically similar sets are considered to be equal, only shape is important. The type of self-similarity is studied: how much does the structure change if we pass to smaller pieces? Even though we are more specific than topologists, our parameters will be more general than those of Euclidean geometry. They are called dimensions or exponents.

The paradigm of classical mathematics is the differential equation. Give me the equation, give me the initial values, and I tell you all details about the system for all times up to infinity. This paradigm is not valid anymore. Even for rather simple differential equations, the tiniest deviation from the initial conditions can completely change the development of the system.

More importantly, nobody wants to care about every detail when the system is complicated. And it is not good to care about too many details of a complex system, because the system can organize itself when the essential parameters are properly regulated. This can be seen in everyday life, for instance in the education of small children. We need a robust but not too detailed description. For fractals, exponents are the appropriate parameters.

7.2 *Box Dimension*

Probably the simplest exponent is box dimension of a fractal F in the plane. Draw a mesh of squares with side length s, and count the number $N(s)$ of squares which intersect F. If you want, you can do this several times, shifting and rotating the mesh, and take $N(s)$ as average. The number $N(s)$ itself, however, is not interesting since size does not matter.

The trick is to do this for different s and *study the function* $s \mapsto N(s)$. When F is a line segment, or a rectifiable curve, then $N(s) \approx \frac{k}{s}$ for some constant k. When F is a rectangle, or, more general, F contains interior points, then $N(s) \approx \frac{k}{s^2}$ for some constant k. Thus a general assumption will be

$$N(s) \approx \frac{k}{s^\beta} \text{ or, equivalently } \log N(s) \approx \log k - \beta \log s \tag{3}$$

for some number $\beta \in [0, 2]$ which is called the *box dimension of* F. Since rectifiable curves have dimension 1 and open sets have dimension 2, and the empty set has dimension 0, the name is justified.

7.3 *How to Continue?*

If you are a theoretical mathematician, you will now look for examples where the approach does not work, will define box dimension as a limit—or better, upper and lower limit so that it always exists and then find classes of sets where upper limit and lower limit coincides. You can also read Hausdorff's beautiful old paper in [2] which presents a mathematically clean dimension concept, using arbitrary sets instead of boxes, and infinite coverings.

If you are a physicist or more applied mathematician, you will look for nice model sets where you can try the method numerically, by determining concrete values $N(s)$, drawing points $(s, N(s))$ into a logarithmic plot and calculating β from a linear regression.

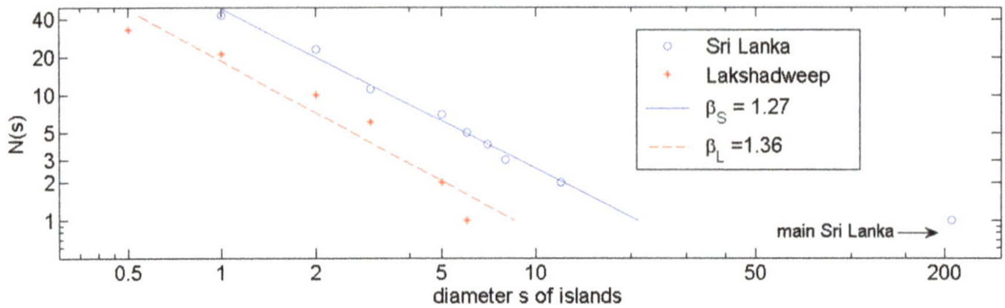

Fig. 9 Exponent of disconnectedness for groups of islands

7.4 *An Illustrative Example*

We conclude by presenting a small example in the physicists' way. We shall not count boxes, but connected components of the fractal F. The number $N(s)$ will be the number of connected components of F with diameter larger than or equal to s. The corresponding β in (3) is some measure of fragmentation of F which could be called *exponent of disconnectedness*. Connected fractals will have $\beta = 0$. Note that β is not defined when F is a Cantor set.

Here we take two maps, from Sri Lanka and from the Lakshadweep islands, and count the number of islands according to diameter (length). Probably my count is not very accurate. You can improve it. For Sri Lanka, I found islands with diameter 210, 12, 8, 7, 6, two times 5 mm, four times 3 mm, twelve times 2 mm and twenty times 1 mm. One millimeter is about 1.65 km in reality, but for the exponent this is not relevant. For Lakshadweep islands, I got diameters 6, 5, four times 3 mm, four times 2 mm, eleven times 1 mm and twelve times $\frac{1}{2}$ mm. Here 1 mm is almost 2 km, but as we said, size does not matter. We draw the values $N(s)$ into the logarithmic plot of Fig. 9 and determine the two regression lines corresponding to Eq. (3).

It turns out that Sri Lanka does not provide a good linear approximation, because of the big main island. If we drop that point, we get a regression line with slope -1.27. The Lakshadweep islands have no mainland, so they have a more fractal appearance. The exponent is $\beta \approx 1.35$, only slightly larger than 1.27. The line does not approximate too well, perhaps due to inaccurate counting. When we neglect the mainland of Sri Lanka, the degree of fragmentation for the two groups of islands is more or less the same.

References

1. Barnsley, M.F.: Fractals Everywhere. Academic, Cambridge (1988)
2. Edgar, G.A.: Classics on Fractals. Addison-Wesley, Reading (1993)
3. Falconer, K.J.: Fractal Geometry: Mathematical Foundations and Applications. Wiley, Chichester (1990)

4. Hutchinson, J.E.: Fractals and self-similarity. Indiana Univ. Math. J. **30**, 713–747 (1981)
5. Mandelbrot, B.B.: The Fractal Geometry of Nature. Freeman, San Francisco (1982)
6. Mandelbrot, B.B.: The Fractalist: Memoir of a Scientific Maverick. Pantheon, New York (2012)
7. Peitgen, H.O., Jürgens, H., Saupe, D.: Chaos and Fractals. Springer, New York (1992)
8. Prusinkiewisz, P., Lindenmayer, A.: The Fractal Beauty of Plants. Springer, New York (1992)
9. Schroeder, M.R.: Fractals, Chaos, Power Laws. Freeman, New York (1991)

Geometry of Self-similar Sets

Christoph Bandt

Abstract Self-similar sets form a well-defined class of fractals which are relatively easy to study. This talk introduces their main features with a lot of examples. We explain the need of a separation condition for the tangential structure. Hausdorff measure is the natural concept of volume. Under certain conditions Hausdorff measures define also the "surface" of the boundary and the interior distance. A number of open problems are mentioned.

Keywords Self-similar measure • Fractal • Hausdorff dimension • Interior distance

1 The Concept of Self-similar Set

Hutchinson's equation. The idea that small pieces of a fractal are similar to the whole figure led Hutchinson [16] to the definition of a mathematically tractable class of fractals. A *self-similar set* in Euclidean $\mathbb{R}^d$ is determined by a finite set of contractive similarity mappings. A map f from $\mathbb{R}^d$ to itself is called contractive if there is a constant $r < 1$ with

$$|f(x) - f(y)| \leq r \cdot |x - y| \qquad \text{for all points } x, y \in \mathbb{R}^d \ , \tag{1}$$

Workshop on Fractals and Wavelets at Rajagiri School, Kochi, India, 10 Nov 2013.

C. Bandt (✉)
Institut für Mathematik und Informatik, Universität Greifswald, 17487 Greifswald, Germany
e-mail: bandt@uni-greifswald.de

C. Bandt et al. (eds.), *Fractals, Wavelets, and their Applications*, Springer Proceedings in Mathematics & Statistics 92, DOI 10.1007/978-3-319-08105-2_2,

Fig. 1 Self-similar sets with finite intersection of pieces

and it is a contractive similarity if equality holds for all x and y. The self-similar set F with respect to $f_1, \ldots, f_m$ is a non-empty compact set F which fulfils the equation

$$F = f_1(F) \cup \ldots \cup f_m(F)\,. \tag{2}$$

It is a beautiful thing that fractal sets can be characterized by such an intuitive equation which specifies how F is composed of similar copies of itself. And there is a rather simple theorem:

Existence and Uniqueness: For any finite set of contractions f_k on $\mathbb{R}^d$, Eq. (2) has a unique solution F in the family of compact non-empty subsets of $\mathbb{R}^d$.

The standard proof uses the fact that $\Phi(E) = \bigcup_{k=1}^{m} f_k(E)$ is a contractive transformation on the space of all compact subsets of $\mathbb{R}^d$, with Hausdorff metric [10, 13, 16]. Below we give a more elementary proof from [1]. First we consider examples. Of course, a square, parallelogram or triangle is self-similar, with n^2 pieces, for arbitrary $n > 1$. Right-angled triangles and rectangles with side ratio $\sqrt{2} : 1$ are self-similar with two pieces. A triangle with angles $30°, 60°, 90°$ [4, Fig. 8] is self-similar with four congruent pieces as well as with three congruent pieces (try it).

Problem 1. Is there any other self-similar set with this property?

Two examples of self-similar sets with two pieces are given in Fig. 1. As an exercise, you may try to describe the mappings f_k which transform the set F to its pieces $F_k = f_k(F)$. The matrix form of a similarity map f is $f(x) = Ax + v$ with a vector $v \in \mathbb{R}^d$ and a matrix $A = rO$ which is the product of an orthogonal $d \times d$ matrix O with the similarity factor r. Since f is contractive, r is smaller than 1. In two dimensions, we better write f as a linear mapping of the complex plane: $f(z) = az + b$ or $f(z) = a\bar{z} + b$ where a, b are complex numbers, and the modulus of a is $|a| = r < 1$. The first formula gives a rotation with angle $\arg a$ around the center $z^* = \frac{b}{1-a}$ which fulfils the fixed-point equation $z^* = f(z^*)$. The second formula gives a reflection or glide reflection at a line which is a bit more difficult to determine. For simplicity, we shall often concentrate on two rotational mappings with fixed points 0 and 1 for which our master student Kirsch wrote a nice program [19]:

$$f_0(z) = az \qquad \text{and} \qquad f_1(z) = bz + 1 - b \qquad \text{with } |a|, |b| < 1\,. \tag{3}$$

Aim of this talk. Self-similar sets seem too artificial to be good models of nature. But they can be treated rigorously, like manifolds in analysis, and their appeal can stimulate our geometric intuition. We shall demonstrate the rich mathematical structure which self-similar sets can possess and mention some of the numerous open problems which arise as soon as we consider the structure a bit more carefully.

2 Addresses and Symbolic Dynamics

Pieces. The fact that a self-similar set consists of smaller and smaller pieces leads to some symbolic notation. Let $I = \{1, \ldots, m\}$. For two mappings, we prefer $I = \{0, 1\}$. Words w from the alphabet I have the form $w = w_1 w_2 \ldots w_n \in I^n$. By substitution of F on the right-hand side of Eq. (2) we can express F as a union of smaller pieces:

$$F = \bigcup_{i,j \in I} f_k f_j(F) \quad \text{and} \quad F = \bigcup_{w \in I^n} f_w(F) \quad \text{for every } n \geq 1,$$

where $f_w = f_{w_1} \cdot \ldots \cdot f_{w_n}$. We shall write $F_k = f_k(F)$ and $F_w = f_w(F)$. The length $n = |w|$ of the word is called the level of the piece F_w.

Points. Since each point $x \in F$ is determined by a nested sequence of pieces $F_{w_1} \supset F_{w_1 w_2} \supset \ldots$, it corresponds to a sequence $s = w_1 w_2 \ldots$ as its *address* [10]. If F is a Cantor set, this correspondence is one-to-one [4, Sect. 4]. However, when x is in the intersection of two pieces F_v, F_w of the same level, then there will be at least two addresses of the point x. Thus in order to get a mathematical function, we have to assign addresses to points, and not conversely. Let $S = I^\infty$ denote the set of all sequences s of symbols from I.

The *address map* $\pi : I^\infty \to F$ is defined as follows.

$$\pi(s) = \lim_{n \to \infty} f_{s_1} \cdot \ldots \cdot f_{s_n}(y) \qquad \text{for } s = s_1 s_2 \ldots \in S \tag{4}$$

where y is an arbitrary point in $\mathbb{R}^d$. A good choice for y is the fixed point z^* of f_1 which is certainly in F. Then we are sure that application of the f_{s_k} will only lead to points of F, because of Eq. (2).

Justification of definition (4). We show that the limit exists and is independent of y. First we note that by compactness of F, the sequence with starting point z^* has a subsequence which converges to some point $x \in F$. For any starting point y, the subsequence will converge to the same x because of (1). The difference on level n,

$$|f_{s_1} \cdot \ldots \cdot f_{s_n}(y) - f_{s_1} \cdot \ldots \cdot f_{s_n}(z)| \leq r^n \cdot |y - z| \tag{5}$$

for $z = z^*$ converges to zero with $n \to \infty$. Here r denotes the maximum of the contraction factors of $f_1, \ldots, f_m$, which is smaller than one.

We now show that the whole sequence with starting value z^* converges to x. Since F is compact, there is a constant c with $|y - z^*| \leq c$ for all $y \in F$. Now let $f_{s_1} \cdot \ldots \cdot f_{s_n}(z^*)$ and $f_{s_1} \cdot \ldots \cdot f_{s_m}(z^*)$ with $m > n$ be two members of our sequence. We define $y = f_{s_{n+1}} \cdot \ldots \cdot f_{s_m}(z^*)$, and use the estimate (5) with $z = z^*$ to see that the difference is smaller than $r^n c$. Thus we have a Cauchy sequence in F which converges. This completes the proof that π is well defined by (4).

Proof of Hutchinson's theorem. So far we have only used the compactness of F, which is an assumption in the definition of self-similar set. We now show that $\pi(S)$ is the unique solution of (2). Equation (2) implies that all points $f_{s_1} \cdot \ldots \cdot f_{s_n}(z^*)$ are in F. By compactness, this holds for their limits, so $\pi(S) \subseteq F$. On the other hand, each point in $x \in F$ can be represented as $f_{s_1} \cdot \ldots \cdot f_{s_n}(y_n)$ for a sequence $s = s_1 s_2 \ldots$ and points $y_n \in F$, by repeated application of Eq. (2) for $y_n, n = 1, 2, \ldots$ With another use of (4) for $y = y_n$, this implies $F \subseteq \pi(S)$ and completes the proof of uniqueness.

We still have to convince ourselves that $F = \pi(S)$ fulfils Hutchinson's equation (2). To this end, we consider the maps $\tau_k : S \to S$ with $\tau_k(s_1 s_2 \ldots) = k s_1 s_2 \ldots$ which shift the symbol k before the sequence, for $k = 1, \ldots, m$. It is obvious that

$$S = \tau_1(S) \cup \ldots \cup \tau_m(S) \quad \text{and} \quad \pi(\tau_k(s)) = f_k(\pi(s)) \quad \text{for all sequences } s\,. \tag{6}$$

This implies that $F = \pi(S)$ is a solution of (2) and concludes the proof of existence.

Remark on topology. For every number $\gamma \in (0, 1)$ we can introduce a metric on S by $\rho(s, t) = \gamma^n$ where n is the length of the common initial word of the sequences s and t. Thus $\rho(s, t) = 1$ if $s_1 \neq t_1$, and $\rho(s, t) = \gamma$ if $s_1 = t_1$ but $s_2 \neq t_2$, etc. With respect to this metric, the τ_k are similarity maps with contraction factor γ, and S is a proper self-similar set, although not a subset of $\mathbb{R}^d$.

Moreover, the metric ρ induces the usual product topology on S for which S is a compact space—in fact a Cantor set, see [4, Sect. 4]. The address map $\pi : S \to F$ is continuous: $\rho(s, t) \leq \gamma^n$ implies

$$|f_{t_1} \cdot \ldots \cdot f_{t_m}(x) - f_{s_1} \cdot \ldots \cdot f_{s_m}(x)| \leq r^n c \qquad \text{for } x \in F \text{ and each } m > n\,.$$

This follows from (5) with $y = f_{t_{n+1}} \cdots f_{t_m}(x)$ and $z = f_{s_{n+1}} \cdots f_{s_m}(x)$. (If we take $\gamma > r$, we even have a Lipschitz map.)

A continuous map from a compact space to a subset of $\mathbb{R}^d$ is always a quotient map. That is, the self-similar set F as a topological space is obtained by identifying

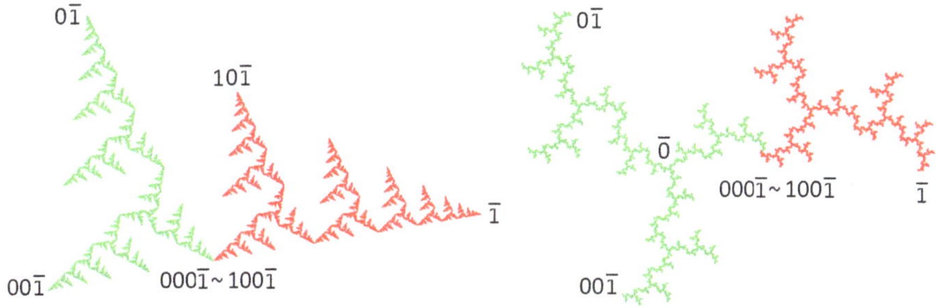

Fig. 2 Addresses of points in two self-similar sets

certain points of S with each other. We write $s \sim t$ for $\pi(s) = \pi(t)$. This is an equivalence relation, and from (6) it follows that

$$s \sim t \text{ implies } ks \sim kt \text{ for all symbols } k \in I \text{ and } ws \sim wt \text{ for all words } w \in I^n .$$

In other words, we need to only define the intersection of the big pieces F_k, then the identification of the smaller pieces will be done automatically. In particular, identification of two points plus self-similarity always yields a connected space [6, 10].

Topological spaces from a simple formula. In cases like Fig. 1 when the pieces F_k intersect only in a few points, the identification can be described by remarkably simple formulas [6, 10, 18]. We demonstrate this for Fig. 2 where $F_0 \cap F_1$ is a single point x for which we can easily identify both addresses. Both pictures show the same topological space, although the geometry is different. The fixed point of mapping f_k has address $kkk\ldots = \overline{k}$. While $\pi(\overline{1})$, the fixed point of f_1, is the "right endpoint of the right-hand piece", the point with address $\overline{0}$ is the branching point which divides the tree into three big components. According to (6), the point $f_w(\pi(\overline{k}))$ has address $w\overline{k}$. Thus the endpoint $\overline{1}$ is mapped by the rotational similarity map f_0 onto the endpoints with addresses $0\overline{1}$ and $00\overline{1}$. The last point is mapped by both f_0 and f_1 to the intersection point x of the pieces. This gives the formula $000\overline{1} \sim 100\overline{1}$ which defines the topology of both spaces. (The example on the left of Fig. 1 can be analyzed similarly, for the example on the right we have to locate the point with address $\overline{01}$. See Sect. 5.)

In this case, $f_0^{-1}(x) = f_1^{-1}(x)$ which means that f_0 and f_1 can be considered as the two inverse branches of a mapping $g : F \to F$. Such topological spaces appear as Julia sets of quadratic mappings. This connection was studied by Kameyama [17]. The following question is open.

Problem 2. For which sequences $s = s_1 s_2 \ldots$ can the topological formula $0s \sim 1s$ be realized by a self-similar set in the plane? Or in three-dimensional space?

How to determine the mappings. When we use the mappings (3), the topological formula allows to determine the parameters a and b for Fig. 2. The fixed points are 0 and 1, so $f_0^3(1) = a^3$ has address $000\overline{1}$, and $f_1 f_0^2(1) = ba^2 + 1 - b$ has address $100\overline{1}$. This leads to the equation

$$b = \frac{1-a^3}{1-a^2} = 1 + \frac{a^2}{1+a}$$

for which two solutions were shown in Fig. 2. Note that the equation only guarantees the existence of an intersection point with prescribed addresses. There can be further intersection points. To avoid them, small a and b must be chosen.

3 The Separation Condition

Although every set of mappings $f_1, \ldots, f_m$ leads to a self-similar set F, it is not so easy to construct examples with interesting geometry. When the contraction factors r_k of the mappings are small ($r_1 + \ldots + r_m < 1$ is sufficient but not necessary) then F is a Cantor set [4, Fig. 1]. When the factors are large, as in Fig. 3, the intersections of pieces will also become large, and the fine structure of F will be lost. Even an apparently small overlap as in Fig. 4 can turn out to be extremely complicated under magnification. Already in 1946 Moran [21] defined the proper separation condition which has to be required for F in order to have reasonable mathematical structure.

Open set condition. There exists an open set U such that

$$f_k(U) \subset U \quad \text{and} \quad f_k(U) \cap f_j(U) = \emptyset \quad \text{for } k, j \in I, k \neq j \; . \tag{7}$$

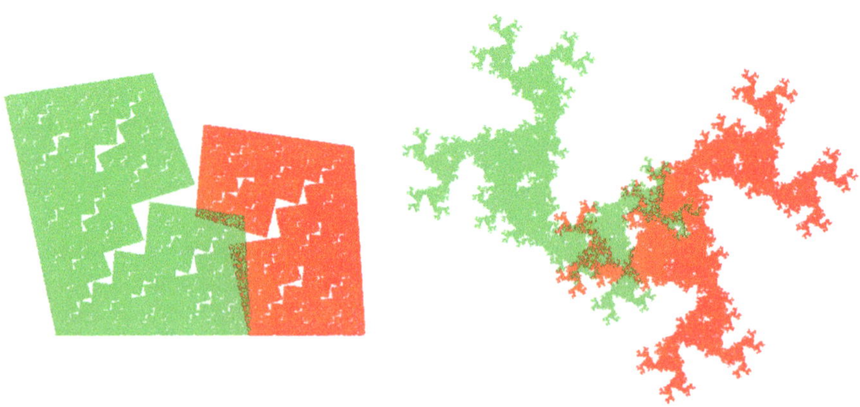

Fig. 3 Self-similar sets with large overlap

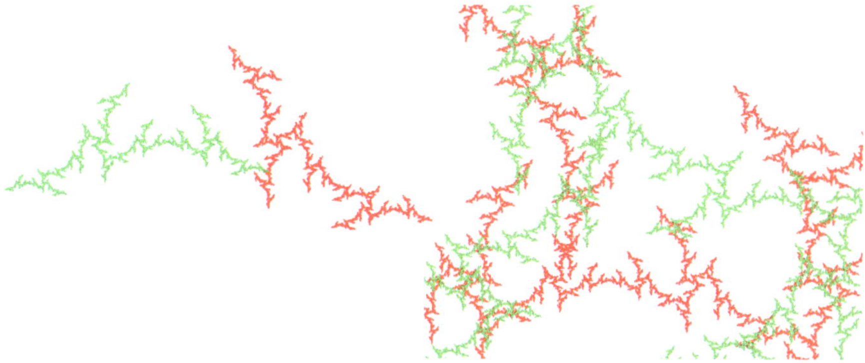

Fig. 4 A small overlap can become large under magnification

The idea is that the pieces $U_k = f_k(U)$ and smaller pieces $U_w = f_w(U)$ will form a disjoint net, as in the construction of a Cantor set in [4, Fig. 2]. However, these open sets can touch at their boundaries. The pieces F_w are subsets of the closure $\overline{U_w}$. They can intersect within the boundaries of the respective open sets which restricts the size of the overlaps.

The condition is a bit mysterious since it does not say how to obtain the open set. U is certainly not unique. If F has interior points we know that $U = \operatorname{int} F$ is a possible open set, and the maximal one. In general there is no maximal U, and U cannot be required to be convex or connected. For some time it was not even clear whether we can require that U intersects F. This was proved by Schief [23] in 1992.

Problem 3. Is the open set condition fulfilled for every self-similar set for which the pieces intersect in finite sets?

For connected sets F in the plane, Bandt and Rao [8] gave an affirmative answer. But when $F \subset \mathbb{R}^2$ is a Cantor set (for example, three pieces, but only two of them intersect), or if $F \subset \mathbb{R}^3$ has two pieces intersecting in a single point, the question remains open.

The neighbor map condition. The open set condition can be replaced by an algebraic condition on the f_k which allows to study separation by computer programs [5]. To this end, Bandt and Graf introduced the concept of a *neighbor map* $h = f_v^{-1} f_w$ where v, w are words from the alphabet I with $v_1 \neq w_1$.

The idea is that h maps the intersection of pieces F_v and F_w to standard size, even for very long words which address tiny pieces. Typically, we take words v, w which describe pieces of about the same size—if the factors of the f_k are all equal, we take words of the same length. f_v^{-1} maps F_v to F, and F_w to a "neighbor set" $h(F)$ which has the same position with regard to F as F_w has to F_v.

The neighbor maps of $f_1, \ldots, f_m$ are similarity maps on $\mathbb{R}^d$. Actually, h is a congruence map if f_v and f_w have the same contraction factor. The natural

Fig. 5 Exact overlaps. *Left*: the golden gasket [11]

concept of convergence of a sequence h_n to a limit map g is pointwise convergence: $h_n(x) \to g(x)$ for all $x \in \mathbb{R}^d$. The following condition for $f_1, \ldots, f_m$ is equivalent to the open set condition [5]:

No sequence of neighbor maps converges to the identity map $id(x) = x$. (8)

Exact overlaps. There are two cases for which the condition does not hold. Either some neighbor map $h = f_v^{-1} f_w$ fulfils $h = id$. This means $f_v = f_w$ and $F_v = F_w$ and $U_v = U_w$ for any open set U, as in Fig. 5. Since two of the pieces do exactly coincide, the open set condition cannot be true. This case is not so bad, however. If this is the only obstacle to (8), we can change notation, consider pieces of different types F^j and replace the Eq. (2) by a system of equations for which Hutchinson's theorem is also valid. See [11] for the golden gasket shown in Fig. 5, and Ngai and Wang [22] for a general proof. For system of equations of type (2), see Barnsley [10, Chap. 10] or Edgar [12]. Many other papers can be found under the keywords "weak separation condition," "recurrent iterated function systems," and "graph-directed constructions".

Problem of tangential structure. Serious overlaps occur when there exist neighbor maps different from id in every neighborhood of the identity map. Figure 4 shows what happens in that case: magnification of the set will reveal an ever increasing clustering of the pieces, cf. [23].

In presence of the open set condition as well as for exact overlapping as in Fig. 5, magnification will not change the view. A set of limit pictures or "tangential views for infinite magnification" can be defined, which do not look essentially different from what we see at large scales [3]. Such fractals can be considered as linear objects of fractal geometry—like lines and planes in Euclidean geometry.

For serious overlaps, this does not remain true. Even the concept of volume defined in the next section seems not to exist. Although the global view of such fractals may look harmless, their mathematical structure is a mess.

Crystallographic fractals. In some cases we can be sure that no serious overlaps can occur, just because the neighbor maps are all contained in a discrete group.

For example, when A is an integer matrix such that all eigenvalues have modulus larger one, and v_k are integer vectors, and $f_k(x) = A^{-1}x + v_k$, then all neighbor maps are translations by integer vectors. Only exact overlaps are possible, and it is easy to give conditions which ensure that there are no overlaps at all [2]. A generalization to crystallographic groups was found by Gelbrich [14]. Thus it is easy to give examples of mappings with open set condition.

4 Measure and Dimension

Uniform distribution. How can we define a volume measure μ, or uniform distribution, on a fractal with three congruent pieces, as shown in Fig. 6? We use the same method as for Lebesgue measure on [0, 1], just replacing intervals by pieces of F. If we want to have a probability measure, each of the three pieces should have measure $\mu(F_k) = \frac{1}{3}$. When the open set condition holds, the measure of the overlap should be zero. The pieces F_{kj} would have measure $\mu(F_{kj}) = \frac{1}{9}$, and all pieces of level n would have measure $\mu(F_w) = 3^{-n}$. Once we have the measure of all pieces, we can define a metric outer measure [13, 20] for all subsets $C \subset F$ as usual, looking at the most efficient coverings of C by pieces F_w:

$$\mu^*(C) = \inf \left\{ \sum_{w \in W} \mu(F_w) \;\middle|\; W \subset I^* \text{ fulfils } \bigcup_{w \in W} F_w \supseteq C \right\} . \tag{9}$$

Here $I^* = \bigcup_{n=1}^{\infty} I^n$ denotes the set of all words from the alphabet I. On the algebra of Borel sets and the larger algebra of measurable sets $C \subset F$, the outer measure μ^* is a measure. Note that for compact sets C, we can restrict ourselves to finite coverings.

Fig. 6 Two modifications of Sierpiński's gasket

The concept of dimension. If there are m pieces instead of 3, we define $\mu(F_w) = m^{-n}$ for words $w = w_1 \dots w_n$. We are going to compare volume and diameter of F_w. The diameter of a set E is $|E| = \sup_{x,y \in E} |x - y|$. We still assume that the pieces are congruent, so the similarity ratio is r for all maps f_k, and r^n for all f_w on level n. Thus a piece F_w of level n has diameter $|F_w| = r^n \cdot |F|$. We assume that $|F| = 1$, otherwise the following conclusion will hold only approximately. Since $m = r^{\log m / \log r}$, we get

$$\mu(F_w) = |F_w|^\alpha \quad \text{for } \alpha = \frac{\log m}{\log 1/r} \tag{10}$$

The number α is called the *dimension of* F. If $F = [0, 1]$ with pieces $[0, \frac{1}{2}]$ and $[\frac{1}{2}, 1]$, then $m = 2, r = \frac{1}{2}$, and $\alpha = 1$. In dimension 1, the measure of an interval equals the diameter. For the unit square with $m = 4$ pieces and $r = \frac{1}{2}$ we get $\alpha = 2$. In two dimensions, the area of a square is the square of its diameter. For cubes in $\mathbb{R}^3$ we get $\alpha = 3$, so our definition is consistent with the ordinary concept of dimension. *Dimension is the exponent which connects volume with diameter.*

Different similarity ratios. Now consider similarity maps f_k with different ratios $r_k, k = 1, \dots, m$. We want to determine the volume $\mu(F_k)$ of the pieces. We put $\mu(F) = |F| = 1$ so that $|F_k| = r_k$. We try with $\mu(F_k) = r_k^\alpha$ and assume that intersections $F_j \cap F_k$ have measure zero. Then

$$\sum_{k=1}^{m} r_k^\alpha = 1 \tag{11}$$

defines the dimension α of the self-similar set F, and the measure is given by $\mu(F_w) = r_w = r_{w_1} \cdots r_{w_n}$ for words $w = w_1 \dots w_n$, and by (9) for arbitrary sets C. Using monotonicity, we can easily prove that (11) has a unique solution $\alpha \in [0, d]$.

Hausdorff dimension. Hausdorff introduced already in 1918 a much more general theory, taking coverings of C with arbitrary sets B_j, and assigning the value $\sum |B_j|^\alpha$ to the covering. Beside (9) we need one more limit for the definition of Hausdorff measure. Moran proved in 1946 that the α-dimensional Hausdorff measure $\mathcal{H}(F)$ of a self-similar set with open set condition is positive and finite. This implies that $\mathcal{H} = c \cdot \mu$ for a positive constant c [5, 13, 20] so that we can use our simplified construction (9). Actually, c is difficult to determine—even for the Sierpiński gasket only estimates are available. For self-similar sets, Hausdorff dimension coincides with box (Minkowski) dimension and with local dimension of the volume measure, so we shall not define these concepts here.

Measure for exact overlaps. When exact overlaps occur, the dimension is strictly smaller than the α in Eq. (11). We take the example of the golden gasket, Fig. 5, to show how a uniform distribution can be defined in the presence of exact overlaps. The similarity of the three mappings is $r = \frac{1}{2}(\sqrt{5} - 1)$, the golden mean. We have $F_{kjj} = F_{jjk}$ for $k \neq j$, that is, any two pieces of first level intersect in a piece of level 3. Now let $\mu(F_w) = r_w^\beta$ and assume that the outer triangular boundaries of the F_w have measure zero. When we calculate $\mu(F)$ we now have to subtract the overlaps:

$$1 = \mu(F) = \sum_{k=1}^{3} \mu(F_k) - \sum_{j<k} \mu(F_{jjk}) = 3r^\beta - 3r^{3\beta} .$$

Thus $\gamma = r^\beta$ is a positive root of the polynomial $3\gamma^3 - 3\gamma + 1$. Since F contains lines, the dimension β must be at least 1, thus $\gamma < r$. This root is $\gamma \approx 0.395$, and the dimension of the golden triangle is $\beta = \frac{\log \gamma}{\log 1/r} \approx 1.93$, rather near to 2. The nice analytic expression $\gamma = \frac{\sin \pi/9}{\sin \pi/3}$ can easily be checked dividing the formula for $\sin 3x$ with $x = \pi/9$ by $\sin 3x = \sqrt{3}/2$. Algebraic numbers often pop up in connection with special self-similar sets.

Dimension and measure in the presence of overlaps. Hochman [15] has recently shown that at least in one dimension, Eq. (11) yields the Hausdorff dimension of the fractal F for many families of mappings f_k, when there are no exact overlaps. When the open set condition fails, the Hausdorff measure of dimension α of every piece will be infinite [5, 23].

Problem 4. Can the f_k be chosen in such a way that the open set condition is not true and there is no exact overlap, but still a uniform probability measure on F can be defined?

5 Further Structure

We now discuss structures from mathematical analysis which exist on certain self-similar sets.

Boundary. We defined a volume measure μ on F. Is there also something like *surface*? What is the boundary of a self-similar set? With the neighbor concept, there is a simple answer. A point $x \in F$ is a boundary point when a potential neighbor $h(F)$ touches F in x. Since the boundary should be a closed set, we define it as $B = \bigcup_h F \cap h(F)$, where h runs over all neighbor maps.

Post-critically finite sets. If the open set condition holds, B is a subset of F with $\mu(B) = 0$. We now define a "surface measure" σ on B. In the case when B is finite, σ will be the point-counting measure. Such cases will be determined by finitely many identifications of eventually periodic addresses, as described in Sect. 2. Such fractals were called *post-critically finite* by Kigami [18], following a similar terminology of Thurston for Julia sets.

On the left of Fig. 1, the formulas for the two intersection points are $010^4\overline{1} \sim 10^2\overline{1}$ and $0^5\overline{1} \sim 10^3\overline{1}$. Deleting leading symbols from these addresses, we get the

Fig. 7 Octagasket and pansy: boundary of linear Cantor sets

six boundary points $10^4\overline{1}$ and $0^k\overline{1}$ with $k = 4, 3, 2, 1, 0$. Similarly, the formula $0^3\overline{01} \sim 10^2\overline{01}$ for the figure on the right gives four boundary points. In Fig. 2 we have 3 boundary points, and on the left of Fig. 6 there are 18 boundary points.

Cantor boundary. Figure 7 shows two cases where the intersections of neighboring pieces are Cantor sets within a line. When we speak of the boundary of F, we mean those Cantor sets where F can meet an outer neighbor in a larger set $f_w^{-1}(F)$. In the octagasket on the left the boundary B is contained in eight line segments. In the pansy example on the right B is contained in three line segments. Each of the linear Cantor boundary sets is a self-similar set with respect to two mappings f_k, f_{k+1} of the original construction. Thus the dimension of B is $\beta = \frac{\log 2}{\log 1/r}$ where the contraction factor r of the f_k is $1/4$ for "pansy" and $1/(2+\sqrt{2})$ for the octagasket.

In both cases, the surface measure σ can be taken as Hausdorff measure of dimension β on the boundary set. A similar result is true whenever the open set condition holds [7]. The right-hand part of Fig. 6 shows a more complicated case where the Cantor set is not contained in a line.

Fractal tiles. When F fulfils the open set condition and the interior of F is nonempty, then $\mathbb{R}^d$ can be tiled by copies of F. In that case, the boundary defined above is indeed the topological boundary $F \setminus \text{int}\, F$. The Lévy curve ([26, 27], see [4, Fig. 6]) is a prominent example with a complicated boundary. If a plane self-similar tile is homeomorphic to a disk, the boundary sets $F \cap h(F)$ will be intervals or singletons, and there are only few neighbor maps. The neighbor concept is very intuitive for tiles since all kinds of neighbors are represented in original size in the tiling. There is no place here to review the vast literature by Lagarias and Wang, Solomyak, Kenyon, Akiyama, Lau and many others on fractal tiles. See [26, 27] for a start.

Minimal separation. Let us assume that F is connected, and let us say that a closed subset C *separates* F if $F \setminus C$ is disconnected. We are looking for separating sets with minimal dimension which we call the separating dimension. For the closure of an open set in $\mathbb{R}^d$, the separating dimension is $d-1$, and each separating set has positive $(d-1)$-dimensional Hausdorff measure.

The boundary of a piece F_w in a self-similar set is always a separating set. However, it need not be minimal as the Sierpiński carpet [13, 18, 25] shows. There the boundary consists of intervals but the separating dimension is $\log 2/\log 3$. The boundary of octagasket and pansy in Fig. 7 has the separating dimension, however. It can be shown that each separating set C fulfils $\sigma(C) > 0$.

For overlapping constructions, as in Fig. 5, the boundary in the above sense is the whole set F. The same holds for the topological boundary. Perhaps we can consider minimal separating sets instead of a boundary. For the example in the middle of Fig. 5, there are finite separating sets (one point will do). So the separating dimension is zero, as for all post-critically finite sets. For the golden triangle on the left, it seems that an altitude A is a minimal separating set. A is a self-similar set with two mappings with contraction factors $r = (\sqrt{5}-1)/2$ and r^3.

Problem 5. Given a self-similar set F with open set condition or only exact overlaps, does there exist a self-similar subset (or graph-directed construction) C which separates F and has the separating dimension?

Geodesics and interior distance. On a manifold, two points x, y can be connected by a geodesic—a rectifiable curve of minimal distance. Do there exist geodesics on fractals? Many self-similar sets contain lines, such that any two points $x, y \in F$ can be connected by a rectifiable curve—a polygonal path consisting of infinitely many line segments. This is true for the Sierpiński triangle and the Sierpiński carpet, for the right-hand part of Fig. 1 and the golden triangle in Fig. 5. The interior distance $\rho(x, y)$ on such fractals can be defined in the same way as on differentiable manifolds. It is *the minimum one-dimensional Hausdorff measure of a connected set C which contains both x and y.* This is called interior distance since we must walk within F. No shortcuts through the complement are allowed.

For post-critically finite sets F, as in Figs. 1, 2, and 6 left, we have the structure of a graph, with vertices on the intersection of pieces, see [18]. Given x and y, we need only compare finitely many possible connecting curves. (More precisely, we first consider the construction on level n, where we definitely have a finite graph, and replace x, y by the nearest boundary points. Then we let n run to infinity. Since the addresses of vertices are eventually periodic, we can stop at finite n.) It turns out that the connecting curves of boundary points form themselves a self-similar set or graph-directed construction [9, 24]. Actually, neighbor maps can directly used to establish the graph-directed construction [7]—by computer if necessary.

We explain the method for Fig. 2 where we have a tree—only one possible connection between any points x, y. Let "base" B the arc connecting the vertices with address $00\overline{1}$ and $\overline{1}$, and "diagonal" D the arc connecting $0\overline{1}$ with $\overline{1}$. In Fig. 8 we see that D is the union of bases of the pieces: $D = f_0(B) \cup f_1(B)$. Moreover, $B = f_{00}(D) \cup f_1(B)$. Putting both equations together, we get

$$B = f_{000}(B) \cup f_{001}(B) \cup f_1(B)$$

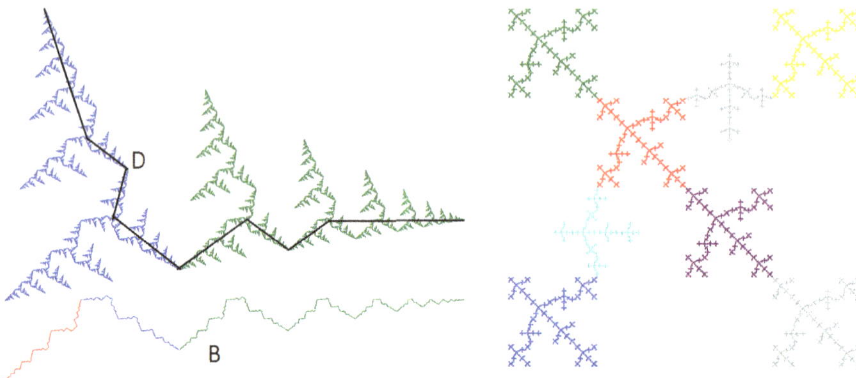

Fig. 8 *Left*: geodesics B and D (approximation) in Fig. 2. *Right*: an example with two different exponents of geodesics

which means that B is a self-similar set with three mappings with contraction factors $r_0^3, r_0^2 r_1, r_1$. The open set condition is inherited from the system $\{f_0, f_1\}$. So B and D have positive finite Hausdorff measure $\tilde{\mu}$ of dimension δ given by (11). For the left of Fig. 2 we have $r_0 = r_1 = r = (\sqrt{5} - 1)/2$ and $r^\delta + 2r^{3\delta} = 1$ which results in $\delta \approx 1.10$. See Fig. 8.

Once we have geodesics connecting the boundary points, we can define arcs connecting arbitrary points x, y as countable unions of copies of the basic arcs. For the theory, it is enough to know that we have a dimension δ of basic geodesics. As the example on the right of Fig. 8 shows, we need an irreducibility condition for the graph of our basic geodesics—otherwise we may have different exponents δ for different directions. When δ is known, we can define the interior distance of $x, y \in F$ as minimal δ-dimensional Hausdorff measure of a connecting set:

$$\rho(x, y) = \inf\{\mu^\delta(C) \mid C \subset F \text{ connected}, x, y \in C\}\,. \tag{12}$$

Problem 6. How to define geodesics when intersections of pieces are infinite? Will they have a self-similar structure? Is there a dimension δ of geodesics?

Remarks on analysis. We have seen that in certain cases we can define not only volume, but also surface and interior distance by means of Hausdorff measures of self-similar constructions. The next thing would be proper analysis: harmonic functions, the Dirichlet problem, heat transfer, eigenvalues of the Laplace operator. For post-critically finite self-similar sets this has been accomplished—see the books of Kigami [18] and Strichartz [25] and their references. One key concept is the resistance metric instead of an interior metric, which takes account of the number of parallel connections from x to y, which may vary on the way. The results for post-critically finite fractals are nearer to analysis of the interval than to analysis in the complex plane. Among others, the vector space of harmonic functions on F is finite-dimensional. For physically meaningful examples it should be infinite-dimensional.

It would be nice to have such examples, even though a general analysis on metric spaces is being developed on the base of Dirichlet forms and heat kernel estimates. This motivates our last problem which will conclude this little survey.

Problem 7. Can harmonic functions and a Laplace operator be constructed for the fractals in Fig. 7?

References

1. Bandt, C.: Self-similar sets 1. Topological Markov chains and mixed self-similar sets. Math. Nachr. **142**, 107–123 (1989)
2. Bandt, C.: Self-similar sets 5. Integer matrices and fractal tilings of $\mathbb{R}^n$. Proc. Am. Math. Soc. **112**, 549–562 (1991)
3. Bandt, C.: Local geometry of fractals given by tangent measure distributions. Monatshefte Math. **133**, 265–280 (2001)
4. Bandt, C.: Introduction to fractals, talk in Kochi, India. 9 Nov 2013
5. Bandt, C., Graf, S.: Self-similar sets 7. A characterization of self-similar fractals with positive Hausdorff measure. Proc. Am. Math. Soc. **114**, 995–1001 (1992)
6. Bandt, C., Keller, K.: Self-similar sets 2. A simple approach to the topological structure of fractals. Math. Nachr. **154**, 27–39 (1991)
7. Bandt, C., Mesing, M.: Self-affine fractals of finite type. Banach Center Publ. **84**, 131–148 (2009)
8. Bandt, C., Rao, H.: Topology and separation of self-similar fractals in the plane. Nonlinearity **20**, 1463–1474 (2007)
9. Bandt, C., Stahnke, J.: Self-similar sets 6. Interior distance on deterministic fractals. Greifswald (1990, preprint)
10. Barnsley, M.F.: Fractals Everywhere, 2nd edn. Academic, Cambridge (1993)
11. Broomhead, D., Montaldi, J., Sidorov, N.: Golden gaskets: variations on the Sierpiński sieve. Nonlinearity **17**, 1455–1480 (2004)
12. Edgar, G.A.: Measure, Topology, and Fractal Geometry. Springer, New York (2008)
13. Falconer, K.J.: Fractal Geometry: Mathematical Foundations and Applications. Wiley, New York (1990)
14. Gelbrich, G.: Crystallographic reptiles. Geometriae Dedicata **51**, 235–256 (1994)
15. Hochman, M.: Self-similar sets, entropy and additive combinatorics (2013). arXiv:1307.6399
16. Hutchinson, J.E.: Fractals and self-similarity. Indiana Univ. Math. J. **30**, 713–747 (1981)
17. Kameyama, A.: Julia sets of postcritically finite rational maps and topological self-similar sets. Nonlinearity **13**, 165–188 (2000)
18. Kigami, J.: Analysis on Fractals. Cambridge University Press, Cambridge (2001)
19. Kirsch, B.: Julia-Mengen linearer Abbildungen. Diplomarbeit, Greifswald (2000)
20. Mattila, P.: Geometry of Sets and Measures in Euclidean Spaces. Cambridge University Press, Cambridge (1999)
21. Moran, P.A.P.: Additive functions of intervals and Hausdorff measure. Proc. Camb. Philos. Soc. **42**, 15–23 (1946)
22. Ngai, S.M., Wang, Y.: Hausdorff dimension of self-similar sets with overlaps. J. Lond. Math. Soc. **63**, 655–672 (2001)
23. Schief, A.: Separation properties for self-similar sets. Proc. Am. Math. Soc. **122**, 111–115 (1994)
24. Strichartz, R.S.: Isoperimetric estimates on Sierpinski gasket type fractals. Trans. Am. Math. Soc. **351**, 1705–1752 (1999)

25. Strichartz, R.S.: Differential Equations on Fractals: A tutorial. Princeton University Press, Princeton (2006)
26. Strichartz, R.S., Wang, Y.: Geometry of self-affine tiles I. Indiana Univ. Math. J. **48**(1), 1–24 (1999)
27. Vince, A.: Self-replicating tiles and their boundary. Discrete Comput. Geom. **21**, 463–476 (1999)

An Introduction to Julia and Fatou Sets

Scott Sutherland

Abstract We give an elementary introduction to the holomorphic dynamics of mappings on the Riemann sphere, with a focus on Julia and Fatou sets. Our main emphasis is on the dynamics of polynomials, especially quadratic polynomials.

Keywords Julia set • Fatou set • Holomorphic dynamical systems • Normal family • Böttcher coordinate • Parabolic orbit • Siegel disk • Cremer point

1 Introduction

In this note, we briefly introduce some of the main elementary ideas in holomorphic dynamics. This is by no means a comprehensive coverage. We do not discuss the Mandelbrot set, despite its inherent relevance to the subject; the Mandelbrot set is covered in companion articles by Devaney [17, 18].

For a more detailed and comprehensive introduction to this subject, the reader is referred to the book [25] by Milnor, the survey article [6] by Blanchard, the text by Devaney [16], or any of several other introductory texts ([4, 11], etc.). References on the history of complex dynamics include [1–3].

These notes were developed from lectures given at the International Conference on Fractals and Wavelets, held at the Rajagiri School of Engineering and Technology in Kerala, India in November, 2013.

S. Sutherland (✉)
Department of Mathematics and Institute for Mathematical Sciences,
Stony Brook University, Stony Brook, NY 11794, USA
e-mail: scott@math.sunysb.edu

C. Bandt et al. (eds.), *Fractals, Wavelets, and their Applications*, Springer Proceedings in Mathematics & Statistics 92, DOI 10.1007/978-3-319-08105-2_3,

Holomorphic dynamics is the study of the iterates of a holomorphic map f on a complex manifold. Classically, this manifold is one of the complex plane $\mathbb{C}$, the punctured plane $\mathbb{C}^*$, or the Riemann sphere $\hat{\mathbb{C}} = \mathbb{C} \cup \{\infty\}$.

In these notes, we shall primarily restrict our attention to the last case, where f is a rational map (in fact, most examples will be drawn from quadratic polynomials $f(z) = z^2 + c$, with z and $c \in \mathbb{C}$).

Based on the behavior of the point z under iteration of f, the Riemann sphere $\hat{\mathbb{C}}$ is partitioned into two sets

- The *Fatou set* $\mathcal{F}_f$ (or merely $\mathcal{F}$), on which the dynamics are tame, and
- The *Julia set* $\mathcal{J}_f$ (or $\mathcal{J}$), where there is sensitive dependence on initial conditions and the dynamics are chaotic.

We shall define these sets more precisely in Sect. 5, but will use these informal definitions for the present to give some intuition.

Given the focus of this conference, we should point out that there is an inverse relationship between the approach of holomorphic dynamics and that of iterated function systems (IFS). More specifically, the Julia set of a rational map $f : \hat{\mathbb{C}} \to \hat{\mathbb{C}}$ corresponds to the attractor for the IFS consisting of $\{g_1, g_2, \ldots, g_d\}$, where the g_i are branches of the inverse of f restricted to a suitable domain. For example, the Julia set of $z^2 - 1$ is the attractor of the IFS $\{w \mapsto \sqrt{w+1}, w \mapsto -\sqrt{w+1}\}$.

2 Linear Maps

Before turning to the dynamics of rational maps, let us first discuss iteration of a single linear map. Much of the theory of holomorphic dynamics doesn't apply in this case, so it is always excluded. However, a brief overview of what happens will be helpful.

The case of iteration of a single linear map $f : \hat{\mathbb{C}} \to \hat{\mathbb{C}}$ is very simple. Such an f is a Möbius transformation of the form

$$f(z) = \frac{az+b}{cz+d} \qquad \text{with } ad - bc \neq 0,$$

which can be viewed as an element of $PSL(2, \mathbb{C})$, and has easily understood dynamics.

Except for the identity and the trivial map $z \mapsto 1/z$, these mappings have two fixed points, counted with multiplicity. If the fixed points are *distinct*, we may make a holomorphic change of coordinates moving one of them to infinity and the other to zero. In this case, the mapping becomes of the form $z \mapsto \lambda z$ with $\lambda \in \mathbb{C}$. See Fig. 1.

- If $|\lambda| \neq 1$, the mapping $z \mapsto \lambda z$ is called *loxodromic* (or *hyperbolic*): under iteration, points move away from one of the fixed points (which is repelling) and toward the other (which is attracting).

Fig. 1 *Left*: under iteration of a loxodromic linear map, points spiral away from one fixed point (not visible) and towards the other. *Center*: an elliptic transformation corresponds to rotation of the Riemann sphere. *Right*: A parabolic transformation corresponds to $z \mapsto 1 + z$. Here points leave infinity from one side (the negative half-plane) and return on the other

- If $|\lambda| = 1$, $z \mapsto \lambda z$ has two *elliptic* fixed points: nonzero finite points orbit around the fixed points at zero and ∞ along an elliptic path. This corresponds to a rotation of the Riemann sphere by the angle $\operatorname{Arg} \lambda$.
- If the two fixed points of f *coincide*, the mapping is conjugate to $z \mapsto z + 1$. Here, the point at infinity is a *parabolic* fixed point: orbits leave from one side, and return on the other.

3 First Examples

Now we turn to our main subject, beginning with some simple examples. First, we remark that in holomorphic dynamics, we only consider mappings f which have degree at least two. The notion of "degree" of a rational mapping of $\hat{\mathbb{C}}$ is unambiguous (unlike in the case of higher-dimensional complex manifolds): the algebraic degree (i.e., the highest power of z in the numerator and denominator of f) and the topological degree (i.e., the number of pre-images of a typical point) coincide.

3.1 *The Map $z \mapsto z^2$*

We begin with an elementary analysis of the simplest rational map, $f(z) = z^2$.

If we write z in polar coordinates with $r = |z|$, $\theta = \operatorname{Arg} z$, then the mapping is $(r, \theta) \mapsto (r^2, 2\theta)$, and it is easy to make the following conclusions.

- There are three fixed points for f: 0, 1, and ∞.
- If $|z| < 1$, then $f^n(z) \to 0$; if $|z| > 1$, then $f^n(z) \to \infty$. Thus, 0 *and* ∞ *are attracting* fixed points (in fact, superattracting; see Definition 5.5). A point

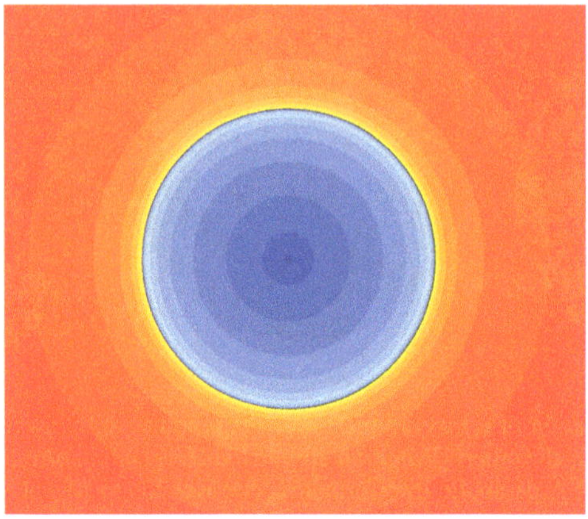

Fig. 2 For $f(z) = z^2$, the Julia set J_f is the unit circle (*black*), and the Fatou set has two components: Bas(∞) (*red* and *orange*) and Bas(0) (*blue*)

p is called an *attracting fixed point* if for z in a neighborhood of p, we have $f^n(z) \to p$.

- If $|z| = 1$, then $|f^n(z)| = 1$ for all n. For this mapping, the unit circle is *forward-* and *backward-invariant*. That is, for each point on the unit circle, the entire forward orbit of that point and all of its preimages are also on the unit circle. Furthermore, every point in a deleted neighborhood of 1 will eventually leave that neighborhood under some iterate of f. The point 1 *is a repelling fixed point*.
- In fact, any small neighborhood of a point z on the unit circle contain points which tend to ∞, points which tend to 0, and points which remain on the unit circle. As we shall see, such behavior characterizes membership in the Julia set.

In the case of $f(z) = z^2$, $\mathcal{J}_f$ is the unit circle, and the Fatou set $\mathcal{F}_f$ has two components: the set of points in the unit disk, which iterate to 0, and those outside the disk, all of which iterate to infinity. (See Fig. 2.)

Even though behavior of f on the Julia set is merely that of angle doubling, there is a surprisingly rich collection of behaviors. For example, there are periodic points of all periods and the pre-images of any point are dense in the unit circle. There are also points whose forward orbit is dense in the circle.

We will refer to the set of points which iterate to ∞ as the *basin of* ∞ and denote it by Bas(∞). More generally, for $p \in \hat{\mathbb{C}}$ we shall use the notation

$$\mathrm{Bas}(p) = \left\{ z \in \hat{\mathbb{C}} \mid f^n(p) \to p \right\}.$$

If p is a periodic point of period m, we can extend this definition as

$$\mathrm{Bas}(p) = \bigcup_{0 \le j < m} \left\{ z \in \hat{\mathbb{C}} \mid f^{n+j}(p) \to p \right\},$$

or, equivalently, as the union of the basins of each point in the orbit of p under f^m.

Because ∞ plays a special role for polynomial maps $f(z)$, it is common and useful to define the *filled Julia set of* f as

$$\mathcal{K}_f = \hat{\mathbb{C}} \setminus \mathrm{Bas}(\infty) = \{z \in \mathbb{C} \mid f^n(z) \text{ is bounded}\}.$$

It is worth noting that for $f(z) = z^2$, we can change coordinates to interchange the roles of ∞ and 0 since $1/f(z) = f(1/z)$.

$$\begin{array}{ccc} \mathrm{Bas}(\infty) & \xrightarrow{f} & \mathrm{Bas}(\infty) \\ \big\downarrow{\scriptstyle 1/z} & & \big\downarrow{\scriptstyle 1/z} \\ \mathrm{Bas}(0) & \xrightarrow{f} & \mathrm{Bas}(0) \end{array}$$

That is, f acting on Bas(0) is *holomorphically conjugate* to f on Bas(∞) via $z \mapsto 1/z$. Note also that $f'(0) = 0$, $f'(1) = 2$, and the derivative at ∞ is also 0 (in the $1/z$ coordinate chart).

3.2 *The Map* $z \mapsto z^2 + \epsilon$

Now let's change things a little, and consider $f(z) = z^2 + \epsilon$. How does this affect the dynamics?

- We still have three fixed points in $\hat{\mathbb{C}}$:
  ∞ (with derivative 0), $\alpha = (1 - \sqrt{1 - 4\epsilon})/2$, and $\beta = (1 + \sqrt{1 - 4\epsilon})/2$.
- For $|z|$ large (i.e., near ∞), we still have $f^n(z) \to \infty$, and (as long as ϵ is not too big), for z near α, we have $f^n(z) \to \alpha$.
- As in the case of z^2, β is a repelling fixed point (since $|f'(\beta)| = |2\beta| > 1$). As such, it lies in the Julia set $\mathcal{J}_f$.
- Furthermore (with a bit more effort), we can show that we have a conformal map ϕ so that

$$\begin{array}{ccc} \mathrm{Bas}_f(\infty) & \xrightarrow{f} & \mathrm{Bas}_f(\infty) \\ \big\downarrow{\scriptstyle \phi} & & \big\downarrow{\scriptstyle \phi} \\ \mathbb{D} & \xrightarrow{z \mapsto z^2} & \mathbb{D} \end{array}$$

However, things are a little more complicated near the fixed points α and β.

In a neighborhood U of the attracting fixed point α, we have a conformal map ϕ so that

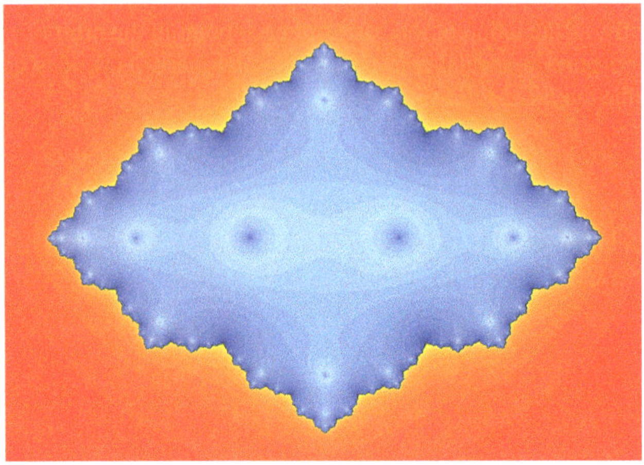

Fig. 3 Julia and Fatou sets for $f(z) = z^2 - 1/2$. As in Fig. 2, the Fatou set $\mathcal{F}_f$ consists of two components shown in *reds* and *blues* and J_f is in *black*, forming the boundary between the two components of $\mathcal{F}_f$

$$\begin{array}{ccc} U & \xrightarrow{\ f\ } & U \\ \big\downarrow \phi & & \big\downarrow \phi \\ \mathbb{C} & \xrightarrow{z \mapsto f'(\alpha) z} & \mathbb{C} \end{array}$$

This neighborhood U cannot include 0, since f is not one-to-one on any neighborhood of 0. This means we cannot hope to extend the local conjugacy above to the whole of $\mathrm{Bas}(\alpha)$, unlike in the case of $\mathrm{Bas}(\infty)$.

One can show that in this case, the Julia set (which is the complement of $\mathrm{Bas}(\alpha)$ and $\mathrm{Bas}(\infty)$) is a Jordan curve with Hausdorff dimension greater than one. See Fig. 3.

4 Some History

The study of the iteration of complex analytic functions began more than a century and a quarter ago. In the 1870s, Schröder [29, 30] investigated the convergence of iterative algorithms for solving equations, with a particular interest in the convergence of Newton's method, which corresponds to the iteration of the function $N_f(z) = z - f(z)/f'(z)$. He discovered that a root ρ of f corresponds to a super-attracting fixed point of N_f; this led him to generalize Newton's method to other numerical methods. In addition, Schröder showed that Newton's method for $f(z) = z^2 - 1$ converges globally in the right half-plane to the root 1, in the left half-plane to -1, and observed sensitive dependence to initial conditions along

Fig. 4 The dynamics of Newton's method for the polynomial $z^3 - 1$. The basins of each of the three roots are colored in shades of *blue*, *green*, and *orange*; the Julia set is the boundary between the colors

the imaginary axis. Later, Cayley independently (and via quite different methods) obtained similar results in [12]. While both Cayley and Schröder had hopes of extending their understanding to higher degree polynomials, they were unable to do so. Figure 4 might give an idea as to the source of some of their difficulties.

Later, Kœnigs [23] greatly extended Schröder's work to further generality, studying the *Schröder functional equation* (SFE) $\phi(f(p)) = f'(p)\phi(p)$ in the neighborhood of a fixed point p. Kœnigs was able to show that the mapping f was locally conjugate to multiplication by its derivative at the fixed point p, in the case where $|f'(p)|$ was not 1 or 0. The more complicated case of $f'(p)$ a root of unity was studied 1897 by Leau [24], and the case of $f'(p) = 0$ was treated by Böttcher in 1904 [8]. (Böttcher was one of the first, if not the first, researchers concerned with developing a general, global theory of iteration of rational maps.) The very difficult cases where $|f'(p)| = 1$ with $f'(p)$ not a root of unity were not understood until the work of Cremer in 1927 [14, 15] and Siegel in 1942 [32]. We will summarize these results in Sect. 8.

In the time immediately after World War I, Fatou [19,20] and Julia [22] laid down the foundations of complex dynamics, looking at the theory of iterated rational functions from a global point of view. Both of them had recently encountered Montel's theory of normal families [26,27] and realized its importance to complex dynamics Fatou and Julia independently proved that the domain of normality must either be empty, or have one, two, or infinitely many components. Each showed that Julia sets are typically fractal, and clearly were able to visualize and understand the very complicated structure of Julia sets, studying and explaining complicated behavior.

The huge amount of interest in the field of iteration of functions of one complex variable, led by the work of Fatou and Julia, continued until the 1930s when it inexplicably faded into obscurity. Although a few important mathematicians worked in this field during that time, it wasn't until the 1980s that it revived, probably due to the advent of accessible computers which enabled others to visualize the extraordinary beauty and complexity that Julia and Fatou obviously understood.

5 Normal Families

The key tool that enabled Fatou and Julia's breakthrough was their realization of the relevance of Montel's work on normal families to the theory of iteration.

Definition 5.1 (Normal Family). Let U be an open subset of the Riemann sphere and $\mathfrak{F} = \{f_i \mid i \in I\}$ be a family of meromorphic functions indexed by I and defined on U with values in $\hat{\mathbb{C}}$. The family $\mathfrak{F}$ is a *normal family* if every sequence f_n contains a subsequence f_{n_j} which converges uniformly on compact subsets of U.

Arzela's theorem gives us an equivalent, and often more useful, condition for checking normality. If X is a metric space with metric d, a family of functions $\{f_i : X \to X\}$ is *equicontinuous* if for every $\epsilon > 0$, there exists $\delta > 0$ so that $d(x, y) < \epsilon$ implies $d(f_i(x), f_i(y)) < \delta$ for all i.

Theorem 5.2 (Arzela). *A family of meromorphic functions* $\left\{f_i : U \to \hat{\mathbb{C}}\right\}$ *is normal if and only if it is equicontinuous on every compact subset of* U.

Corollary 5.3. *If a family of holomorphic functions* $\{f_i : U \to \mathbb{C}\}$ *is locally uniformly bounded, then it is a normal family.*

The concept of normal families enables us to define the Julia set $\mathcal{J}_f$ and the Fatou set $\mathcal{F}_f$.

Definition 5.4. A point z is in the *Fatou set* for f if there is a neighborhood U of z such that the family of iterates $\{f^n|_U\}$ is normal. The complement of the Fatou set is called the *Julia set*.

If p is a periodic point of period n, the *multiplier* λ of the periodic orbit is $\lambda = (f^n)'(p)$; by the chain rule, this is the product of the derivatives of f along the periodic orbit.

Definition 5.5. A periodic orbit p with multiplier λ is

- *superattracting* if $\lambda = 0$,
- *attracting* if $0 < |\lambda| < 1$,
- *neutral* if $|\lambda| = 1$, and
- *repelling* if $|\lambda| > 1$.

The next result follows easily from the definitions and Arzela's theorem.

Proposition 5.6. *If* p *is an attracting or superattracting periodic point, then* $\mathrm{Bas}(p) \subset \mathcal{F}$.
If p *is a repelling periodic point, then* $p \in \mathcal{J}$.

The Fatou set is sometimes called "the domain of normality" or "the domain of equicontinuity."

Corollary 5.7. *The Fatou set* $\mathcal{F}$ *is an open set which is completely invariant. That is, if* $z \in \mathcal{F}$*, then* $f(z) \in \mathcal{F}$ *and* $f^{-1}(z) \subset \mathcal{F}$.
The Julia set $\mathcal{J}$ *is a completely invariant and compact set in* $\hat{\mathbb{C}}$.

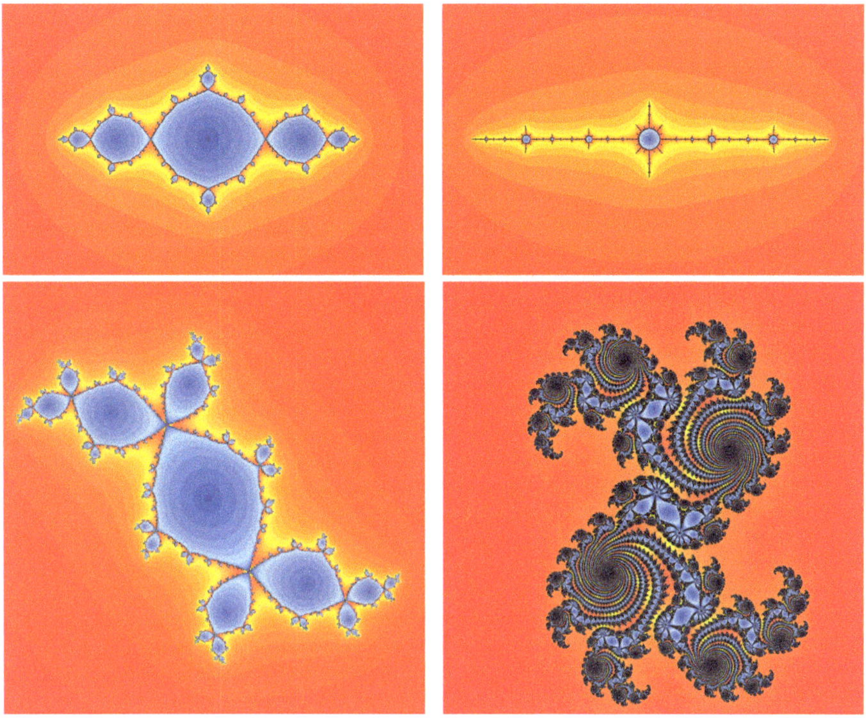

Fig. 5 Pictured are J_f (in *black*) and $\mathcal{F}_f$ (in *reds* and *blues*) for several examples. (*top left*) "The basilica" corresponds to $f(z) = z^2 - 1$, and has an attracting period 2 orbit. (*top right*) "The airplane," which has an attracting period 3 orbit. (*bottom left*) "Douady's rabbit," also has an attracting period 3 orbit. (*bottom right*) A map with an attracting period 72 orbit

Proof. The fact that $\mathcal{F}$ is open follows immediately from the definition. We have the invariance of $\mathcal{F}_f$ since f is a continuous, open mapping. Since J is the complement of a completely invariant, open set, its invariance and compactness follows. □

A few examples of $\mathcal{J}$ and $\mathcal{F}$ for polynomials with attracting periodic points are shown in Fig. 5.

While the previous examples have all had attracting periodic orbits, for many polynomials, all finite periodic orbits are repelling. As examples, consider the mappings $f(z) = z^2 - 2$ and $f(z) = z^2 + i$.

Since infinity is always an attracting fixed point for a polynomial, we have $\mathrm{Bas}(\infty) \subset \mathcal{F}$. But if all orbits in $\mathbb{C}$ are repelling, there can be no other components of $\mathcal{F}$. That is, the filled Julia set is equal to the Julia set, i.e. $\mathcal{K} = \mathcal{J}$. In such a situation, if $\mathcal{J}$ is connected it will be a *dendrite*. See Fig. 6.

In the case of a polynomial where the Julia set is not connected, it consists of infinitely many components. We will examine this case shortly.

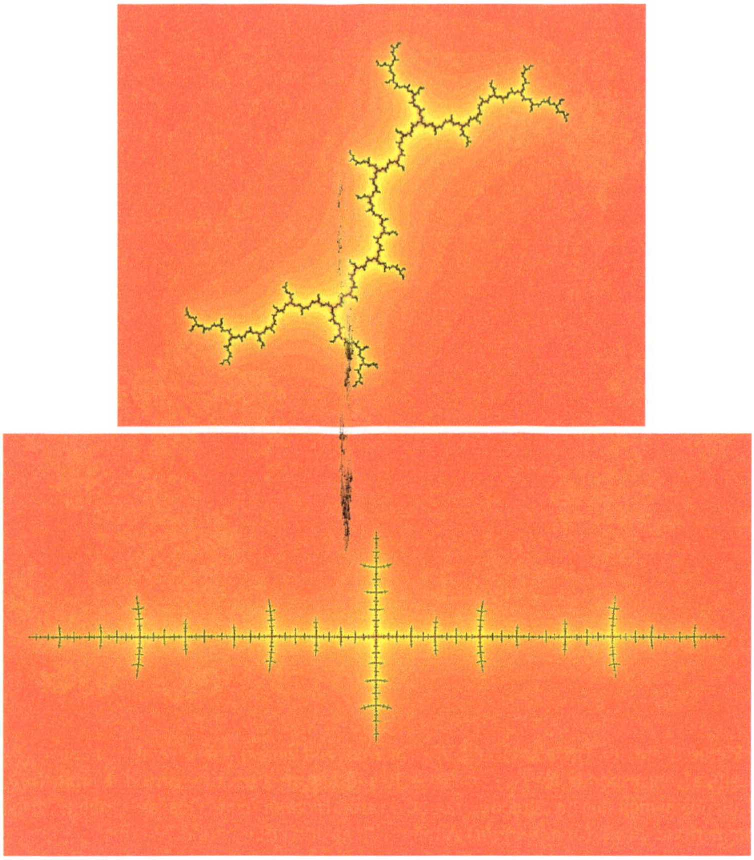

Fig. 6 The Julia sets for $z^2 - i$ and the limit of period doubling ($z^2 - 1.4011552\ldots$), which are connected but have no attracting periodic orbits

6 The Local Theory

As mentioned in Sect. 4, the behavior in a neighborhood of a fixed point or periodic point was, in most cases, well known to Fatou, Julia, and their predecessors. In this section, we briefly address those situations.

6.1 Böttcher coordinates

In the case where $f(z)$ is a polynomial, we have $\mathrm{Bas}(\infty) \subset \mathcal{F}$. Furthermore, since polynomials have no poles, $\mathrm{Bas}(\infty)$ is a completely invariant subset of the Fatou set.

For $|z|$ large, $f(z)$ is conjugate to $w \mapsto w^d$, where d is the degree of f. Let U be a neighborhood of ∞; there is a holomorphic map ϕ so that

$$\begin{array}{ccc} U & \xrightarrow{\ f\ } & U \\ \Big\downarrow \phi & & \Big\downarrow \phi \\ \hat{\mathbb{C}} \smallsetminus \overline{\mathbb{D}} & \xrightarrow{w \mapsto w^d} & \hat{\mathbb{C}} \smallsetminus \overline{\mathbb{D}}, \end{array}$$

where d is the degree of the polynomial f.[1]

We now explain how to extend this conjugacy to a larger neighborhood. Choose r large so that $A_0 = \phi^{-1}\left(\overline{\mathbb{D}}_{r^d} \smallsetminus \mathbb{D}_r\right)$ is contained in U. Here $\mathbb{D}_r$ denotes the disk of radius r centered at the origin. Now define $A_1 = f^{-1}(A_0)$. As long as there is no critical point of f in A_0, the map $f : A_1 \to A_0$ will be a d-fold covering map.

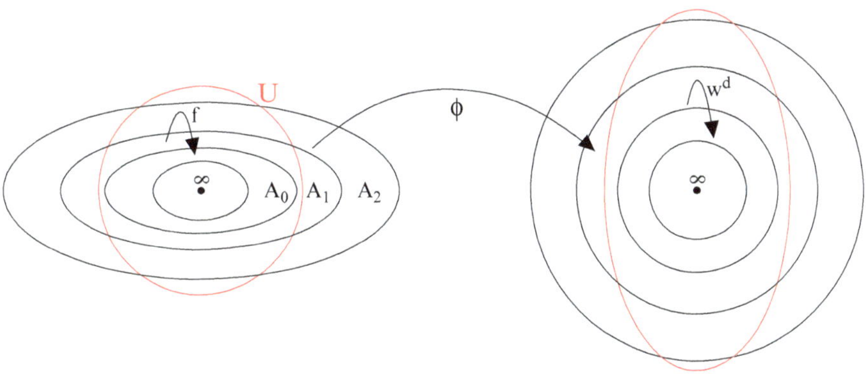

We can extend the conjugacy ϕ to $U \cup A_1$ by setting $\phi(z) = \phi(f(z))^{1/d^k}$, where we take care to choose the appropriate branches of the inverses. There is no problem doing this, since both maps are covering maps. This process can be continued inductively as long as A_k does not contain a critical point of f. This gives us a set of coordinates on $\{\infty\} \cup \bigcup_{k=0}^{\infty} A_k$; these are usually called *Böttcher coordinates*.

The preimages of curves of constant radius under ϕ are called *equipotentials*; the preimage of the radial line $te^{2\pi i\theta}$ $(t > r_0)$ is called the *external ray of angle* θ (where θ is measured in *turns*); we denote it by $\mathcal{R}_\theta$, and use $\mathcal{R}_\theta(t)$ as a parameterization, with $\phi(\mathcal{R}_\theta(t)) = te^{2\pi i\theta}$.

If the orbits of all critical points for the polynomial f are bounded, these coordinates will extend to all of $\mathrm{Bas}(\infty)$; see Figs. 7 and 8. Even in the case where a finite critical value lies in $\mathrm{Bas}(\infty)$, equipotentials and all but a countable number of external rays can be defined on the whole of $\mathrm{Bas}(\infty)$, as we shall see shortly.

If $\lim_{t\to 1} \mathcal{R}_\theta(t)$ exists, we say that the ray $\mathcal{R}_\theta$ *lands* at a point z in the Julia set. If all rays land, then the Julia set must be locally connected (since then we have ϕ^{-1} extending as a continuous mapping from $\overline{\mathbb{D}}$ to $\overline{\mathrm{Bas}(\infty)}$).

[1]More precisely, we set $\phi(z) = \left(\lim_{k\to\infty} f^k(z)\right)^{1/d^k}$.

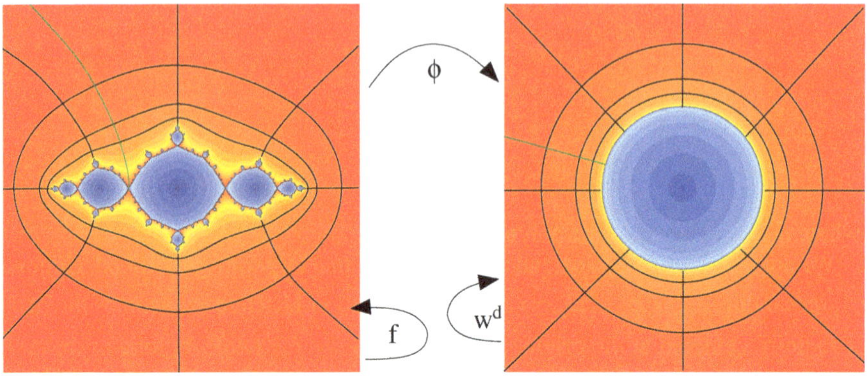

Fig. 7 Böttcher coordinates on Bas(∞), with equipotential lines and external rays shown

Fig. 8 On Bas(∞), preimages of the annulus A_0 (as described earlier) are shown, where points z for which $f^n(z) \in A_0$ has positive imaginary part are colored in shades of *red*, and those where $\text{Im}(f^n(z)) < 0$ are colored *green*. This coloring enables one to read off equipotentials and external angles of the form $p/2^n$

It should be apparent that the previous construction can be immediately adapted to a bounded Fatou component in the case of a fixed (or periodic) point p with $f'(p) = 0$. In this situation, the mapping will be conjugate to w^k, where k is the smallest integer such that the kth derivative $f^{(k)}(p)$ is nonzero ($k \geq 2$). In this case, the corresponding ray $\mathcal{R}$ is called an *internal ray*.

For all points $z \in \text{Bas}(\infty)$, the *escape rate* ($|\phi(z)|$) is well defined, since we have

$$|\phi(z)| = \left(\lim_{k\to\infty} |f^k(z)|\right)^{1/d^k},$$

and the dth root is unambiguous for non-negative numbers.

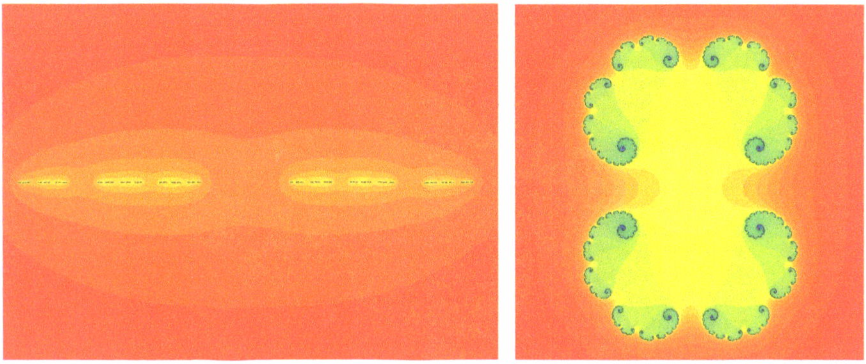

Fig. 9 Two quadratic Julia sets homeomorphic to the Cantor set

This enables us to define the *Green's function* for the filled Julia set $\mathcal{K}_f$ (sometimes also called the *canonical potential function*) as

$$G_f(z) = \begin{cases} \log|\phi(z)| & z \notin \mathcal{K}_f \\ 0 & z \in \mathcal{K}_f \end{cases}.$$

One sees easily that G_f is continuous everywhere and harmonic, and that $G_f(f(z)) = dG_f(z)$.

The level curves $G_f(z) = \text{constant} > 0$ are *equipotentials for* f; observe that f sends one equipotential to another equipotential. While we won't go into details here, one can use G_f to define a measure on the Julia set which corresponds to the harmonic measure on K_f.

Observe that if there is a critical point which lies in $\text{Bas}(\infty)$, we cannot extend the conjugacy ϕ without ambiguity to the entire basin of infinity, since there will be at least two rays which land at the critical point. (Of course, we should not expect to be able to extend the conjugacy, since w^d has no nonzero finite critical points. Instead, the conjugacy will correspond to a Blaschke product.)

By analytically continuing the inverse of ϕ we can define external rays and equipotentials on the whole of $\text{Bas}(\infty)$, although rays which contain preimages of the critical point(s) will not be disjoint.

Consider $f(z) = z^2 + c$ with $c < -2$. In this case, $f^n(0) \to \infty$, so $0 \in \text{Bas}(\infty)$. The equipotential $G_f(z) = \log|c|$ will be an (analytic) circle, but its preimage $G_f(z) = \log|c|/2$ will branch at the origin, making a figure-8 shape. Note that the two preimages of the ray passing through c must cross at the origin (since it is a critical point), although all other rays extend without ambiguity to the interior of the figure-8. Of course, there will be two more rays which branch at the pre-image of 0, and so on. See Fig. 9.

Observe the close similarity to the classical construction of a Cantor set: the Julia set must lie in the interior of the nested figure-8 shapes. If these figure-8s contract to points, the result will be homeomorphic to a Cantor set.

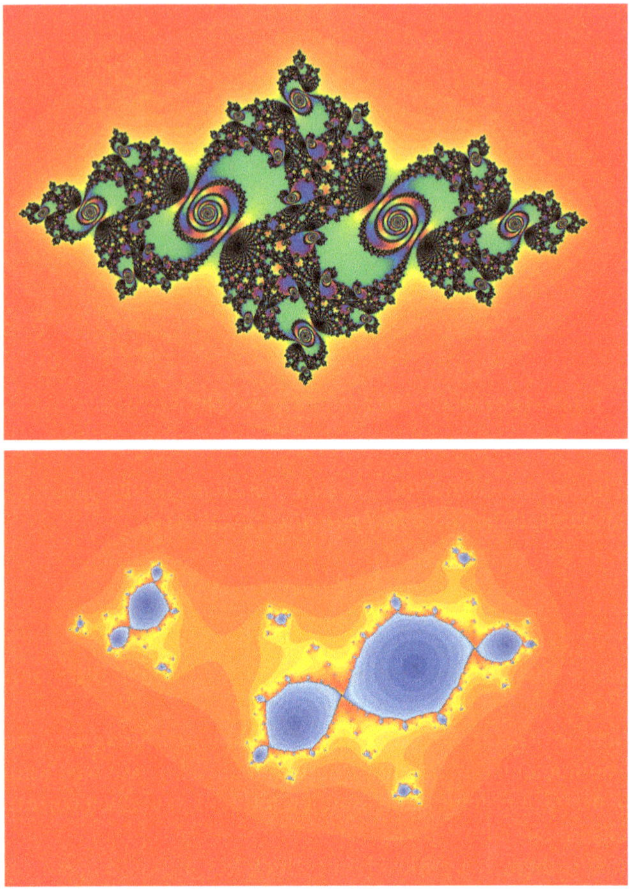

Fig. 10 Another quadratic Cantor Julia set, and a cubic with only one critical point which escapes

This is indeed what happens if *all* the critical points of f lie in the basin of ∞. For quadratic polynomials, since there is only one critical point, we have the following.

Lemma 6.5 (Dichotomy for Quadratic Polynomials). *Let $f(z)$ be a quadratic polynomial with critical value c.*

- *If the orbit of c is unbounded, then $\mathcal{J}_f$ is homeomorphic to a Cantor set.*
- *If the orbit of c is bounded, then $\mathcal{J}_f$ is connected.*

Using this dichotomy, one can define the *Mandelbrot set* as

$$\mathcal{M} = \left\{c \in \mathbb{C} \mid \text{the orbit of 0 under } z \mapsto z^2 + c \text{ is bounded}\right\}.$$

Of course, for a polynomial of degree three or higher there can be an intermediate case where the Julia set consists of infinitely many components but is not a Cantor set; see Fig. 10.

7 Montel's Theorem and Its Consequences

One powerful tool to understanding the Julia set $\mathcal{J}$ is Montel's theorem.

Theorem 7.1 (Montel). *Let $\mathfrak{F}$ be a family of meromorphic functions defined on a domain U. If there exist three points $a, b, c \in \hat{\mathbb{C}}$ so that*

$$\{a, b, c\} \cap \left(\bigcup_{f \in \mathfrak{F}} f(U) \right) = \emptyset,$$

then $\mathfrak{F}$ is a normal family.

Corollary 7.2. *Let $z \in \mathcal{J}_f$. Then for any neighborhood U of z, the exceptional set*

$$\mathcal{E}_U = \hat{\mathbb{C}} \setminus \bigcup_{n>0} f^n(U)$$

contains at most two points.

The points in $\mathcal{E}_U$ are called *exceptional points.*

Note that for a polynomial $p(z)$, ∞ is an exceptional point. We have $p^{-1}(\infty) = \{\infty\}$. Furthermore, since ∞ is a superattracting point, $\mathrm{Bas}(\infty) \subset \mathcal{F}$, and the Julia set $\mathcal{J}$ lies in a bounded region of $\mathbb{C}$. Since p has no poles, $p(\mathbb{C}) = \mathbb{C}$; thus ∞ is an exceptional point.

Theorem 7.3. *Suppose $z \in \mathcal{J}_f$, and let $E_z = \bigcup E_U$, where U ranges over all neighborhoods of z. Then*

- *If $\mathcal{E}_z$ contains exactly one point, then f is conjugate to a polynomial.*
- *If $\mathcal{E}_z$ contains two points, then either f is conjugate to z^d or $1/z^d$, where d is the degree of f.*

In both cases, E_z does not depend on the choice of $z \in \mathcal{J}$, and $\mathcal{E}_z \subset \mathcal{F}$.

The proof follows from conjugating f by a Möbius transformation which moves $\mathcal{E}_z$ to $\{\infty\}$ (in the first case) or $\{0, \infty\}$ in the second. An easy calculation finishes the proof.

Corollary 7.4. *If $\mathcal{J}_f$ has nonempty interior, then $\mathcal{J}_f$ is the entire Riemann sphere $\hat{\mathbb{C}}$.*

Proof. Suppose there is a domain $U \subset \mathcal{J}$. Since $\mathcal{J}$ is forward invariant,

$$\bigcup_{n>0} f^n(U) \subset \mathcal{J} \quad \text{and} \quad \bigcup_{n>0} f^n(U) = \hat{\mathbb{C}} \setminus \mathcal{E}_u.$$

But $\mathcal{J}$ is closed and $\mathcal{E}_u$ contains at most two points, so $J = \hat{\mathbb{C}}$. □

The map

$$L(z) = \frac{(z^2+1)^2}{4z(z^2-1)}$$

is known as a *Lattès' example*. $L(z)$ corresponds to multiplication by 2 on the torus $\mathbb{T}^2$ via the Weierstrass $\wp$ function. Furthermore, the Julia set of L is all of $\hat{\mathbb{C}}$. For further details, a good reference is [25]. Other examples of rational maps where $\mathcal{J}$ is the entire Riemann sphere can be found in [28].

Corollary 7.5. *If z is any point in the Julia set of f, then preimages of z are dense in $\mathcal{J}_f$:*

$$\mathcal{J}_f = \overline{\bigcup_{n\geq 0} f^{-n}(z)}.$$

Proof. Observe that for any $w \in \hat{\mathbb{C}}$ which is not an exceptional point, the preimages of w accumulate on the Julia set:

$$\mathcal{J}_f \subset \overline{\bigcup f^{-n}(w)}.$$

This follows since for any $z \in \mathcal{J}$ and any neighborhood U of z, we have $w \in f^n(U)$ for some n.

Since $\mathcal{J}$ is a completely invariant set and $\mathcal{J}$ is closed, we also have

$$\overline{\bigcup f^{-n}(z)} \subset \mathcal{J} \quad \text{for any } z \in \mathcal{J}.$$

□

Theorem 7.6. *The Julia set is a perfect set. That is, it has no isolated points.*

Proof. Let $z \in \mathcal{J}$. We can find another point $a \in \mathcal{J}$ so that $f^n(a) = z$ for some n, but a is not a forward image of z. (If z is not periodic, any inverse of z will do. If z is periodic of period m, consider $g = f^m$, and choose a so that $g(a) = z$ but $a \neq z$.)

Now let U be a neighborhood of z. Since $a \in \mathcal{J}$, a is not an exceptional point, so there is a $k > 0$ so that $f^k(U)$ contains a.

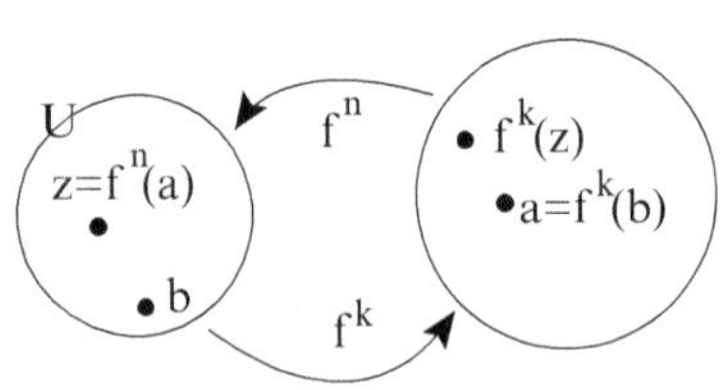

Let b be another point of U for which $f^k(b) = a$. Then $b \neq z$ since z is not in the forward orbit of a. Since $\mathcal{J}$ is completely invariant, a and b are in $\mathcal{J}$. Thus, every point z of $\mathcal{J}$ is an accumulation point of other points of $\mathcal{J}$. □

We now show that the Julia set is the closure of the repelling periodic points.

Lemma 7.7. $\mathcal{J} \subset \overline{\{\text{periodic points}\}}$.

Proof. We give the proof for polynomials $p : \mathbb{C} \to \mathbb{C}$; a similar idea works for rational functions.

Let J_0 be the Julia set with any critical values of p removed, and let $w \in J_0$. Since J_0 contains no critical values, there is a neighborhood U of w and a local inverse $S : U \to \mathbb{C} \setminus U$. Let

$$g_n(z) = \frac{p^n(z) - z}{S(z) - z}, \quad n > 0.$$

Observe that g_n is bounded and takes the values 0 and 1 only on periodic orbits z. Consequently, the family $\{g_n\}$ is normal if and only if $\{p_n\}$ is normal.[2]

Since $w \in \mathcal{J}$, the family $\{p_n\}$ and hence $\{g_n\}$ cannot be normal on any neighborhood of w. But unless there is a sequence of periodic orbits accumulating to w, the family $\{g_n\}$ omits the values 0, 1, and ∞ on a neighborhood of w, contradicting Montel's theorem. □

To prove the inclusion the other way, we show that there are only finitely many non-repelling orbits. Then the previous lemma, coupled with the fact that $\mathcal{J}$ is a perfect set, will give the result.

Let $\mathcal{B}(p)$ denote the *immediate basin of* p, that is, the connected component of $\text{Bas}(p)$ which contains p.

Theorem 7.8. *If p is an attracting fixed point, then $\mathcal{B}(p)$ contains a critical value.*

Proof. Suppose there is no critical value in $\mathcal{B}(p)$. Then if U is a simply connected neighborhood of p contained in $\mathcal{B}(p)$, we can construct a local inverse S_1 for $f|_U$ so that $S_1(p) = p$.

Observe that $U_1 = S_1(U)$ is a simply connected subset of $\mathcal{B}(p)$, so we can repeat the process using U_1 to construct S_2. Indeed, since there is no critical value in $\mathcal{B}(p)$, it can be repeated infinitely often, giving a family $\{S_k\}$ on U, which is normal (by Montel's theorem) since $S_k(U) \subset \mathcal{B}(p) \subset \mathcal{F}$.

But p is a repelling fixed point for each S_k, so the family cannot be normal. □

We can extend this result to the a count on attracting periodic orbits: if p is of period k, apply the theorem to f^k.

[2] To adapt this proof to a rational function $R(z)$, instead one sets J_0 to be the Julia set with infinity, all critical values, and all poles of R removed. Then, for a neighborhood U of a point $w \in \mathcal{K}$, let S_1, S_2, and S_3 be three local inverses of R^2 and take

$$g_n(z) = \left(\frac{R^n(z) - S_1(z)}{R^n(z) - S_2(z)}\right)\left(\frac{S_3(z) - S_2(z)}{S_3(z) - S_1(z)}\right).$$

The family $\{g_n\}$ is normal if and only if $\{R^n\}$ is normal.

From Theorem 7.8, we get an upper bound of $2d-2$ on the number of attracting orbits of a rational map f of degree d. Fatou and Julia showed (by means of a perturbation argument) that the number of neutral cycles is at most $4d-4$. However, in 1987 Shishikura [31] used quasiconformal surgery techniques to give the sharp bound of at most $2d-2$ non-repelling orbits for any rational map of degree d. This bound is commonly called the *Fatou-Shishikura inequality*. Applying the Fatou-Shishikura inequality (or the larger Fatou/Julia bound) gives the following.

Theorem 7.9. *The Julia set $\mathcal{J}_f$ is the closure of the repelling periodic orbits of f.*

Proof. To prove this, we merely combine the fact that $\mathcal{J} \subset \overline{\{\text{periodic points}\}}$ with the fact that there are only finitely many non-repelling periodic orbits. Since $\mathcal{J}$ is a perfect set, each point in it must be the accumulation of some sequence of repelling periodic points. □

8 Neutral Periodic Orbits

We have seen that repelling periodic orbits always lie in the Julia set, and that attracting orbits lie in the Fatou set. If p is part of a periodic cycle with multiplier on the unit circle, p may either lie in the Julia set or in the Fatou set.

Theorem 8.1. *Let p be a fixed point with multiplier $f'(p) = \lambda$, $|\lambda| = 1$. Then $p \in \mathcal{F}$ if and only if the SFE $\phi(f(z)) = \lambda\phi(z)$ has an analytic solution in a neighborhood of p.*

Proof. If SFE has a solution in a neighborhood of p, then $p \in \mathcal{F}$ follows immediately.

Conversely, suppose $p \in \mathcal{F}$ and let U be the maximal domain such that $p \in U$ and $U \subset \mathcal{F}$. Since U is disjoint from J, its complement in $\hat{\mathbb{C}}$ contains more than three points, so the universal cover $\tilde{U}$ is conformally equivalent to $\mathbb{D}$.

Thus, we have a cover $\varphi : \mathbb{D} \to U$ with $\varphi(0) = p$, and we can lift f to $\tilde{f} : \mathbb{D} \to \mathbb{D}$ with $\tilde{f}(0) = 0$. Since $|\tilde{f}'(0)| = |\lambda| = 1$, then by the Schwarz lemma, $\tilde{f}(z) = \lambda z$. The mapping φ is the solution ϕ to SFE. □

Corollary 8.2. *If p is a periodic point with multiplier λ a root of unity, then $p \in \mathcal{J}$.*

Proof. Replacing f by an iterate if necessary, we may assume p is a fixed point. Suppose $\lambda^k = 1$ and $p \in \mathcal{F}$. Then if SFE has a solution ϕ in some neighborhood U of p, we have $\phi \circ f^k \circ \phi^{-1}$ is the identity on U. But since f is analytic, it is the identity on $\hat{\mathbb{C}}$, and thus f is of degree 1, a contradiction. □

Theorem 8.3 (Fatou–Leau Flower Theorem). *Suppose $\lambda^n = 1$ with $\lambda^j \neq 1$ for $1 < j < n$. Let $f(z) = \lambda z + a_2 z^2 + \ldots$ be analytic in a neighborhood of the origin. If $f^n \neq \mathrm{Id}$, there is an integer k for which there are nk attracting petals bounded by analytic curves which are tangent at the origin. The union of the petals is forward*

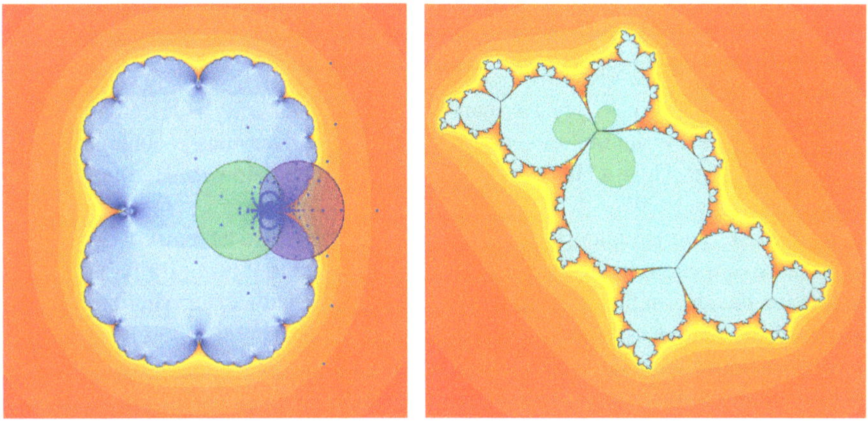

Fig. 11 On the *left* is the "cauliflower" ($z \mapsto z^2 + 1/4$) with a cardioid-shaped attracting petal in *green*, a repelling petal in *violet*, and the forward orbits of several points shown. At *right* is the "fat rabbit" with attracting petals indicated in *green*

invariant, and any orbit in a petal is asymptotic to the origin. Any compact set within a petal converges uniformly to the origin under f^n.

In addition, there are nk *repelling petals*; in each, every orbit eventually leaves the petal (alternatively, in each petal there is a branch of the inverse for which the petal is attracting). See Fig. 11.

8.1 Rotation Domains

In 1942, Siegel [32] found a full measure subset of the unit circle Λ such that whenever the multiplier at p is in Λ, the SFE has a solution, so $p \in \mathcal{F}$.

Theorem 8.4. *Let p be a fixed point for f with multiplier $\lambda = e^{2\pi i\theta}$, where θ is irrational. Suppose also there exist constants a and b such that*

$$\left|\theta - \frac{p}{q}\right| > \frac{a}{q^b} \qquad \text{for all rationals } \frac{p}{q}.$$

Then SFE has a solution and $p \in \mathcal{F}$.

The condition above is roughly that "θ is badly approximated by rationals." For example, the golden mean $\frac{1+\sqrt{5}}{2}$ is such a number, as is any number for which the terms in its continued fraction expansion are bounded. In this case, the topological disk around p on which f is conjugate to an irrational rotation is called a *Siegel disk*.

The Julia set of a polynomial with a Siegel disk can be extremely complicated, despite the fact that the dynamics on the Siegel disk is conjugate to a "simple" rigid rotation. For example, there are quadratic polynomials with Siegel disks whose Julia set is a non-computable set (see [9]), as well as Julia sets with no interior but nonzero Lebesgue measure (see Sect. 10).

Related to Siegel disks are Herman rings. A *Herman ring* is a domain which is conformally equivalent to an annulus, on which the dynamics is conjugate to an irrational rotation. Thus, a Herman ring is a subset of the Fatou set. Herman showed their existence in 1979.

Siegel disks and Herman rings are collectively referred to as *rotation domains.*

Because of the maximum principle, Herman rings cannot occur for polynomials. Shishikura has shown that the degree of a rational map with a Herman ring must be at least three. For any odd degree at least 3, we can construct a Blaschke product f which sends the unit circle to itself by an orientation-preserving diffeomorphism with any desired rotation number ρ. If ρ is Diophantine (i.e., badly approximated by rationals), then f will have a Herman ring.

For example, the map $e^{2\pi i t} z^2 \dfrac{z-4}{1-4z}$, with $t \approx 0.6151732$, has a Herman ring with rotation number $(\sqrt{5}-1)/2$. See Fig. 12.

8.2 Cremer Points

The condition of a Diophantine rotation number for the existence of a Siegel disk is sufficient, but not sharp. For the map $z^2 + \lambda z$, if λ is *Brjuno* (i.e., if the convergents p_n/q_n of λ satisfy $\sum(\log q_{n+1})/q_n < \infty$), then the map will have a Siegel disk about zero.

Yoccoz (see [21,35]) showed that this condition is sharp for quadratic polynomials: if it fails to hold, the map cannot be linearized in a neighborhood of the fixed point. Indeed, it has the *small cycles property*: every neighborhood of the origin contains infinitely many periodic orbits.

A point with this property is called a *Cremer point*, and is of necessity in the Julia set.

We know of no good means of making a picture of a map with a Cremer point, although topological models exist and such Julia sets are always computable [5] when they have empty interior (which is necessarily true for quadratics). Julia sets with Cremer points are not locally connected sets [33] and can have positive Lebesgue measure (see Sect. 10). See [7] for more details and further references.

9 Fatou–Sullivan Classification of Fatou Components

In 1983, Sullivan [33] classified the possibilities for a Fatou component which is eventually periodic. While most of these possibilities were known to Fatou and Julia, Sullivan's work completed the classification.

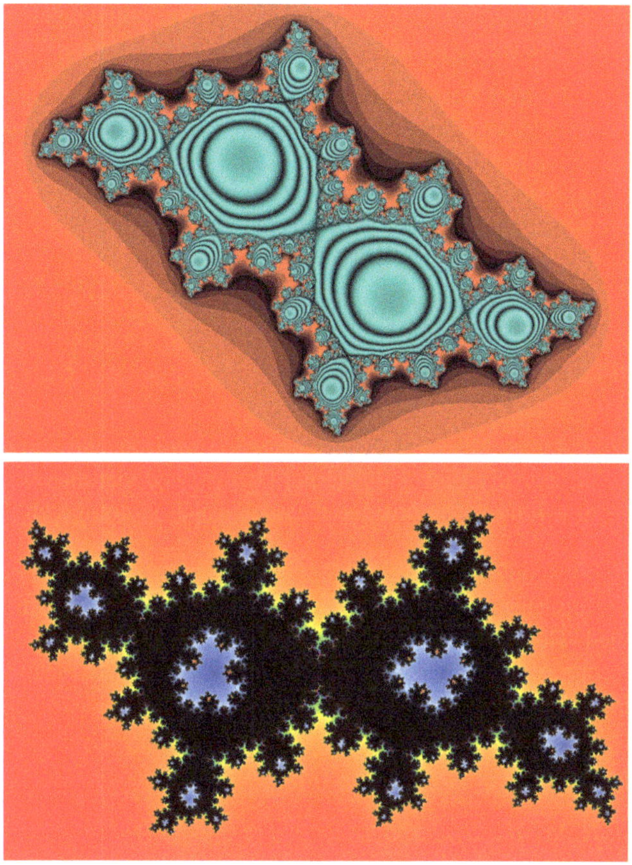

Fig. 12 On the *left* is the Julia set of a polynomial which has a Siegel disk with a rotation number of the golden mean (the Siegel disk and its preimages are shown in *blue*; the *light–dark bands* indicate orbits within the Siegel disk). On the *right* is a rational map with a Hermann ring with a golden-mean rotation (Bas(∞) is in *reds* and *yellows*, Bas(0) in shades of *blue*, with the Hermann ring and its preimages shown in *black*)

Let U be an eventually periodic such a component of the Fatou set. Then, after passing to an iterate, we can view U as being fixed. There are only four possibilities:

(1) U is the immediate basin of an attracting (or super-attracting) point p.
(2) U is the immediate basin of one petal of a parabolic point.
(3) U is a Siegel disk.
(4) U is a Herman ring.

Furthermore, Sullivan [34] showed that every Fatou component is eventually periodic, that is, there are *no wandering domains* for rational maps.[3]

Theorem 9.1 (Sullivan's Non-wandering Theorem). *Every Fatou component U for a rational map is eventually periodic. That is, there exist integers $n \geq 0$ and $p \geq 1$ so that the forward image $f^n(U)$ is mapped onto itself by f^p. In particular, it follows that every Fatou component is either a branched covering or a biholomorphic copy of some periodic Fatou component, which necessarily belongs to one of the four types described above.*

10 A Julia Set of Positive Measure

Earlier, we noted that either $\mathcal{J}$ has no interior or it is the entire Riemann sphere. However, recently Buff and Chéritat [10] showed that there is a quadratic polynomial with a Julia set of positive measure which is not the entire Riemann sphere.

They did this by constructing a sequence of perturbations of Siegel disks with an increasingly complicated boundary, such that the loss of measure in the filled Julia set is controlled. The Siegel disks become "digitated," with deep channels entering towards the center of the disk. Informative pictures of the process can be found on Arnaud Chéritat's web page [13].

More precisely, Buff and Chéritat showed the following.

Theorem 10.1. *Let $P_\alpha(z) = e^{2\pi i \alpha} z + z^2$. Then*

- *there exists α such that P_α has a fixed point of Cremer type and $\mathcal{J}_{P_\alpha}$ has positive measure;*
- *there exists β such that P_β has a Siegel disk and $\mathcal{J}_{P_\beta}$ has positive measure.*

11 Conclusion

In this brief note, I hope I have given you some idea of the beauty and complexity related to holomorphic dynamics in one variable. Of course, due to limited space many details and interesting, relevant topics were omitted. I hope I have inspired the reader to follow up and learn more about this vibrant and exciting area of mathematics.

[3]This result is not valid for transcendental maps, since $z + \sin(2\pi z)$ has a wandering domain.

References

1. Alexander, D.S.: A History of Complex Dynamics: From Schröder to Fatou and Julia. Aspects of Mathematics, vol. 24. Vieweg, Heidelberg (1994)
2. Alexander, D.S., Iavernaro, F., Rosa, A.: Early Days in Complex Dynamics: A History of Complex Dynamics in One Variable During 1906–1942. History of Mathematics, vol. 38. American Mathematical Society/London Mathematical Society, Providence/London (2012)
3. Audin, M.: Fatou, Julia, Montel, the Great Prize of Mathematical Sciences of 1918, and Beyond. Springer, Berlin (2011)
4. Beardon, A.: Iteration of Rational Functions: Complex Analytic Dynamical Systems. Graduate Texts in Mathematics, vol. 132. Springer, New York (1991)
5. Binder, I., Braverman, M., Yampolsky, M.: Filled Julia sets with empty interior are computable. J. Found. Comput. Math. **7**, 405–416 (2007)
6. Blanchard, P.R.: Complex analytic dynamics on the Riemann sphere. Bull. Am. Math. Soc. **11**, 85–141 (1984)
7. Blokh, A., Buff, X., Chéritat, A., Oversteegen, L.: The solar Julia sets of basic quadratic Cremer polynomials. Ergodic Theory Dyn. Syst. **30**, 51–65 (2010)
8. Böttcher, L.E.: The principal laws of convergence of iterates and their applications to analysis (Russian). Izv. Kazan. Fiz.-Mat. Obshch. **14**, 155–234 (1904)
9. Braverman, M., Yampolsky, M.: Computability of Julia Sets. Algorithms and Computation in Mathematics, vol. 23. Springer, Berlin (2006)
10. Buff, X., Chéritat, A.: Quadratic Julia sets with positive area. Ann. Math. **176**, 673–746 (2012)
11. Carleson, L., Gamelin, T.: Complex Dynamics. Springer, New York (1993)
12. Cayley, A.: Applications of the Newton-Fourier method to an imaginary root of an equation. J. Math. Pur. Appl. **16**, 179–185 (1879)
13. Chéritat, A.: Galerie II: Holomorphic Dynamics and Complex Analysis. http://www.math.univ-toulouse.fr/~cheritat/GalII/galery.html (retrieved Jan. 2014)
14. Cremer, H.: Zum zentrumproblem. Math. Ann. **98**, 151–163 (1927)
15. Cremer, H.: Über die Schrödersche funktionalgleichung und das Schwarzsche eckenabbildungsproblem. Leipziger Berichte **84**, 291–324 (1932)
16. Devaney, R.L.: An Introduction to Chaotic Dynamical Systems. Studies in Nonlinearity, 2nd edn. Westview Press, Boulder (2003)
17. Devaney, R.L.: Complex geometry of the Mandelbrot set. In: A. Sanayei, I. Zelinka, and O. E. Rössler, (eds.) Interdisciplinary Symposium on Complex Systems 2013, Prague, pp. 3–8. Emergence, Complexity and Computation 8, Springer-Verlag (2014). http://math.bu.edu/people/bob/papers/prague.pdf
18. Devaney, R.L.: Parameter planes for complex analytic maps. In: Bandt, C., Falconer, K.J., Barnsley, M., Devaney, R., Kannan, V., Vinod Kumar, P.B. (eds.) Fractals, Wavelets, and their Applications, Chap. 4. Springer, New York (2014)
19. Fatou, P.: Sur les substitutions rationnelles. Comptes Rendus de l'Académie des Sciences de Paris **164**, 806–808 (1917); **165**, 992–995 (1917)
20. Fatou, P.: Sur les équations fonctionnelles. Bull. Soc. Math. France **47**, 161–271 (1919)
21. Hubbard, J.H.: Local connectivity of Julia sets and bifurcation loci: three theorems of J.-C. Yoccoz. In: Goldberg, L.R., Phillips, A.V. (eds.) Topological Methods in Modern Mathematics: A Symposium in Honor of John Milnor's 60th Birthday, pp. 467–511. Publish or Perish, Houston (1993)
22. Julia, G.: Mémoire sur l'iteration des fonctions rationnelles. J. de Mathématiques Pures et Appliquées **8**, 47–245 (1918)
23. Kœnigs, G.: Recherches sur les intégrales de certaines équations fonctionnelles. Ann. Sci. Éc. Norm. Sup. **3**, 1–41 (1884)
24. Leau, L.: Étude sur les equations fonctionelles à une ou plusièrs variables. Ann. Fac. Sci. Toulouse **11**, 1–110 (1897)

25. Milnor, J.: Dynamics in One Complex Variable: Introductory Lectures. Annals of Mathematics Studies, 3rd edn., vol. 160. Princeton University Press, Princeton (2006)
26. Montel, P.: Sur les familles de fonctions analytiques qui admettent des valeurs exceptionnelles dans un domaine. Ann. Sci. École Norm. Sup. **29**, 487–535 (1912)
27. Montel, P.: Lecons sur les familles normales de fonctions analytiques et leurs applications. Gauthier-Villars, Paris (1927)
28. Rees, M.: Ergodic rational maps with dense critical point forward orbit. Ergodic Theory Dyn. Syst. **4**, 311–322 (1984)
29. Schröder, E.: Über unendlich viele algorithmen zur auflösung der gleichungen. Math. Ann. **2**, 317–365 (1870)
30. Schröder, E.: Über iterirte funktionen. Math. Ann. **3**, 296–322 (1871)
31. Shishkura, M.: On the quasiconformal surgery of rational functions. Ann. Sci. Ecole Norm. Sup. **20**, 1–29 (1987)
32. Siegel, C.L.: Iterations of analytic functions. Ann. Math. **43**, 607–612 (1942)
33. Sullivan, D.: Conformal Dynamical Systems. Lecture Notes in Mathematics, vol. 1007. Springer, Berlin (1983)
34. Sullivan, D.: Quasiconformal homeomorphisms and dynamics. I. Solution of the Fatou-Julia problem on wandering domains. Ann. Math. **122**(3), 401–418 (1985)
35. Yoccoz, J.C.: Théorème de Siegel, nombres de Bruno et polynômes quadratiques. Petits diviseurs en dimension 1. Astérisque **231**, 3–88 (1995). Soc. Math. France

Parameter Planes for Complex Analytic Maps

Robert L. Devaney

Abstract In this paper we describe the structure of the parameter planes for certain families of complex analytic functions. These families include the quadratic polynomials $z^2 + c$, the exponentials $\lambda \exp(z)$, and the family of rational maps $z^n + \lambda/z^n$. These are, in a sense, the simplest polynomial, transcendental, and rational families, as each has essentially one critical orbit.

In this paper we give a brief overview of the structure of the parameter plane for three different families of complex analytic maps, namely quadratic polynomials (the Mandelbrot set), singularly perturbed rational maps, and the exponential family. The goal is to show how these objects allow us to understand almost completely the different dynamical behaviors that arise in these families as well as the accompanying bifurcations.

Keywords Mandelbrot set • Julia set • Singular Perturbations

AMS Subject Classification (2010): Primary 37F10; Secondary 37F45

1 The Mandelbrot Set

The Mandelbrot set $\mathcal{M}$ is one of the most interesting and beautiful objects in all of mathematics. Amazingly, it arises as the parameter plane for the seemingly simple quadratic family $P_c(z) = z^2 + c$. See Fig. 1. This is a picture in the c-plane (the parameter plane) that describes the fate of the orbit of the only critical point for

R.L. Devaney (✉)
Department of Mathematics, Boston University, 111 Cummington Mall, Boston, MA 02215, USA
e-mail: bob@bu.edu

C. Bandt et al. (eds.), *Fractals, Wavelets, and their Applications*, Springer Proceedings in Mathematics & Statistics 92, DOI 10.1007/978-3-319-08105-2_4,

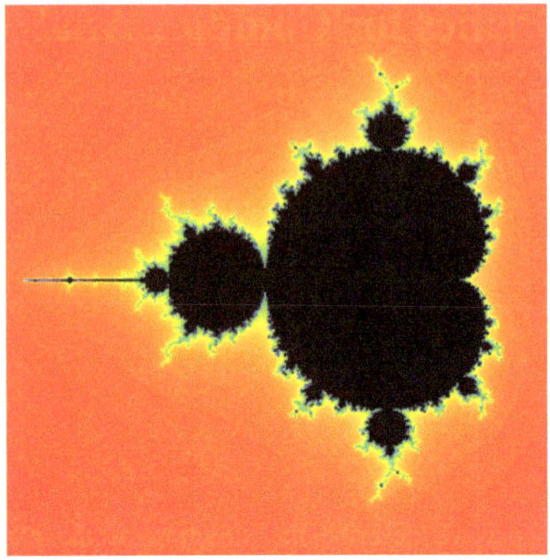

Fig. 1 The Mandelbrot set. *Colored points* are c-values for which the orbits of 0 escape to ∞; *black points* are c-values for which this does not happen. So the Mandelbrot set is the *black region* in these images

this family, namely 0. If the orbit of 0 does not tend to ∞, then the corresponding parameter c lies in $\mathcal{M}$ and we color this point black. If the orbit does escape to ∞, then c is not in the Mandelbrot set and we color c according to how quickly the orbit of 0 reaches the exterior of a large disk surrounding the origin (with red points escaping fastest, followed in order by orange, yellow, green, blue, and violet).

In complex dynamics, the object of central interest in the dynamical plane is the *Julia set*. For the family P_c, there is an open neighborhood of ∞ in the Riemann sphere consisting of points whose orbits tend to ∞. The set of all points whose orbits tend to ∞ is called the basin of ∞. Then the Julia set, denoted by $J(P_c)$ is the boundary of this basin. There are other equivalent definitions of $J(P_c)$. For example, it is known that $J(P_c)$ is also the closure of the set of repelling periodic points of P_c. As a consequence, we see that the Julia set is the chaotic set, for arbitrarily close to any point in $J(P_c)$, we have points whose orbits are periodic and other points whose orbits tend to ∞. In fact, via Montel's Theorem, given any point in the Julia set, then any open neighborhood of this point, no matter how small, is eventually mapped over the entire complex plane, minus at most one point. So the family of iterates of P_c on the Julia set is very sensitive to initial conditions. The *filled Julia set* is, by definition, the set of all points whose orbits do not tend to ∞. So $J(P_c)$ is also the boundary of the filled Julia set. The *Fatou set* is then the complement of $J(P_c)$ in the Riemann sphere.

The natural question is: Why are we interested in the fate of the orbit of the critical point? Well, in short, the critical orbit "knows it all" in complex dynamics. In particular, for the family P_c, if the orbit of 0 tends to ∞, then the Julia set of P_c is a Cantor set. If the orbit of 0 does not escape to ∞, then $J(P_c)$ is a connected set. So there are only two possible types of Julia sets for P_c: those that

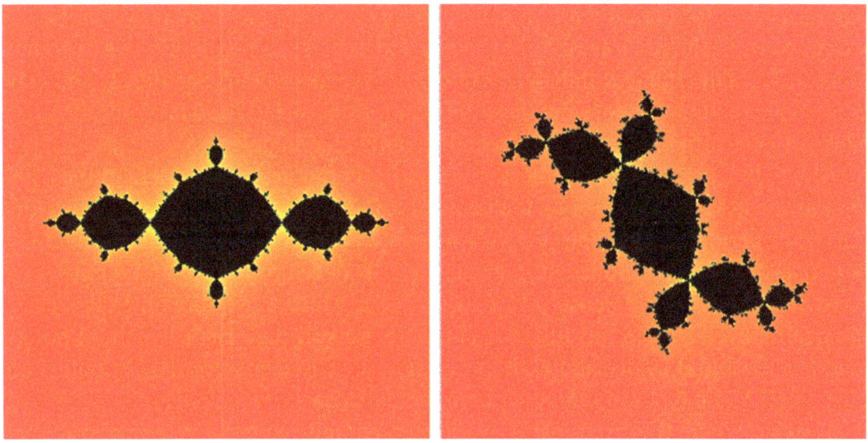

Fig. 2 The Julia sets for $z^2 - 1$ (the basilica) and $z^2 - 0.12 + .75i$ (the Douady rabbit). The filled Julia sets are the *black regions*, so the Julia sets here are the boundaries between the *black* and *colored regions*

consist of uncountably many point components, and those that consist of exactly one component. There are no Julia sets for quadratic polynomials that consist of 2 or 20 or 200 components.

The large black open regions visible in the Mandelbrot set are regions for which P_c has an attracting cycle of some given period. It is known that, if P_c has an attracting cycle, then the orbit of the critical point must tend to this cycle. Hence there can be at most one attracting cycle for a quadratic polynomial. For example, any c-value drawn from the central cardioid has an attracting fixed point. For c in the large open disk just to the left of this cardioid, P_c has an attracting 2-cycle. We therefore call this the period 2-bulb. And, for c in the northernmost and southernmost bulbs off the main cardioid, P_c has an attracting cycle of period 3, so these are the period 3-bulbs. Such open disks are called *hyperbolic components*, since it is known that P_c must then be hyperbolic on the Julia set, i.e., in some suitable metric, P_c is everywhere expanding.

As c moves from one hyperbolic component to another, the map undergoes a bifurcation. The simplest part of this bifurcation is the fact that we move from having an attracting cycle of some period when we are in one hyperbolic component to having an attracting cycle of some other period in the subsequent hyperbolic component. But, in fact, much more happens: the topology of the Julia sets changes dramatically. For example, if we move from the main cardioid to the period-2 bulb, the Julia set, which is just a simple closed curve when c is in the main cardioid, becomes a "basilica" when c is in the period 2-bulb. What happens is a repelling 2-cycle that lies in $J(P_c)$ when c is in the cardioid suddenly merges with the attracting fixed point and thereby makes it neutral when the parameter reaches the boundary of the cardioid. So two points in $J(P_c)$ become identified to one point. Meanwhile, infinitely many pairs of preimages of this point also become identified. This is what accounts for the infinitely many "pinch-points" visible in the basilica. Or, as we move from the main cardioid to the period 3-bulbs, a period 3-cycle becomes identified and the Julia set transforms into the "Douady rabbit." See Fig. 2. You may construct an animation to view these bifurcations by using the Mandelbrot Movie Maker applet at the website http://math.bu.edu/DYSYS/applets.

Along the boundaries of these hyperbolic components is where things get complicated. At each c-value on the boundary, P_c has a neutral cycle, i.e., a periodic point z of period n for which $(P_c^n)'(z) = \exp(2\pi i\theta)$. As c winds once around the boundary of this hyperbolic component, θ winds once around the unit circle. As a consequence, there is a dense set of such c's for which θ is rational. In this case, the neutral cycle lies in the Julia set but there are still regions in which all points tend to the neutral cycle (although these regions no longer surround the points on the cycle). These types of periodic points are called *parabolic points*.

The case where θ is irrational is much more complicated. If θ is highly irrational (i.e., "far" from rationals), then there is an open disk around each point on the cycle on which P_c^n is conjugate to the irrational linear rotation of angle θ. These disks are called Siegel disks. When θ is close to rationals, the structure of the Julia set near this cycle is still not completely understood. This is one of the major open problems in complex dynamics. See [16] for details.

A natural question is how do we understand how all of the bulbs and other smaller Mandelbrot sets are arranged in $\mathcal{M}$. Amazingly, if we zoom in to any portion of the boundary of the Mandelbrot set, it turns out that this zoom is very different from any other zoom that is non-symmetric with respect to $c \mapsto \overline{c}$. More importantly, with a keen eye for geometry, one can deduce exactly where in the boundary of $\mathcal{M}$ this zoom is, and, more importantly, what the corresponding dynamical behavior in the associated bulb is. It turns out that there are several different geometric and dynamical ways to understand the structure of these bulbs. First we will look at this geometrically, and then, using complex analysis, we will indicate how to prove this.

For simplicity, let's concentrate on the bulbs attached to the main cardioid. How do we know what their period is? One way is easy: look at the bulb. There is an antenna attached to this bulb. This antenna has a junction point from which a certain number of spokes emanate. The number of these spokes tells us exactly what the period is. For example, in Fig. 3, we display two bulbs having periods 5 and 7. Note that this is the exact number of antennas hanging off the junction point in the antenna of each bulb.

There is another way to read off the periods of these bulbs. Choose a parameter from the interior of a period n bulb and plot the corresponding filled Julia set. There is a central disk in these filled Julia sets that surrounds the origin. Then there are exactly $n - 1$ smaller disks that join this main disk at certain junction points. For example, in Fig. 2, we see that the rabbit has two "ears" attached to the central disk and the period of this bulb is $2 + 1 = 3$. Similarly, the basilica has just 1 ear and the period here is $1 + 1 = 2$. In Fig. 4, we display Julia sets from the above period 5 and period 7 bulbs, and we see the same phenomenon. The Mandelbrot/Julia Set applet

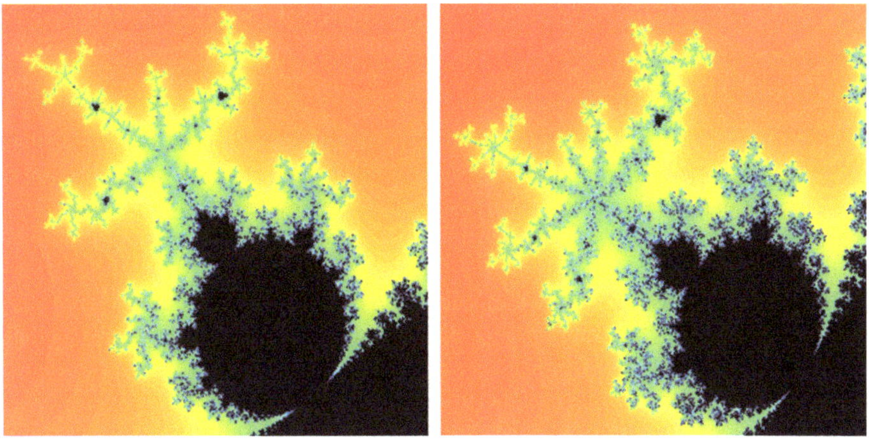

Fig. 3 Period 5 and 7 bulbs hanging off the main cardioid

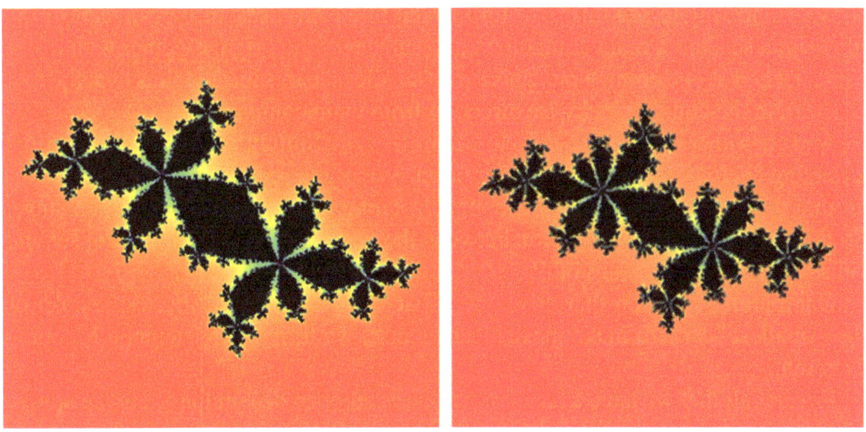

Fig. 4 Julia sets drawn from the above period 5 and 7 bulbs hanging off the main cardioid. Note that there are 4 and 6 "ears" hanging off the central disks of these filled Julia sets

at the website http://math.bu.edu/DYSYS/applets allows you to view and zoom in on the Mandelbrot and Julia sets of P_c to see more examples of these phenomena.

Now let us turn to the arrangement of the bulbs around the main cardioid. Recall that, on the boundary of the main cardioid, P_c has a fixed point whose derivative is given by $\exp(2\pi i\theta)$. Then a little algebra shows that a parametrization of the boundary of this main cardioid is given by

$$c = c(\theta) = \frac{e^{2\pi i\theta}}{2} - \frac{e^{4\pi i\theta}}{4}.$$

So when $\theta = 0$, $c = 1/4$ and we are at the cusp of the main cardioid; when $\theta = 1/2$, $c = -3/4$, and we are at the point where the period 2-bulb meets the main cardioid. In general, when θ is a rational number p/q in lowest terms, the corresponding c-value lies at the meeting point (also called the root point) of the main cardioid and a period q-bulb which we now call the p/q-bulb. So we see that the bulbs are arranged around the main cardioid in the exact order of the rational numbers. In particular, we can count exactly how many period q-bulbs are there. For example, there are 6 period 7-bulbs and 4 period 10-bulbs touching the main cardioid.

But there are several other geometric and dynamical ways to understand this. Look at the period five bulb in Fig. 3. We call the spoke of the antenna that extends down to the bulb from the junction point the *principal spoke*. Note that the "shortest" spoke (that is not the principal spoke) is located $2/5$ of a turn in the counterclockwise direction from the principal spoke. And this bulb is exactly the $2/5$-bulb. In that same figure, we also see that the period 7-bulb is, in fact, the $3/7$-bulb.

A second way to see this is to turn to the filled Julia set. In Fig. 4, each of the filled Julia sets has a main component that surrounds the origin together with $q-1$ ears attached at one point. Note where the "smallest" ear is located; it is exactly p/q of a turn in the counterclockwise direction from main component.

And then there is a third way to read off p/q. Simply plot the points on the attracting cycle of period q in the Fatou set. What you see is that this cycle moves around the ears and the main component, rotating by p/q of a turn at each stage. So there is a very nice connection between the geometry of the Mandelbrot set and Julia sets and the dynamics of P_c.

One natural question that arises is: What is meant by the "shortest" spoke or the "smallest" ear? To make these ideas precise, we turn to the Riemann Mapping Theorem.

First recall that we have a basin of ∞ that is an open disk in the Riemann sphere whenever c is chosen to lie in $\mathcal{M}$. Call this basin B_c. Then it is known that we can construct and analytic homeomorphism ϕ_c that takes B_c to the open unit disk $\mathbb{D}$ and maps ∞ to 0. Moreover, ϕ_c conjugates P_c on B_c with the simple map $z \mapsto z^2$ on $\mathbb{D}$. That is,

$$\phi_c(P_c(z)) = (\phi_c(z))^2.$$

In particular, the map z^2 takes the straight ray of angle θ given by $te^{i\theta}$ for $0 < t < 1$ to the ray $te^{i2\theta}$, the ray of angle 2θ. Then the preimage under ϕ_c^{-1} of the straight ray of angle θ in B_c is called the external ray of angle θ, and P_c interchanges these external rays just as $z \mapsto z^2$ interchanges the straight rays.

Now it is a fact that, when P_c has an attracting cycle, each of these external rays lands at a unique point in $J(P_c)$. The reason for this is that P_c is hyperbolic on $J(P_c)$ and consequently the Julia set is locally connected. When c is chosen from the main cardioid, each external ray has a unique landing point, so this says that P_c is conjugate to $z \mapsto z^2$ on its Julia set. But, when c lies in other bulbs, certain of

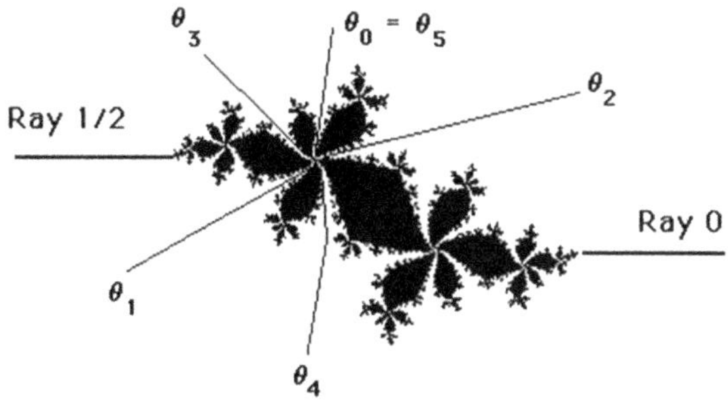

Fig. 5 The orbit $\theta_0 \to \theta_1 \to \theta_2 \ldots$ of the rays landing at the fixed point $J(P_c)$ when c is in the 2/5 bulb

these rays land at the same point. It is true that the external ray of angle 0 always lands at a particular fixed point in $J(P_c)$ and this is the only such ray landing at this point. Similarly, the external ray of angle $1/2$ is the unique ray landing at the preimage of this fixed point. But, when c is chosen from the p/q-bulb, there is a fixed point that lies on the boundary of the main Fatou component containing 0 and is the connection point for the q basins of the attracting cycle. Now there are exactly q rays that land at this fixed point. Moreover, P_c must interchange these rays just as above, by angle-doubling.

So, for example, when c is in the $2/5$-bulb, there must be five rays of angle $\theta_0, \ldots, \theta_4$ that land on this fixed point. And they must be mapped around just as P_c interchanges the ears in the Julia set, so

$$\theta_0 \mapsto \theta_1 \mapsto \theta_2 \mapsto \theta_3 \mapsto \theta_4 \mapsto \theta_0 \ldots.$$

Then a little computation shows that $\theta_0 = 9/31$, so that $\theta_1 = 18/31$, $\theta_2 = 5/31$, $\theta_3 = 10/31$, and $\theta_4 = 20/31$. In similar fashion, the external rays that land at the fixed point on the main component of a Julia set when c is in other p/q-bulbs may also be calculated.

So, how do we determine the size of the "ears" on these Julia sets? Using what is called *harmonic measure*, we define the size of the ears just to be the difference of the angles of the two landing external rays that separate this ear from the other components containing the attracting cycle. So, in the $2/5$ case, we see that the smallest ear is contained between the external rays of angles $\theta_0 = 9/31$ and $\theta_3 = 10/31$, so this ear has "size" $10/31 - 9/31 = 1/31$, whereas all the other ears are larger (Fig. 5).

Now how do we determine the size of the spokes of the antennas in the Mandelbrot set? We use essentially the same technique, but now in the parameter plane. Using a celebrated result of Douady and Hubbard [11], there is a similar "uniformization" of the exterior of $\mathcal{M}$ in the Riemann sphere which again maps ∞ to 0. Let $\mathcal{C}$ denote this external region. To construct this map, for each $c \in \mathcal{C}$, we have that the critical value c for P_c now lies in B_c. So we can consider the function

$\Phi(c) = \phi_c(c)$. Just as in the previous case, Φ is now an analytic homeomorphism that takes $\mathcal{C}$ onto $\mathbb{D}$. So again we have external rays, but now they are in the parameter plane. It is known that all rational rays land at a unique point on the boundary of the Mandelbrot set. Some land at root points of bulbs or cusp points on the cardioids of small copies of the Mandelbrot set. Others land at the endpoints of the spokes of the antennas or at the junction points. And one can use similar techniques as above to determine exactly where certain of these external rays land. See [3,8]. For example, it is known that if an external ray of angle θ lands at the root point of a period q-bulb, then the angle θ must have period q under angle-doubling. So, for example, the two rays landing at the root point of the period 2-bulb must be $1/3$ and $2/3$. The rays landing at the northern period 3-bulb are $1/7$ and $2/7$ and at the southern period 3-bulb are $5/7$ and $6/7$. So the rays $3/7$ and $4/7$ must also land on a period 3-bulb that is somewhere to the left of the northern and southern period 3-bulbs. Indeed, as is well known, there is a small Mandelbrot set lying along the negative real axis whose main cardioid contains parameters for which there is an attracting 3-cycle. So these two external rays both land at the cusp of this main cardioid.

One curious fact that relates to the Farey tree involves the size of the bulbs hanging off the main cardioid. To begin, we think of the root point of the main cardioid as being the cusp at $c = 1/4$. Then we call the main cardioid the $0/1$-bulb. Which is the largest bulb between the root points of the $0/1$ and $1/2$-bulbs (in, say, the upper portion of $\mathcal{M}$)? It is clearly the $1/3$-bulb. And note that $1/3$ is obtained from the previous two fractions by *Farey addition*, i.e., adding the numerators and adding the denominators

$$\frac{0}{1}\text{“}+\text{"}\frac{1}{2} = \frac{1}{3}.$$

Similarly, the largest bulb between the $1/3$ and $1/2$-bulbs is the $2/5$-bulb, again given by Farey addition. As above, we again measure the size of these bulbs by determining the interval of external rays that land on the bulb. So the size of the period 3-bulb is $2/7 - 1/7 = 1/7$ while the $2/5$-bulb has size $1/31$, as seen in Fig. 6. Note that the $2/5$-bulb is the largest bulb between the $1/2$ and $1/3$-bulbs. Then this process continues. The largest bulb between the $2/5$ and $1/2$-bulb is the $3/7$-bulb and the largest bulb between the $2/5$ and $1/3$-bulbs is the $3/8$-bulb and so on along the “Farey tree [4]”.

One of the most interesting and important open problems in complex dynamics is the question of whether or not the boundary of the Mandelbrot set is locally connected. If this is the case, then all of the external rays land at unique points along the boundary of $\mathcal{M}$. As a consequence, we would understand everything about the Mandelbrot set. However, it is not at all clear that this boundary is locally connected. Think about the period one-millionth bulb—the antenna here has a million spokes! And as the denominators of p/q get larger, the antenna structure also becomes even more “complex.” It is true that the size of these bulbs gets smaller as q increases, so it is possible that the boundary is locally connected. However, a result

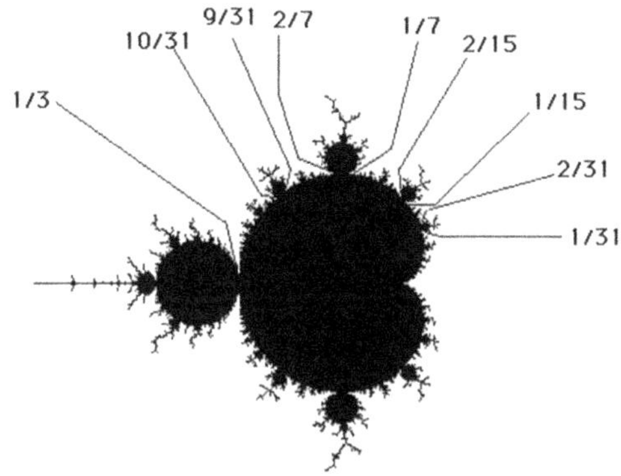

Fig. 6 Rays landing on the Mandelbrot set

of Shishikura [19] shows that the boundary of $\mathcal{M}$ has Hausdorff dimension 2, so, indeed, this boundary is pretty "crazy." Furthermore, a result of Buff and Chéritat [2] shows that Julia sets of P_c that contain fixed points that are close to rationals have positive Lebesgue measure, something that also indicates that things are getting quite complicated along the boundary of $\mathcal{M}$.

2 Singularly Perturbed Rational Maps

We now consider a very different type of map, namely rational maps of the form

$$F_\lambda(z) = z^n + \frac{\lambda}{z^n}$$

where $n \geq 2$. While the degree of these maps can be quite large, there is really only one "free" critical orbit just as in the case of quadratic polynomials. Indeed, one checks easily that there are $2n$ critical points given by $\lambda^{1/2n}$. However, there are only two critical values $\pm 2\sqrt{\lambda}$; n of the critical points map to one critical value and the other critical points map to the second critical value. But, when n is even, both critical values then map to the same point, whereas, if n is odd, we have $F_\lambda(-z) = -F_\lambda(z)$, so the two critical values have orbits that are symmetric under $z \mapsto -z$. We call this the free critical orbit, since ∞ and 0 are also critical points, but ∞ is fixed and 0 is mapped by F_λ onto ∞.

Just as in the case of $z^2 + c$, the point at ∞ is an attracting fixed point when $n \geq 2$, so we have an immediate basin of attraction B_λ of ∞ that lies in the Fatou set. Also, 0 is a pole, so there is a neighborhood of 0 that is mapped into B_λ. If the component of the Fatou set containing 0 is disjoint from B_λ, we denote this set by T_λ and call it the trap door since any orbit that eventually ends up in B_λ must pass

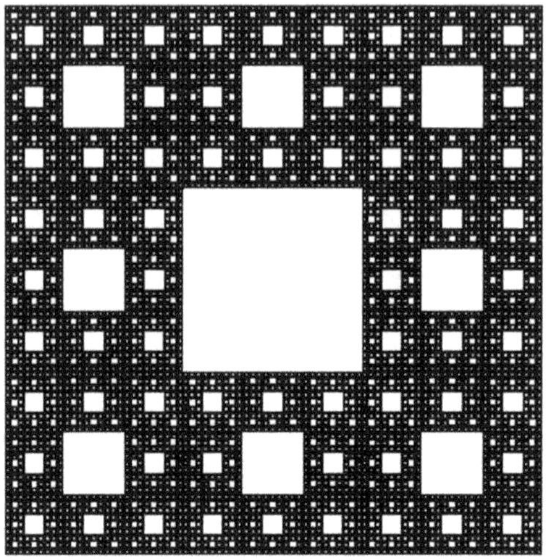

Fig. 7 The Sierpinski carpet

through T_λ. This follows since F_λ maps both B_λ and T_λ n-to-1 onto B_λ and the map F_λ has degree $2n$.

Unlike the quadratic polynomial case, where we had only one possibility for the structure of the Julia set when the critical orbit escapes, here we have an escape trichotomy. As shown in [10],

1. If the critical values lie in B_λ, then $J(F_\lambda)$ is a Cantor set;
2. If the critical values lie in T_λ, then $J(F_\lambda)$ is a Cantor set of simple closed curves;
3. In all other cases, the Julia set is connected. If the critical orbit enters B_λ at iteration 2 or later, then $J(F_\lambda)$ is a Sierpinski curve.

The second result here is due to McMullen [15]. Incidentally, case 2 does not occur when $n = 2$; indeed, the situation when $n = 2$ is very different from (and much more complicated than) the case $n > 2$ [6, 7].

A Sierpinski curve is any planar set that is homeomorphic to the well-known Sierpinski carpet fractal displayed in Fig. 7. These sets are important for three reasons. First, by a result due to Whyburn [20], there is a topological characterization of any such set: any planar set that is compact, connected, nowhere dense, locally connected, and has the property that any pair of complementary domains are bounded by simple closed curves that are pairwise disjoint is necessarily homeomorphic to the carpet. Second, as proved by Sierpini, the carpet is a universal plane continuum: any planar, one-dimensional, compact curve can be homeomorphically manipulated to fit inside the carpet. And finally, Sierpinski curves occur all the time as Julia sets for rational maps.

In Fig. 8, we display the parameter planes (the λ-planes) for the cases where $n = 3$ and $n = 4$. In both cases, the external region is where the Julia sets are

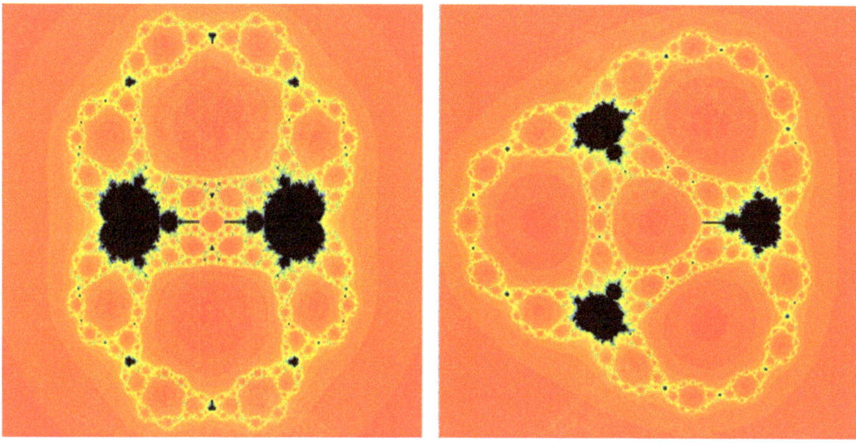

Fig. 8 The parameter planes when $n = 3$ and $n = 4$

Cantor sets; this is the Cantor set locus. The central disk surrounding the origin contains parameters whose Julia sets are Cantor sets of simple closed curves; we call this region the McMullen domain. All of the other red regions contain parameters whose Julia sets are Sierpinski curves; these are Sierpinski holes.

The arrangement of the Sierpinski holes in the parameter plane is fairly well understood. It is known that there are exactly $(n-1)(2n)^{\kappa-3}$ Sierpinski holes with escape time κ (the number of iterates it takes for the critical orbits to enter B_λ). Each Sierpinski hole contains parameters for which the corresponding maps all have conjugate dynamics on their Julia sets. However, most of the maps drawn from different Sierpinski holes have very different dynamics. In fact, only parameters drawn from Sierpinski holes that are symmetric under either complex conjugation or rotation by an $(n-1)^{\text{st}}$ root of unity have conjugate dynamics. Then it follows that, when n is odd, there are exactly $(2n)^{\kappa-3}$ conjugacy classes of maps drawn from Sierpinski holes. When n is even, there are $(2n)^{\kappa-3}/2 - 2^{\kappa-4}$ such holes. The discrepancy between n odd and even arises because there are no Sierpinski holes along the negative axis when n is odd, whereas there are such holes when n is even. So, when n is odd, there are exactly $2(n-1)$ Sierpinski holes in each conjugacy class, but when n is even, certain conjugacy classes have only $n-1$ Sierpinksi holes. See [9]. In Fig. 9, we display four different Sierpinski curve Julia sets drawn from the family when $n = 2$. All of these Julia sets are homeomorphic, but it turns out that all have very different dynamics.

One way that Sierpinski curve Julia sets have non-conjugate dynamics occurs when the escape times are different. If F_λ is a map with a Sierpinski curve Julia set for which the critical orbits escape to ∞, then the Fatou components that contain the critical points are the only ones that have boundaries that are mapped 2 to 1 onto their images. So if F_μ has a different escape time, then F_λ cannot be conjugate to F_μ since the boundaries of the escape components containing the critical points

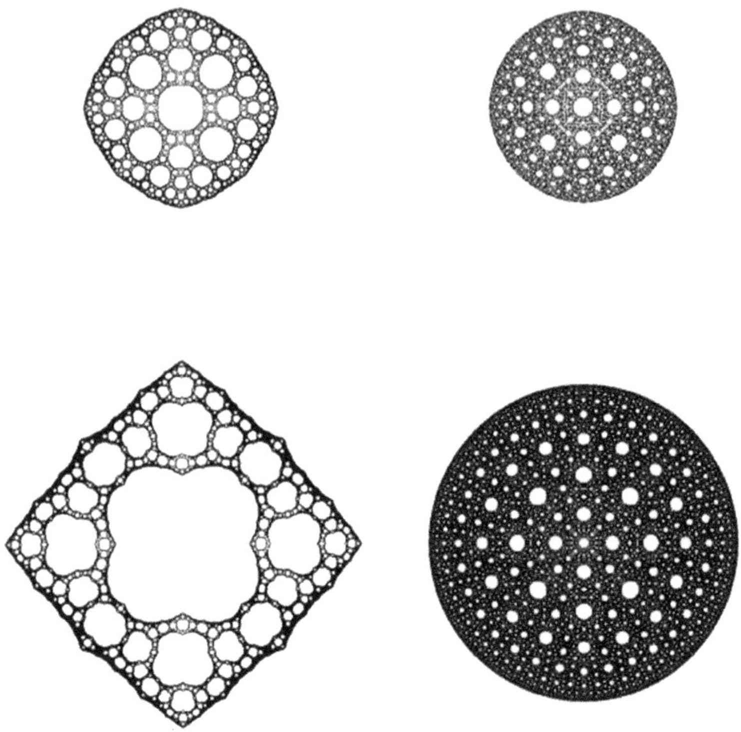

Fig. 9 The Julia sets for various values of λ when $n = d = 2$

would have to be mapped to each other. For escape time Julia sets with the same escape times, many still have non-conjugate dynamics. Proving this involves using Thurston's Theorem [9]. Moreno Rocha [17] has recently produced a dynamical invariant that explains why two such maps have non-conjugate dynamics.

As mentioned earlier, the case $n = 2$ is very different from the case $n > 2$. One reason for this is apparent in Fig. 9. Note that, as $\lambda \to 0$, the Julia sets of F_λ seem to converge to the unit disk. Of course, when $\lambda = 0$, we have the very simple map $F_0(z) = z^2$ for which the Julia set is just the unit circle. By Montel's Theorem, if the Julia set ever contains an open set in the plane, then it must be the entire plane. So here we see Julia sets getting closer and closer to the unit disk as $\lambda \to 0$, but, when $\lambda = 0$, things change dramatically. This is why these maps are called singular perturbations.

It is known that there are infinitely many small copies of the Mandelbrot set in each of these parameter planes. Certain of the Mandelbrot sets extend out to the boundary of the Cantor set locus while others do not. In these "buried" Mandelbrot sets, Julia sets drawn from the main cardioids are also Sierpinski curves. Since these Julia sets have an attracting cycle of some given period, the dynamics on these Sierpinski curves is quite different than on the escaping Sierpinski curves described above. For parameters in the main cardioids of the Mandelbrot sets that touch the boundary of the Cantor set locus, the structure of the Julia sets is quite different. See [1] for details.

Unlike the Mandelbrot set, these parameter planes have much simpler boundaries. Indeed the boundaries of the Cantor set locus, the McMullen domain, and all of the Sierpinski holes are known to be simple closed curves [18] (when $n > 2$). Of course, as mentioned above, there are also infinitely many small copies of the Mandelbrot set included in these sets, so the full structure in the parameter plane is still at least as complicated as the Mandelbrot set.

3 Complex Exponential Maps

In this final section, we consider another, very different, family of maps, the complex exponential family, $E_\lambda(z) = \lambda \exp(z)$. These are entire transcendental maps, so ∞ is no longer an attracting fixed point. Rather, ∞ is an essential singularity. For the exponential maps there is no longer a critical point. However, 0 is an asymptotic value (the only one), and hence this point plays the same role as the critical points did for the previous two families. A point z is an *asymptotic value* if there is a curve $\gamma(t)$ which tends to the essential singularity as $t \to \infty$ but whose image tends to z as $t \to \infty$. Any curve whose real part tends to $-\infty$ has this property for E_λ.

Because E_λ has an essential singularity at ∞, the Julia set has one slightly different definition. In the previous cases, the Julia set was the boundary of the set of points whose orbits tend to ∞. Now the Julia set, $J(E_\lambda)$, is the *closure* of the set of points that escape to ∞. So any point whose orbit tends to ∞ is now in the Julia set.

When $\lambda > 0$, the dynamical behavior on the real axis is pretty simple. The graph of E_λ shows that there is a simple saddle-node bifurcation when $\lambda = 1/e$. See Fig. 10. When $\lambda < 1/e$, there is an attracting fixed point a_λ and a repelling fixed point r_λ in $\mathbb{R}^+$. All points to the left of r_λ in $\mathbb{R}$ have orbits that tend to a_λ and hence do not lie in the Julia set, while the half-line $[r_\lambda, \infty)$ is in $J(E_\lambda)$. In fact, all points in $\mathbb{C}$ to the left of the vertical line through r_λ lie in the Fatou set. To see this, let x_λ be the point in $\mathbb{R}$ for which $E'_\lambda(x_\lambda) = 1$. So x_λ lies in the open interval (a_λ, r_λ). Then the vertical line through x_λ is mapped infinitely often around a circle centered at the origin which includes a_λ in its interior. Thus the open half-plane to the left of this vertical line is contracted inside the disk bounded by this circle. Hence, by the Contraction Mapping Principle, all points in this open half-plane have orbits that tend to a_λ. Then one checks easily that all points to the left of the vertical line through r_λ eventually map inside this half-plane as well and so are also in the Fatou set.

When $\lambda = 1/e$, the two fixed points a_λ and r_λ merge and again all points to the left of this now neutral fixed point in $\mathbb{R}$ do not lie in the Julia set, while the fixed point and all points to the right of it in $\mathbb{R}$ again do lie in $J(E_\lambda)$. A similar argument

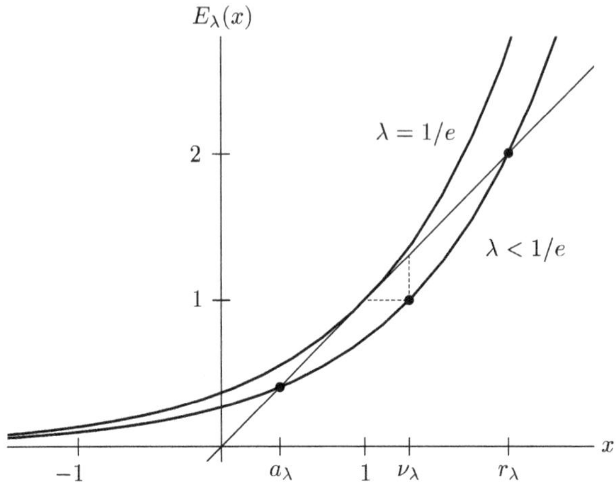

Fig. 10 The graphs of E_λ for $\lambda = 1/e$ and $\lambda < 1/e$

as above then says that all points to the left of a vertical line through this fixed point in $\mathbb{C}$ also tend to the neutral fixed point and so are in the Fatou set.

When $\lambda > 1/e$, the fixed points in $\mathbb{R}$ disappear (they actually become complex), and now all points in $\mathbb{R}$ tend to ∞ under iteration and so lie in $J(E_\lambda)$. So it looks like the Julia set undergoes an abrupt change when λ increases through $1/e$. In fact, much more happens: a result of Goldberg and Keen [12] states that if the orbit of the asymptotic value 0 tends to ∞, then the $J(E_\lambda)$ is the entire complex plane. So, for $\lambda \leq 1/e$, the Julia set is contained in the right half-plane, but, as soon as $\lambda > 1/e$, the Julia set becomes the entire complex plane.

Interestingly, no new periodic points are born as λ increases through $1/e$; all of the periodic points simply migrate continuously but do so in a way that they suddenly become dense in the plane when $\lambda > 1/e$. This is quite an interesting bifurcation!

In Fig. 11, we display the Julia set for a value of $\lambda \in (0, 1/e)$. Black points are in the basin of attraction of a_λ and colored points escape to ∞. So the colored region is the Julia set. It appears that the Julia set contains open strips that tend off to ∞, but, By Montel's Theorem, this cannot happen. In fact, the Julia set in this case is a *Cantor bouquet*, a collection of uncountably many smooth curves which tend off to ∞ in the right half-plane and each of which has a distinguished endpoint. See [5]. These curves are called "hairs" and all points (except the endpoints) have orbits that tend to ∞ and so are in the Julia set. For example, one hair is the subset of the real axis given by (r_λ, ∞); the endpoint is then the fixed point r_λ. Since the bounded orbits must lie in the set of endpoints, we have that the repelling periodic points must lie in the set of endpoints. Therefore this set is much more intricate than it at first seems: these endpoints must be everywhere dense in the Julia set. An interesting result of Mayer [14] shows that the only points that are accessible from the Fatou set are these endpoints; there is no curve contained in the Fatou set that limits on

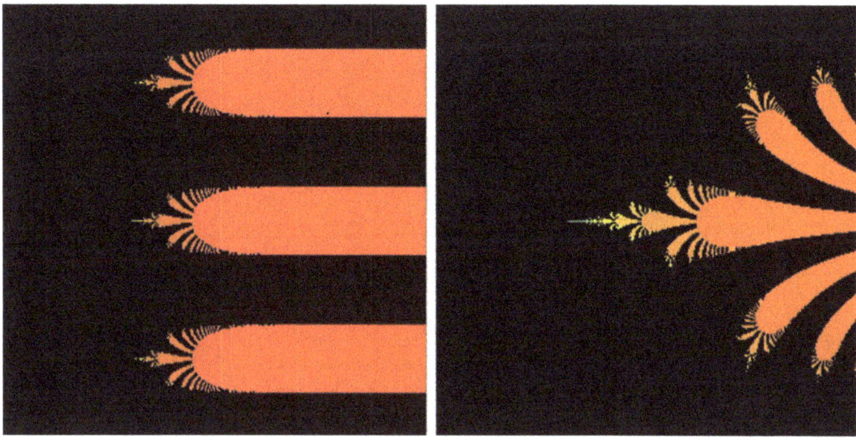

Fig. 11 The Julia set for $E_{0.3}$ and a magnification along the real axis

Fig. 12 The Julia set for $E_{0.6}$ near the real axis

any single point in the hairs. Moreover, a result of Karpinska [13] shows that the Hausdorff dimension of the set of all points on the hairs is 1 whereas the Hausdorff dimension of the supposedly much smaller set of endpoints is 2!

In Fig. 12 a portion of the Julia set for $\lambda = 0.6$ is displayed; here $J(E_{0.6}) = \mathbb{C}$. The two spirals actually converge down to the pair of repelling fixed points that appear after a_λ and r_λ coalesce and disappear off the real line.

As in the case of the other families discussed in this paper, we now turn briefly to the parameter plane for the complex exponential. In Fig. 13 we display a portion of this parameter plane and a magnification near the origin. The cardioid shaped region is where E_λ has an attracting fixed point. The cusp of this cardioid is the

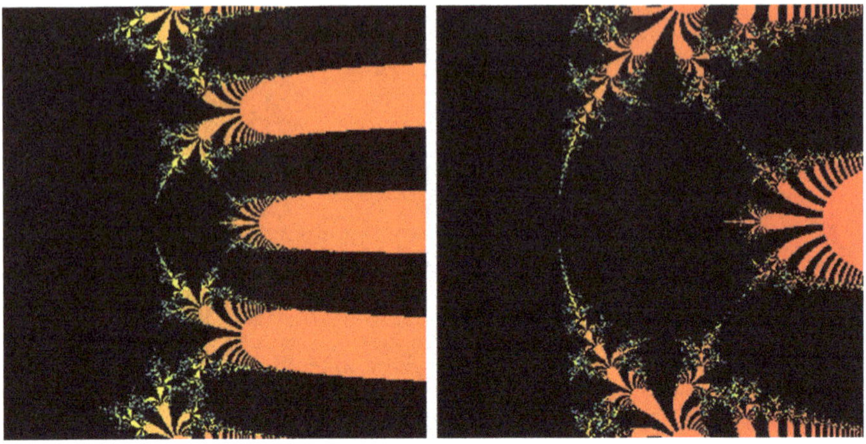

Fig. 13 The parameter plane for E_λ

parameter $\lambda = 1/e$. The large black region to the left of the cardioid actually extends to ∞ in the left half-plane and contains parameters for which E_λ has an attracting cycle of period 2. Hanging off the cardioid are strips that all tend to ∞ in the right half-plane and contain parameters for which there is an attracting cycle of some period greater than 2. The two largest strips are regions where E_λ has an attracting cycle of period 3.

As in the case of the Julia sets, the colored regions contain parameters for which the orbit of 0 tends to ∞ and so the Julia set for these parameters is the entire complex plane. Again as in the dynamical plane, these colored regions are really curves. For example, one such curve is the interval in $\mathbb{R}^+$ given by $(1/e, \infty)$.

Acknowledgements This work was partially supported by grant #208780 from the Simons Foundation.

References

1. Blanchard, P., Çilingir, F., Cuzzocreo, D., Devaney, R.L., Look, D.M., Russell, E.D.: Checkerboard Julia sets for rational maps. Int. J. Bifurcat. Chaos **23**, 48–60 (2013)
2. Buff, X., Chéritat, A.: The Yoccoz function continuously estimates the size of Siegel disks. Ann. Math. **164**, 265–312 (2006)
3. Devaney, R.L.: The fractal geometry of the Mandelbrot set: II. How to add and how to count. Fractals **3**(4), 629–640 (1995). See also http://math.bu.edu/DYSYS/FRACGEOM2/FRACGEOM2.html
4. Devaney, R.L.: The Mandelbrot set, the Farey tree, and the Fibonacci sequence. Am. Math. Mon. **106**, 289–302 (1999)
5. Devaney, R.L.: Complex exponential dynamics. In: Handbook of Dynamical Systems, vol. 3, pp. 125–224. Elsevier, Amsterdam (2010)

6. Devaney, R.L.: Dynamics of $z^n + C/z^n$; Why the case $n = 2$ is crazy. In: Conformal Dynamics and Hyperbolic Geometry. Contemporary Mathematics, vol. 573, pp. 49–65. AMS, Providence (2012)
7. Devaney, R.L.: Singular perturbations of complex polynomials. Bull. Am. Math. Soc. **50**, 391–429 (2013)
8. Devaney, R.L., Moreno Rocha, M.: Geometry of the antennas in the Mandelbrot set. Fractals **10**, 39–46 (2002)
9. Devaney, R.L., Pilgrim, K.M.: Dynamic classification of escape time Sierpinski curve Julia sets. Fundam. Math. **202**, 181–198 (2009)
10. Devaney, R.L., Look, D.M., Uminsky, D.: The escape trichotomy for singularly perturbed rational maps. Indiana Univ. Math. J. **54**, 1621–1634 (2005)
11. Douady, A., Hubbard, J.: Étude Dynamique des Polynômes Complexes. Partie I. Publ. Math. D'Orsay **84–02**, 287–343 (1984)
12. Goldberg, L., Keen, L.: A finiteness theorem for a dynamical class of entire functions. Ergodic Theory Dyn. Syst. **6**, 183–192 (1986)
13. Karpinska, B.: On the accessible points in the Julia sets for entire functions. Fundam. Math. **180**, 89–98 (2003)
14. Mayer, J.: An explosion point for the set of endpoints in the Julia set of $\lambda \exp(z)$. Ergodic Theory Dyn. Syst. **10**, 177–184 (1990)
15. McMullen, C.: Automorphisms of rational maps. In: Holomorphic Functions and Moduli, vol. 1. Mathematical Sciences Research Institute Publications, 10. Springer, New York (1988)
16. Milnor, J.: Dynamics in One Complex Variable. Princeton University Press, Princeton (2006)
17. Moreno Rocha, M.: A combinatorial invariant for escape time Sierpinski rational maps. Fundam. math. **222**, 99–130 (2013)
18. Qiu, W., Roesch, P., Wang, X., Yin, Y.: Hyperbolic components of McMullen maps. Ann. Sci. École Norm. Sup. Paris
19. Shishikura, M.: On the quasiconformal surgery of rational functions. Ann. Sci. École Norm. Sup. Paris **20**, 1–29 (1987)
20. Whyburn, G.T.: Topological characterization of the Sierpinski curve. Fundam. Math. **45**, 320–324 (1958)

Measure Preserving Fractal Homeomorphisms

Michael F. Barnsley, Brendan Harding, and Miroslav Rypka

Abstract The basic theory of fractal transformations is recalled. For a fractal homeomorphism generated by a pair of affine iterated function systems (IFSs), a condition under which the transformation is measure (i.e. area, volume, etc.) preserving is established. Then three families of fractal homeomorphisms, two of them entirely new, generated by pairs of affine IFSs, are introduced. It is proved that they admit subfamilies that preserve n-dimensional Lebesque measure, where n is 2 or 3. Several examples are illustrated and applications to computer aided design and manufacture, via three-dimensional printing, are envisaged.

Keywords Fractal • Iterated function system

1991 Mathematics Subject Classification. 28A80

1 Introduction

The main goal of this paper is to introduce three special families of fractal homeomorphisms, two of them unexpected, generated by pairs of affine iterated function systems (IFSs). Using a new result (Theorem 3.1) we prove that these families admit subfamilies that preserve n-dimensional Lebesgue measure, where n

---

M.F. Barnsley (✉) • B. Harding
Australian National University, Canberra, ACT 0200, Australia
e-mail: michael.barnsley@anu.edu.au; Harding@anu.edu.au

M. Rypka
Australian National University, Canberra, ACT 0200, Australia

Palacky University Olomouc, Olomouc, 77146, Czech Republic
e-mail: miroslav.rypka01@upol.cz

C. Bandt et al. (eds.), *Fractals, Wavelets, and their Applications*, Springer Proceedings in Mathematics & Statistics 92, DOI 10.1007/978-3-319-08105-2_5,

is 2 or 3. We show explicitly how to construct these transformations and we present some examples. Applications to computer aided design and manufacture, via three-dimensional printing, are envisaged.

A measure preserving fractal homeomorphism is a type of fractal transformation. Fractal transformations have attracted our attention because they lead to innovative practical tools that may be used to model rough phenomena, such as digital images and biological objects. In place of using recursive structures and iteration to model rough objects, we consider instead the use of recursively generated transformations to model rough objects, (see, e.g., [7, 12]). Fractal transformations provide us with new tools in fractal theory (see, for example, [3–6, 8, 10, 11, 13]). They are generated using matched pairs of affine, projective, or bi-affine IFSs.

Affine IFSs have been used for image compression [2], where image objects, digital images for example, are represented by attractors of IFSs. Our goal, using fractal transformations, is different: we aim to relate simple objects to complicated objects via algorithmically simple (but geometrically complicated) transformations.

Why is the topic of this paper interesting? Measure preserving fractal homeomorphisms might be used to model area and volume preserving deformations of incompressible objects, for example to represent metamorphic rocks in geology, such as those described in, e.g., [17]; also, they are related to area preserving piecewise affine maps [15] used in image animation. But our interest is driven by other factors. (i) Area and volume preserving affine fractal homeomorphisms were a surprise to us: for a while we did not know how to construct them, and we had begun to suspect that they did not exist—if they did exist, then they would be in some sense both rare and natural. (ii) Area preserving affine fractal homeomorphisms applied to digital images produce visually appealing transformed images with applications to multimedia, for example [14]. (iii) Three-dimensional volume preserving affine fractal homeomorphisms can be used to deform malleable materials such as clay, metal, and plastic, to make new shapes out of familiar ones, and thus lend themselves to a role in computer aided design and manufacture, as we illustrate in Figs. 12 and 13.

In Sect. 2 we summarize the key ideas needed to understand fractal transformations, including basic notions related to IFSs. In Sect. 3 we establish a new result, Theorem 3.1, that gives a condition under which measure preserving fractal homeomorphisms would exist. Then in Sect. 4 we show that they do exist: we construct the explicit families of two- and three-dimensional measure preserving fractal homeomorphisms that lie at the core of this work, and give examples.

2 Point-Fibred IFSs

We recall some basic ideas concerning IFSs, code space, and fractal homeomorphisms. We follow terminology and results presented in the review article [11].

Definition 2.1. An *IFS* is a topological space X together with a finite set of continuous functions $f_n : X \to X, n = 1, 2, \ldots, N$.

We use the notation

$$\mathcal{F} = \{X; f_1, f_2, \ldots, f_N\}$$

to denote an IFS. Throughout we assume that X is a (complete) metric space with metric d. Let $H = H(X)$ denote the collection of nonempty compact subsets of X and let d_H denote the Hausdorff metric induced by d. The metric space (H, d_H) is complete. We define the Hutchinson operator $F : H \to H$ by

$$F(S) = \bigcup_{f \in \mathcal{F}} f(S)$$

for all $S \in H$. For $S \subset X$, define $F^0(S) = S$ and let $F^k(S)$ denote the k-fold composition of F applied to S, namely, the union of $f_{i_1} \circ f_{i_2} \circ \cdots \circ f_{i_k}(S)$ over all finite words $i_1 i_2 \cdots i_k$ of length k.

Definition 2.2. An *attractor* of the IFS $\mathcal{F}$ is a set $A \in H(X)$ such that there is an open set $U \subset X$ such that $A \subset U$ and $\lim_{k\to\infty} F^k(S) = A$, for all $S \in H$ with $S \subset U$, where the limit is with respect to the Hausdorff metric on H. The *basin* $B(A)$ of an attractor A of the IFS $\mathcal{F}$ is the largest open set U such that the latter assertion holds.

Since X is a metric space, we have $F : H(A) \to H(A)$ is continuous (w.r.t. the d_H) so it is necessarily true that if A is an attractor, then $F(A) = A$.

An IFS $\mathcal{F}$ on a metric space (X, d) is *contractive* if there is a metric $\hat{d}$, inducing the same topology on X as the metric d, with respect to which the functions in $\mathcal{F}$ are contractions. If $\mathcal{F}$ is a contractive IFS on a nonempty complete metric space (X, d), then $\mathcal{F}$ has a unique attractor A and the basin of A is X.

If $X = \mathbb{R}^M$ and the functions in $\mathcal{F}$ are affine functions, represented in the form $f(x) = Lx + a$, where L is an $M \times M$ matrix and $a \in \mathbb{R}^M$, then $\mathcal{F}$ is called an *affine IFS*. A fundamental result is that an affine IFS possesses an attractor if and only if there is a metric, inducing the same topology on $\mathbb{R}^M$ as the Euclidean metric, such that $\mathcal{F}$ is contractive, for precise information, see [1].

Let $\Sigma = \{1, 2, \ldots, n\}^\infty$ denote the set of all infinite sequences $\sigma \in \Sigma$, $\sigma = \sigma_1\sigma_2\sigma_3\cdots$, where $\sigma_i \in \{1, 2, \ldots, N\}$ for all $i = 1, 2, \ldots$ We have that (Σ, d_Σ), which we refer to as *code space*, is a compact metric space, where, for $\sigma \neq \omega$,

$$d_\Sigma(\sigma, \omega) = 2^{-\min\{k \in \mathbb{N}:\ \sigma_k \neq \omega_k\}}.$$

We will denote by $\Sigma_{\mathcal{F}}^k$ the space of "first" k-tuples from $\Sigma_{\mathcal{F}}$. For $\sigma|_k := \sigma_1\sigma_2\ldots\sigma_k \in \Sigma_{\mathcal{F}}^k$ we write $f_{\sigma|_k} = f_{\sigma_1} \circ f_{\sigma_2} \circ \cdots \circ f_{\sigma_k}$.

An attractor A of $\mathcal{F}$ is said to be *point-fibred* if there is a continuous *coding map* $\pi : \Sigma \to A$ that is well defined by

$$\pi(\sigma) = \lim_{k\to\infty} f_{\sigma_1} \circ f_{\sigma_2} \circ \cdots \circ f_{\sigma_k}(x)$$

independently of $x \in B(A)$. The coding map is uniformly continuous in σ and onto. If $\mathcal{F}$ is contractive, then it possesses a unique point-fibred attractor. If A is an attractor of an affine IFS, then is point-fibred.

The *set of addresses* of a point $x \in A$ is defined to be

$$\pi^{-1}(\{x\}) := \{\omega \in \Sigma : x \in \pi(\omega)\}.$$

If A is a point-fibred attractor of $\mathcal{F}$, then the coding map $\pi : \Sigma \to A$ assigns a point in the attractor A of $\mathcal{F}$ to each infinite string. A *section* of a coding map $\pi : \Sigma \to A$ is a function $\tau : A \to \Sigma$ such that $\pi \circ \tau$ is the identity on A. For $x \in A$, the string $\tau(x)$ is referred to as the *address* of x with respect to the section τ. The set $\Sigma_\tau := \tau(A)$ is called the *address space* of the section τ.

Properties of sections of coordinate maps are discussed in [12, Sect. 3].

Here it is assumed that all IFSs have point-fibred attractors, each of which possesses a coding map. Given two point-fibred IFSs $\mathcal{F}$ and $\mathcal{G}$ with respective attractors $A_{\mathcal{F}}$ and $A_{\mathcal{G}}$, a fractal transformation is a mapping $h : A_{\mathcal{F}} \to A_{\mathcal{G}}$ that maps a given point in $A_{\mathcal{F}}$ to the point in $A_{\mathcal{G}}$ with the same address. We assume that $\mathcal{F}$ and $\mathcal{G}$ have the same number of functions, and let $\pi_{\mathcal{F}}$ and $\pi_{\mathcal{G}}$ be the respective coordinate maps.

Definition 2.3. A transformation $h : A_{\mathcal{F}} \to A_{\mathcal{G}}$ is called a *fractal transformation* if

$$h = \pi_{\mathcal{G}} \circ \tau_{\mathcal{F}}$$

for some shift invariant section $\tau_{\mathcal{F}}$ of $\mathcal{F}$. If h is a homeomorphism, then h is called a *fractal homeomorphism*.

The following result is proved in [10].

Proposition 2.4. *If a fractal transformation $h = \pi_{\mathcal{G}} \circ \tau_{\mathcal{F}} : A_{\mathcal{F}} \to A_{\mathcal{G}}$ is a homeomorphism, then there exists a shift invariant section $\tau_{\mathcal{G}}$ of $\pi_{\mathcal{G}}$ such that the following diagram commutes:*

$$\begin{array}{ccc} A_{\mathcal{F}} & \underset{h}{\rightarrow} & A_{\mathcal{G}} \\ \tau_{\mathcal{F}} \searrow & & \swarrow \tau_{\mathcal{G}} \\ & \Sigma & \end{array}$$

Conversely, if there exist sections $\tau_{\mathcal{F}}$ and $\tau_{\mathcal{G}}$ and a homeomorphism h such that the above diagram commutes, then $h = \pi_{\mathcal{G}} \circ \tau_{\mathcal{F}}$ and $h^{-1} = \pi_{\mathcal{F}} \circ \tau_{\mathcal{G}}$.

An IFS $\mathcal{F} = \{X; f_1, f_2, \ldots, f_N\}$ is said to be *injective* if the map $f_i : X \to X$ is injective, for $i = 1, 2, \ldots, N$.

Definition 2.5. For an IFS $\mathcal{F}$ with attractor A, a *mask* is a partition $M = \{M_i, 1 \le i \le N\}$ of A such that $M_i \subseteq f_i(A)$ for all $f_i \in \mathcal{F}$. Given an injective IFS $\mathcal{F}$ and a mask M, consider the function $T : A \to A$ defined by

$$T(x) := f_i^{-1}(x)$$

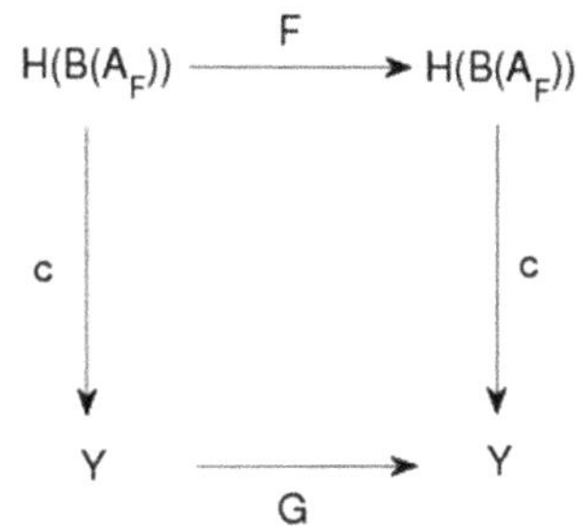

Fig. 1 Semiconjugacy of IFSs

when $x \in M_i$. The *itinerary* $\tau_M(x)$ of a point $x \in A$ is the string $\omega_0\,\omega_1\,\omega_2 \cdots \in \Sigma$, where ω_k is the unique integer $1 \leq \omega_k \leq N$ such that

$$T^k(x) \in M_{\omega_k}.$$

A particular mask for an IFS with N functions and attractor A is defined by $M_1 = f_1(A)$ and

$$M_k = f_k(A) \setminus \bigcup_{i=1}^{k-1} f_i(A),$$

for $k = 2, 3, \ldots, N$. This is called the *tops mask.*

The following theorem states that any shift invariant section is constructed from a mask.

Theorem 2.6. *Let $\mathcal{F}$ be a contractive and injective IFS.*

(1) *If M is a mask, then τ_M is a shift invariant section of π.*
(2) *If τ is a shift invariant section of π, then $\tau = \tau_M$ for some mask M.*

The section from a tops mask is given by

$$\tau(x) = \max \pi^{-1}(x),$$

where the maximum is with respect to the lexicographic order on Σ.

The relationship of attractors of IFSs under continuous mappings is captured in Proposition 2.7 and Lemma 2.8.

Proposition 2.7. *Let $A_{\mathcal{F}}$ be an attractor of IFS $\mathcal{F} = \{X; f_1, f_2, \ldots, f_N\}$. Let Y be a topological space. Let $c\colon X \to Y$ be continuous, let $g_i\colon Y \to Y$ be continuous and such that $cf_i = g_i c$ for all $i = 1, 2, \ldots, N$. Then the IFS $\mathcal{G} := \{Y, g_1, g_2, \ldots, g_N\}$ possesses the invariant set $A_{\mathcal{G}} = cA_{\mathcal{F}}$, and the diagram in Fig. 1 is commutative.*

Proof. Observe that the attractor $A_{\mathcal{F}}$ of $\mathcal{F}$ has the following property,

$$A_{\mathcal{F}} = \bigcup_{i=1}^{N} f_i(A_{\mathcal{F}}).$$

Since $A_{\mathcal{F}}$ is compact and $cf_i = g_i c$ then

$$cA_{\mathcal{F}} = c\bigcup_{i=1}^{N} f_i(A_{\mathcal{F}}) = \bigcup_{i=1}^{N} cf_i(A_{\mathcal{F}}) = \bigcup_{i=1}^{N} g_i c(A_{\mathcal{F}}).$$

This means that $cA_{\mathcal{F}} = \cup g_i(cA_{\mathcal{F}})$ is an invariant set of $\mathcal{G}$.

Notice also that

$$cF = Gc,$$

where $F(A) = \cup_{i=1}^{N} f_i(A)$.

Let $D \in H(B_{\mathcal{F}})$, then

$$\lim_{k\to\infty} F^k(D) = A_{\mathcal{F}}.$$

It follows that

$$c\lim_{k\to\infty} F^k(D) = cA_{\mathcal{F}}.$$

We can write that

$$c\lim_{k\to\infty} F^k(D) = \lim_{k\to\infty} cFF^{k-1}(D) = \lim_{k\to\infty} GcF^{k-1}(D)$$

and similarly

$$\lim_{k\to\infty} GcFF^{k-2}(D) = \lim_{k\to\infty} G^2cF^{k-2}(D).$$

Finally we obtain

$$c\lim_{k\to\infty} F^k(D) = \lim_{k\to\infty} G^k c(D) = c(A_{\mathcal{F}}) = A_{\mathcal{G}}.$$

□

Consider, for example, two IFSs $\mathcal{S} = \{\{1,2,\dots,N\}^{\infty}; s_1, s_2, \dots, s_N\}$ such that $s_n(\sigma) = n\sigma \forall \sigma \in \Sigma, n \in \{1,2,\dots,N\}$ and $\mathcal{G}' = \{X; g_1, g_2, \dots, g_N\}$. Then the continuous mapping is obviously $c = \pi_{\mathcal{G}'}$ (see Fig. 2). Replacing continuous map c with homeomorphism h we obtain the following.

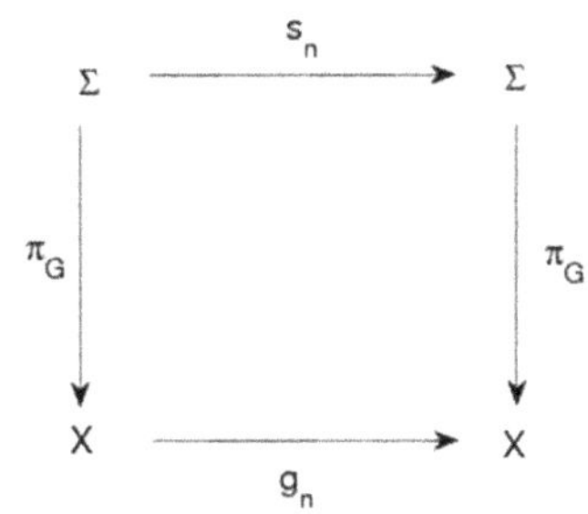

Fig. 2 Semiconjugacy between the symbolic IFS $\{\Sigma; s_1, s_2, \ldots, s_N\}$ acting on code space, its attractor, and the IFS $\mathcal{G} = \{X; g_1, g_2, \ldots, g_N\}$ acting on its attractor. The latter is a factor of the former

Lemma 2.8. *Let $A_{\mathcal{F}}$ be an attractor of IFS $\{X \subset \mathbb{R}^M, f_1, f_2, \ldots, f_N\}$ with basin $B(A_{\mathcal{F}})$ and let $h\colon X \to Y \subset \mathbb{R}^M$ be a homeomorphism. Then $A_{\mathcal{G}} := h(A_{\mathcal{F}})$ is an attractor of $\mathcal{G} = \{Y; hf_1h^{-1}, hf_2h^{-1}, \ldots, hf_Nh^{-1}\}$. All topological properties of the two systems are shared; in particular*

(1) $\dim(A_{\mathcal{F}}) = \dim A_{\mathcal{G}}$
(2) *$A_{\mathcal{F}}$ is point-fibred if and only if $A_{\mathcal{G}}$ is point-fibred*
(3) $B(A_{\mathcal{G}}) = h(B(A_{\mathcal{F}}))$

Next, we will be concerned with the opposite problem. Given two IFSs, we look for conditions sufficient for the existence of a homeomorphism between their attractors.

In what follows we assume that all attractors are point-fibred. This is because here we are concerned exclusively with attractors of affine IFSs and their children, obtained by topological conjugation, as in Lemma 2.8.

Definition 2.9. Let $\mathcal{F}$ and $\mathcal{G}$ be IFSs with attractors $A_{\mathcal{F}}$ and $A_{\mathcal{G}}$, respectively. A transformation $T\colon A_{\mathcal{F}} \to A_{\mathcal{G}}$ is called *a fractal transformation* if it can be written in the form

$$T_{\mathcal{F}\mathcal{G}} = \pi_{\mathcal{G}} \circ \tau_{\mathcal{F}}$$

where $\tau_{\mathcal{F}}\colon A_{\mathcal{F}} \to \Sigma_{\mathcal{F}}$ is a shift invariant section for $\mathcal{F}$ and $\pi_{\mathcal{G}}\colon \Sigma_{\mathcal{G}} \to A_{\mathcal{G}}$ is the coding map for $\mathcal{G}$. A fractal transformation $T_{\mathcal{F}\mathcal{G}}$ is called a *fractal homeomorphism* when it is a homeomorphism between the attractors $A_{\mathcal{F}}$ and $A_{\mathcal{G}}$.

Theorem 2.10 gives conditions under which a fractal transformation is either continuous or a homeomorphism.

Theorem 2.10 ([12, Theorem 3.4, p. 6]). *Let $\mathcal{F}$ and $\mathcal{G}$ be IFSs, as above. Let $\tau_{\mathcal{F}}\colon A_{\mathcal{F}} \to \Sigma$ be a section for $\mathcal{F}$ and let $\pi_{\mathcal{G}}\colon \Sigma_{\mathcal{G}} \to A_{\mathcal{G}}$ be the coding map for $\mathcal{G}$. The fractal transformation $T_{\mathcal{F}\mathcal{G}} = \pi_{\mathcal{G}} \circ \tau_{\mathcal{F}}$ has the following properties:*

(1) *If, for all $\sigma, \omega \in \overline{\Sigma_{\mathcal{F}}}$, $\pi_{\mathcal{F}}(\sigma) = \pi_{\mathcal{F}}(\omega) \Rightarrow \pi_{\mathcal{G}}(\sigma) = \pi_{\mathcal{G}}(\omega)$, then the fractal transformation $T_{\mathcal{F}\mathcal{G}}$ is continuous.*

(2) *If, for all* $\sigma, \omega \in \overline{\Sigma_{\mathcal{F}}}$, $\pi_{\mathcal{F}}(\sigma) = \pi_{\mathcal{F}}(\omega) \Leftrightarrow \pi_{\mathcal{G}}(\sigma) = \pi_{\mathcal{G}}(\omega)$, *and in addition* $\Sigma_{\mathcal{F}}$ *is the address space for* $\mathcal{G}$, *then* $T_{\mathcal{FG}}$ *is a homeomorphism and* $T_{\mathcal{FG}}^{-1} = T_{\mathcal{GF}}$.

The following lemma follows immediately from Theorem 2.10 and Lemma 2.8.

Lemma 2.11. *If* $A_{\mathcal{F}}$ *is an attractor of IFS* $\mathcal{F} = \{\mathbb{R}^n, f_1, f_2, \ldots, f_N\}$ *and* $\mathcal{G} = \{\mathbb{R}^n, hf_1h^{-1}, hf_2h^{-1}, \ldots, hf_Nh^{-1}\}$, *then* $T_{\mathcal{FG}} := h|A_{\mathcal{F}}$ *is a fractal homeomorphism.*

Remark 1. If the conditions in Theorem 2.10 (2) are satisfied, we say that $A_{\mathcal{F}}$ and $A_{\mathcal{G}}$ have the same address structure.

3 Lebesgue Measure Preserving Fractal Homeomorphism

A pair of IFSs $\mathcal{F}$ and $\mathcal{G}$ with respective attractors $A_{\mathcal{F}}$ and $A_{\mathcal{G}}$, satisfying conditions in Theorem 2.10 (2), yields a fractal homeomorphism. It is, in general, quite difficult to find pairs of affine IFSs that generate fractal homeomorphisms. Some such families of pairs, say $\mathcal{F}$ and $\mathcal{G}$ with attractors $A_{\mathcal{F}}$ and $A_{\mathcal{G}}$, respectively, have been established [6], in $\mathbb{R}^2$ but, until this paper, no truly three-dimensional "non trivial" examples had been reported. Why are we interested in affine systems? Affine transformations in $\mathbb{R}^n$ play a special role: they preserve ratios of n-dimensional Lebesgue measures of subsets of sets upon which they act. As an analogy, if subsets are defined in terms of colours, then affine transformations preserve histograms. Using this property, Theorem 3.1 asserts that if a fractal homeomorphism $T_{\mathcal{FG}}: A_{\mathcal{F}} \to A_{\mathcal{G}}$ is generated by a pair of IFSs $\mathcal{F}$ and $\mathcal{G}$ with the property

$$\mathcal{L}(f_i(A_{\mathcal{F}})) = \mathcal{L}(g_i(A_{\mathcal{G}})) \forall i \in \{1, 2, \ldots, N\}, \tag{1}$$

then Lebesgue n-dimensional measure is invariant under $T_{\mathcal{FG}}$.

In Sect. 4 we will establish that Theorem 3.1 is far from vacuous, by exhibiting interesting families of pairs of affine IFSs.

Theorem 3.1. *Let* $A_{\mathcal{F}}$ *and* $A_{\mathcal{G}}$ *be respective attractors of affine IFSs* $\mathcal{F} = \{\mathbb{R}^n, f_1, f_2, \ldots, f_N\}$ *and* $\mathcal{G} = \{\mathbb{R}^n, g_1, g_2, \ldots, g_N\}$, *such that* $T_{\mathcal{FG}} = \pi_{\mathcal{G}} \circ \tau_{\mathcal{F}}: A_{\mathcal{F}} \to A_{\mathcal{G}}$ *is a homeomorphism, and such that* $\mathcal{L}(A_{\mathcal{F}}) > 0$, *where* $\mathcal{L}$ *is* n*−dimensional Lebesgue measure. If*

$$\mathcal{L}(f_i(A_{\mathcal{F}})) = \mathcal{L}(g_i(A_{\mathcal{G}})) \text{ for all } i = 1, 2, \ldots, N$$

then

$$\mathcal{L}(S) = \mathcal{L}(T_{\mathcal{FG}}(S)) \forall S \subset A_{\mathcal{F}}.$$

Proof. Let the assumptions of the theorem hold. Observe that since affine transformations scale uniformly and $\mathcal{L}(f_i(A_{\mathcal{F}})) = \mathcal{L}(g_i(A_{\mathcal{G}}))$ for $i = 1, 2, \dots N$, then

$$\mathcal{L}(f_{\sigma|_k}(A_{\mathcal{F}})) = \mathcal{L}(g_{\sigma|_k}(A_{\mathcal{G}}))$$

for all $\sigma|_k \in \Sigma^k_{\mathcal{F}}, k \geq 1$.

In order to prove the equation

$$\mathcal{L}(S) = \mathcal{L}(T_{\mathcal{F}\mathcal{G}}(S))$$

we will show two inequalities with the help of Vitali covering (see, e.g., [16, p. 128]),

$$\mathcal{L}(S) \leq \mathcal{L}(T_{\mathcal{F}\mathcal{G}}(S)) + \delta$$

and

$$\mathcal{L}(T_{\mathcal{F}\mathcal{G}}(S)) \leq \mathcal{L}(S) + \delta$$

for any $\delta > 0$.

First, we will prove that $S \subset A_{\mathcal{F}}$, then $\{f_{\sigma|_k}(A_{\mathcal{F}}) : \sigma|_k \in \Sigma^k_{\mathcal{F}}, k \geq 1\}$ is a Vitali covering of S. Let us pick $x \in S$. Since $S \subset A_{\mathcal{F}}$, there exists $\sigma \in \Sigma_{\mathcal{F}}$ such that $\pi_{\mathcal{F}}(\sigma) = x$. Hence,

$$\pi_{\mathcal{F}}(\sigma) = \lim_{k\to\infty} f_{\sigma|_k}(y)$$

is independent of y. In particular, we have $x \in f_{\sigma|_k}(A_{\mathcal{F}})$ for all $k = 1, 2, 3, \dots$ and $\{x\} = \lim_{k\to\infty} f_{\sigma|_k}(A_{\mathcal{F}})$. Therefore, for any $\epsilon > 0$ there exists $k \geq 1$ such that $\text{diam}(f_{\sigma|_k}(A_{\mathcal{F}})) < \epsilon$ and hence $\{f_{\sigma|_k}(A_{\mathcal{F}}) \subset A_{\mathcal{F}} : \sigma^k \in \Sigma^k_{\mathcal{F}}, k \geq 1\}$ is a Vitali covering of S.

Hence, we may apply the Vitali covering lemma. For any $\delta > 0$, we can find finitely many disjoint sets $\{B_1, B_2, \dots, B_r\} \subset \{f_{\sigma|_k}(A_{\mathcal{F}}) \colon \sigma|_k \in \Sigma^k_{\mathcal{F}}, k \geq 1\}$ such that

$$\mathcal{L}(S) \leq \delta + \sum_{j=1}^{r} \mathcal{L}(B_j).$$

Furthermore, these sets may be chosen such that $B_j \subset S$ for $j = 1, 2, \dots, r$.

Now, consider the sets $T_{\mathcal{F}\mathcal{G}}(B_j)$, for $j = 1, 2, \dots, r$. Since $T_{\mathcal{F}\mathcal{G}}$ is a homeomorphism then

$$B_j \subset S \Rightarrow T_{\mathcal{F}\mathcal{G}}(B_j) \subset T_{\mathcal{F}\mathcal{G}}(S)$$

and the sets $T_{\mathcal{F}\mathcal{G}}(B_j)$ are disjoint. Observe that

$$\mathcal{L}(T_{\mathcal{F}\mathcal{G}}(B_j)) = \mathcal{L}(B_j)$$

since $B_j = f_{\sigma|_k}(A_{\mathcal{F}})$ for some $\sigma|_k \in \Sigma_{\mathcal{F}}^k, k \geq 1$ and $T_{\mathcal{F}\mathcal{G}}(B_j) = g_{\sigma|_k}(A_{\mathcal{G}})$. We arrive to

$$L(S) \leq \delta + \sum_{j=1}^{r} \mathcal{L}(B_j) = \delta + \sum_{j=1}^{r} \mathcal{L}(T_{\mathcal{F}\mathcal{G}}(B_j)) = \delta + \mathcal{L}\left(\bigsqcup_{j=1}^{r} T_{\mathcal{F}\mathcal{G}}(B_j)\right).$$

This means

$$\mathcal{L}(S) \leq \delta + \mathcal{L}(T_{\mathcal{F}\mathcal{G}}(S)).$$

The second inequality may be obtained in the similar way. Performing the previous construction for $T_{\mathcal{F}\mathcal{G}}(S) \subset A_{\mathcal{G}}$, we find that

$$\mathcal{L}(T_{\mathcal{F}\mathcal{G}}(S)) \leq \delta + \mathcal{L}(S).$$

Therefore, we have that

$$|\mathcal{L}(S) - \mathcal{L}(T_{\mathcal{F}\mathcal{G}}(S))| \leq \delta$$

for any $\delta > 0$.

We arrive to

$$\mathcal{L}(S) = \mathcal{L}(T_{\mathcal{F}\mathcal{G}}(S)).$$

□

Corollary 1. *Let $A_{\mathcal{F}}$ and $A_{\mathcal{G}}$ be respective attractors of affine IFSs $\mathcal{F} = \{\mathbb{R}^n, f_1, f_2, \ldots, f_N\}$ and $\mathcal{G} = \{\mathbb{R}^n, g_1, g_2, \ldots, g_N\}$, such that $T_{\mathcal{F}\mathcal{G}} = \pi_{\mathcal{G}} \circ \tau_{\mathcal{F}} \colon A_{\mathcal{F}} \to A_{\mathcal{G}}$ is a homeomorphism, and such that $\mathcal{L}(A_{\mathcal{F}}) > 0$, where $\mathcal{L}$ is $n-$dimensional Lebesgue measure. If*

$$\mathcal{L}(f_i(A_{\mathcal{F}}))/\mathcal{L}(A_{\mathcal{F}}) = \mathcal{L}(g_i(A_{\mathcal{G}}))/\mathcal{L}(A_{\mathcal{G}}) \text{ for all } i = 1, 2, \ldots, N,$$

then

$$\mathcal{L}(S)/\mathcal{L}(A_{\mathcal{F}}) = \mathcal{L}(T_{\mathcal{F}\mathcal{G}}(S))/\mathcal{L}(A_{\mathcal{G}}) \forall S \subset A_{\mathcal{F}}.$$

Proof. We will consider the image of $A_{\mathcal{G}}$ under a similitude on $\mathbb{R}^n$. Let $\mathcal{F}, \mathcal{G}$ and $S \subset \mathbb{R}^n$ satisfy the assumptions of the corollary. Consider the similitude $h \colon Y \to Y$,

$$h(y) = \sqrt[n]{\frac{\mathcal{L}(A_{\mathcal{F}})}{\mathcal{L}(A_{\mathcal{G}})}}\, y$$

which is an affine homeomorphism. Observe that $\mathcal{L}(A_{\mathcal{H}}) = \mathcal{L}(A_{\mathcal{F}})$, where $A_{\mathcal{H}}$ is an attractor of the IFS $\mathcal{H} = \{\mathbb{R}^n, hg_1h^{-1}, hg_2h^{-1}, \ldots, hg_Nh^{-1}\}$ given by Lemma 2.8. The existence of fractal homeomorphism $h \circ T_{\mathcal{FG}}\colon A_{\mathcal{F}} \to A_{\mathcal{H}}$ follows from Lemma 2.11. Hence $\mathcal{L}(S) = \mathcal{L}(h \circ T_{\mathcal{FG}}(S))$. Since h is affine, it follows that

$$\mathcal{L}(S)/\mathcal{L}(A_{\mathcal{F}}) = \mathcal{L}(h \circ T_{\mathcal{FG}}(S))/\mathcal{L}((A_{\mathcal{H}})) = \mathcal{L}(T_{\mathcal{FG}}(S))/\mathcal{L}(A_{\mathcal{G}}).$$

□

4 Existence of Nontrivial (Measure Preserving) Fractal Homeomorphisms in $\mathbb{R}^2$ and $\mathbb{R}^3$

In this section we exhibit three families of pairs of IFSs that generate fractal homeomorphisms. We also show that sub-families can be chosen so that the corresponding fractal homeomorphism is measure preserving. The first family acts on $\mathbb{R}^2$ and is already known, but the formal application of Theorem 3.1 to it is new. The second and third families acting in $\mathbb{R}^3$ are new in all aspects.

Example 4.1. This example of an area preserving fractal homeomorphism was suggested but not established in [6, pp. 292–293]. Let ABC be an equilateral triangle in $\mathbb{R}^2$. Let $\alpha \in (0,1)$ and a, b, c denote points on BC, AC, and AB, respectively, such that $|Bc|/|AB| = |Ca|/|BC| = |Ab|/|AC| = \alpha$. Let $f_1 : \mathbb{R}^2 \to \mathbb{R}^2$ denote the unique affine transformation such that

$$f_1(ABC) = caB,$$

which means $f_1(A) = c$, $f_1(B) = a$ and $f_1(A) = B$. Similarly, let

$$f_2(ABC) = Cab,\ f_3(ABC) = cAb,\ f_4(ABC) = cab.$$

Then the attractor of $\mathcal{F} = \{\mathbb{R}^2, f_1, f_2, f_3, f_4\}$ is the filled triangle ABC (see Fig. 3).

Next, let us consider points $r \in AB, q \in AC, p \in BC$, such that $|Br|/|AB| = |Cq|/|AC| = |Ap|/|BC| = 1 - \alpha$. Let $\mathcal{G}$ denote the IFS $\{\mathbb{R}^2; g_1, g_2, g_3, g_4\}$ such that

$$g_1(ABC) = rpB, g_2(ABC) = Cpq, g_3(ABC) = rAq, g_4(ABC) = rpq.$$

Observe that the attractor $A_{\mathcal{G}}$ is again the filled triangle ABC and $\mathcal{L}(f_i(ABC)) = \mathcal{L}(g_i(ABC))$ for any $i = 1, 2, 3, 4$.

In [6, p. 300] it is proven that the address structures of both attractors are the same. Hence, there exists a fractal homeomorphism $T_{\mathcal{FG}}\colon ABC \to ABC$. Since for any $i = 1, 2, 3, 4$ we have that $\mathcal{L}(f_i(ABC)) = \mathcal{L}(g_i(ABC))$, it follows from Theorem 3.1 that the homeomorphism $T_{\mathcal{FG}}$ is area preserving, that is $\mathcal{L}(S) = \mathcal{L}(T_{\mathcal{FG}}(S))$ for any set $S \subset ABC$.

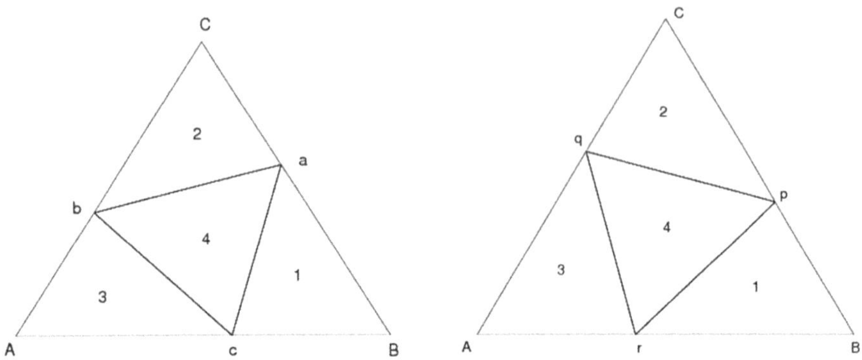

Fig. 3 The two self-affine tilings of the filled triangle ABC shown here generate a measure preserving fractal homeomorphism under appropriate conditions (see Sect. 4 Example 4.1)

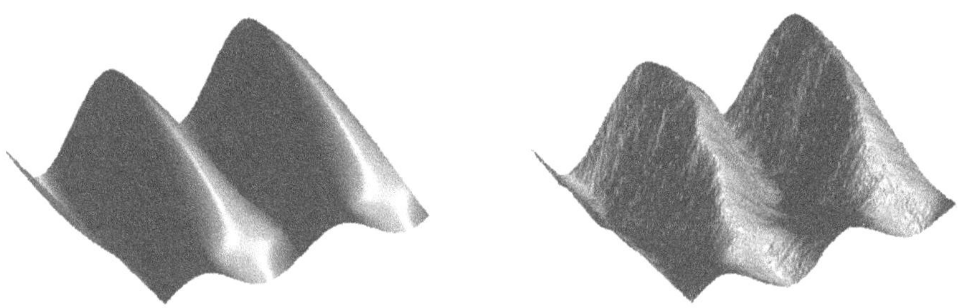

Fig. 4 Transformation of a surface obtained by applying a two-dimensional area preserving homeomorphism to the (x, y) coordinates of a surface

We note the following. We can represent a surface over the equilateral triangle in the obvious manner, regarding it as the graph of a function $f : ABC \to \mathbb{R}$. Then we can apply $T_{\mathcal{FG}}$ to ABC, yielding a new surface, the graph of $f_1 : T_{\mathcal{FG}}(ABC) \subset \mathbb{R}^2 \to \mathbb{R}$, defined by

$$f_1(x) = f(T_{\mathcal{FG}}^{-1}(x)).$$

Clearly, if $T_{\mathcal{FG}}$ is sufficiently extreme, the new surface will be rough. An example of such transformation is shown in Fig. 4.

Remark 2. Since the Lebesgue measure is preserved it follows that

$$\int_S f d\mathcal{L} = \int_{T_{\mathcal{FG}}(S)} f T_{\mathcal{FG}}^{-1} d\mathcal{L}$$

for any measure preserving fractal homeomorphism $T_{\mathcal{FG}}$.

Fig. 5 Transformation of a picture obtained by applying an area preserving fractal homeomorphism (Example 4.1)

Figure 5 illustrates application of a fractal homeomorphism to a picture. Any picture may be regarded as a map $P : S \subset \mathbb{R}^2 \to C$, where C is a colour space. Applying a homeomorphism $T_{\mathcal{F}\mathcal{G}}$, we obtain a map $P_1 : T_{\mathcal{F}\mathcal{G}}(S) \subset \mathbb{R}^2 \to C$,

$$P_1(x) = PT_{\mathcal{F}\mathcal{G}}^{-1}(x).$$

Area preserving homeomorphism ensures the same amount of each colour on the original and transformed picture.

Next, we will provide examples of a volume preserving fractal homeomorphism.

Example 4.2. Let us denote by A, B, C, D the vertices of a regular tetrahedron. Let us consider $\alpha \in (0,1)$ and points $a \in BC, b \in AC, c \in AB$ such that $|aC|/|BC| = |bA|/|AC| = |cB|/|AB| = \alpha$. Let $\mathcal{F}$ denote IFS $\{\mathbb{R}^3; f_1, f_2, f_3, f_4\}$, where $f_1(ABCD) = cbDA, f_2(ABCD) = cBDa, f_3(ABCD) = CbDa, f_4(ABCD) = cbDa$ as illustrated by Fig. 6. Notice that an attractor of $\mathcal{F}$ is filled tetrahedron $ABCD$. Let us denote by $\mathcal{G}$ the IFS $\{\mathbb{R}^3; g_1, g_2, g_3, g_4\}$ such that $g_1(ABCD) = rqDA, g_2(ABCD) = rBDp, g_3(ABCD) = CqDp, g_4(ABCD) = rqDp$, where $r \in AB, q \in AC, p \in BC$ and $|pC|/|BC| = |qA|/|AC| = |rB|/AB = 1 - \alpha$. Again, attractor $A_{\mathcal{G}}$ is the filled tetrahedron $ABCD$. For the details of the IFS see Fig. 6, where we write $f_i(ABCD) = A'B'C'D'$ for any $i = 1, \ldots, 4$.

The fact that this example possesses a unique point-fibred attractor is quite subtle: the key point is that the maps are all non-antipodal, see [1].

In the same way as in [6] for the previous example, it may be shown that the attractors have the same address structure, and hence, by Theorem 2.10, $T_{\mathcal{F}\mathcal{G}}$ is a fractal homeomorphism. Since $\mathcal{L}(f_i(ABCD)) = \mathcal{L}(g_i(ABCD)), i = 1, 2, 3, 4$, it follows by Theorem 3.1 that $T_{\mathcal{F}\mathcal{G}}$ is volume preserving. For any set $S \subset ABCD$ we have $\mathcal{L}(S) = \mathcal{L}(T_{\mathcal{F}\mathcal{G}}(S))$, as illustrated, for example, in Figs. 7 and 8.

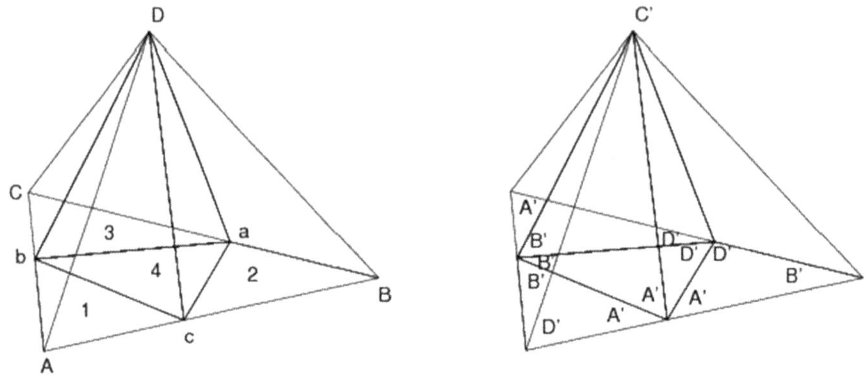

Fig. 6 A pair of self-affine tilings of a tetrahedron used to generate a volume preserving fractal homeomorphism, as explained in Sect. 4 Example 4.2

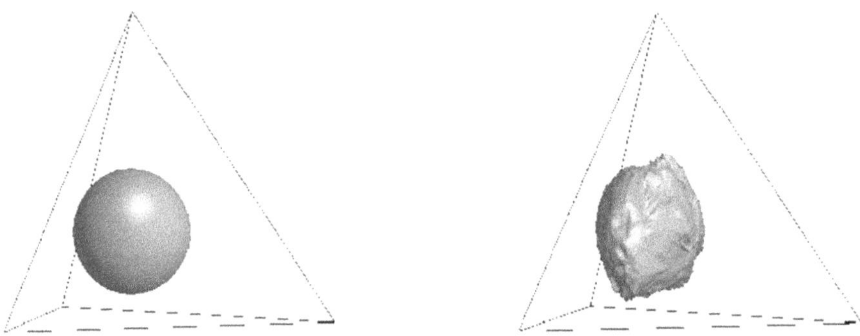

Fig. 7 Ball, original and transformed under a volume preserving fractal homeomorphism of the type explained in Sect. 4 Example 4.2

Fig. 8 Ball, original and transformed by an Example 4.2 type volume preserving fractal homeomorphism

Remark 3. For the visualization of transformed sets and images we use the chaos game and perpixel algorithm. These are described in detail in [9].

Example 4.3. Finally, we will introduce a second tiling of a tetrahedron which also generates volume preserving fractal homeomorphisms. Consider the regular tetrahedron $ABCD$ $A = (0,0,0), B = (1,0,0), C = (1/2, \sqrt{3}/2, 0), D = (1/2, \sqrt{3}/6, \sqrt{2/3})$. Given six parameters $\alpha_{AB}, \alpha_{AC}, \alpha_{AD}, \alpha_{BC}, \alpha_{BD}, \alpha_{CD}$ we can split the regular tetrahedron into the eight smaller tetrahedrons characterized by the following sets of vertices

$$\{A, A + \alpha_{AB}(B - A), A + \alpha_{AC}(C - A), A + \alpha_{AD}(D - A)\},$$

$$\{B, A + \alpha_{AB}(B - A), B + \alpha_{BC}(C - B), B + \alpha_{BD}(D - B)\},$$

$$\{C, A + \alpha_{AC}(C - A), B + \alpha_{BC}(C - B), C + \alpha_{CD}(D - C)\},$$

$$\{D, A + \alpha_{AD}(D - A), B + \alpha_{BD}(D - B), C + \alpha_{CD}(D - C)\},$$

$$\{A + \alpha_{AB}(B - A), A + \alpha_{AC}(C - A), A + \alpha_{AD}(D - A), B + \alpha_{BC}(C - B)\},$$

$$\{A + \alpha_{AC}(C - A), A + \alpha_{AD}(D - A), B + \alpha_{BC}(C - B), C + \alpha_{CD}(D - C)\},$$

$$\{A + \alpha_{AB}(B - A), A + \alpha_{AD}(D - A), B + \alpha_{BC}(C - B), B + \alpha_{BD}(D - B)\},$$

$$\{A + \alpha_{AD}(D - A), B + \alpha_{BC}(C - B), B + \alpha_{BD}(D - B), C + \alpha_{CD}(D - C)\}.$$

Let us consider an affine IFS which maps a regular tetrahedron to itself by mapping it to the eight smaller tetrahedrons described above (see Fig. 9). Such an IFS is characterized by six α parameters (assuming we match up all of the vertices in a particular way). In particular we note that if $\alpha_{AB} = \alpha_{AC} = \alpha_{AD} = \alpha_{BC} = \alpha_{BD} = \alpha_{CD} = 0.5$ then the eight affine mappings map the regular tetrahedron to eight smaller tetrahedrons which all have the same volume. Furthermore, four of the mappings are similitudes. From now on we will denote an IFS of the above form by $F(\alpha_{AB}, \alpha_{AC}, \alpha_{AD}, \alpha_{BC}, \alpha_{BD}, \alpha_{CD})$ with each of the six parameters in the open interval $(0, 1)$. For instance, the IFS $F(0.5, 0.5, 0.5, 0.5, 0.5, 0.5)$ is the special case mentioned above.

Proposition 4.4. *The homeomorphism between two IFSs $\mathcal{F}, \mathcal{G}$ which map the tetrahedron into itself as described above is a volume preserving if*

$$\mathcal{F} = \mathcal{F}\left(a, \frac{1}{2}, \frac{1}{2}, \frac{1}{2}, \frac{1}{2}, 1 - a\right) \text{ and } \mathcal{G} = \mathcal{G}\left(\frac{1}{2}, a, \frac{1}{2}, \frac{1}{2}, 1 - a, \frac{1}{2}\right)$$

or

$$\mathcal{F} = \mathcal{F}\left(a, 1 - a, \frac{1}{2}, \frac{1}{2}, a, 1 - a\right) \text{ and } \mathcal{G} = \mathcal{G}\left(1 - a, a, \frac{1}{2}, \frac{1}{2}, 1 - a, a\right)$$

for some $a \in (0, 1)$

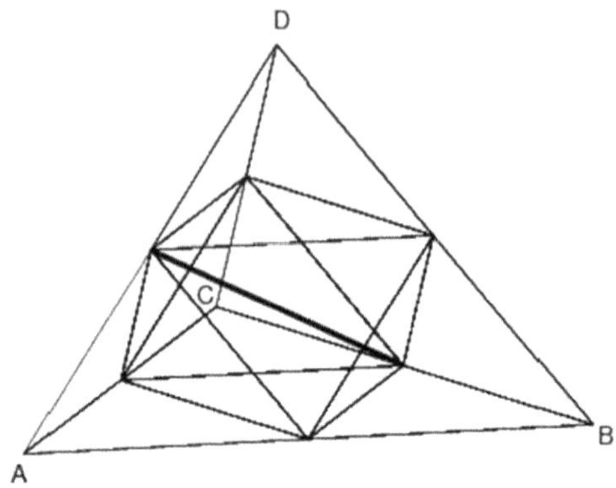

Fig. 9 A diagram that represents the eight tetrahedron used to generate the family of volume preserving fractal homeomorphisms explained in Sect. 4 Example 4.3

Proof. In order to show that the homeomorphism is measure preserving, we will investigate the volumes of $f_i(ABCD)$ and express them with the help of parameters $\alpha_{AB}, \alpha_{AC}, \alpha_{AD}, \alpha_{BC}, \alpha_{BD}, \alpha_{CD}$. The volume of the tetrahedron $ABCD$ is given by

$$V(ABCD) = \frac{|(D - A) \cdot ((B - A) \times (C - A))|}{6}.$$

Note that the volume is unchanged under permutation the ordering of the vertices. Let $\alpha_{AB}, \alpha_{AC}, \alpha_{AD}, \alpha_{BC}, \alpha_{BD}, \alpha_{CD} \in (0, 1)$, and define points by

$$\begin{aligned} P_{AB} &= A + \alpha_{AB}(B - A), & P_{AC} &= A + \alpha_{AC}(C - A), \\ P_{AD} &= A + \alpha_{AD}(D - A), & P_{BC} &= B + \alpha_{BC}(C - B), \\ P_{BD} &= B + \alpha_{BD}(D - B), & P_{CD} &= C + \alpha_{CD}(D - C). \end{aligned}$$

Now the volume of $AP_{AB}P_{AC}P_{AD}$ is given by

$$\begin{aligned} V(AP_{AB}P_{AC}P_{AD}) &= \frac{|(P_{AD} - A) \cdot ((P_{AB} - A) \times (P_{AC} - A))|}{6} \\ &= \frac{|\alpha_{AD}(D - A) \cdot (\alpha_{AB}(B - A) \times (\alpha_{AC}(C - A)))|}{6} \\ &= \alpha_{AB}\alpha_{AC}\alpha_{AD}\frac{|(D - A) \cdot ((B - A) \times (C - A))|}{6} \\ &= \alpha_{AB}\alpha_{AC}\alpha_{AD}V(ABCD). \end{aligned}$$

Similarly we have

$$\begin{aligned}
V(BP_{AB}P_{BC}P_{BD}) &= \frac{|(P_{BD}-B)\cdot((P_{AB}-B)\times(P_{BC}-B))|}{6}\\
&= (1-\alpha_{AB})\alpha_{BC}\alpha_{BD}\frac{|(D-B)\cdot((A-B)\times(C-B))|}{6}\\
&= (1-\alpha_{AB})\alpha_{BC}\alpha_{BD}V(ABCD),\\
V(CP_{AC}P_{BC}P_{CD}) &= \frac{|(P_{CD}-C)\cdot((P_{AC}-C)\times(P_{BC}-C))|}{6}\\
&= (1-\alpha_{AC})(1-\alpha_{BC})\alpha_{CD}\frac{|(D-C)\cdot((A-C)\times(B-C))|}{6}\\
&= (1-\alpha_{AC})(1-\alpha_{BC})\alpha_{CD}V(ABCD),\\
V(DP_{AD}P_{BD}P_{CD}) &= \frac{|(P_{CD}-D)\cdot((P_{AD}-D)\times(P_{BD}-D))|}{6}\\
&= (1-\alpha_{AD})(1-\alpha_{BD})(1-\alpha_{CD})\frac{|(C-D)\cdot((A-D)\times(B-D))|}{6}\\
&= (1-\alpha_{AD})(1-\alpha_{BD})(1-\alpha_{CD})V(ABCD).
\end{aligned}$$

Let us study the remaining four cases. Note that

$$\begin{aligned}
P_{AB}-P_{BC} &= A+\alpha_{AB}(B-A)-B-\alpha_{BC}(C-B)\\
&= (1-\alpha_{AB})(A-B)-\alpha_{BC}(C-B)\\
P_{AC}-P_{BC} &= A+\alpha_{AC}(C-A)-B-\alpha_{BC}(C-B)\\
&= (A-B)+\alpha_{AC}(C-A)-\alpha_{BC}(C-B)\\
P_{AD}-P_{BC} &= A+\alpha_{AD}(D-A)-B-\alpha_{BC}(C-B)\\
&= (A-B)+\alpha_{AD}(D-A)-\alpha_{BC}(C-B),
\end{aligned}$$

and therefore

$$\begin{aligned}
&(P_{AB}-P_{BC})\times(P_{AC}-P_{BC})\\
&= ((1-\alpha_{AB})(A-B)-\alpha_{BC}(C-B))\times((A-B)+\alpha_{AC}(C-A)-\alpha_{BC}(C-B))\\
&= (1-\alpha_{AB})\alpha_{AC}(A-B)\times(C-A)-(1-\alpha_{AB})\alpha_{BC}(A-B)\times(C-B)\\
&-\alpha_{BC}(C-B)\times(A-B)-\alpha_{BC}\alpha_{AC}(C-B)\times(C-A)\\
&= [-(1-\alpha_{AB})\alpha_{AC}+(1-\alpha_{AB})\alpha_{BC}-\alpha_{BC}+\alpha_{AC}\alpha_{BC}](B-A)\times(C-A)\\
&= [-(1-\alpha_{AB})\alpha_{AC}+(\alpha_{AC}-\alpha_{AB})\alpha_{BC}](B-A)\times(C-A).
\end{aligned}$$

Further, we note this is perpendicular to the plane containing A, B, and C. It follows that

$$
\begin{aligned}
&(P_{AD} - P_{BC}) \cdot ((B - A) \times (C - A)) \\
&= ((A - B) + \alpha_{AD}(D - A) - \alpha_{BC}(C - B)) \cdot ((B - A) \times (C - A)) \\
&= \alpha_{AD}(D - A) \cdot ((B - A) \times (C - A)),
\end{aligned}
$$

and hence

$$
\begin{aligned}
V(P_{BC} P_{AB} P_{AC} P_{AD}) &= \frac{|(P_{AD} - P_{BC}) \cdot ((P_{AB} - P_{BC}) \times (P_{AC} - P_{BC}))|}{6} \\
&= |-(1 - \alpha_{AB})\alpha_{AC}\alpha_{AD} + (\alpha_{AC} - \alpha_{AB})\alpha_{AD}\alpha_{BC}| \\
&\cdot |(D - A) \cdot ((B - A) \times (C - A)|/6 \\
&= |(1 - \alpha_{AB})\alpha_{AC}\alpha_{AD} + (\alpha_{AB} - \alpha_{AC})\alpha_{AD}\alpha_{BC}| V(ABCD).
\end{aligned}
$$

Let us consider $P_{BC} P_{CD} P_{AC} P_{AD}$. Observe that we can reuse some of the previous quantities, in addition we also have

$$
\begin{aligned}
P_{CD} - P_{BC} &= C + \alpha_{CD}(D - C) - B - \alpha_{BC}(C - B) \\
&= (1 - \alpha_{BC})(C - B) + \alpha_{CD}(D - C).
\end{aligned}
$$

Furthermore, we can rewrite $P_{AC} - P_{BC}$ as

$$
\begin{aligned}
P_{AC} - P_{BC} &= C + (1 - \alpha_{AC})(A - C) - C - (1 - \alpha_{BC})(B - C) \\
&= (1 - \alpha_{AC})(A - C) - (1 - \alpha_{BC})(B - C).
\end{aligned}
$$

It follows

$$
\begin{aligned}
&(P_{CD} - P_{BC}) \times (P_{AC} - P_{BC}) \\
&= ((1 - \alpha_{BC})(C - B) + \alpha_{CD}(D - C)) \times ((1 - \alpha_{AC})(A - C) - (1 - \alpha_{BC})(B - C)) \\
&= (1 - \alpha_{AC})(1 - \alpha_{BC})(C - B) \times (A - C) + (1 - \alpha_{AC})\alpha_{CD}(D - C) \times (A - C) \\
&- (1 - \alpha_{BC})\alpha_{CD}(D - C) \times (B - C).
\end{aligned}
$$

In the same way we obtain

$$
\begin{aligned}
&(P_{AD} - P_{BC}) \cdot ((P_{CD} - P_{BC}) \times (P_{AC} - P_{BC})) \\
&= (1 - \alpha_{AC})\alpha_{AD}(1 - \alpha_{BC})(D - A) \cdot ((C - B) \times (A - C)) \\
&+ (1 - \alpha_{AC})\alpha_{CD}((A - B) - \alpha_{BC}(C - B)) \cdot ((D - C) \times (A - C)) \\
&- (1 - \alpha_{BC})\alpha_{CD}((A - B) + \alpha_{AD}(D - A)) \cdot ((D - C) \times (B - C)) \\
&= (1 - \alpha_{AC})\alpha_{AD}(1 - \alpha_{BC})(D - A) \cdot ((C - B) \times (A - C))
\end{aligned}
$$

$$
\begin{aligned}
&+ (1-\alpha_{AC})\alpha_{CD}(A-B)\cdot((D-C)\times(A-C)) \\
&- (1-\alpha_{AC})\alpha_{BC}\alpha_{CD}(C-B)\cdot((D-C)\times(A-C)) \\
&- (1-\alpha_{BC})\alpha_{CD}(A-B)\cdot((D-C)\times(B-C)) \\
&- \alpha_{AD}(1-\alpha_{BC})\alpha_{CD}(D-A)\cdot((D-C)\times(B-C)).
\end{aligned}
$$

We now note that

$$
\begin{aligned}
(D-A)\cdot((C-B)\times(A-C)) &= (D-A)\cdot((C-B)\times(A-D+D-C)) \\
&= (D-A)\cdot((C-D+D-B)\times(D-C)) \\
&= -(A-D)\cdot((B-D)\times(C-D)), \\
(A-B)\cdot((D-C)\times(A-C)) &= (A-B)\cdot((A-D)\times(C-D)) \\
&= (A-D+D-B)\cdot((A-D)\times(C-D)) \\
&= (D-B)\cdot((A-D)\times(C-D)) \\
&= (A-D)\cdot((B-D)\times(C-D)), \\
(C-B)\cdot((D-C)\times(A-C)) &= (C-D+D-B)\cdot((D-C)\times(A-C)) \\
&= (D-B)\cdot((D-C)\times(A-C)) \\
&= (D-B)\cdot((A-D)\times(C-D)) \\
&= (A-D)\cdot((B-D)\times(C-D)), \\
(A-B)\cdot((D-C)\times(B-C)) &= (A-B)\cdot((B-D)\times(C-D)) \\
&= (A-D+D-B)\cdot((B-D)\times(C-D)) \\
&= (A-D)\cdot((B-D)\times(C-D)), \\
(D-A)\cdot((D-C)\times(B-C)) &= (D-A)\cdot((D-C)\times(B-D+D-C)) \\
&= (D-A)\cdot((D-C)\times(B-D)) \\
&= -(A-D)\cdot((B-D)\times(C-D)).
\end{aligned}
$$

Therefore

$$
\begin{aligned}
&(P_{AD}-P_{BC})\cdot((P_{CD}-P_{BC})\times(P_{AC}-P_{BC})) \\
&= [-(1-\alpha_{AC})\alpha_{AD}(1-\alpha_{BC}) + (1-\alpha_{AC})\alpha_{CD} \\
&- (1-\alpha_{AC})\alpha_{BC}\alpha_{CD} - (1-\alpha_{BC})\alpha_{CD} \\
&+ \alpha_{AD}(1-\alpha_{BC})\alpha_{CD}](A-D)\cdot((B-D)\times(C-D))
\end{aligned}
$$

and hence the volume is given by

$$\begin{aligned} V(P_{BC}P_{CD}P_{AC}P_{AD}) = V(ABCD)| &-\alpha_{AD}+\alpha_{AD}\alpha_{BC}+\alpha_{AC}\alpha_{AD} \\ &-\alpha_{AC}\alpha_{AD}\alpha_{BC}+\alpha_{AD}\alpha_{CD} \\ &-\alpha_{AC}\alpha_{CD}+\alpha_{AC}\alpha_{BC}\alpha_{CD} \\ &-\alpha_{AD}\alpha_{BC}\alpha_{CD}|\,. \end{aligned}$$

Calculations for $P_{BC}P_{AB}P_{BD}P_{AD}$ and $P_{BC}P_{BD}P_{CD}P_{AD}$ are similar to the first two. For $P_{BC}P_{AB}P_{BD}P_{AD}$ we first note that

$$\begin{aligned} P_{BD}-P_{BC} &= B+\alpha_{BD}(D-B)-B-\alpha_{BC}(C-B) \\ &= \alpha_{BD}(D-B)-\alpha_{BC}(C-B)\,. \end{aligned}$$

and hence

$$\begin{aligned} &(P_{AB}-P_{BC})\times(P_{BD}-P_{BC}) \\ &= ((1-\alpha_{AB})(A-B)-\alpha_{BC}(C-B))\times(\alpha_{BD}(D-B)-\alpha_{BC}(C-B)) \\ &= (1-\alpha_{AB})\alpha_{BD}(A-B)\times(D-B)-(1-\alpha_{AB})\alpha_{BC}(A-B)\times(C-B) \\ &-\alpha_{BC}\alpha_{BD}(C-B)\times(D-B)\,. \end{aligned}$$

It follows that

$$\begin{aligned} &(P_{AD}-P_{BC})\cdot((P_{AB}-P_{BC})\times(P_{BD}-P_{BC})) \\ &= -(1-\alpha_{AB})\alpha_{BC}\alpha_{BD}(C-B)\cdot((A-B)\times(D-B)) \\ &-(1-\alpha_{AB})\alpha_{AD}\alpha_{BC}(D-A)\cdot((A-B)\times(C-B)) \\ &-\alpha_{BC}\alpha_{BD}((A-B)+\alpha_{AD}(D-A))\cdot((C-B)\times(D-B)) \\ &= (1-\alpha_{AB})\alpha_{BC}\alpha_{BD}(A-B)\cdot((C-B)\times(D-B)) \\ &-(1-\alpha_{AB})\alpha_{AD}\alpha_{BC}(A-B)\cdot((C-B)\times(D-A)) \\ &-\alpha_{BC}\alpha_{BD}(A-B)\cdot((C-B)\times(D-B)) \\ &-\alpha_{AD}\alpha_{BC}\alpha_{BD}(D-A)\cdot((C-B)\times(D-B)) \\ &= [(1-\alpha_{AB})\alpha_{BC}\alpha_{BD}-(1-\alpha_{AB})\alpha_{AD}\alpha_{BC} \\ &-\alpha_{BC}\alpha_{BD}+\alpha_{AD}\alpha_{BC}\alpha_{BD}](A-B)\cdot((C-B)\times(D-B))\,, \end{aligned}$$

and therefore

$$V(P_{BC}P_{AB}P_{BD}P_{AD}) = V(ABCD)|(\alpha_{AD}-\alpha_{AB})\alpha_{BC}\alpha_{BD}-(1-\alpha_{AB})\alpha_{AD}\alpha_{BC}|\,.$$

Finally we calculate the volume for $P_{BC}P_{BD}P_{CD}P_{AD}$.

$$\begin{aligned}
&(P_{BD}-P_{BC})\times(P_{CD}-P_{BC})\\
&=(\alpha_{BD}(D-B)-\alpha_{BC}(C-B))\times(\alpha_{CD}(D-C)+(1-\alpha_{BC})(C-B))\\
&=\alpha_{BD}\alpha_{CD}(D-B)\times(D-C)+(1-\alpha_{BC})\alpha_{BD}(D-B)\times(C-B)\\
&-\alpha_{BC}\alpha_{CD}(C-B)\times(D-C)\\
&=[\alpha_{BD}\alpha_{CD}-(1-\alpha_{BC})\alpha_{BD}-\alpha_{BC}\alpha_{CD}](B-D)\times(C-D)\,.
\end{aligned}$$

Noting that

$$\begin{aligned}
(A-B)\cdot((B-D)\times(C-D)) &= (A-D)\cdot((B-D)\times(C-D))\\
(D-A)\cdot((B-D)\times(C-D)) &= -(A-D)\cdot((B-D)\times(C-D))\\
(C-B)\cdot((B-D)\times(C-D)) &= 0\,,
\end{aligned}$$

it follows that

$$\begin{aligned}
&(P_{AD}-P_{BC})\cdot((P_{BD}-P_{BC})\times(P_{CD}-P_{BC}))\\
&=[\alpha_{BD}\alpha_{CD}-(1-\alpha_{BC})\alpha_{BD}-\alpha_{BC}\alpha_{CD}\\
&-\alpha_{AD}\alpha_{BD}\alpha_{CD}+\alpha_{AD}(1-\alpha_{BC})\alpha_{BD}+\alpha_{AD}\alpha_{BC}\alpha_{CD}](A-D)\cdot((B-D)\times(C-D))\,.
\end{aligned}$$

Therefore we have

$$\begin{aligned}
V(P_{BC}P_{BD}P_{CD}P_{AD}) = V(ABCD)|&\alpha_{BD}\alpha_{CD}-(1-\alpha_{BC})\alpha_{BD}-\alpha_{BC}\alpha_{CD}\\
&-\alpha_{AD}\alpha_{BD}\alpha_{CD}\\
&+\alpha_{AD}(1-\alpha_{BC})\alpha_{BD}+\alpha_{AD}\alpha_{BC}\alpha_{CD}|\,.
\end{aligned}$$

In order to finish the proof it suffices to substitute the parameters and to verify that the volumes match for $\mathcal{F}$ and $\mathcal{G}$.

We except that the proposition admits additional families of solutions. However, finding these is not so simple, for a general solution we need to reduce the following simultaneous equations

$$\begin{aligned}
1:&\alpha_{AB}\alpha_{AC}\alpha_{AD}\\
&=\beta_{AB}\beta_{AC}\beta_{AD}\\
2:&(1-\alpha_{AB})\alpha_{BC}\alpha_{BD}\\
&=(1-\beta_{AB})\beta_{BC}\beta_{BD}\\
3:&(1-\alpha_{AC})(1-\alpha_{BC})\alpha_{CD}\\
&=(1-\beta_{AC})(1-\beta_{BC})\beta_{CD}
\end{aligned}$$

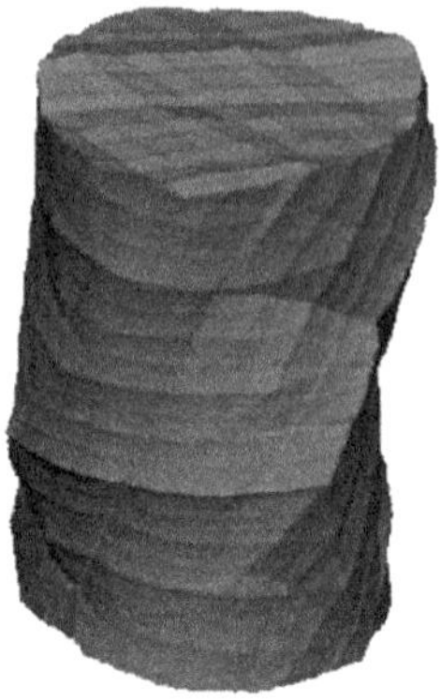

Fig. 10 Two illustrations of volume preserving fractal homeomorphisms (Example 4.3 type) applied to a cylinder

$$
\begin{aligned}
4 :&(1-\alpha_{AD})(1-\alpha_{BD})(1-\alpha_{CD})\\
&=(1-\beta_{AD})(1-\beta_{BD})(1-\beta_{CD})\\
5 :&(1-\alpha_{AB})\alpha_{AC}\alpha_{AD}+(\alpha_{AB}-\alpha_{AC})\alpha_{AD}\alpha_{BC}\\
&=(1-\beta_{AB})\beta_{AC}\beta_{AD}+(\beta_{AB}-\beta_{AC})\beta_{AD}\beta_{BC}\\
6 :&-\alpha_{AD}+\alpha_{AD}\alpha_{BC}+\alpha_{AC}\alpha_{AD}-\alpha_{AC}\alpha_{AD}\alpha_{BC}+\alpha_{AD}\alpha_{CD}\\
&-\alpha_{AC}\alpha_{CD}+\alpha_{AC}\alpha_{BC}\alpha_{CD}-\alpha_{AD}\alpha_{BC}\alpha_{CD}\\
&=-\beta_{AD}+\beta_{AD}\beta_{BC}+\beta_{AC}\beta_{AD}-\beta_{AC}\beta_{AD}\beta_{BC}+\beta_{AD}\beta_{CD}\\
&\quad-\beta_{AC}\beta_{CD}+\beta_{AC}\beta_{BC}\beta_{CD}-\beta_{AD}\beta_{BC}\beta_{CD}\\
7 :&(\alpha_{AD}-\alpha_{AB})\alpha_{BC}\alpha_{BD}-(1-\alpha_{AB})\alpha_{AD}\alpha_{BC}\\
&=(\beta_{AD}-\beta_{AB})\beta_{BC}\beta_{BD}-(1-\beta_{AB})\beta_{AD}\beta_{BC}\\
8 :&\alpha_{BD}\alpha_{CD}-(1-\alpha_{BC})\alpha_{BD}-\alpha_{BC}\alpha_{CD}-\alpha_{AD}\alpha_{BD}\alpha_{CD}\\
&+\alpha_{AD}(1-\alpha_{BC})\alpha_{BD}+\alpha_{AD}\alpha_{BC}\alpha_{CD}\\
&=\beta_{BD}\beta_{CD}-(1-\beta_{BC})\beta_{BD}-\beta_{BC}\beta_{CD}-\beta_{AD}\beta_{BD}\beta_{CD}\\
&\quad+\beta_{AD}(1-\beta_{BC})\beta_{BD}+\beta_{AD}\beta_{BC}\beta_{CD}
\end{aligned}
$$

□

Further examples of fractal homeomorphisms on pictures and three-dimensional objects are illustrated in Figs. 10, 11, 12 and 13.

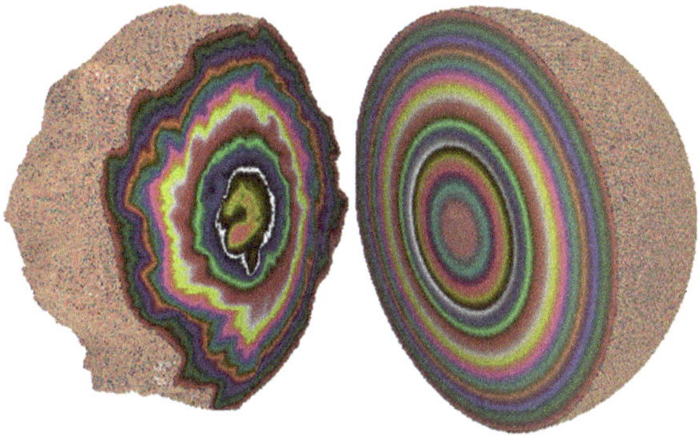

Fig. 11 Illustration of a tri-affine fractal homeomorphism, after and before, applied to a nested family of coloured hemispherical shells, see [9]

Fig. 12 Before and after an Example 4.1 type area preserving fractal homeomorphism

Fig. 13 Photograph of a bowl and saltshaker, both physical objects having been produced by a three-dimensional printer applied to Example 4.2 type volume preserving fractal homeomorphisms of simple three-dimensional computer graphical objects

References

1. Atkins, R., Barnsley, M., Wilson, D.C., Vince, A.: A characterization of point-fibred affine iterated function systems. Topol. Proc. **38**, 189–211 (2010)
2. Barnsley, M.F.: Fractal functions and interpolation. Constr. Approx. **2** , 303–329 (1986)
3. Barnsley, M.F.: Fractal image compression. Not. Am. Math. Soc. **43** , 657–662 (1996)
4. Barnsley, M.F.: Theory and applications of fractal tops. In: Levy-Vehel, J., Lutton, E. (eds.) Fractals in Engineering: New Trends in Theory and Applications. Springer, London (2005)
5. Barnsley, M.F.: SuperFractals. Cambridge University Press, Cambridge (2006)
6. Barnsley, M.F.: Transformations between self-referential sets. Am. Math. Mon. **116**, 291–304 (2009)
7. Barnsley, M.F.: The life and survival of mathematical ideas. Not. Am. Math. Soc. **57**, 10–22 (2010)
8. Barnsley, M.F.: Fractals Everywhere. Academic, Boston (1988); 2nd edn., Morgan Kaufmann (1993); 3rd edn., Dover Publications (2012)
9. Barnsley, M.F., Harding, B.: Three-Dimensional Fractal Homeomorphisms. World Scientific, in a volume dedicated to Benoit B. Mandelbrot (to appear)
10. Barnsley, M.F., Vince, A.: Fractal homeomorphism for bi-affine iterated function systems. Int. J. Appl. Nonlinear Sci. **1**, 3–19 (2012)
11. Barnsley, M.F., Vince, A.: Developments in fractal geometry. Bull. Math. Sci. **3**, 299–348 (2013)
12. Barnsley, M.F., Harding, B., Igudesman, K.: How to transform and filter images using iterated function systems. SIAM J. Imaging Sci. **4**(4), 1001–1028 (2011)
13. Barnsley, M.F., Harding, B., Vince, A.: The entropy of a special overlapping dynamical system. Ergodic Theory Dyn. Syst. (2012). doi: 10.1017/etds.2012.140
14. FrangoCamera, iPad (2012) and iPhone (2013) applications. www.frangostudio.com
15. Saalfeld, A.: Area-preserving piecewise affine mappings. In: Proceedings of the Seventeenth Annual Symposium on Computational Geometry. ACM, New York (2001)
16. Stein Elias, M., Shakarchi, R.: Real Analysis: Measure Theory, Integration, and Hilbert Spaces. Princeton University Press, Princeton (2005)
17. Touret, J.L.R.: Fluids in metamorphic rocks. Lithos **55**(1), 1–25 (2001)

The Dimension Theory of Almost Self-affine Sets and Measures

Károly Simon

Abstract A self-affine IFS $\mathcal{F} = \{f_i(x) = A_i x + t_i\}_{i=1}^m$ is a finite list of contracting affine maps on $\mathbb{R}^d$, for some $d \geq 1$. The attractor of $\mathcal{F}$ is

$$\Lambda = \bigcap_{n=1}^{\infty} \bigcap_{i_1,\dots,i_n} f_{i_1} \circ \cdots \circ f_{i_n}(B), \tag{0.1}$$

where B is a sufficiently large ball centered at the origin. In most cases we cannot compute the dimension of Λ. However, if we add an independent additive random error to each f_{i_k} in (0.1) then the dimension of this random perturbation (called almost self-affine system) is almost surely the so-called affinity dimension of the original deterministic system. The dimension theory of almost self-affine sets and measures were described in Jordan et al. (Commun. Math. Phys. 270(2):519–544, 2007). The multifractal analysis of almost self-affine measures has been studied in some recent papers (Falconer, Nonlinearity 23:1047–1069, 2010; Barral and Feng, Commun. Math. Phys. 318(2):473–504, 2013). In the second part of this note I give a survey of this field but first we review some results related to the dimension theory of self-affine sets.

Keywords Random fractals • Hausdorff dimension • Processes in random environment

2000 Mathematics Subject Classification. 37H15, 37C45

K. Simon (✉)
Institute of Mathematics, Technical University of Budapest,
H-1529 B.O. Box 91, Budapest, Hungary
e-mail: simonk@math.bme.hu

C. Bandt et al. (eds.), *Fractals, Wavelets, and their Applications*, Springer Proceedings in Mathematics & Statistics 92, DOI 10.1007/978-3-319-08105-2_6,

1 Introduction

Let A_i be $d \times d$ non-singular matrices with $\|A_i\| < 1$ and $t_i \in \mathbb{R}^d$ for $i = 1, \ldots, m$. Let

$$\mathcal{F} := \{f_i\}_{i=1}^m = \{A_i \cdot x + t_i\}_{i=1}^m . \tag{1}$$

We study the attractor Λ of the IFS $\mathcal{F}$ which is defined as follows: let $B = B(0, r)$ be a sufficiently large closed ball such that: $f_i(B) \subset B$ for all $i = 1, \ldots, m$. We set

$$\Lambda := \bigcap_{n=0}^{\infty} \bigcup_{i_1,\ldots,i_n} f_{i_1} \circ \cdots \circ f_{i_n}(B). \tag{2}$$

Then Λ is the unique nonempty compact set satisfying

$$\Lambda = \bigcup_{i=1}^{m} f_i(\Lambda) \tag{3}$$

is called self-affine set.

Our understanding of the dimension theory of self-affine sets is far from complete. This is so even in the diagonal case that is when all of the matrices A_i, $i = 1, \ldots, m$ are diagonal. Most of the results about the dimension theory of self-affine sets can be divided in to two fields:

(a) The cylinders are aligned in some ways similar to the ones in Fig. 1. Namely, for example let $d = 2$ and $Q := [0, 1]^2$, all matrices A_i are diagonal and the sets $f_i(Q) = \{A_i x + t_i : x \in Q\}$ for $i = 1, \ldots, m$ are the gray squares in Fig. 1. The attractors generated by these three IFS in Fig. 1 are (from left to right) members of the families called Bedford–McMullen carpets [21], Lalley–Gatzouras carpets [19], and Barański carpets [1].
(b) The geometric position of the cylinders is general.

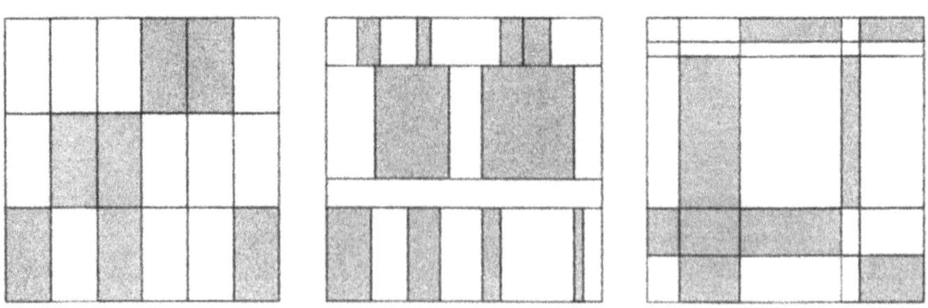

Fig. 1 Bedford–McMullen, Gatzouras–Lalley, and Barański carpets (in this order). This figure is from [1]

A good account about the case (a) is available in forthcoming book by Bishop and Peres. Although there are very important recent developments related to the multifractal analysis of the Bedford–McMullen carpet in this note we only review some results related to case (b) that is when the geometric position of the cylinders is general.

2 Self-affine Sets and Measures

The following classical theorem of Falconer [5] was improved by Solomyak [30].

Theorem 1 (Falconer and Solomyak). *Let $\mathcal{F}$ be a self-affine IFS of the form* (1). *If $\|A_i\| < \frac{1}{2}$ holds for all $i = 1,\dots,m$ then for Lebesgue almost all $(t_1,\dots,t_m) \in \mathbb{R}^{d\cdot m}$ the Hausdorff dimension and the box dimension of the attractor of $\mathcal{F}$ is the same and is equal to the affinity dimension defined below.*

2.1 The Affinity Dimension

The affinity dimension is a generalization of the similarity dimension [7]. It gives the best guess for the Hausdorff dimension of a self-affine attractor based only on the linear parts A_i of the maps f_i. Recall that for a $t > 0$ the t-dimensional Hausdorff measure of the attractor is

$$\mathcal{H}^t(\Lambda) = \sup_{\delta\to 0} \inf \left\{ \boxed{\sum_{i=1}^{\infty} |E_i|^t} : \Lambda \subset \bigcup_{i=1}^{\infty} E_i, |E_i| < \delta \right\}, \tag{4}$$

where $|E_i|$ is the diameter of the set E_i.

Assume that we are in $\mathbb{R}^3$ and B in the definition of the attractor is actually $Q = [0,1]^3$. Then the most natural cover of Λ is the cover by the n-cylinders $f_{\mathbf{i}}(Q)$, $\mathbf{i} = (i_1,\dots,i_n)$. If we are looking for the most economic cover of Λ [the one that minimizes the boxed sum in (4)] without taking into consideration the translation parts t_i of f_i then we need to cover each n-cylinder $f_{\mathbf{i}}(Q)$ individually. For simplicity here we assume that $f_{\mathbf{i}}(Q)$ is a box with side lengths

$$\alpha_k := \alpha_k(A_{\mathbf{i}}) = \alpha_k(\underbrace{A_{i_1}\cdots A_{i_n}}_{A_{\mathbf{i}}}), \tag{5}$$

where $\alpha_k(A_{\mathbf{i}})$ the k-th biggest singular value of $A_{\mathbf{i}} = A_{i_1}\cdots A_{i_n}$.

In Fig. 2 we indicated the three sensible ways to cover $f_{\mathbf{i}}(Q)$. The contributions of these coverings of $f_{\mathbf{i}}(Q)$ to $\sum_{i=1}^{\infty} |E_i|^t$ are

$$\alpha_1^t, \qquad \frac{\alpha_1}{\alpha_2}\alpha_2^t = \alpha_1\alpha_2^{t-1}, \qquad \frac{\alpha_1}{\alpha_3}\frac{\alpha_2}{\alpha_3}\alpha_3^t = \alpha_1\alpha_2\alpha_3^{t-2}$$

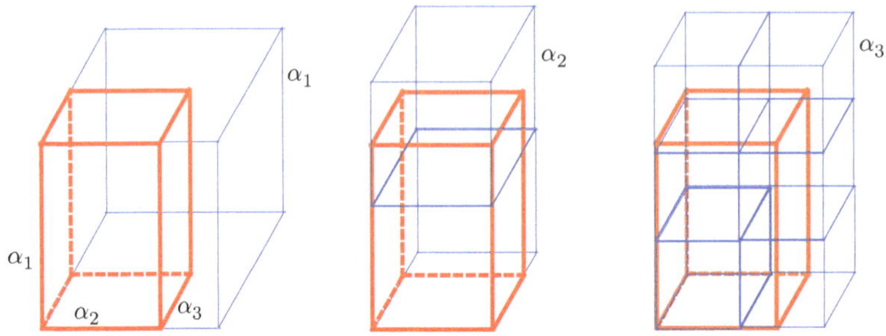

Fig. 2 Three ways of covering the *red* cylinder

respectively. Observe that for a $0 \leq t \leq d$

$$\min_i \alpha_1 \cdots \alpha_i \alpha_{i+1}^{t-i} = \alpha_1 \cdots \alpha_{[t]} \alpha_{[t]+1}^{t-[t]}. \tag{6}$$

Therefore to estimate $\mathcal{H}^t(\Lambda)$, we estimate the boxed sum $\sum_{i=1}^{\infty} |E_i|^t$ in (4) with the $[t]+1$-th cover in Fig. 2. By this cover, the contribution of the n-th cylinder $f_{\mathbf{i}}[0,1]$ to the boxed sum in (4) is

$$\varphi^t(A_{\mathbf{i}}) = \alpha_1 \cdots \alpha_{[t]} \alpha_{[t]+1}^{t-[t]}.$$

This $\varphi^t(A_{\mathbf{i}})$ is called singular value function (for $A_{\mathbf{i}} = A_{i_1} \cdots A_{i_n}$).

Definition 2. For $\mathcal{A} := (A_1, \ldots, A_m) \in (GL_d(\mathbb{R}))^m)$ and $s \geq 0$ the subadditive topological pressure function is: $P(\mathcal{A}, \cdot) : [0, \infty) \to \mathbb{R}$

$$P(\mathcal{A}, t) := \lim_{n \to \infty} \frac{1}{n} \log \left(\sum_{|\mathbf{i}|=n} \varphi^t(A_{\mathbf{i}}) \right). \tag{7}$$

Fix $\mathcal{A}$. The map $t \mapsto P(\mathcal{A}, t)$ is strictly decreasing and

$$P(\mathcal{A}, 0) = m > 0, \quad \lim_{t \to \infty} P(\mathcal{A}, t) = -\infty.$$

Hence there is a unique zero of $P(\mathcal{A}, t)$.

Definition 3. The affinity dimension $s(\mathcal{A})$ of the self-affine IFS

$$\mathcal{F} := \{f_i\}_{i=1}^m = \{A_i \cdot x + t_i\}_{i=1}^m$$

is defined as the zero of $P(\mathcal{A}, s)$.

Using earlier results of Falconer and Solan [9], Feng and Shmerkin [11] proved recently that

Theorem 4 (Feng and Shmerkin (2012)). $(\mathcal{A}, t) \mapsto P(\mathcal{A}, t)$ *is continuous.*

So, the affinity dimension is also continuous.

2.2 *When Does* $\dim(\Lambda) = s(\mathcal{A})$ *Hold for All Translations?*

Falconer–Solomyak Theorem says that the Hausdorff dimension is equal to the affinity dimension:

(a) if all contractions are stronger $1/2$,
(b) only for $\mathcal{L}eb_{m \cdot d}$ almost all $(t_1, \cdots, t_m) \in \mathbb{R}^{m \cdot d}$.

It is easy to see that the bound $1/2$ is sharp. On the other hand, there are some results which give further conditions under which the assertion of Falconer–Solomyak Theorem remains valid. For example in $\mathbb{R}^2$, Hueter and Lalley [13] proved that this is the case assuming that the smaller singular value is greater than the square of the bigger one (the so-called one-bunched property holds) and the self-map of S^1 generated by $A_1^{-1}, \ldots, A_m^{-1}$ preserves the negative quadrant and satisfies a kind of strong separation property. More precisely:

Theorem 5 (Hueter and Lalley). *Given a self-affine IFS* $\mathcal{F}$ *on* $\mathbb{R}^2$*. The linear parts of the maps of* $\mathcal{F}$ *are* $\mathcal{A} = \{A_1, \ldots A_i\}$*. We assume that*

(1) $(\alpha_1(A_i))^2 < \alpha_2(A_i) \leq \alpha_1(A_i)$ *for all* $i = 1, \ldots, m$*. (1-bunched property.)*
(2) Let Q_2 *be the closed second quadrant of* $\mathbb{R}^2 \setminus (0,0)$*. Then we assume that for every* $i = 1, \ldots, m$*:* $A_i^{-1} Q_2 \subset \mathrm{int}(Q_2)$ *and* $A_i^{-1}(Q) \bigcap A_j^{-1}(Q) = \phi$ *if* $i \neq j$*.*
(3) $\det(A_i) > 0$*.*
(4) Open set V *such that for the closure of the sets* $\overline{A_1 V}, \ldots, \overline{A_m V}$ *are pairwise disjoint subsets of* V*.*

Then

$$\dim_{\mathrm{H}} \Lambda = \dim_{\mathrm{B}} \Lambda = s(\mathcal{F})$$

We remark that the conditions of Hueter and Lalley Theorem can hold only if $\dim_{\mathrm{H}} \Lambda < 1$. If we require only that $\dim_{\mathrm{B}} \Lambda = s(\mathcal{A})$ we may have a chance to apply another theorem of Falconer [6].

Theorem 6 (Falconer). *Assume that*

(a) $\mathcal{F}$*, defined as in* (1)*, satisfies the Open Set Condition (see* [7]*) with a connected open set.*
(b) *For some* $c > 0$ *we have* $\mathcal{L}eb_{d-1}\mathrm{proj}_{\Pi}(\Lambda) > c$ *for all* $d - 1$ *dimensional subspaces* Π*.*

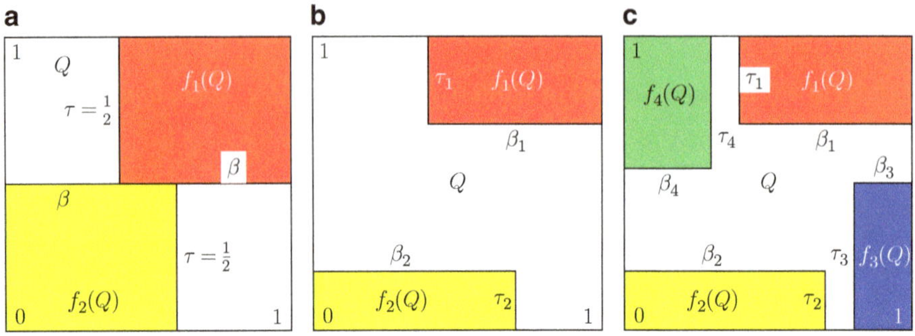

Fig. 3 The self-affine IFS: the diagonal case. (**a**) PU-set. (**b**) Generalized PU-set. (**c**) Generalized four corner set

Then

$$\dim_B(\Lambda) = s(\mathcal{F}). \tag{8}$$

We remark that assumption (b) in the previous theorem holds if Λ has a connected component that is not contained in any straight line. Känmäki and Shmerkin [17] proved a theorem having the same conclusion as Theorem 6 for the so-called self-affine sets of Kakeya type but without assuming any separation properties.

Käenmäki [16] proved that under the conditions of Theorem (1) there is an ergodic self-affine measure with maximal dimension:

Theorem 7 (Käenmäki). *Let $\mathcal{F}$ be a self-affine IFS of the form* (1)*. Assume that $\|A_i\| < \frac{1}{2}$ hods for all $i = 1, \dots, m$. Then there exists an ergodic measure μ on the symbolic space $\Sigma := \{1, \dots, m\}^{\mathbb{N}}$ such that for Lebesgue almost all $(t_1, \dots, t_m) \in \mathbb{R}^{d \cdot m}$ we have*

$$\dim_H(\Lambda) = s(\mathcal{F}) = \dim_H(\Pi_* \mu). \tag{9}$$

2.3 *The Diagonal Case*

Here we assume that all of the matrices A_i, $i = 1, \dots, m$ are diagonal. The first result in this direction is due to Przytycki and Urbanski [25].

Example 8. See Fig. 3a. Przytycki and Urbanski introduced a one-parameter family of self-affine IFS on the plane

$$\mathcal{F}_\beta := \{f_1(x) = A_1 \cdot x + t_1\}_{i=1}^{2}, \quad \Lambda_\beta \text{ is the attractor of } \mathcal{F}_\beta \tag{10}$$

where

$$A_1 = A_2 = \begin{bmatrix} \beta & 0 \\ 0 & \frac{1}{2} \end{bmatrix}, \ t_1 = (0,0), \ t_2 = \left(1-\beta, \frac{1}{2}\right)$$

That is for each parameter $\beta \in \left(\frac{1}{2}, 1\right)$ the fixed point of f_1 and f_2 are the opposite corners of the unit square $Q = [0,1]^2$ as shown in Fig. 3a. Its attractor Λ_β is called PU-set.

Pollicott and Weiss [24], and Neunhauserer [23] considered a generalization of this system that we call Generalized PU system:

Example 9. See Fig. 3b.

$$\mathcal{F}_{\boldsymbol{\beta},\boldsymbol{\tau}} := \left\{ f_1(x) = \begin{bmatrix} \beta_1 & 0 \\ 0 & \tau_1 \end{bmatrix} \cdot x + \begin{bmatrix} 1-\beta_1 \\ 1-\tau_1 \end{bmatrix}, f_2(x) = \begin{bmatrix} \beta_2 & 0 \\ 0 & \tau_2 \end{bmatrix} \cdot x \right\} \tag{11}$$

where $\boldsymbol{\beta} := (\beta_1, \beta_2)$, $\boldsymbol{\tau} := (\tau_1, \tau_2)$ satisfying $0 < \beta_i, \tau_i$, $i = 1, 2$, $\beta_1 + \beta_2 > 1$ and $\tau_1 + \tau_2 < 1$.

Finally B. Bárány considered the generalized four corner set:

Example 10 (Generalized Four Corner Set). (See Fig. 3c.) Let $\boldsymbol{\beta} := (\beta_1, \ldots, \beta_4)$ and $\boldsymbol{\tau} := (\tau_1, \ldots, \tau_4)$

$$\mathcal{F}_{\boldsymbol{\beta},\boldsymbol{\tau}} := \left\{ f_i(x) = \begin{bmatrix} \beta_i & 0 \\ 0 & \tau_i \end{bmatrix} \cdot x + t_i, \right\}_{i=1}^{4} \tag{12}$$

where t_i is chosen in such a way that the fixed point of $f_1, \ldots, f_4$ is $(1,1), (0,0), (1,0), (0,1)$, respectively. The parameters $\boldsymbol{\beta}, \boldsymbol{\tau}$ are chosen so that $f_i(Q) \cap f_j(Q) = \emptyset$ for $i \neq j$ and $Q = [0,1]^2$.

2.3.1 Przytycki and Urbanski Theorem

To get an upper bound for the box dimension of the attractor Λ_β of the Przytycki Urbaski system (Example 8) we follow the argument related to Fig. 2. Clearly the projection to the x-axis of the attractor $\mathrm{proj}_x(\Lambda_\beta)$ contains $[0,1]$ (since for all n we have $\mathrm{proj}_x\left(\bigcup_{i_1\ldots i_n} f_{i_1\ldots i_n}(Q)\right) \supset [0,1]$). Hence the argument related to Fig. 2 suggests that the most natural system of covers of Λ_β is as follows: for an arbitrary n first we cover Λ_β by the level n cylinders: $f_{i_1\ldots i_n}(Q)$ (which are translated copies of $[0,\beta^n] \times [0, 2^{-n}]$), then to get the desired cover of Λ we cover each of these 2^n rectangles by $\lfloor (2\beta)^n \rfloor + 1$ squares of side length β^n. This cover consists of altogether $2^n \times (2\beta)^n$ squares of size 2^{-n}. This yields the trivial upper bound

$$\overline{\dim}_{\mathrm{B}}(\Lambda) \leq 2 + \frac{\log \beta}{\log 2}. \tag{13}$$

To prove that the right-hand side of (13) is also a lower bound on the Hausdorff dimension of Λ, Przytycki and Urbanski considered the natural measure μ of the IFS $\mathcal{F} = \{f_1, f_2\}$ which is the unique probability measure supported by Λ satisfying

$$\mu(H) = \frac{1}{2}\mu(f_1^{-1}(H)) + \frac{1}{2}\mu(f_2^{-1}(H)) \text{ for all Borel set } H \subset Q. \tag{14}$$

Using that $\dim_{\mathrm{H}}(\mu) \leq \dim_{\mathrm{H}}(\Lambda)$, to get a lower bound on $\dim_{\mathrm{H}}(\Lambda)$ it is enough to compute $\dim_{\mathrm{H}}(\mu)$.

The Ledrappier–Young [20] formula for the dimension of invariant measures (Sect. 4.2) would yield the required conclusion but this formula was proved only for measures which are ergodic and invariant for a smooth diffeomorphism (in the time when [25] was prepared) as opposed to μ which is the invariant measure of a self-affine IFS. So, Przytycki and Urbanski verified that the Ledrappier–Young formula was valid for μ defined in (13). This Ledrappier–Young formula yields that

$$\dim_{\mathrm{H}}(\mu) = 2 + \frac{\log \beta}{\log 2} \cdot \dim_{\mathrm{H}}(\mu_{x,\beta}), \tag{15}$$

where $\mu_{x,\beta}$ is the invariant measure with respect to the IFS

$$\mathcal{S} = \{S_1(x) = \beta \cdot x, S_2(x) = \beta x + 1 - \beta\}.$$

Note that $\mathcal{S}$ is the orthogonal projection to the x axis of the original system $\mathcal{F}$. That is

$$\mu_{x,\beta}(A) = \frac{1}{2}\mu_{x,\beta}\left(S_1^{-1}(A)\right) + \frac{1}{2}\mu_{x,\beta}\left(S_2^{-1}(A)\right) \text{ for all Borel } A \subset [0,1]. \tag{16}$$

It happened only in 2009 when Feng and Hu [10] verified that Ledrappier and Young formula holds for self-affine IFS in the diagonal case. Formulas (14) and (15) imply the first part of the following theorem

Theorem 11 (Prytyzcki, Urbanski).

(1) If $\dim_{\mathrm{H}}(\mu_{x,\beta}) = 1$ *holds, then*

$$\dim_{\mathrm{H}} \Lambda = \dim_{\mathrm{B}} \Lambda = 2 + \frac{\log \beta}{\log 2}. \tag{17}$$

(2) If β *is the reciprocal of a PV number, then*

$$\dim_{\mathrm{H}} \Lambda < \dim_{\mathrm{B}} \Lambda = 2 + \frac{\log \beta}{\log 2}.$$

It is a very hard open problem to characterize those β for which $\dim_{\mathrm{H}}(\mu_{x,\beta}) = 1$ holds. Actually this question was asked by P. Erdős in 1930s. A very important achievement was made by Solomyak [29] in 1995:

Theorem 12 (Solomyak). *Let μ_β be the invariant measure of the self-similar IFS on the line*

$$\mathcal{S}_\beta := \{S_1(x) = \beta \cdot x, S_2(x) = \beta x + 1 - \beta\}. \tag{18}$$

Then μ_β is absolute continuous with respect to the Lebesgue measure with L^2 density for Lebesgue almost all $\beta \in \left(\frac{1}{2}, 1\right)$.

Here we remark that at least the absolute continuity part of Solomyak Theorem has been extended by Shmerkin [28, Theorem 1.1] to all but a set of zero Hausdorff dimension $\beta \in \left(\frac{1}{2}, 1\right)$. For $\beta < \frac{1}{2}$ the measure μ_β is supported by a Cantor set. So, μ_β is singular in this case. The following extraordinarily important recent result of Hochman [12, Theorem 1.8] implies that $\dim_{\mathrm{H}} \mu_\beta = 1$ for all but a set of Hausdorff dimension zero of $\beta \in \left(\frac{1}{2}, 1\right)$. Since this theorem can be used in many related more general examples in combination with the Ledrappier–Young formula here we give further details about it.

2.3.2 Hochman Theorem

Let $I \subset \mathbb{R}$ be a compact parameter interval and $m \geq 2$. For every parameter $t \in I$ given a self-similar IFS on the line:

$$\Phi_t := \{\varphi_{i,t}(x) = r_i(t) \cdot (x - a_i(t))\}_{i=1}^m,$$

where

$$r_i : I \to (-1, 1) \setminus \{0\} \text{ and } a_i : I \to \mathbb{R}$$

are real analytic functions. Let Π_t be the natural projection from $\Sigma := \{1, \ldots, m\}^{\mathbb{N}}$ to the attractor Λ_t of Φ_t. For every probability vector $\mathbf{p} := (p_1, \ldots, p_m)$ the associated self-similar measure is

$$\nu_{\mathbf{p},t} := (\Pi_t)_*(\mathbf{p}^{\mathbb{N}}).$$

Its similarity dimension is defined by

$$\dim_{\mathrm{S}}(\nu_{\mathbf{p},t}) := \frac{\sum_{i=1}^m p_i \log p_i}{\sum_{i=1}^m p_i \log r_i(t)}$$

The similarity dimension of Λ_t is the solution $s(t)$ of

$$r_1^{s(t)}(t) + \cdots + r_m^{s(t)}(t) = 1.$$

We say that a parameter $t \in I$ is exceptional if either $\dim_H \Lambda_t < \min\{1, s(t)\}$ or there exists a probability vector $\mathbf{p} := (p_1, \ldots, p_m)$ such that $\dim_H(\nu_{\mathbf{p},t}) < \min\{1, \dim_S(\nu_{\mathbf{p},t})\}$

Theorem 13 (Hochman). *Assume that*

$$\textit{if } \Pi_t(\mathbf{i}) = \Pi_t(\mathbf{j}) \textit{ holds for } \textbf{all } t \in I \textit{ then } \mathbf{i} = \mathbf{j}.$$

Then both the Hausdorff and the packing dimension of the set of exceptional parameters are equal to 0.

Combining this with Theorem 24 we obtain that (17) holds for all but a set of Hausdorff dimension zero $\beta \in \left(\frac{1}{2}, 1\right)$.

2.3.3 A Generalization with Different Contraction Ratios

In the Przytycki–Urbanski Theorem above the idea was

(a) Use Ledrappier–Young formula [20] to express the dimension of the self-affine natural measure with the dimension of the projection of the natural measure.
(b) The projected measure is a self-similar measure with overlapping cylinders. Apply Solomyak or Hochman Theorem to compute its dimension.

To generalize Przytycki–Urbanski Theorem, a similar plan was carried out for the case of different contraction ratios (see the IFS on Fig. 3b) by Neunhäuserer [23].

More precisely, first we define two open triangles $T_\beta, T_\tau \subset [0,1]^2$. Their product $T_{\beta,\tau}$ will be the parameter space. Let

$$T_\beta := \left\{\boldsymbol{\beta} = (\beta_1, \beta_2) \in (0,1)^2 : \beta_1 + \beta_2 > 1\right\}$$

and

$$T_\tau := \{\boldsymbol{\tau} = (\tau_1, \tau_2) : \tau_1 + \tau_2 < 1\}$$

For an arbitrary $(\boldsymbol{\beta}, \boldsymbol{\tau}) \in T_{\beta,\tau} := T_b \times T_\tau$ let $\Lambda_{\boldsymbol{\beta},\boldsymbol{\tau}}$ be the attractor of $\mathcal{F}_{\boldsymbol{\beta},\boldsymbol{\tau}}$ defined in (11).

Using a similar argument as above one can easily see that the upper box dimension of the attractor satisfies:

$$\overline{\dim}_B(\Lambda_{\boldsymbol{\beta},\boldsymbol{\tau}}) \le t + 1, \tag{19}$$

where t is the solution of

$$\beta_1 \tau_1^t + \beta_2 \tau_2^t = 1. \tag{20}$$

This equation naturally defines a Bernoulli measure $\{p, 1-p\}^{\mathbb{N}}$ with $p = \beta_1 \tau_1^t$ on the symbolic space $\Sigma := \{1,2\}^{\mathbb{N}}$. Let $\mu^{\boldsymbol{\beta},\boldsymbol{\tau}}$ be its push forward measure to the attractor by the natural projection $\Pi : \Sigma \to \Lambda_{\boldsymbol{\beta},\boldsymbol{\tau}}$,

$$\Pi(\mathbf{i}) := \lim_{n\to\infty} f_{i_1\ldots i_n}(0), \tag{21}$$

where $\mathbf{i} = (i_1, i_2, \ldots)$ and $f_{i_1\ldots i_n} := f_{i_1} \circ \cdots \circ f_{i_n}$. That is for

$$\mu^{\boldsymbol{\beta},\boldsymbol{\tau}}(H) = p \cdot \mu^{\boldsymbol{\beta},\boldsymbol{\tau}}(f_1^{-1}(H)) + (1-p) \cdot \mu^{\boldsymbol{\beta},\boldsymbol{\tau}}(f_2^{-1}(H)) \tag{22}$$

for all Borel set $H \subset Q$. The natural measure for $\mathcal{F}_{\boldsymbol{\beta},\boldsymbol{\tau}}$ is $\mu^{\boldsymbol{\beta},\boldsymbol{\tau}}$. To estimate its dimension we consider its projection to the direction of weaker contraction that is to the x-axis. The projected measure $\mu_x^{\boldsymbol{\beta},\boldsymbol{\tau}}$ is the invariant measure of the projected IFS

$$\mathcal{S}_\beta := \{\beta_2 \cdot x, \beta_1 \cdot x + (1-\beta_1)\}, \text{ with probabilities } (p, 1-p). \tag{23}$$

Using the Ledrappier–Young formula (see Sect. 4.2) we obtain that

$$\dim_{\mathrm{H}}(\mu^{\boldsymbol{\beta},\boldsymbol{\tau}}) = 1 + t \text{ iff } \dim_{\mathrm{H}}(\mu_x^{\boldsymbol{\beta},\boldsymbol{\tau}}) = 1. \tag{24}$$

The heart of the matter is the application of Hochman Theorem mentioned above. This together with (24) and (19) yields that

Theorem 14 (Neunhauserer (Improved by Hochman Theorem)). *Fix an arbitrary $\boldsymbol{\tau} \in T_\tau$. Then for $\mathcal{L}eb_2$-almost all $\boldsymbol{\beta} \in T_\beta$ we have*

$$\dim_{\mathrm{H}}(\mu^{\boldsymbol{\beta},\boldsymbol{\tau}}) = 1 + t, \tag{25}$$

consequently,

$$\dim_{\mathrm{B}}(\Lambda) = \dim_{\mathrm{H}}(\Lambda) = 1 + t. \tag{26}$$

2.3.4 Four-Corner Sets

Here we show a family of self-affine IFS for which even the box dimension of the typical element is smaller than its affinity dimension.

Bárány [3] studied the so-called generalized four-corner set which is the attractor of a self-affine IFS like the one in Fig. 3c. We consider the IFS $\mathcal{F}_{\boldsymbol{\beta},\boldsymbol{\tau}}$, defined in (12). The only assumption made is that the rectangles in Fig. 3c are disjoint. More precisely, let

$$T := \left\{(\boldsymbol{\beta}, \boldsymbol{\tau}) \in (0,1)^4 : f_i(Q) \cap f_j(Q) = \emptyset \text{ for } i \neq j\right\},$$

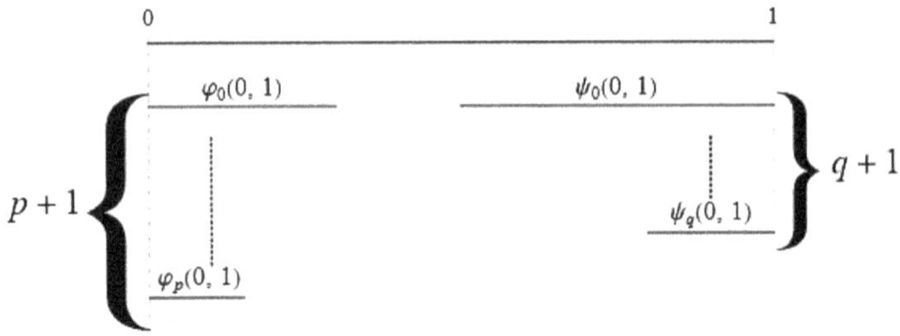

Fig. 4 IFS with some functions sharing the same fixed point. The figure is from [3]

where $Q = [0,1]^2$. In the sequel we always assume that $(\boldsymbol{\beta}, \boldsymbol{\tau}) \in T$. The generalized four corner set is the attractor $\Lambda_{\boldsymbol{\beta},\boldsymbol{\tau}}$ of $\mathcal{F}_{\boldsymbol{\beta},\boldsymbol{\tau}}$. To compute its box dimension we consider the two IFS which are the projections of $\mathcal{F}_{\boldsymbol{\beta},\boldsymbol{\tau}}$ to the coordinate axis:

$$\mathcal{F}_{\boldsymbol{\beta}}^1 := \left\{f_i^1(x) := \beta_i x + 1 - \beta_i\right\}_{i=1,3} \cup \left\{ f_j^1(x) := \beta_j x \right\}_{j=2,4} \tag{27}$$

and

$$\mathcal{F}_{\boldsymbol{\tau}}^2 := \left\{f_i^2(y) := \tau_i y + 1 - \tau_i\right\}_{i=1,4} \cup \left\{ f_j^2(y) := \tau_j y \right\}_{j=2,3}. \tag{28}$$

Clearly, for $k = 1, 2$ there are two-two functions in $\mathcal{F}^k$, both sharing the same fixed point zero and also two-two functions which share the fixed point 1. For this reason the box (and the Hausdorff dimension) for the corresponding attractor $\Lambda_{\boldsymbol{\beta}}^1$ and $\Lambda_{\boldsymbol{\tau}}^2$ are smaller than the minimum of their similarity dimension and 1.

As in the previous two examples we have to find the dimension of the projected attractors. In order to do that Bárány [3] proved the following more general theorem about a $p+q$-parameter family of IFS on the line containing functions which share the same fixed point (Fig. 4).

Theorem 15 (Bárány).

Let $p, q \in \mathbb{N}$ such that at least one of them is positive. Let $\Lambda_{\boldsymbol{\gamma}}$ be the attractor of the IFS

$$\mathcal{S} := \{\varphi_i(x) := \gamma_{i,\ell} x\}_{i=0}^p \cup \left\{\psi_j(x) = \gamma_{j,r} x + 1 - \gamma_{j,r}\right\}_{j=0}^q$$

Without loss of generality we may assume that $\gamma_{0,\ell} = \max \gamma_{i,\ell}$ and $\gamma_{0,r} = \max \gamma_{j,r}$. We define $\hat{s} = \hat{s}(\mathcal{F})$ as the solution of

$$\prod_{i=0}^{p} \left(1 - \gamma_{i,\ell}^{\hat{s}}\right) + \prod_{j=0}^{q} \left(1 - \gamma_{j,r}^{\hat{s}}\right) = 1. \tag{29}$$

Then for $\mathcal{L}eb_{p+q}$ *almost all* $(\gamma_{1,\ell}, \dots, \gamma_{p,\ell}, \gamma_{1,r}, \dots, \gamma_{q,r}) \in (0, \gamma_{0,\ell}) \times (0, \gamma_{0,r})$ *we have*

(a) $\dim_{\mathrm{H}}(\Lambda_{\boldsymbol{\gamma}}) = \min\{1, \hat{s}\}$
(b) *If* $\hat{s} > 1$ *then* $\mathcal{L}eb(\Lambda_{\boldsymbol{\gamma}}) > 0$.

Observe that $\hat{s}$ is smaller than the similarity dimension of the system $\mathcal{S}$. If $\gamma_{0,\ell} + \gamma_{0,r} > 1$, then $\Lambda_{\boldsymbol{\gamma}} = [0, 1]$, so in this case the assertion is trivial. We compute the box dimension of the four-corner set $\Lambda_{\boldsymbol{\beta},\boldsymbol{\tau}}$ for typical $(\boldsymbol{\beta}, \boldsymbol{\tau})$ using the previous theorem for the two projected IFS $\mathcal{F}^1_\beta$ and $\mathcal{F}^2_\tau$. Namely,

Theorem 16 (Bárány). *Assume that* $\boldsymbol{\beta}, \boldsymbol{\tau} \in T$. *Let*

$$s_1 := \hat{s}(\mathcal{F}^1_\beta) \text{ and } s_2 := \hat{s}(\mathcal{F}^2_\tau)$$

be defined as in (29). *Moreover, let* d_1, d_2 *be the solution of*

$$\sum_{i=1}^{4} \beta_i^{\min\{1,s_1\}} \cdot \tau_i^{d_1 - \min\{1,s_1\}} = 1 \text{ and } \sum_{i=1}^{4} \tau_i^{\min\{1,s_2\}} \cdot \beta_i^{d_2 - \min\{1,s_2\}} = 1.$$

Then for $\mathcal{L}eb_4$*-almost all* $\boldsymbol{\beta}, \boldsymbol{\tau} \in T$ *we have*

$$\dim_{\mathrm{B}}(\Lambda_{\boldsymbol{\beta},\boldsymbol{\tau}}) = \max\{d_1, d_2\} < s(\mathcal{F}_{\boldsymbol{\beta},\boldsymbol{\tau}}), \tag{30}$$

where $s(\mathcal{F}_{\boldsymbol{\beta},\boldsymbol{\tau}})$ *is the affinity dimension (see Definition 3).*

2.4 *The Dimension of Self-affine Measures*

Given the self-affine IFS $\mathcal{F} := \{f_i(x) = A_i \cdot x + t_i\}_{i=1}^m$ on $\mathbb{R}^d$, where $\mathcal{A} := (A_1, \dots, A_m) \in (GL_d(\mathbb{R}))^m)$ and $\|A_i\| < 1$ for all $i = 1, \dots, m$. We denote the attractor by Λ. Recall that the affinity dimension and the pressure for this system was defined in Sect. 2.1. The natural projection Π from the symbolic space $\Sigma := \{1, \dots m\}^{\mathbb{N}}$ to the attractor Λ is defined by $\Pi(\mathbf{i}) := \lim_{n\to\infty} f_{i_1\dots i_n}(0)$, where the $\mathbf{i} = (i_0, i_1, \dots)$. Let $\mathcal{E}$ be the set of ergodic invariant measures on Σ.

2.4.1 Thermodynamical Formalism

Let $\nu \in \mathcal{E}$. To define the Lyapunov exponents $\lambda_1(\nu) \geq \dots \geq \lambda_d(\nu)$ of ν, first we introduce the map $A : \Sigma \to GL_d(\mathbb{R})$,

$$A(\mathbf{i}) := A_{i_0}^T.$$

Consider the stationary process given by the measure ν and

$$\left\{P_n(A,\mathbf{i}) := A_{i_{n-1}}^T \cdots A_{i_0}^T\right\}_{n=1}^{\infty} \tag{31}$$

Using Oseledec Theorem for this process we obtain the Lyapunov exponents (see [18, Sect. 1.5])

$$0 > \lambda_1(\nu) \geq \lambda_2(\nu) \geq \cdots \geq \lambda_d(\nu). \tag{32}$$

For a $\nu \in \mathcal{E}$ we can think of the k-th biggest Laypunov exponent $\lambda_k(\nu)$ as the limit

$$\lambda_k(\nu) = \lim_{n\to\infty} \frac{1}{n} \log \alpha_k(A_{\mathbf{i}|_n}) \text{ for } \nu \text{ a.a. } \mathbf{i} \in \Sigma, \tag{33}$$

where $\alpha_k(A_{\mathbf{i}|_n})$ is the k-th biggest singular value of the n-fold matrix product $A_{\mathbf{i}|_n} = A_{i_1} \cdots A_{i_n}$. For a $\mu \in \mathcal{E}$, following Kaenmaki [16], we define the t-energy of μ for $t \geq 0$ by

$$\Lambda_\mu(t) := \lim_{n\to\infty} \frac{1}{n} \sum_{|\mathbf{i}|=n} \mu(\mathbf{i}) \log \varphi^t(A_\mathbf{i}). \tag{34}$$

We write h_μ for the entropy of μ. The following variational principle is due to Kaenmaki [16]:

Theorem 17 (Kaenmaki). *For $t \geq 0$ we have*

$$P(\mathcal{A}, t) = \sup_{\mu\in\mathcal{E}} \left\{h_\mu + \Lambda_\mu(t)\right\} \tag{35}$$

where the corresponding topological pressure function $P(\mathcal{A}, t)$ was defined in (7).

For a given $t \geq 0$, the ergodic measures where the supremum in (35) $\mu \in \mathcal{E}$ are attained are called t-equilibrium measures. Kaenmaki proved [16, Theorem 2.6] that for every $t \geq 0$ such a t-equilibrium measure always exists. However, T. Jordan pointed out that the t-equilibrium measures are not necessarily unique. The same idea appeared in the paper of Rams [26] who studied the same problem for horseshoes.

Now we derive another formula for $\Lambda_\mu(t)$. Fix a $\mu \in \mathcal{E}$ and a $t \geq 0$. We write $\ell := \lfloor t \rfloor$ in the argument below. In the first step of the argument we will use the Sub-additive Ergodic Theorem [31] for the sub-additive function sequence $\psi_n(\mathbf{i}) := \log \varphi^t(A_{\mathbf{i}_n})$:

$$\begin{aligned}
\Lambda_\mu(t) &= \lim_{n\to\infty} \frac{1}{n} \log \varphi^t(A_{\mathbf{i}|n}) \qquad \mu\text{-a.a.}\mathbf{i} \in \Sigma \\
&= \lim_{n\to\infty} \left(\sum_{k=1}^{\ell} \frac{1}{n} \log \alpha_k(A_\mathbf{i}) + (t-\ell)\frac{1}{n} \log \alpha_{\ell+1}(A_\mathbf{i}) \right) \\
&= \sum_{k=1}^{\ell} \lambda_k(\mu) + (t-\ell)\lambda_{\ell+1}(\mu).
\end{aligned} \tag{36}$$

That is $t \mapsto \Lambda_\mu(t)$ is the blue function in Fig. 5.

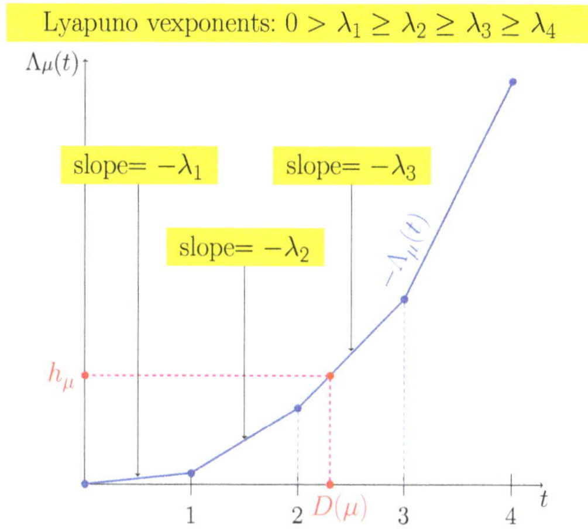

Fig. 5 The definitions of $D(\mu)$ and $\Lambda_\mu(t)$ in $\mathbb{R}^4$

We will see later that under some conditions, generically (in some sense, specified later) the Hausdorff dimension of the push forward measure $\Pi_* \nu$ is the Lyapunov dimension $D(\nu)$ of $\nu \in \mathcal{E}$.

Definition 18 (Lyapunov Dimension of Ergodic Measures). If

$$k := \max\{i : 0 < h_\nu + \lambda_1(\nu) + \cdots + \lambda_i(\nu)\} \leq d - 1, \tag{37}$$

then we define the Lyapunov dimension of ν:

$$D(\nu) := k + \frac{h_\nu + \lambda_1(\nu) + \cdots + \lambda_k(\nu)}{-\lambda_{k+1}(\nu)};$$

If $h_\nu + \lambda_1(\nu) + \cdots + \lambda_d(\nu) > 0$, then we define

$$D(\nu) := d \cdot \frac{h_\nu}{-(\lambda_1(\nu) + \cdots + \lambda_d(\nu))},$$

where h_ν is the entropy of the measure ν.

It is immediate from (36) that

Fact 19 *For a $\mu \in \mathcal{E}$, an alternative way to define $D(\mu)$ is as follows:*

$$-\Lambda_\mu(D(\mu)) = h_\mu. \tag{38}$$

This is shown in Fig. 5.

Recall that Falconer–Solomyak Theorem stated that the dimension of a typical self-affine set is equal to the affinity dimension, if all contractions are stronger than $1/2$. The corresponding theorem for the dimension of ergodic measures was proved in [15].

Theorem 20 (Jordan, Pollicott, and Simon). *Fix an arbitrary* $\mathcal{A} := (A_1, \ldots, A_m) \in (GL_d(\mathbb{R}))^m)$ *satisfying* $\|A_i\| < \frac{1}{2}$ *for all* $i = 1, \ldots, m$. *For* $\mathbf{t} := (t_1, \ldots, t_m) \in \mathbb{R}^{m \cdot d}$ *we define*

$$\mathcal{F}^{\mathbf{t}} := \left\{ f_1^{\mathbf{t}}(x) = A_1 \cdot x + t_1, \ldots, f_m^{\mathbf{t}}(x) := A_m \cdot x + t_m \right\}, \tag{39}$$

and the corresponding natural projection $\Pi^{\mathbf{t}} : \Sigma \to \Lambda^{\mathbf{t}}$ *is*

$$\Pi^{\mathbf{t}}(\mathbf{i}) := \lim_{n \to \infty} f_{i_1 \ldots i_n}^{\mathbf{t}}(0). \tag{40}$$

Then for $\mathcal{L}eb_{md}$ *almost all* $\mathbf{t}$ *we have*

$$\dim_{\mathrm{H}}(\Pi_*^{\mathbf{t}}(\nu)) = D(\nu). \tag{41}$$

3 Dimension Theory of Almost Self-affine Sets and Measures

In this section we always consider some families of function systems that are derived from a given self-affine IFS (like its random perturbations). We conclude that whenever this family satisfies the so-called self-affine Hölder- and transversality conditions (see conditions (a) and (b) in Sect. 3.1.2) then we have a complete understanding about the dimension theory of a typical element of the family. The canonical example of the systems for which these conditions hold, are the so-called almost-self affine system. These are random perturbations of a deterministic self-affine system.

Namely, to have a chance to get rid of the rather restrictive condition $\|A_i\| < \frac{1}{2}$ in Theorems 1 and 20, we consider the so-called almost self-affine sets instead of the self-affine one. The almost self-affine sets are random sets which can be considered as a random perturbation of deterministic self-affine sets like the one in (2). Starting from a deterministic self-affine system like (1) in the construction (2) of the attractor Λ on the right-hand side of (2) we add an independent additive random error to each function which appears. A version of the resulted random set which is called almost self-affine set in the case of the golden gasket is presented in Fig. 6.

For the sake of the reader first we summarize both the definition and the results in a more intuitive way.

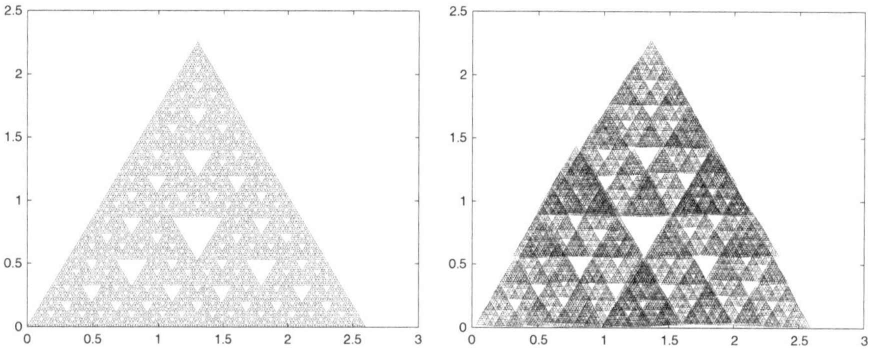

Fig. 6 The golden gasket with and without random errors in the translation

3.1 Self-affine Transversality Condition

In this subsection we introduce two conditions among which the self-affine transversality condition is the more important one. The self-affine Hölder condition basically always holds.

3.1.1 The Definition and Results Presented Heuristically

These definitions were introduced by Jordan et al. in [15]. We consider a one parameter family of self-affine IFS $\{f_i^u\}_{i=1}^m$, on $\mathbb{R}^d$, where the parameter $u \in U$ a compact set which is endowed with a Borel measure $\mathcal{M}$. We introduce a so-called *self-affine Hölder condition and self-affine transversality condition*. Assuming these for $\mathcal{M}$-a.a.parameter $u \in U$ we have a complete understanding of the dimension theory of such a system:

- The Hausdorff dimension of the attractor is minimum of the dimension of the ambient space d and the affinity dimension.
- If the affinity dimension $s(\mathcal{F}) > d$, then $\mathcal{L}eb_d(\Lambda) > 0$.
- The dimension of the push forward of an ergodic measure μ is $\min\{d, D(\mu)\}$.
- If $D(\mu) > d$, then the push forward measure of μ is absolute continuous.

Then we need to know which system satisfy both the self-affine Hölder and transversality conditions. Actually we know a number such families including

- Self-affine systems having all contractions stronger than $1/2$
- Almost self-affine systems.
- Some graph directed systems motivated by fractal image recognition [15, Sect. 6].

We get these results by assuming only that $\|A_1\| < 1$ (instead of $\|A_1\| < 1/2$) in Theorems 1 and 20 but, on the other hand, our system is a random system.

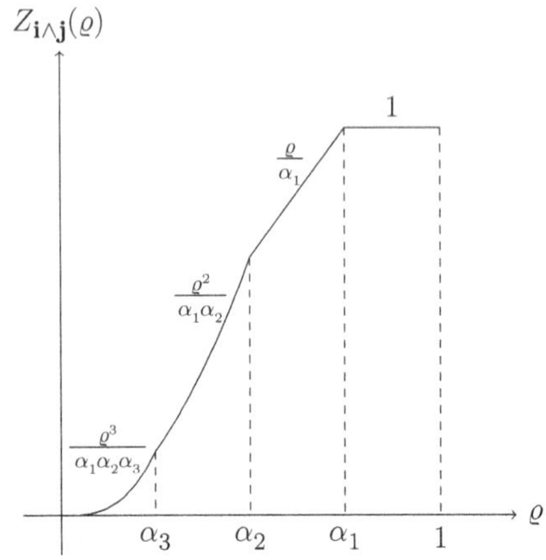

Fig. 7 The function $\rho \to Z_{\mathbf{i}\wedge\mathbf{j}}(\rho)$, $\alpha_k := \alpha_k(\mathbf{i} \wedge \mathbf{j})$ in $\mathbb{R}^3$

3.1.2 The Precise Definitions

We are given a compact σ-invariant $\Omega \subset \Sigma$. (Mostly $\Omega = \Sigma := \{1,\ldots,m\}^{\mathbb{N}}$ except when we use graph directed construction.) We assume that we are also given a continuous map $\Pi : U \times \Omega \to \mathbb{R}^d$ which for a fixed u gives us the natural projection from $\Pi^u : \Omega \to \Lambda^u$, where Λ^u is the attractor for parameter u.

(a) *Self-affine Hölder condition:* There exists a constant $K > 0$ such that for every $u \in U$ and $\mathbf{i} \in \Omega$ and for every $n \in \mathbb{N}$ we can find an isometry $G = G(\mathbf{i}, n) : \mathbb{R}^d \to \mathbb{R}^d$ with

$$\Pi^u(\mathbf{i}|n) \subset K \cdot G\left([0, \alpha_1(\mathbf{i}|n)] \times \cdots \times [0, \alpha_d(\mathbf{i}|n)]\right) \tag{42}$$

where $\mathbf{i}|n$ denotes the truncation of $\mathbf{i} \in \Sigma$ to a word of length n.

(b) *Self-affine transversality condition:* There is a constant $C > 0$ (independent of $\mathbf{i}, \mathbf{j}$) such that for all $\mathbf{i}, \mathbf{j} \in \Omega$, $\mathbf{i} \neq \mathbf{j}$ we have

$$\mathcal{M}\left\{u \in U : |\Pi^u(\mathbf{i}) - \Pi^u(\mathbf{j})| < \rho\right\} < C \cdot Z_{\mathbf{i}\wedge\mathbf{j}}(\rho), \tag{43}$$

where

$$Z_{\mathbf{i}\wedge\mathbf{j}}(\rho) := \prod_{k=1}^{d} \min\left\{\rho, \alpha_k(\mathbf{i} \wedge \mathbf{j})\right\}. \tag{44}$$

See Fig. 7.

These conditions were motivated by the Generalized projection scheme of Solomyak [30].

3.2 *Almost Self-affine Systems*

Given the self-affine IFS on $\mathbb{R}^n$ by $\mathcal{F} := \{f_i\}_{i=1}^m = \{A_i \cdot x + t_i\}_{i=1}^m$, where A_i are supposed to be contracting for all $i = 1, \ldots, m$ and as usual we write $\mathcal{A} := \{A_1, \ldots, A_m\}$. We assume that with each application of the functions from the given IFS we make a random additive error Y. We assume that these errors have distribution η, where η is an absolutely continuous distribution with bounded density supported on a compact disk D which is centered at the origin and can be chosen to be arbitrarily small. Let $\mathcal{T}$ be the m-array tree. For an $\mathbf{i}_n := (i_0, \ldots, i_{n-1}) \in \{1, \ldots, m\}^n$ let

$$f_{\mathbf{i}_n}^{\mathbf{y}_{\mathbf{i}_n}} := (f_{i_0} + y_{i_0}) \circ (f_{i_1} + y_{i_0 i_1}) \circ \cdots \circ (f_{i_{n-1}} + y_{i_0 \ldots i_{n-1}}), \tag{45}$$

and we assume that

$$\mathbf{y}_{\mathbf{i}_n} := (y_{i_0}, y_{i_0 i_1}, \ldots, y_{i_0 \ldots i_{n-1}}) \in \underbrace{D \times \cdots \times D}_{n}$$

are i.i.d. with distribution η. Let

$$\mathbf{y} := \{\mathbf{y}_{\mathbf{i}_n}\}_{\mathbf{i}_n \in \{1,\ldots,m\}^*} .$$

The attractor is defined by

$$\Lambda^{\mathbf{y}} := \bigcap_{n=0}^{\infty} \bigcup_{\mathbf{i}_n} f_{\mathbf{i}_n}^{\mathbf{y}_{\mathbf{i}_n}}(B), \tag{46}$$

where B is a large ball centered at the origin.

So, for an $\mathbf{i} = (i_0, \ldots, i_k, \ldots) \in \underbrace{\{1, \ldots, m\}^{\mathbb{N}}}_{\Sigma}$

$$\Pi^{\mathbf{y}}(\mathbf{i}) := \lim_{n \to \infty} \Big(\mathbf{t}_{i_0} + \sum_{k=1}^{n} A_{i_0, \ldots, i_{k-1}} \cdot \mathbf{t}_{i_k} \tag{47}$$
$$+ y_{i_0} + \sum_{k=1}^{n} A_{i_0, \ldots, i_{k-1}} \cdot y_{i_0 \ldots i_k} \Big).$$

The black part is the deterministic part and the yellow part is the random part. Finally, on $D^{\mathcal{T}}$ we define the infinite product measure:

$$\mathbb{P} := \eta \times \cdots \times \eta \cdots .$$

It was proved in [15, Lemma 5.1] that

Lemma 21. *The family of almost self-affine maps defined above satisfy both the self-affine Hölder and Transversality conditions.*

Hence we obtain

Theorem 22 (Jordan, Pollicott, and Simon). *Let ν be an ergodic measure on Σ. For $\mathbb{P}$-almost all $\mathbf{y} \in D^{\infty}$ we have:*

(1) If $s(\mathcal{A}) \leq d$, then $\dim_{\mathrm{H}} \Lambda^{\mathbf{y}} = s(\mathcal{A})$;
(2) If $s(\mathcal{A}) > d$, then $\mathcal{L}eb_d(\Lambda^{\mathbf{y}}) > 0$.
(3) $\dim_{\mathrm{H}} \Pi_*^{\mathbf{y}}(\nu) = \min\{d, D(\nu)\}$.
(4) If $D(\nu) > d$, then $\Pi_*^{\mathbf{y}}(\nu) \ll \mathcal{L}eb_d$.
(5) The measure $\Pi_*^{\mathbf{y}}(\nu)$ is exact dimensional:

$$\lim_{r \downarrow 0} \frac{\log \Pi_*^{\mathbf{y}}(\nu) B(x,r)}{\log r} = D(\nu) \text{ for } \Pi_*^{\mathbf{y}} \text{ a.a. } x.$$

T. Jordan observed that last statement follows by a little modification of a proof of [15, Theorem 4.3].

4 Recent Developments

4.1 Non-compactly Supported Random Perturbations

In a very recent paper Jordan and Jurga [14] considered a construction which is similar to the almost-self affine case but here the random perturbations are non-compactly supported. That is the distribution η of additive the random error Y is not supported by a disk. In this case it is already a question how to define the attractor because (46) clearly does not work. A definition of the attractor which is equivalent to (46) when η is compactly supported and may work also in the general case is:

$$\Lambda^{\mathbf{y}} := \{\Pi^{\mathbf{y}}(\mathbf{i}) : \mathbf{i} \in \Sigma\}.$$

Theorem 23. *Assume that η decays super-polynomially, (e.g., η is the normal distribution) then assertions (1) and (2) of Theorem 22 hold.*

4.2 Multifractal Analysis

Exciting recent papers of Falconer [8], Barral and Feng [4] obtained partial results about the multifractal analysis of almost self-affine systems. A conference proceedings paper (International Conference on Fractals and Related Topics, 2012, Hong Kong) will be published by K. Falconer about this topic.

Appendix

This section is devoted to the Ledrappier–Young formula for self-affine IFS in the diagonal case on the plane. This was proved by Feng and Hu [10] in 2009 in full generality. Given the self-affine IFS on $\mathbb{R}^2$:

$$\mathcal{F}_{\boldsymbol{\beta},\boldsymbol{\tau}} := \left\{ f_i(x) := \begin{bmatrix} \beta_i & 0 \\ 0 & \tau_i \end{bmatrix} \cdot x + t_i \right\}, \text{ where } 0 < \tau_i, \beta_i < 1,\ i = 1, \dots, m. \tag{48}$$

The projection on the plane to the x-axis is denoted by proj_x. As always in this paper Π is the natural projection from the symbolic space $\Sigma := \{1, \dots, m\}^{\mathbb{N}}$ to the attractor Λ defined by (21). Let μ be an ergodic measure on Σ. We define the push forward measures of μ by Π and $\mathrm{proj}_x \circ \Pi$:

$$\nu := \Pi_* \mu \text{ and } \nu_x := (\mathrm{proj}_x \circ \Pi)_* \mu. \tag{49}$$

Let us consider the Lyapunov exponents $0 > \lambda_1(\mu) \geq \lambda_2(\mu)$. Since the matrices A_i are diagonal, the corresponding eigenspaces [18, Sect. 1.5] are the coordinate axes. Moreover, between the two Lyapunov exponents $\lambda_1(\mu), \lambda_2(\mu)$ one of them is $\lambda_x(\mu)$ the other one is $\lambda_y(\mu)$, where

$$\lambda_x(\mu) := \lim_{n \to \infty} \frac{1}{n} \log(\beta_{i_1} \cdots \beta_{i_n}) = \sum_{i=1}^{m} \mu([i]) \log \beta_i, \tag{50}$$

and

$$\lambda_y(\mu) := \lim_{n \to \infty} \frac{1}{n} \log(\tau_{i_1} \cdots \tau_{i_n}) = \sum_{i=1}^{m} \mu([i]) \log \tau_i, \tag{51}$$

where the limits above are meant to be limits for μ-almost all $\mathbf{i} \in \Sigma$. With Ledrappier–Young formula, we can express $\max\left(\dim_{\mathrm{H}}(\nu_x), \dim_{\mathrm{H}}(\nu_y)\right)$ by $\dim_{\mathrm{H}}(\nu_x)$. The following Theorem is an immediate corollary of Feng–Hu [10, Theorems 2.8 and 2.11]:

Theorem 24 (Ledrappier–Young Formula for IFS). *Assume that the IFS $\mathcal{F}_{\boldsymbol{\beta},\boldsymbol{\tau}}$ satisfies the strong separation property (that is all cylinders are disjoint). Further, we assume that $\lambda_x(\mu) \geq \lambda_y(\mu)$. That is $\lambda_1(\mu) = \lambda_x(\mu)$ and $\lambda_2(\mu) = \lambda_y(\mu)$. Then*

$$\dim_{\mathrm{H}}(\nu) = \frac{h_\mu}{-\lambda_2(\mu)} + \left(1 - \frac{\lambda_1(\mu)}{\lambda_2(\mu)}\right) \cdot \dim_{\mathrm{H}}(\nu_x). \tag{52}$$

From this we get

Corollary 25. *Under the assumption of the Theorem 24,*

$$D(\nu) = \dim_{\mathrm{H}}(\nu) \iff \dim_{\mathrm{H}}(\nu_x) = \min\left\{1, \frac{h_\mu}{-\lambda_1(\mu)}\right\}. \tag{53}$$

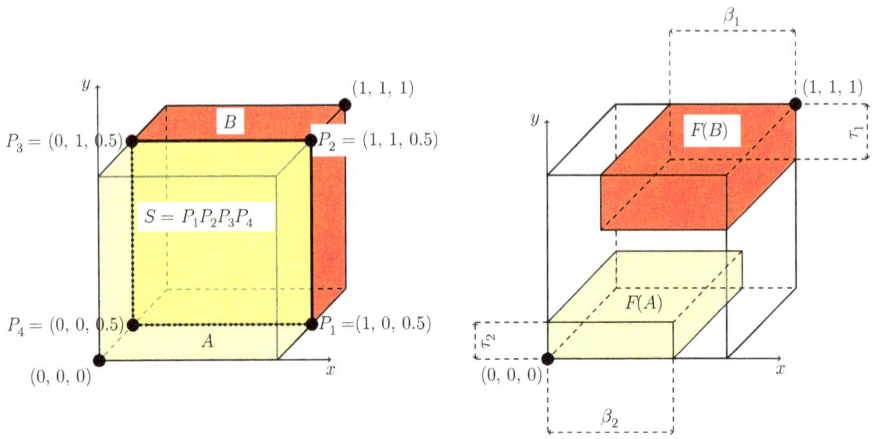

Fig. 8 A is the box with opposite vertices $(0,0,0)$ and P_2. B is the box with opposite vertices P_4 and $(1,1,1)$

Ledrappier–Young [20, p. 545] proved a formula for the dimension of measures that are ergodic and invariant with respect to some C^2 diffeomorphisms of a compact C^∞ manifold without boundary. In order to see how to apply this for self-affine IFS on the plane, in the diagonal case, we associate a baker map in dimension three in a natural way. Namely, if the IFS is the generalized PU set of Example 9, then the associated baker map is the map which appears in Fig. 8. In this case (using the definition of Example 9)

$$F(x,y,z) = \begin{cases} (\beta_2 x, \tau_2 y, 2z), & \text{if } (x,y,z) \in A; \\ (\beta_1 x + 1 - \beta_1, \tau_1 y + 1 - \tau_1, 2z - 1), & \text{if } (x,y,z) \in B. \end{cases} \tag{54}$$

In the general case the corresponding baker map in dimension three is defined analogously. This is a map with singularities (see the square S in Fig. 8). If there were no singularities in the associated baker map, then we would apply Ledrappier–Young formula (see [20, p. 545, (i)–(iii)]) for the baker map, and that would immediately imply that (52) holds. I explain everything in the special case of Example 9 and the corresponding baker map F [defined in (54) and in Fig. 8]. Let $\hat{\Lambda}$ be the attractor of the baker map. The natural projection $\hat{\Pi} := \Sigma_2 \to \hat{\Lambda}$ is defined for an $\mathbf{i} = (\dots i_{-2}, i_{-1}, i_0, i_1, i_2, \dots) \in \Sigma_2$ by

$$\hat{\Pi}(\cdots i_{-2} i_{-1}; i_0 i_1 \cdots) = \left(\lim_{n\to\infty} f_{i_{-1}} \circ \cdots \circ f_{i_{-n}}(0,0), \sum_{n=0}^{\infty} \frac{i_n}{2^{n+1}} \right).$$

Let $\hat{\mu}$ be the canonical ergodic measure on $\Sigma_2 := \{1,\dots,m\}^{\mathbb{Z}}$ which corresponds to μ and we write $\hat{\nu} := (\hat{\Pi})_* \hat{\mu}$ for the push forward measure on $\hat{\Lambda}$. The Lyapunov exponents of $\hat{\nu}$ are

$$\lambda_0 = \log 2, \lambda_1 = \lambda_x(\mu) \text{ and } \lambda_2 = \lambda_y(\mu),$$

where $\lambda_x(\mu), \lambda_y(\mu)$ where defined in (50) and (51).

Then the decomposition $\bigoplus_{i=0}^{2} E_i$ corresponding to the Lyapunov exponents $\lambda_0 > \lambda_1 \geq \lambda_2$ are the lines through the point (x, y, z) which are parallel to the z, x, and y axes in this order. We call them unstable, stable and strong stable directions and denote them by $E_0(x, y, z)$, $E_1(x, y, z)$ and $E_2(x, y, z)$, respectively. The unstable manifold $W^u(x, y, z)$ of a point $(x, y, z) \in \hat{\Lambda}$ is $E_0(x, y, z)$, the stable manifold is $W^1(x, y, z) := E_1(x, y, z) \times E_2(x, y, z)$ and the strong stable manifold is $W^2(x, y, z) := E_2(x, y, z)$. Observe that

$$\hat{\Lambda} = \Lambda \times [0, 1], \quad \hat{\nu} = \nu \times [0, 1]. \tag{55}$$

Further, for every $(x, y, z) \in \hat{\Lambda}_2$ let $\hat{\nu}^k$ be the conditional measure of $\hat{\nu}$ to $W^k(x, y, z)$, $k = 0, 1, 2$. Observe that by (55) we have

$$\hat{\nu}^1(x, y, z) = \hat{\nu}^1(x', y', z'), \text{ for any } (x, y, z), (x', y'z') \in \hat{\Lambda} \tag{56}$$

and this measure can be naturally identified by ν. This and formulas [20, p. 545 (i)–(iii)] applied to the system $(\hat{\Lambda}, \hat{\nu}, F)$ would imply immediately that Theorem 24 holds if we knew that formulas [20, p. 545 (i)–(iii)] can be applied not only for the ergodic invariant measures invariant for C^2 diffeomorphisms of C^∞ manifolds but also for the system $(\hat{\Lambda}, \hat{\nu}, F)$ which has singularities.

So, we need to prove that the presence of singularities do not influence the way to compute the dimension of the invariant measure.

This (in a much higher generality) was done first by Feng and Hu [10]. They proved this (beside other important things) with making all the proofs from sketch, basically following the idea of Ledrappier–Young's proof. The proofs in [10] are completely self-contained in a sense that they do not refer back to Ledrappier–Young [20]. In particular they do not use the baker map above.

On the other hand, Jörg Neunhäuserer showed in his Ph.D. thesis [22, p. 30] that it follows from [27, Sect. 4] that Ledrappier–Young formula can be applied for self-affine IFS in the diagonal case. His argument was that the associated baker map has Lyapunov exponents and Lyapunov charts almost everywhere. Using this and invertibility, it follows from Schmeling and Troubetzkoy [27] that Theorem 24 holds. This idea was worked out completely with all of the details in a more general case than that of considered by Neunhauserer in the M.Sc. thesis of B. Bárány [2, pp. 21–38] providing an alternative proof to Theorem 24. In this proof, Bárány checked only those details which are different in the case of the IFS and referred to Ledrappier–Young [20] at all steps of the proof which were the same. In either way, there is no short proof for Theorem 24.

Acknowledgements The research was supported by the grant OTKA # K 104745.

References

1. Barański, K.: Hausdorff dimension of the limit sets of some planar geometric constructions. Adv. Math. **210**(1), 215–245 (2007)
2. Bárány, B.: Dimension theory of non-conformal attractors (2008) http://www.math.bme.hu/~balubs/
3. Bárány, B.: Dimension of the generalized 4-corner set and its projections. Ergodic Theory Dyn. Syst. **32**(4), 1190 (2011)
4. Barral, J., Feng, D.J.: Multifractal formalism for almost all self-affine measures. Commun. Math. Phys. **318**(2), 473–504 (2013)
5. Falconer, K.J.: The hausdorff dimension of self-affine fractals. In: Mathematical Proceedings of the Cambridge Philosophical Society, vol. 103, pp. 339–350. Cambridge University Press, Cambridge (1988)
6. Falconer, K.J.: The dimension of self-affine fractals ii. In: Mathematical Proceedings of the Cambridge Philosophical Society, vol.111, pp. 169–179. Cambridge University Press, Cambridge (1992)
7. Falconer, K.J.: Fractal Geometry: Mathematical Foundations and Applications. Wiley, New York (2007)
8. Falconer, K.J.: Generalised dimensions of measures on almost self-affine sets. Nonlinearity **23**, 1047–1069 (2010)
9. Falconer, K.J., Sloan, A.: Continuity of subadditive pressure for self-affine sets. Real Anal. Exchange **34**(2), 413–428 (2008)
10. Feng, D.J., Hu, H.: Dimension theory of iterated function systems. Commun. Pure Appl. Math. **62**(11), 1435–1500 (2009)
11. Feng, D.J., Shmerkin, P.: Non-conformal repellers and the continuity of pressure for matrix cocycles Geometric and Functional Analysis, 1–28, Springer (2013)
12. Hochman, M.: On self-similar sets with overlaps and inverse theorems for entropy. Preprint (2012). arXiv:1212.1873
13. Hueter, I., Lalley, S.P.: Falconer's formula for the hausdorff dimension of a self-affine set in r2. Ergodic Theory Dyn. Syst. **15**(01), 77–97 (1995)
14. Jordan, T., Jurga, N.: Self-affine sets with non-compactly supported random perturbations. Preprint (2013). arXiv:1310.0944
15. Jordan, T., Pollicott, M., Simon, K.: Hausdorff dimension for randomly perturbed self affine attractors. Commun. Math. Phys. **270**(2), 519–544 (2007)
16. Käenmäki, A.: On natural invariant measures on generalized iterated function systems. Ann. Acad. Sci. Fenn. Math. **29**, 419–458 (2004)
17. Käenmäki, A., Shmerkin, P.: Overlapping self-affine sets of kakeya type. Ergodic Theory Dyn. Syst. **29**(3), 941–965 (2009)
18. Krengel, U., Brunel, A.: Ergodic Theorems, De Gruyter Studies in Mathematics, vol. 59. W. de Gruyter, Cambridge Univ Press, Berlin (1985)
19. Lalley, S.P., Gatzouras, D.: Hausdorff and box dimensions of certain self-affine fractals. Indiana Univ. Math. J. **41**, 533 (1992)
20. Ledrappier, F., Young, L.S.: The metric entropy of diffeomorphisms: part ii: relations between entropy, exponents and dimension. Ann. Math. **122**(3), 540–574 (1985)
21. McMullen, C.: The hausdorff dimension of general sierpiński carpets. Nagoya Math. J. **96**, 1–9 (1984)
22. Neunhauserer, J.: Dimensional theoretical properties of some affine dynamical systems. Ph.D. thesis, Free University Berlin (1999)
23. Neunhauserer, J.: Properties of some overlapping self-similar and some self-affine measures. Acta Math. Hung. **92**, 143–161 (2001)
24. Pollicott, M., Weiss, H.: The dimensions of some self-affine limit sets in the plane and hyperbolic sets. J. Stat. Phys. **77**(3–4), 841–866 (1994)

25. Przytycki, F., Urbański, M.: On the hausdorff dimension of some fractal sets. Stud. Math. **93**(2), 155–186 (1989)
26. Rams, M.: Measures of maximal dimension for linear horseshoes. Real Anal. Exchange **31**(1), 55–62 (2005)
27. Schmeling, J., Troubetzkoy, S.: Scaling properties of hyperbolic measures. Preprint (1998)
28. Shmerkin, P.: On the exceptional set for absolute continuity of bernoulli convolutions. Preprint (2013). arXiv:1303.3992
29. Solomyak, B.: On the random series $\pm\lambda$ n (an erdös problem). Ann. Math. **142**(3), 611–625 (1995)
30. Solomyak, B.: Measure and dimension for some fractal families. In: Mathematical Proceedings of the Cambridge Philosophical Society, vol. 124, pp. 531–546. Citeseer (1998)
31. Walters, P.: An Introduction to Ergodic Theory, vol. 79. Springer, New York (2000)

Countable Alphabet Non-autonomous Self-affine Sets

Mariusz Urbański

Abstract We extend Falconer's formula from Falconer (Math. Proc. Camb. Philos. Soc. 103:339–350, 1988) by identifying the Hausdorff dimension of the limit sets of almost all contracting affine iterated function systems to the case of an infinite alphabet, non-autonomous choice of iterating matrices, and time-dependent random choice of translations.

Keywords Self-affine sets • Hausdorff dimension • Non-autonomous systems

1 Introduction

In the seminal paper [1], given k contracting matrices $A_1, A_2, \ldots, A_k$, Ken Falconer has provided a close formula which gives the Hausdorff dimension of the limit sets of the iterated function system

$$\mathcal{S}_a = \{\mathbb{R}^q \ni x \mapsto A_i x + a_i\}_{i=1}^k$$

for Lebesgue almost every vector $a = (a_i)_{i=1}^k \in \mathbb{R}^{qk}$. In our article we extend Falconer's result in several directions.

- We allow k to be infinite; instead of Lebesgue measure we then consider appropriately defined product measure with infinitely many factors.
- Being in the iterating process we allow all the matrices A_i to depend on the time, i.e. making a new composition at a step n, we take the contracting matrices from an entirely new collection $A_1^{(n)}, \ldots, A_k^{(n)}$.

M. Urbański (✉)
Department of Mathematics, University of North Texas,
Denton, TX 76203-1430, USA
e-mail: urbanski@unt.edu

C. Bandt et al. (eds.), *Fractals, Wavelets, and their Applications*, Springer Proceedings in Mathematics & Statistics 92, DOI 10.1007/978-3-319-08105-2_7,

- We choose the vectors $(a_i)_{i=1}^k$ randomly according to some random process.

Roughly speaking we have either a finite or countable infinite alphabet E, the system $\mathcal{S}_a$ consists now of maps

$$\phi_e^{(n,a)}(x) = A_e^{(n)}x + a_e, \ \ e \in E,$$

we have also a measurable transformation $\theta : X \to X$ preserving some Borel probability measure X, and smooth transformations $S_x : G^E \to G^E$ $(G \subset \mathbb{R}^q)$, $x \in X$, with some additional technical properties. Each point $x \in X$ generates a non-autonomous iterative scheme

$$\phi_{\omega_1}^{(1,a)} \circ \phi_{\omega_2}^{(2,S_x(a))} \circ \ldots \circ \phi_{\omega_n}^{(n,S_x^n(a))} \circ \ldots, \ \ \omega \in E^{\mathbb{N}},$$

where

$$S_x^n := S_x \circ S_{\theta(x)} \circ \ldots \circ S_{\theta^{n-1}(x)}.$$

This determines (see (2) for a rigorous definition) the limit set $J_{(x,a)}$, and our main result identifies the Hausdorff dimension of $J_{(x,a)}$ for m–a.e. $x \in X$ and "Lebesgue"–a.e. $a \in G^E$. We do this by introducing the Falconer dimension FD($\mathcal{S}$), which depends only on matrices $A_e^{(n)}$, $e \in E, n \in \mathbb{N}$, and is independent of the maps $S_x : G^E \to G^E$. We prove the following main result

Theorem 1.1. *If $\mathcal{S}$ is an affine scheme on $\mathbb{R}^q$, then*

$$\mathrm{HD}\big(J_{(x,a)}\big) = \min\{q, \mathrm{FD}(\mathcal{S})\}.$$

for m-a.e. $x \in X$ and λ_G^E-a.e. $a \in G^E$.

which is Theorem 5.3 from the last section of our paper. We would like to add that another extension of Falconer's result, incorporating a different randomizing procedure, was treated in [3].

2 Affine Schemes

Fix E, a countable set, either finite or infinite; it will be called an alphabet in the sequel. Fix an integer $q \geq 1$ and two real numbers $\kappa, \xi \in (0,1)$. For every $n \geq 1$ and every $e \in E$ let $A_e^{(n)} : \mathbb{R}^q \to \mathbb{R}^q$ be an invertible linear map with

$$\big\|A_e^{(n)}\big\| \leq \kappa \ \text{ and } \ \Big\|\big(A_e^{(n)}\big)^{-1}\Big\| \leq \xi^{-1}. \tag{1}$$

Let $G \subset \mathbb{R}^q$ be a bounded Borel subset of $\mathbb{R}^q$ with positive Lebesgue measure. Let λ_G be the normalized (so that $\lambda_G(G) = 1$) q-dimensional Lebesgue measure on G and let λ_G^E be the corresponding infinite product measure on G^E. This measure is uniquely determined by the requirement that

$$\lambda_G^E\left(\prod_{a\in\Gamma} F_a \times G^{E\setminus F}\right) = \prod_{a\in\Gamma} \lambda_G(F_a)$$

for every finite subset Γ of E and all Borel sets $F_a \subset G$, $a \in \Gamma$. Denote by R_G the largest radius $r > 0$ such that $G \subset B(0, r)$. For every $n \geq 1$, every $e \in E$, and every $a \in G^E$ consider the maps $\phi_e^{(n,a)} : \mathbb{R}^q \to \mathbb{R}^q$ given by respective formulas

$$\phi_e^{(n,a)}(x) = A_e^{(n)}x + a_e.$$

Since all the maps $A_e^{(n)}$ are uniform linear contractions and since the set G is bounded, there exists B, a sufficiently large closed ball in $\mathbb{R}^q$ centered at the origin, such that

$$\phi_e^{(n,a)}(B) \subset B$$

for all $n \geq 1$, all $e \in E$, and all $a \in G^E$. Let $l_E^\infty(\mathbb{R}^q)$ be the Banach space of all bounded functions from E to $\mathbb{R}^q$, endowed with the supremum norm, i.e.

$$||a||_\infty = \sup\{||a_e|| : e \in E\}.$$

Of course, G^E is a subset of $l_E^\infty(\mathbb{R}^q)$. Let $(X, \mathcal{F}, m)$ be a probability space and let $\theta : X \to X$ be an invertible measurable map preserving the measure m. For every $x \in X$ let $S_x : G^E \to G^E$ be a map for which there exists a bounded convex open set $\hat{G} \subset \mathbb{R}^q$ with the following properties.

(p1) $G \subset \hat{G}$ and $\operatorname{dist}(G, \mathbb{R}^q \setminus \hat{G}) > 0$; then $G^E \subset \operatorname{Int}_{l_E(\mathbb{R}^q)}(\hat{G}^E)$.
(p2) There exists a continuous map $\hat{S}_x : \hat{G}^E \to \hat{G}^E$ such that
(p3) $\hat{S}_x$ is differentiable throughout $\operatorname{Int}_{l_E(\mathbb{R}^q)}(\hat{G}^E)$.
(p4)

$$||DS_x||_\infty := \sup\{||D_a\hat{S}_x|| : a \in G^E\} < \infty$$

and

$$\beta := \operatorname{ess\,sup}\{||DS_x||_\infty : x \in X\} < \infty$$

is so small that

$$\kappa\beta < 1/3.$$

(p5) For m-a.e. $x \in X$ there exists a Borel probability measure μ_x on G^E equivalent (with bounded Radon–Nikodym derivatives) to λ_G^E such that

$$\mu_{\theta(x)} = \mu_x \circ S_x^{-1}.$$

Note that if the space X is a singleton, then we are talking about one mapping $S : G^E \to G^E$ (and its extension $\tilde{S} : \tilde{G}^E \to \tilde{G}^E$) preserving a Borel probability measure μ on G^E equivalent (with bounded Radon–Nikodym derivatives) to λ_G^E. This of course comprises the case of S being the identity map on G^E. This case is referred to as translation deterministic. $S = \mathrm{Id}_{G^E}$ was a part of Falconer's setup in [1]. He was also assuming that the alphabet E is finite and the linear contractions $A_e^{(n)} : \mathbb{R}^q \to \mathbb{R}^q$ are independent of n. We do not assume any of these. Now, the collection of maps

$$\left\{\phi_e^{(n,a)} : \mathbb{R}^q \to \mathbb{R}^q : n \geq 1,\, a \in G^E,\, e \in E\right\}$$

along with the map $\theta : X \to X$ and described above maps $S_x : G^E \to G^E, x \in X$, are referred to as an affine scheme $\mathcal{S}$. We classify affine schemes as follows.

(1) Autonomous if the affine contractions $A_e^{(n)} : \mathbb{R}^q \to \mathbb{R}^q$ are independent of n.
(1') Finitely autonomous if $\mathcal{S}$ is autonomous and the alphabet E is finite.
(2) Non-autonomous if $\mathcal{S}$ is not autonomous.
(3) Of dynamically deterministic type if the maps $S_x : G^E \to G^E, x \in X$, are independent of $x \in X$. Then the action $\theta : X \to X$ is irrelevant, and we may assume without loss of generality that X is a singleton.
(4) Deterministic if $\mathcal{S}$ is of dynamically deterministic type and $S : G^E \to G^E$ is the identity map on G^E.
(5) Of dynamically random type if $\mathcal{S}$ is not of dynamically deterministic type, meaning that $S_x : G^E \to G^E$ do depend on $x \in X$.
(6) A Falconer scheme if $\mathcal{S}$ is finitely autonomous and of dynamically deterministic type.

From now on $\mathcal{S}$ is an arbitrary affine scheme. As in the introduction, for every integer $k \geq 1$ and every $x \in X$ let

$$S_x^k := S_x \circ S_{\theta(x)} \circ \ldots \circ S_{\theta^{k-1}(x)}.$$

Given $n \geq 1$, $\omega \in E^n$, and $a \in G^E$, we define the maps

$$A_\omega := A_{\omega_1}^{(1)} \circ A_{\omega_2}^{(2)} \circ \ldots \circ A_{\omega_n}^{(n)} : R^q \to \mathbb{R}^q$$

and

$$\phi_\omega^{(x,a)} := \phi_{\omega_1}^{(1,a)} \circ \phi_{\omega_2}^{(2,S_x(a))} \circ \ldots \circ \phi_{\omega_n}^{(n,S_x n-1(a))} : B \to B.$$

Note that A_ω is the linear part of the affine map $\phi_\omega^{(x,a)}$. For every infinite word $\omega \in E^{\mathbb{N}}$ and every integer $n \geq 1$ we put

$$\omega|_n := \omega_1\omega_2 \dots \omega_n.$$

Then $\left(\phi_{\omega|_n}^{(x,a)}(B)\right)_{n=1}^{\infty}$ is a descending sequence of nonempty compact subsets of B and

$$\operatorname{diam}\left(\phi_{\omega|_n}^{(x,a)}(B)\right) \leq \operatorname{diam}(B)\kappa^n.$$

So, the intersection

$$\bigcap_{n=1}^{\infty} \phi_{\omega|_n}^{(x,a)}(B)$$

is a singleton, and we denote its only element by $\pi_{(x,a)}(\omega)$. So, for every $x \in X$ and every $a \in G^E$ we have defined the projection map

$$\pi_{(x,a)} : E^{\mathbb{N}} \to B.$$

Slightly more generally, given any integer $k \geq 1$, we consider the maps

$$\phi_\omega^{(x,a;k)} := \phi_{\omega_1}^{(k,a)} \circ \phi_{\omega_2}^{(k+1,S_x(a))} \circ \dots \circ \phi_{\omega_n}^{(k+n-1,S_x^{n-1}(a))} : B \to B.$$

and the corresponding projections

$$\pi_{(x,a)}^k : E^{\mathbb{N}} \to B.$$

In particular,

$$\pi_{(x,a)}^1 = \pi_{(x,a)}.$$

The set

$$J_{(x,a)} := \pi_{(x,a)}(E^{\mathbb{N}}) \subset B \subset \mathbb{R}^q \tag{2}$$

is called the limit set (or the attractor) of the affine scheme $\mathcal{S}$ at the point (x, a). Our goal is to determine the Hausdorff dimensions of these limit sets. Indeed, we will show that these dimensions are equal for m-almost all $x \in X$ and λ_G^E-almost all $a \in G^E$, and the resulting common value is directly expressible in terms of the sequence alone $\left(\{A_e^{(n)} : e \in E\}\right)_{n=1}^{\infty}$.

3 The Singular Value Function

Let $A : \mathbb{R}^q \to \mathbb{R}^q$ be an invertible linear contraction and let

$$1 > \alpha_1 \geq \alpha_1 \geq \ldots \geq \alpha_q > 0$$

be the square roots of (necessarily positive) eigenvalues of the self-adjoint map $A^*A : \mathbb{R}^q \to \mathbb{R}^q$. Geometrically, the numbers $\alpha_1, \ldots \alpha_q$ are the lengths of the (mutually perpendicular) principal semi-axes of $A(\overline{B}(0,1))$, where $\overline{B}(0,1)$ is the closed ball in R^q centered at 0 and of radius 1. These numbers are called singular values of the map $A : \mathbb{R}^q \to \mathbb{R}^q$. Following Falconer [1] we define

$$\alpha^t(A) := \alpha_1\alpha_2 \ldots \alpha_{k-1}\alpha_k^{t-(k-1)}$$

if $0 \leq t \leq q$, where k is the least integer greater than or equal to s, i.e. $k-1 < t \leq k$, and

$$\alpha^t(A) := (\alpha_1\alpha_2 \ldots \alpha_{k-1}\alpha_q)^{t/q}$$

if $t > q$. Denote by $L_*(\mathbb{R}^q)$ the set of all invertible linear contractions from $\mathbb{R}^q$ onto itself. Note that $L_*(\mathbb{R}^q)$ is closed under the compositions of maps. We quote from [1] the following two lemmas.

Lemma 3.1. *For each $t \geq 0$ the function $L_*(\mathbb{R}^q)$ by $(0, +\infty)$ is submultiplicative, meaning that*

$$\alpha^t(AC) \leq \alpha^t(A)\alpha^t(C)$$

for all $A, C \in L_*(\mathbb{R}^q)$.

and

Lemma 3.2. *Given a non-integral real number $0 < t < q$ and a real number $R > 0$ there exists a constant $c < +\infty$ (depending on all of then q, t, and R) such that*

$$\int_{\overline{B}(0,1)} \frac{d\,\lambda_q(x)}{||Ax||^t} \leq \frac{c}{\alpha^t(A)}$$

for all $A \in L_*(\mathbb{R}^q)$, where λ_q denotes q-dimensional Lebesgue measure on $\mathbb{R}^q$.

4 Falconer Dimension

Let $\mathcal{S}$ be an affine scheme. Fix $t \geq 0$. Define the metric $\rho_F^{(t)}$ on $E^{\mathbb{N}}$ as follows.

$$\rho_F^{(t)}(\omega, \tau) := \begin{cases} \alpha^t\left(A_{\omega\wedge\tau}\right) & \text{if } \quad \omega \neq \tau \\ 0 & \text{if } \quad \omega = \tau. \end{cases}$$

To check that $\rho_F^{(t)}$ is a metric indeed only triangle inequality requires an argument. For this take also $\gamma \in E^{\mathbb{N}}$. Then $|\omega \wedge \tau| \geq \min\{|\omega \wedge \gamma|, |\tau \wedge \gamma|\}$. Assume without loss of generality that $\omega \wedge \tau \geq |\omega \wedge \gamma|$. Then $\omega \wedge \tau = (\omega \wedge \gamma)\theta$ with some $\theta \in E^*$, say $\theta \in E^k$. Denote $n := |\omega \wedge \tau|$. We then have

$$\begin{aligned}
\rho_F^{(t)}(\omega, \tau) &= \alpha^t\left(A_{\omega\wedge\tau}\right) = \alpha^t\left(A_{(\omega\wedge\gamma)\theta}\right) \leq \alpha^t\left(A_{\omega\wedge\gamma}\right)\alpha^t\left(A_{\theta_1}^{(n+1)}A_{\theta_2}^{(n+1)}\dots A_{\theta_k}^{(n+1)}\right) \\
&\leq \alpha^t\left(A_{\omega\wedge\gamma}\right) = \rho_F^{(t)}(\omega, \gamma) \\
&\leq \max\{\rho_F^{(t)}(\omega, \gamma), \rho_F^{(t)}(\gamma, \tau)\}.
\end{aligned}$$

So, $\rho_F^{(t)}$ is a metric indeed, in fact we have proved the following.

Proposition 4.1. *For every $t \geq 0$, $\rho_F^{(t)}$ is an ultra-metric on $E^{\mathbb{N}}$.*

Let H_F^t be the 1-dimensional Hausdorff measure on $E^{\mathbb{N}}$ generated by the metric $\rho_F^{(t)}$. Of course if $s < t$ and $\mathrm{H}_F^s(E^{\mathbb{N}}) < +\infty$, then $\mathrm{H}_F^t(E^{\mathbb{N}}) = 0$. Therefore,

$$\inf\{t \geq 0 : \mathrm{H}_F^t(E^{\mathbb{N}}) = 0\} = \sup\{t \geq 0 : \mathrm{H}_F^t(E^{\mathbb{N}}) = +\infty\}.$$

Call this common number the Falconer dimension of the scheme $\mathcal{S}$ and denote it by $\mathrm{FD}(\mathcal{S})$. Note that it in fact depends only on the sequence $\left(\{A_e^{(n)} : e \in E\}\right)_{n=1}^{\infty}$ and is entirely independent of the vectors a_e, $e \in E$, or the maps $S_x : G^E \to G^E$.

We now define an auxiliary dimension $\mathrm{FD}_*(\mathcal{S})$. For every $l \geq 1$ and every set $\Gamma \subset E^{\mathbb{N}}$ define

$$\mathrm{F}_l^t(\Gamma) := \inf\left\{\sum_{\omega \in \mathcal{A}_l} \alpha^t(A_\omega)\right\},$$

where the infimum is taken over the family $\mathcal{A}_l$ of all countable covers of Γ by cylinders $[\omega]$ of length $\geq l$. The sequence $\left(\mathrm{F}_l^t(\Gamma)\right)_{l=1}^{\infty}$ is monotone increasing, and therefore the following limit

$$\mathrm{F}^t(\Gamma) = \lim_{l\to\infty} F_l^t(\Gamma)$$

exists and is equal to

$$\sup\{\mathrm{F}_l^t(\Gamma) : l \geq 1\}.$$

Note that if $s < t$ and $\mathrm{F}^s(E^{\mathbb{N}}) < +\infty$, then $\mathrm{F}^t(E^{\mathbb{N}}) = 0$. Therefore,

$$\inf\{t \geq 0 : \mathrm{F}^t(E^{\mathbb{N}}) = 0\} = \sup\{t \geq 0 : \mathrm{F}^t(E^{\mathbb{N}}) = +\infty\}$$

Denote this common number by $\mathrm{FD}_*(\mathcal{S})$. Note that as in the case of $\mathrm{FD}(\mathcal{S})$ it in fact depends only on the sequence $\left(\{A_e^{(n)} : e \in E\}\right)_{n=1}^{\infty}$ and is entirely independent of the vectors a_e, $e \in E$, or the maps $S_x : G^E \to G^E$. We shall prove the following.

Proposition 4.2. *If* $\mathcal{S}$ *an affine scheme and*

$$\varliminf_{e\to\infty} \left\|A_e^{(n)}\right\| > 0$$

for all $n \geq 1$, *then*

$$\mathrm{FD}_*(\mathcal{S}) = \mathrm{FD}(\mathcal{S}).$$

Proof. Obviously,

$$\mathrm{FD}(E^{\mathbb{N}}) \leq \mathrm{FD}_*(E^{\mathbb{N}}).$$

In order to prove the opposite inequality fix $\delta > 0$ and consider $\mathcal{A}$, an arbitrary cover of $E^{\mathbb{N}}$ by sets of diameters (with respect to $\rho_F^{(t)}$) $\leq \delta$. For every $\Gamma \in \mathcal{A}$ let $\omega_\gamma \in E_A^*$ be a longest word such that

$$\Gamma \subset [\omega_\Gamma].$$

Then of course

$$\mathrm{diam}_{\rho_F^{(t)}}(\Gamma) \leq \mathrm{diam}_{\rho_F^{(t)}}\big([\omega_\Gamma]\big) \tag{3}$$

but, more importantly for us at the moment, there exist two elements $\beta, \gamma \in \Gamma$ such that $\beta|_{|[\omega_\Gamma]|+1} \neq \gamma|_{|[\omega_\Gamma]|+1}$. As also $\beta|_{|[\omega_\Gamma]|} = \gamma|_{|[\omega_\Gamma]|}$, we thus get

$$\mathrm{diam}_{\rho_F^{(t)}}(\Gamma) \geq \rho_F^{(t)}(\beta, \gamma) = \alpha^t\left(A_{\beta\wedge\gamma}\right) = \alpha^t\left(A_{\omega_\Gamma}\right) = \mathrm{diam}_{\rho_F^{(t)}}\big([\omega_\Gamma]\big).$$

Along with (3) this yields

$$\mathrm{diam}_{\rho_F^{(t)}}\big([\omega_\Gamma]\big) = \mathrm{diam}_{\rho_F^{(t)}}(\Gamma) \tag{4}$$

Hence $\{[\omega_\Gamma]\}_{\Gamma\in\mathcal{A}}$ is also a cover of $E^{\mathbb{N}}$ by sets with diameter (with respect to the metric $\rho_F^{(t)}$) $\leq \delta$. Therefore, we are done since, by our hypothesis, $\sup\{|\omega_\Gamma| : \gamma \in \mathcal{A}\}$ converges to zero if $\delta \to 0$. □

As an immediate consequence of this proposition we get the following.

Corollary 4.3. *If* $\mathcal{S}$ *is a finitely autonomous affine scheme, then*

$$\mathrm{FD}_*(\mathcal{S}) = \mathrm{FD}(\mathcal{S}).$$

We also define

$$\underline{\mathrm{P}}_{\mathcal{S}}(t) := \varliminf_{n\to\infty} \frac{1}{n} \log \sum_{|\omega|=n} \alpha^t(A_\omega),$$

and call $\underline{\mathrm{P}}_{\mathcal{S}}(t)$ the lower topological pressure of the affine scheme $\mathcal{S}$ at the parameter t. Let

$$\theta_{\mathcal{S}}^- := \inf\{t \geq 0 : \underline{\mathrm{P}}_{\mathcal{S}}(t) < +\infty\}$$

and

$$\theta_{\mathcal{S}}^+ := \inf\{t \geq 0 : \underline{\mathrm{P}}_{\mathcal{S}}(t) = -\infty\}.$$

Since for $0 \leq s < t$, we have $\alpha^t(A_\omega) \leq \alpha_1^{t-s}(A_\omega)\alpha^s(A_\omega) \leq \kappa^{t-s}\alpha^s(A_\omega)$, we immediately get the following.

Proposition 4.4. *If $\mathcal{S}$ is an affine scheme, then*

(a) *the function $[0,+\infty) \ni t \mapsto \underline{\mathrm{P}}_{\mathcal{S}}(t) \in [-\infty,+\infty]$ is monotone decreasing,*
(b) *the function $(\theta_{\mathcal{S}}^-, \theta_{\mathcal{S}}^+) \ni t \mapsto \underline{\mathrm{P}}_{\mathcal{S}}(t)$ is strictly decreasing.*

Proposition 4.5. *If $\mathcal{S}$ is a finitely autonomous affine scheme, then the following numbers are equal.*

(a) $\mathrm{FD}(\mathcal{S})$,
(b) $\mathrm{FD}_*(\mathcal{S})$
(c) $\inf\{t \geq 0 : \underline{\mathrm{P}}_{\mathcal{S}}(t) \leq 0\}$,
(d) $\inf\{t \geq 0 : \sum_{\omega\in E^*} \alpha^t(A_\omega) < +\infty\} = \sup\{t \geq 0 : \sum_{\omega\in E^*} \alpha^t(A_\omega) = +\infty\}$.

Proof. Because of Corollary 4.3 it suffices to prove that the numbers in (b), (c), and (d) are all equal. Indeed, if $s < t$ and $\sum_{\omega\in E^*} \alpha^s(A_\omega) < +\infty$, then $\inf\{t \geq 0 : \sum_{\omega\in E^*} \alpha^t(A_\omega) < +\infty\}$. Therefore, the equality in (d) is proved. The equality of numbers in (c) and (d) is a direct consequence of Proposition 4.4(b). Now, if $\Gamma :=: \sum_{\omega\in E^*} \alpha^t(A_\omega) < +\infty$, then for every $l \geq 1$, $\sum_{\omega\in E^l} \alpha^t(A_\omega) \leq \Gamma$, and therefore $F_l^t(E^{\mathbb{N}}) \leq \Gamma$. Consequently, $F^t(E^{\mathbb{N}}) \leq \Gamma < +\infty$, and so (b)≤(d). The implication (c)≤(b) requires the system $\mathcal{S}$ to be finitely autonomous and is established in [1].

□

The proof of the following lemma is an adaptation of the proof of Lemma 3 in [3].

Lemma 4.6. *If $\mathcal{S}$ is an affine scheme and $F^t(E^{\mathbb{N}}) = +\infty$, then there exist a finite Borel measure ν on $E^{\mathbb{N}}$ and a constant $C > 0$ such that*

$$\nu([\omega]) \leq C\alpha^t(A_\omega)$$

for all $\omega \in E^*$.

Proof. Because of Proposition 4.1 it follows from Theorem 57(c) in [5] that there exists a compact set $\Gamma \subset E^{\mathbb{N}}$ such that

$$0 < \mathrm{H}^t_F(\Gamma) < +\infty.$$

Since $\operatorname{diam}_{\rho_F^{(t)}}([\omega]) = \alpha^t(A_\omega)$, the proof is thus completed by invoking Theorem 8.17 in [4]. □

5 Main Theorem: The Proof

The proof of our main theorem, Theorem 1.1 will consist of several lemmas. We start with the following.

Lemma 5.1. *Let $\mathcal{S}$ be an affine scheme acting on $\mathbb{R}^q$. Let $0 < t < q$ be a non-integral number. Then there exists a constant $C \in (0, +\infty)$ such that*

$$\int_X \int_{G^E} \frac{d\lambda_G^E(a)\, dm(x)}{||\pi_{(x,a)}(\omega) - \pi_{(x,a)}(\tau)||^t} \leq \frac{C}{\alpha^t\big(A_{\omega\wedge\tau}\big)}$$

for all $\omega, \tau \in E^{\mathbb{N}}$ with $\omega \neq \tau$.

Proof. Let

$$\rho := \omega \wedge \tau$$

and let $k := |\omega \wedge \tau| < +\infty$. Let $\omega' := \sigma^k(\omega)$ and $\tau' := \sigma^k(\tau)$. Then

$$\begin{aligned}
I(\omega, \tau) &:= \int_X \int_{G^E} \frac{d\mu_x(a)\, dm(x)}{||\pi_{(x,a)}(\omega) - \pi_{(x,a)}(\tau)||^t} \\
&= \int_X \int_{G^E} \frac{d\mu_x(a)\, dm(x)}{\left\|\phi_\rho^{(a)}\big(\pi^{(k+1)}_{\theta^k(x), S_x^k(a)}(\omega')\big) - \phi_\rho^{(a)}\big(\pi^{(k+1)}_{\theta^k(x), S_x^k(a)}(\tau')\big)\right\|^t} \\
&= \int_X \int_{G^E} \frac{d\mu_{\theta^{-k}(x)}(a)\, dm(x)}{\left\|\phi_\rho^{(S_x^{-k}(a))}\big(\pi^{(k+1)}_{(x,a)}(\omega')\big) - \phi_\rho^{(S_x^{-k}(a))}\big(\pi^{(k+1)}_{(x,a)}(\tau')\big)\right\|^t} \qquad (5)\\
&= \int_X \int_{G^E} \frac{d\mu_{\theta^{-k}(x)}(a)\, dm(x)}{\left\|A_\rho\big((\pi^{(k+1)}_{(x,a)}(\omega')\big) - \pi^{(k+1)}_{(x,a)}(\tau')\big)\right\|^t} \\
&\asymp \int_X \int_{G^E} \frac{d\mu_x(a)\, dm(x)}{\left\|A_\rho\big((\pi^{(k+1)}_{(x,a)}(\omega')\big) - \pi^{(k+1)}_{(x,a)}(\tau')\big)\right\|^t}.
\end{aligned}$$

Now,

$$\pi_{(x,a)}^{(k+1)}(\omega')) - \pi_{(x,a)}^{(k+1)}(\tau') = a_{\omega_1'} - a_{\tau_1'} + F(a),$$

where $F : G^E \to \mathbb{R}^q$ is given by the formula:

$$F(a) := \sum_{j=1}^{\infty} A_{\omega'|j^{(k+1)}}\big((S_x^j(a))_{\omega'_{j+1}}\big) - \sum_{j=1}^{\infty} A_{\tau'|j^{(k+1)}}\big((S_x^j(a))_{\tau'_{j+1}}\big). \tag{6}$$

Now consider the product measure

$$\ell_{\omega_1'} := \lambda_q \otimes \prod_{e\in E\setminus\{\omega_1'\}} = \lambda_q \otimes \lambda_G^{E\setminus\{\omega_1'\}}$$

on $\mathbb{R}^q$, where, we recall, λ_q is the q-dimensional Lebesgue measure on $\mathbb{R}^q$. Let $H : \hat{G}^E \to \mathbb{R}^q \times G^{E\setminus\{\omega_1'\}}$ be given by the following formula:

$$H(a)_j := \begin{cases} a_{\omega_1'} - a_{\tau_1'} + F(a) & \text{if } j = \omega_1' \\ a_j & \text{if } j \neq \omega_1'. \end{cases} \tag{7}$$

We shall prove the following.

Claim 1. The map $H : \hat{G}^E \to \mathbb{R}^q \times G^{E\setminus\{\omega_1'\}}$ is injective.

Proof. Suppose that $H(a') = H(a)$. Then immediately $a'_e = a_e$ for all $e \in E \setminus \{\omega_1'\}$. Since $\tau_1' \neq \omega_1'$, this entails $a'_{\tau_1'} = a_{\tau_1'}$. So,

$$F(a') - F(a) = a_{\omega_1'} - a'_{\omega_1'}.$$

It then follows from (6), (p4), linearity of the maps $A_{\omega'|j^{(k+1)}}$ and $A_{\tau'|j^{(k+1)}}$, and Q-quasi-convexity of $\hat{G}$, that

$$\begin{aligned} &||a' - a||_\infty = \\ &= ||a'_{\omega_1'} - a_{\omega_1'}|| = ||F(a') - F(a)|| \\ &= \Big\| \sum_{j=1}^{\infty} A_{\omega'|j^{(k+1)}}\big((S_x^j(a'))_{\omega'_{j+1}}\big) - (S_x^j(a))_{\omega'_{j+1}}\big) - \end{aligned}$$

$$
\begin{aligned}
&- \sum_{j=1}^{\infty} A_{\tau'|j^{(k+1)}}\big((S_x^j(a'))_{\tau'_{j+1}}) - (S_x^j(a))_{\tau'_{j+1}}\big)\Big\| \\
&\le \sum_{j=1}^{\infty} \kappa^j ||(S_x^j(a'))_{\omega'_{j+1}}) - S_x^j(a))_{\omega'_{j+1}}|| + \sum_{j=1}^{\infty} \kappa^j ||(S_x^j(a'))_{\tau'_{j+1}}) - S_x^j(a))_{\tau'_{j+1}}|| \\
&\le 2\sum_{j=1}^{\infty} \kappa^j ||S_x^j(a') - S_x^j(a)||_\infty \\
&\le 2\sum_{j=1}^{\infty} \kappa^j \beta ||a' - a||_\infty \\
&= 2Q\kappa\beta(1-\kappa\beta)^{-1}||a' - a||_\infty \\
&< ||a' - a||_\infty,
\end{aligned}
\tag{8}
$$

where the last equality followed from the assumption (see (p4)) that $\kappa\beta < 1/3$. This contradiction finishes the proof of Claim 1. □

In the same vein let us prove now the existence and estimate the norm of the partial derivative $D_{\omega'_1} F(a)$ at every point $a \in G^E$. Indeed, it again follows from (6), (p4), and linearity of both $A_{\omega'|j^{(k+1)}}$ and $A_{\tau'|j^{(k+1)}}$, that

$$
\begin{aligned}
&||D_{\omega'_1} F(a)|| = \\
&= \Big\| \sum_{j=1}^{\infty} A_{\omega'|j^{(k+1)}} \circ D_{\omega'_1}(p_{\omega'_{j+1}} \circ S_x^j)(a) - \sum_{j=1}^{\infty} A_{\omega'|j^{(k+1)}} \circ D_{\omega'_1}(p_{\tau'_{j+1}} \circ S_x^j)(a) \Big\| \\
&\le 2\sum_{j=1}^{\infty} \kappa^j ||DS_x^j||_\infty \le 2\sum_{j=1}^{\infty} (\kappa\beta)^j \\
&= \frac{2\kappa\beta}{1-\kappa\beta},
\end{aligned}
\tag{9}
$$

i.e. $D_{\omega'_1} F(a)$ exists and (9) holds. So, because of the special form (7), we now conclude that the map $H : \hat{G}^E \to \mathbb{R}^q \times G^{E\setminus\{\omega'_1\}}$ is non-singular with respect to the measure $\ell_{\omega'_1}$, and its Jacobian is given by the formula

$$
J_H^*(a) = |\det(\mathrm{Id}_{\mathbb{R}^q} + D_{\omega'_1} F(a))| \ge (1 - ||D_{\omega'_1} F(a)||)^q \ge \left(1 - \frac{2\kappa\beta}{1-\kappa\beta}\right)^q.
$$

So, if we consider the measure $\ell_{\omega'_1}$ on $H(G^E)$ but the measure λ_G^E on G^E, then $J_{H^{-1}}(a)$, the corresponding Jacobian of the map $H^{-1} : H(G^E) \to G^E$ is

$$J_{H^{-1}}(b) = \frac{1}{\lambda_q(G)} J^*_{H^{-1}}(b) \le \gamma := \left(\lambda_q(G)\left(1 - \frac{2\kappa\beta}{1-\kappa\beta}\right)^q\right)^{-1}$$

for all $b \in H(G^E)$. Therefore, we can single out the inner integral in (5) to get

$$\begin{aligned} I_x(\omega,\tau) &:= \int_{H^{-1}(H(G^E))} \frac{d\lambda_G^E(a)}{\left\| A_\rho\left(\left(\pi_{(x,a)}^{(k+1)}(\omega')\right) - \pi_{(x,a)}^{(k+1)}(\tau')\right)\right\|^t} \\ &= \int_{H(G^E)} \frac{J_{H^{-1}}(b)}{||A_\rho((b)_{\omega_1'})||^t}\, d\ell_{\omega_1'}(b) \\ &\le \gamma \int_{H(G^E)} \frac{d\ell_{\omega_1'}(b)}{||A_\rho((b)_{\omega_1'})||^t} \\ &= \gamma \int_{p_*(H(G^E))} \int_{p_{\omega_1'}(p_*^{-1}(b))} \frac{d\lambda_q(y)}{||A_\rho(y)||^t}\, d\lambda_G^{E\setminus\{\omega_1'}(b), \end{aligned}$$

where $p_* : (\mathbb{R}^q)^E \to (\mathbb{R}^q)^{E\setminus\{\omega_1'\}}$ is the canonical projection onto $(\mathbb{R}^q)^{E\setminus\{\omega_1'\}}$, i.e. $p_*(b_e)_{e\in E} = (b_e)_{e\in E\setminus\{\omega_1'\}}$, and, we recall, $p_{\omega_1'} : (\mathbb{R}^q)^E \to \mathbb{R}^q$ is the canonical projection onto ω_1'th coordinate. Now, if $a \in G^E$, then

$$\begin{aligned} ||(H(a))_{\omega_1'}|| &= ||a_{\omega_1'} - a_{\tau_1'} + F(a)|| \le ||a_{\omega_1'}|| + ||a_{\tau_1'}|| + ||F(a)|| \\ &\le 2R_G + R_G\kappa(1-\kappa)^{-1} \\ &= (2 + \kappa(1-\kappa)^{-1})R_G, \end{aligned}$$

where the estimate $||F(a)|| \le R_G\kappa(1-\kappa)^{-1}$ is a simplification of the calculation from (8). Therefore, for every $b \in p_*(H(G^E))$, we have that $p_{\omega_1'}(p_*^{-1}(b)) \subset B(0, (2+\kappa(1-\kappa)^{-1})R_G)$. So, by virtue of Lemma 3.2, there exists a constant $C > 0$ such that

$$\begin{aligned} I_x(\omega,\tau) &\le \gamma \int_{p_*(H(G^E))} \int_{\overline{B}(0,(2+\kappa(1-\kappa)^{-1})R_G)} \frac{d\lambda_q(y)}{||A_\rho(y)||^t}\, d\lambda_G^{E\setminus\{\omega_1'\}}(b) \\ &\le \frac{C}{\alpha^t(A_\rho)} \int_{p_*(H(G^E))} d\lambda_G^{E\setminus\{\omega_1'\}} \\ &= \frac{C}{\alpha^t(A_\rho)} \lambda_G^{E\setminus\{\omega_1'\}}\big(p_*(H(G^E))\big) \\ &\le \frac{C}{\alpha^t(A_\rho)}. \end{aligned}$$

Therefore,

$$\begin{aligned}\int_X \int_{G^E} \frac{d\lambda_G^E(a)\,dm(x)}{||\pi_{(x,a)}(\omega) - \pi_{(x,a)}(\tau)||^t} &\asymp \int_X \int_{G^E} \frac{d\mu_x(a)\,dm(x)}{||\pi_{(x,a)}(\omega) - \pi_{(x,a)}(\tau)||^t} \\ &\asymp \int_X I_x(\omega,\tau)\,dm(x) \\ &\le \frac{C}{\alpha^t\left(A_{\omega\wedge\tau}\right)} \int_X dm(x) \\ &= \int_X I_x(\omega,\tau)\,dm(x).\end{aligned}$$

The proof of our lemma is complete. □

The proof of the following proposition goes, with almost no changes, as the proof of Proposition 5.1 in [1]

Proposition 5.2. *If $\mathcal{S}$ is an affine scheme and $H_F^t(E^{\mathbb{N}}) < +\infty$, then $H^t(J_{(x,a)}) < +\infty$ for all $x \in X$ and all $a \in G^E$.*

Proof. Begin in the same way as in the proof of Proposition 4.2. Fix $\delta > 0$ and consider $\mathcal{A}$, an arbitrary cover of $E^{\mathbb{N}}$ by sets of diameters (with respect to the metric $\rho_F^{(t)}$) $\le \delta$. For every $\Gamma \in \mathcal{A}$ let $\omega_\gamma \in E_A^*$ be a longest word such that

$$\Gamma \subset [\omega_\Gamma].$$

Then of course

$$\operatorname{diam}_{\rho_F^{(t)}}(\Gamma) \le \operatorname{diam}_{\rho_F^{(t)}}\left([\omega_\Gamma]\right) \tag{10}$$

but, more importantly for us at the moment, there exist two elements $\beta, \gamma \in \Gamma$ such that $\beta|_{|[\omega_\Gamma]|+1} \neq \gamma|_{|[\omega_\Gamma]|+1}$. As also $\beta|_{|[\omega_\Gamma]|} = \gamma|_{|[\omega_\Gamma]|}$, we thus get

$$\operatorname{diam}_{\rho_F^{(t)}}(\Gamma) \ge \rho_F^{(t)}(\beta,\gamma) = \alpha^t\left(A_{\beta\wedge\gamma}\right) = \alpha^t\left(A_{\omega_\Gamma}\right) = \operatorname{diam}_{\rho_F^{(t)}}\left([\omega_\Gamma]\right).$$

Along with (10) this yields

$$\operatorname{diam}_{\rho_F^{(t)}}\left([\omega_\Gamma]\right) = \operatorname{diam}_{\rho_F^{(t)}}(\Gamma) \tag{11}$$

Hence $\{[\omega_\Gamma]\}_{\Gamma\in\mathcal{A}}$ is also a cover of $E^{\mathbb{N}}$ by sets with diameter (with respect to the metric $\rho_F^{(t)}$) $\le \delta$. Therefore, for all $x \in X$ and all $a \in G^E$ we have that

$$J_{(x,a)} \subset \bigcup_{\Gamma\in\mathcal{A}} \phi_{\omega_\Gamma}^{(x,a)}(B).$$

But each set $\phi_{\omega\Gamma}^{(x,a)}(B)$ is contained in a rectangular box with sides of length

$$2\text{diam}(B)\alpha_1(A_\omega),\ 2\text{diam}(B)\alpha_2(A_\omega), \dots, 2\text{diam}(B)\alpha_q(A_\omega).$$

If k is the least integer greater than or equal to t, then each such box can be divided into at most

$$\left(4\text{diam}(B)\frac{\alpha_1(A_\omega)}{\alpha_k(A_\omega)}\right)\cdot\left(4\text{diam}(B)\frac{\alpha_2(A_\omega)}{\alpha_k(A_\omega)}\right)\cdot\ldots\cdot\left(4\text{diam}(B)\frac{\alpha_2(A_\omega)}{\alpha_k(A_\omega)}\right)\cdot \\ \cdot (4\text{diam}(B))^{q-k+1}$$

rectangular cubes with sides of length α_k, that is of diameter $\sqrt{q}\alpha_k$. Therefore, fixing $\eta > 0$, there exists, because of (1) and (11), $\delta_\eta > 0$ such that $\text{diam}\big(\phi_{\omega_\gamma}(B)\big) \leq \eta$ for all $\Gamma \in \mathcal{A}$. Hence,

$$\begin{aligned}
&\mathrm{H}_\eta^t\big(J_{(x,a)}\big) \leq \\
&\leq \sum_{\Gamma\in\mathcal{A}}\left(4\text{diam}(B)\frac{\alpha_1(A_\omega)}{\alpha_k(A_\omega)}\right)\cdot\left(4\text{diam}(B)\frac{\alpha_2(A_\omega)}{\alpha_k(A_\omega)}\right)\cdot\ldots\cdot\left(4\text{diam}(B)\frac{\alpha_2(A_\omega)}{\alpha_k(A_\omega)}\right)\cdot \\
&\qquad \cdot (4\text{diam}(B))^{q-k+1}(\sqrt{q}\alpha_k)^t \\
&\preceq \sum_{\Gamma\in\mathcal{A}} \alpha_1(A_{\omega\Gamma})\alpha_2(A_{\omega\Gamma})\ldots\alpha_{k-1}(A_{\omega\Gamma})\alpha_k^{t-(k-1)}(A_{\omega\Gamma}) \\
&\preceq \sum_{\Gamma\in\mathcal{A}} \alpha^t(A_{\omega\Gamma}).
\end{aligned}$$

Therefore,

$$\mathrm{H}^t\big(J_{(x,a)}\big) \preceq \mathrm{H}_F^t(E^{\mathbb{N}}).$$

So,

$$\mathrm{H}^t\big(J_{(x,a)}\big) = \lim_{\eta\to 0}\mathrm{H}_\eta^t\big(J_{(x,a)}\big) \preceq \lim_{\eta\to 0}\mathrm{H}_F^t(E^{\mathbb{N}}) < +\infty.$$

The proof is complete. □

Now we can prove the main theorem of our paper.

Theorem 5.3. *If $\mathcal{S}$ is an affine scheme on $\mathbb{R}^q$, then*

$$\text{HD}\big(J_{(x,a)}\big) = \min\{q, \text{FD}(\mathcal{S})\}.$$

for m-a.e. $x \in X$ and λ_G^E-a.e. $a \in G^E$.

Proof. Because of the previous proposition we only have to prove that

$$\mathrm{HD}\big(J_{(x,a)}\big) = \min\{q, \mathrm{FD}(\mathcal{S})\}.$$

for m-a.e. $x \in X$ and λ_G^E-a.e. $a \in G^E$. Indeed, fix a non-integral number $0 < s < \min\{q, \mathrm{FD}(\mathcal{S})\}$. Take then arbitrary $0 < s < t < \min\{q, \mathrm{FD}(\mathcal{S})\}$. So, $F^t(E^{\mathbb{N}}) = +\infty$, and, by Lemma 4.6, there is a finite Borel measure ν on $E^{\mathbb{N}}$ such that

$$\nu([\omega]) \le C\alpha^t\big(A_\omega\big) \tag{12}$$

for all $\omega \in E^*$. Applying Lemma 5.1, formula (12), and the observation that $\nu \otimes \nu$ does not charge the diagonal, we get

$$\begin{aligned}
I &:= \int_X \int_{E^{\mathbb{N}}} \int_{E^{\mathbb{N}}} \int_{G^E} \frac{dm(x)\, d\lambda_G^E(a)\, d\nu(\omega)\, d\nu(\tau)}{||\pi_{(x,a)}(\omega) - \pi_{(x,a)}(\tau)||^s} \le \int_{E^{\mathbb{N}}} \int_{E^{\mathbb{N}}} \frac{d\nu(\omega)\, d\nu(\tau)}{\alpha^s\big(A_{\omega\wedge\tau}\big)} \\
&\preceq \sum_{n=0}^{\infty} \sum_{|\gamma|=n} \iint_{\substack{\omega,\tau\in E^{\mathbb{N}} \\ \omega\wedge\tau=\gamma}} \alpha^s\big(A_\gamma\big)^{-1} d\nu(\omega)\, d\nu(\tau) \\
&= \sum_{n=0}^{\infty} \sum_{|\gamma|=n} \alpha^s\big(A_\gamma\big)^{-1} \nu \otimes \nu(A_\gamma) \\
&\le \sum_{n=0}^{\infty} \sum_{|\gamma|=n} \alpha^s\big(A_\gamma\big)^{-1} \nu^2([\gamma]) \\
&\le \sum_{n=0}^{\infty} \sum_{|\gamma|=n} \alpha^s\big(A_\gamma\big)^{-1} \alpha^t\big(A_\gamma\big) \nu([\gamma]).
\end{aligned} \tag{13}$$

Now, with k being the least integer greater than or equal to s and l being the least integer greater than or equal to t, we get

$$\begin{aligned}
\alpha^t\big(A_\gamma\big)\alpha^s\big(A_\gamma\big)^{-1} &= \alpha_1(A_\gamma)\alpha_2(A_\gamma)\alpha_{k-1}(A_\gamma)\ldots\alpha_k(A_\gamma)\alpha_{k+1}(A_\gamma)\ldots\alpha_{l-1}(A_\gamma)\alpha_l(A_\gamma)^{t-l+1} \\
&\qquad \alpha_1(A_\gamma)^{-1}\ldots\alpha_{k-1}^{-1}(A_\gamma)\alpha_k(A_\gamma)^{-s+k-1} \\
&= \alpha_k(A_\gamma)^{k-s}\alpha_{k+1}(A_\gamma)\ldots\alpha_{l-1}(A_\gamma)\alpha_l(A_\gamma)^{t-l+1}.
\end{aligned}$$

Since $t - l + 1 \ge 0$ and since $k - s > 0$, we further get

$$\alpha^t\big(A_\gamma\big)\alpha^s\big(A_\gamma\big)^{-1} \le \alpha_k(A_\gamma)^{k-s} \le ||A_\gamma||^{k-s} \le \kappa^{(k-s)|\gamma|}.$$

Hence, we can continue (13) as follows.

$$I \leq C \sum_{n=0}^{\infty} \kappa^{(k-s)n} \sum_{|\gamma|=n} \nu([\gamma]) = C \sum_{n=0}^{\infty} \kappa^{(k-s)n} = C(1-\kappa^{k-s})^{-1} < +\infty.$$

Hence, for m-a.e. $x \in X$ and λ_G^E-a.e. $a \in G^E$, we have that

$$I_{(x,a)} := \int_{E^{\mathbb{N}}} \int_{E^{\mathbb{N}}} \frac{d\nu(\omega)\, d\nu(\tau)}{||\pi_{(x,a)}(\omega) - \pi_{(x,a)}(\tau)||^s} < +\infty.$$

This means that

$$\int_{J_{(x,a)}} \int_{J_{(x,a)}} \frac{d(\nu \circ \pi_{(x,a)}^{-1})(z)\, d(\nu \circ \pi_{(x,a)}^{-1})(\xi)}{||z-\xi||^s} < +\infty,$$

and this in turn (see [2], comp. [4]) implies that $\mathrm{HD}(J_{(x,a)}) \geq s$. Thus, $\mathrm{HD}(J_{(x,a)}) \geq \min\{q, \mathrm{FD}(\mathcal{S})\}$, and the proof is finished. □

References

1. Falconer, K.: The Hausdorff dimension of self-affine fractals. Math. Proc. Camb. Philos. Soc. **103**, 339–350 (1988)
2. Falconer, K.: The Geometry of Fractal Sets. Cambridge University Press, Cambridge (1990)
3. Jordan, T., Pollicott, M., Simon, K.: Hausdorff dimension for randomly perturbed self affine attractors. Commun. Math. Phys. **270**, 519–544 (2007)
4. Mattila, P.: Geometry of Sets and Measures in Euclidean Spaces: Fractals and Rectifiability. Cambridge Studies in Advanced Mathematics, vol. 44. Cambridge University Press, Cambridge (1995)
5. Rogers, C. A.: Hausdorff Measures, Cambridge University Press, Cambridge (1998)

On Transverse Hyperplanes to Self-similar Jordan Arcs

Andrey Tetenov

Abstract We consider self-similar Jordan arcs γ in $\mathbb{R}^d$, different from a line segment and show that they cannot be projected to a line bijectively. Moreover, we show that the set of points $x \in \gamma$, for which there is a hyperplane, intersecting γ at the point x only, is nowhere dense in γ.

Keywords Self-similar set • Zipper • Projection • Rigidity theorems • Transverse foliation

MSC classification: Primary 28A80

1 Introduction

The first examples of self-similar fractals which appeared in the beginning of twentieth century were the constructions of self-similar curves with predefined geometrical properties [11, 13]. Though the study of geometrical properties of self-similar curves is so close to historical origins of fractal geometry, some of their elementary geometric properties were established only in recent times.

For example, it was a common opinion that self-similar curves have no tangent at any of their points. But in 2005 Kravchenko [7] found that there are self-affine curves which are differentiable everywhere and therefore have a tangent at any of their points. In 2011 the problem of differentiability for self-affine curves with two generators and the problem of existence of tangent subspaces for self-similar sets found their exhaustive solution in the paper of Bandt and Kravchenko [3].

A. Tetenov (✉)
Gorno-Altaisk State University, 649000 Gorno-Altaisk, Russia
e-mail: atet@mail.ru

C. Bandt et al. (eds.), *Fractals, Wavelets, and their Applications*, Springer Proceedings in Mathematics & Statistics 92, DOI 10.1007/978-3-319-08105-2_8,

In this note we study the projections of self-similar Jordan arcs in $\mathbb{R}^d$ to the real line along families of parallel hyperplanes. Analysing the case when there is a bijective projection of a self-similar Jordan arc γ to a straight line segment, we show that this is possible only when the arc is a straight line segment itself.

Theorem 1. *Let γ be a self-similar Jordan arc in $\mathbb{R}^d$. Suppose there is such hyperplane σ, that for any $x \in \gamma$ the parallel copy of σ passing through the point x intersects γ only once, then γ is a straight line segment.*

Really, we prove a much more general statement, in which transverse hyperplanes $\sigma(x)$ at different points x of γ need not be parallel to each other, and transversality is understood in the sense of Definition 6:

Theorem 2. *Let γ be a self-similar Jordan arc in $\mathbb{R}^d$. Suppose there is a dense subset $D \subset \gamma$ such that for any $x \in D$ there is a hyperplane σ, which is weakly transverse to γ at the point x, then γ is a straight line segment.*

The proof is based on a simple and almost obvious observation (Theorem 5) that the invariant set of a multizipper of similarity dimension 1 is always a collection of straight line segments. We prove it in Sect. 3.

The author is thankful to V.V. Aseev for numerous fruitful discussions of the topic.

2 Preliminaries

We give some definitions needed in current paper. Some of them are slightly different from generally accepted ones, but they are best fit for our further argument.

2.1 Self-similar Arcs

A contraction similarity S in $\mathbb{R}^d$ is a map of the form $S(x) = q \cdot O(x - x_0) + x_0$, where x_0 is the fixed point of S, $q \in (0, 1)$ is the contraction ratio, and O is the orthogonal transformation called the *orthogonal part* of S.

Let $\mathcal{S} = \{S_1, \ldots, S_m\}$ be a system of contraction similarities in $\mathbb{R}^d$. A compact set K is called the *invariant set* or the *attractor* of the system $\mathcal{S}$, if $K = \bigcup\limits_{i=1}^{m} S_i(K)$. If this invariant set is an arc γ we call γ a *self-similar arc* defined by the system $\mathcal{S}$.

We denote the semigroup generated by $S_1, \ldots, S_m$, by $G(\mathcal{S})$.

2.2 Directed Multigraphs

A *directed multigraph* (or *digraph*) Γ is defined by a set of *vertices* $V(\Gamma)$, a set of *edges* $E(\Gamma)$ and maps $\alpha, \omega : E(\Gamma) \to V(\Gamma)$. Here $\alpha(e)$ is the *beginning* of the edge e and $\omega(e)$ is its *end*.

By E_{uv} we denote the set of all edges $e \in E$ for which $\alpha(e) = u, \omega(e) = v$, and by $E_u = \bigcup\limits_{v \in V} E_{uv}$—the set of all edges with the starting point at u.

To make the further argument more convenient, the set V will be supposed to be equal to $\{1, 2, \dots, n\}$, where $n = \#V$. In this case $u \in V$ means the same as $1 \le u \le n$. We also denote the numbers $\#E_{uv}$ by m_{uv} and $\#E_u$ by m_u.

A *path* σ from a vertex $\alpha(e_1) = u$ to $\omega(e_n) = v$ in a digraph Γ is a sequence of edges $\sigma = e_1 e_2 \dots e_n$, with $\omega(e_i) = \alpha(e_{i+1})$ for every $1 \le i \le n-1$. The set of all paths σ of the length n with the beginning u and the end v is denoted by $E_{uv}^{(n)}$ and $E_{uv}^{(*)} = \bigcup\limits_{n=1}^{\infty} E_{uv}^{(n)}$ is the set of all paths from u to v.

A digraph Γ is *strongly connected* if for every two vertices u and v it has a path from u to v.

2.3 Graph-Directed Systems of Contraction Similarities

A *graph-directed system of contraction similarities* $\mathcal{S}$ with *structural graph* $\Gamma = \langle V, E, \alpha, \omega \rangle$ is a finite collection of metric spaces $\{X_v\}_{v \in V}$, together with a collection of contraction similarities $\{S_e : X_{\omega(e)} \to X_{\alpha(e)}\}_{e \in E}$.

We denote the contraction ratios of the similarities by $q_e = Lip(S_e)$.

Throughout this paper all the spaces X_u will be different copies of the same space $\mathbb{R}^d$ for certain d.

A graph-directed system of similarities $\mathcal{S}$ is called *regular*, if its structural graph Γ is strongly connected.

A finite collection of compact subsets $\{K_v\}_{v \in V}$ is called the *invariant set*, or the *attractor* of the system $\mathcal{S}$, if for every $v \in V$

$$K_u = \bigcup_{\alpha(e)=u} S_e(K_{\omega(e)}). \tag{1}$$

The sets $\{K_u\}_{u \in V}$ are called the *components of the attractor* of the system $\mathcal{S}$.

We use the following definition of a similarity dimension of graph-directed system of similarities [4, 10]:

Definition 3. Let $\mathcal{S}$ be a regular graph-directed system of similarities with a structure graph $\Gamma = \langle V, E, \alpha, \omega \rangle$. For each positive real number s, let $\mathbf{B(s)}$ be the matrix (with rows and columns indexed by V) with entry $\mathbf{B_{uv}}(s) = \sum\limits_{e \in E_{uv}} q_e^s$ in row

u column v. Let $\Phi(s) = r(\mathbf{B}(s))$ be the spectral radius of $\mathbf{B}(s)$. The unique solution $s_1 \geq 0$ of $\Phi(s) = 1$ is the similarity dimension of the system $\mathcal{S}$.

3 Multizippers of Similarity Dimension 1

A method of construction of self-similar curves, used by many authors [6, 9, 11] was studied in 2002 by Aseev [1] as a zipper construction. This construction proved to be an efficient tool in the investigation of geometrical properties of self-similar curves and continua [2]. Its graph-directed version was introduced by the author in 2006 and was called a multizipper construction; it gives a complete description of self-similar Jordan arcs in $\mathbb{R}^d$ [12, Theorem 4.1]:

Theorem 4. *Let $\mathcal{S}$ be a regular graph directed system of similarities in $\mathbb{R}^d$ with Jordan attractor $\vec{\gamma}$. If one of the components γ_u of the attractor $\vec{\gamma}$ is different from a straight line segment, then there is a multizipper $\mathcal{Z}$ such that the set of the components of the attractor of $\mathcal{Z}$ contains each of the arcs γ_u.*

Consider a graph-directed system $\mathcal{Z}$ of similarities with structural graph Γ, which satisfies the following conditions:

MZ1. In each of the spaces $X_u, u \in V$, a chain of points $\{z_0^{(u)}, \ldots, z_{m_u}^{(u)}\}$, is specified. These chains are defined in such a way that

$$\|z_i^{(u)} - z_{i-1}^{(u)}\| < \|z_{m_v}^{(v)} - z_0^{(v)}\| \text{ for any } u, v \in V, i = 1, \ldots, m_u.$$

MZ2. There is a bijection ϵ from the set of all pairs $\{(u, i), u \in V, 1 \leq i \leq m_u\}$ to the set E.

MZ3. For any pair (u, i), the map S_e, corresponding to the edge $e = \epsilon(u, i)$ with $v = \omega(e)$, sends two-point set $\{z_0^{(v)}, z_{m_v}^{(v)}\}$ to the set $\{z_{i-1}^{(u)}, z_i^{(u)}\}$.

The graph-directed system $\mathcal{Z}$, satisfying the conditions **MZ1–MZ3** is called a *multizipper with structural graph Γ and node points $z_i^{(u)}$.*

Let $L^{(u)}$ be the polygonal line specified by the sequence $\{z_0^{(u)}, z_1^{(u)}, \ldots, z_{m_u}^{(u)}\}$ of the nodes of the multizipper $\mathcal{Z}$. Denote the distance $||z_{m_u}^{(u)} - z_0^{(u)}||$ by l_u. Observe that if $S_e(\{z_0^{(v)}, z_{m_v}^{(v)}\}) = \{z_{i-1}^{(u)}, z_i^{(u)}\}$, then $||z_{i-1}^{(u)} - z_i^{(u)}|| = q_e l_v$. So, the length of the polygonal line $L^{(u)}$ is equal to $\sum_{v=1}^{n} \sum_{e \in E_{uv}} q_e l_v$.

Theorem 5. *Let $\mathcal{Z}$ be a regular self-similar multizipper whose similarity dimension is 1. Then all the components $\gamma^{(u)}$ of its invariant set are line segments.*

Proof. Suppose there is a component $\gamma^{(u)}$ of the attractor of $\mathcal{Z}$, which is not a line segment. Since $\mathcal{Z}$ is regular, for any $v \in V$ there is a path $\sigma = e_1 \dots e_k \in E_{vu}^{(*)}$, so the similarity $S_\sigma = S_{e_1} \cdot \dots \cdot S_{e_k}$ maps the arc $\gamma^{(u)}$ to a subarc of $\gamma^{(v)}$. Therefore, each $\gamma^{(v)}$ is also different from a straight line.

Then, choosing appropriate refinement of the multizipper $\mathcal{Z}$, we may suppose that all the polygonal lines $L^{(u)}$ are different from a straight line. For each component $\gamma^{(u)}$ we have:

$$\gamma^{(u)} = \bigcup_{v=1}^{n} \bigcup_{e \in E_{uv}} S_e(\gamma^{(v)}).$$

The similarity dimension of the multizipper $\mathcal{Z}$ is equal to such value of a parameter s, that the spectral radius of the matrix $\mathbf{B(s)}$ whose entries are $B_{uv}(s) = \sum\limits_{e \in E_{uv}} q_e^s$, is equal to 1.

So, the spectral radius of the matrix $\mathbf{B(1)}$ with entries $B_{uv}(1) = \sum\limits_{e \in E_{uv}} q_e$ is equal to 1.

Since all the polygonal lines $L^{(u)}$ are not straight lines, they obey the inequality

$$l_u < \sum_{v=1}^{n} \sum_{e \in E_{uv}} q_e l_v = (\mathbf{B(1)}\vec{l})_u.$$

Therefore, for a vector $\vec{l} = (l_1, \dots, l_n)$ and for the matrix $\mathbf{B(1)}$ we have the inequality

$$\min_{1 \le u \le n} \frac{(\mathbf{B}\vec{l})_u}{l_u} > 1.$$

The structural graph of the system $\mathcal{Z}$ is strongly connected. Then the matrix $\mathbf{B(1)}$ is a positive irreducible matrix. According to [5, Remark 4, Sect. 2, Chap. XIII] its spectral radius is equal to

$$r = \max_{\vec{l} \neq 0} \min_{1 \le u \le n} \frac{(\mathbf{B}\vec{l})_u}{l_u}.$$

So, if $r = 1$, then for any $\vec{l}$, there is such u, that $\dfrac{(\mathbf{B}\vec{l})_u}{l_u} \le 1$.

The contradiction shows that all $L^{(u)}$ are straight line segments, so all $\gamma^{(u)}$ are straight line segments too.■

4 Theorem on Transverse Hyperplanes

4.1 *Jordan Arcs and Transverse Hyperplanes*

Let $\gamma : [0,1] \to \mathbb{R}^d$ be a Jordan arc in $\mathbb{R}^d$. For any point $x = \gamma(t)$ we define the half-open subarcs $\gamma_x^+ = \gamma((t,1])$ and $\gamma_x^- = \gamma([0,t))$.

Let $x, y \in \gamma$, and $y \in \bar{\gamma}_x^+$. We denote the open subarc $\gamma_x^+ \cap \gamma_y^-$ by (x, y) and $\bar{\gamma}_x^+ \cap \bar{\gamma}_y^-$ by $[x, y]$.

A hyperplane containing the origin 0 is denoted by σ, while $V^+(\sigma)$ and $V^-(\sigma)$ are open half-spaces, defined by σ. A hyperplane parallel to σ and containing x is denoted by $\sigma(x)$ or $\sigma + x$. The open half-spaces defined by $\sigma(x)$ are denoted by $V^+(\sigma, x)$ and $V^-(\sigma, x)$ or $V^+(\sigma) + x$ and $V^-(\sigma) + x$.

Definition 6. We say a hyperplane σ is *weakly transverse* to the arc γ at the point x, if $\gamma_x^+ \subset \bar{V}^+(\sigma, x)$, $\gamma_x^- \subset \bar{V}^-(\sigma, x)$.

We say a hyperplane σ is *transverse* to the arc γ at the point x, if $\gamma_x^+ \subset V^+(\sigma, x)$, $\gamma_x^- \subset V^-(\sigma, x)$.

4.2 *The Cones Q^+ and Q^-*

By $Q^+(x, y)$ (respectively, $Q^-(x, y)$) we denote the intersection of all closed half-spaces $\bar{V}^+(\sigma, z)$ (resp. $\bar{V}^-(\sigma, z)$) corresponding to the hyperplanes $\sigma(z)$, weakly transverse to γ at the points $z \in [x, y]$. These sets are convex and closed and they satisfy the relations

$$\gamma^+(y) \subset Q^+(x, y) \quad \text{and } \gamma^-(x) \subset Q^-(x, y).$$

Taking $x = y$ we come to the sets $Q^+(x)$ ($Q^-(x)$) which are the intersections of all closed half-spaces $\bar{V}^+(\sigma, x)$ ($\bar{V}^-(\sigma, x)$) corresponding to hyperplanes $\sigma(x)$, weakly transverse to γ at the point x. We can also consider the set $Q^+(x) \cup Q^-(x)$ as the intersection of all unions $Q_i^+ \cup Q_i^-$ of pairs of convex closed cones symmetric with respect to x which satisfy relations $\gamma^+(x) \subset Q_i^+$ and $\gamma^-(x) \subset Q_i^-$.

Lemma 7. *Let γ be a Jordan arc in $\mathbb{R}^n$. Suppose a sequence of points $x_n \in \gamma$ converges to a point x_0, while a sequence of hyperplanes σ_n, weakly transverse to γ at points x_n, converges to a hyperplane σ_0. Then σ_0 is weakly transverse to γ at the point x_0.*

Proof. For any n, $\bar{\gamma}_{x_n}^+ \subset V^+(\sigma_n, x_n)$. Since $x_n \to x_0$, σ_n converge to σ_0 if and only if $\sigma_n(x_n)$ converge to $\sigma_0(x_0)$. Taking the closed half-spaces, corresponding to $\sigma_n(x_n)$, we get $\lim\limits_{n\to\infty} \bar{V}^+(\sigma_n, x_n) = \bar{V}^+(\sigma_0, x_0)$. At the same time, $\lim\limits_{n\to\infty} \bar{\gamma}_{x_n}^+ = \bar{\gamma}_{x_0}^+$. Therefore, $\bar{\gamma}_{x_0}^+ \subset \bar{V}^+(\sigma_0, x_0)$. The same way we get $\bar{\gamma}_{x_0}^- \subset \bar{V}^-(\sigma_0, x_0)$. ∎

Denote by $\Sigma(x)$ the set of all hyperplanes, weakly transverse to the arc γ at the point $x \in \gamma$. This set is a compact subset of $\mathbf{RP^d}$. It follows from the Lemma 7, that

$$\Sigma(x) \supset \limsup_{y \to x, y \in \gamma} \Sigma(y).$$

This inclusion implies that the cones $Q^+(x)$ and $Q^-(x)$ satisfy the following semicontinuity condition:

Lemma 8. *Let γ be a Jordan arc in $\mathbb{R}^d$ and $x \in \gamma$. Then,*

$$Q^+(x) \subset \liminf_{y \to x, y \in \gamma} Q^+(y).$$

Proof. Since

$$Q^+(x) = \bigcap_{\sigma \in \Sigma(x)} \bar{V}^+(\sigma, x),$$

using basic properties of upper and lower limits [8, Sect. 29], we can write

$$Q^+(x) = (\bigcup_{\sigma \in \Sigma(x)} V^-(\sigma, x))^{\mathsf{c}} \subset (\limsup_{y \to x, y \in \gamma} \bigcup_{\sigma \in \Sigma(y)} V^-(\sigma, y))^{\mathsf{c}} =$$

$$= \liminf_{y \to x, y \in \gamma} (\bigcup_{\sigma \in \Sigma(y)} V^-(\sigma, y))^{\mathsf{c}} = \liminf_{y \to x, y \in \gamma} Q^+(y). \blacksquare$$

Lemma 9. *Let γ be a self-similar Jordan arc. If for any $x \in \gamma$ there is a hyperplane, weakly transverse to γ at the point x, then there is a hyperplane σ, which is transverse to γ at any point $x \in \gamma$.*

Proof. Suppose the affine hull of γ is $\mathbb{R}^d$ so it is not contained in a hyperplane.

Take some $\delta > 0$.

Consider the family of all the cones $A = \{Q^+(x), x \in \gamma\}$. Taking the parallel copy of each cone $Q^+(x)$ having the vertex at the center 0 of the unit ball $B \subset \mathbb{R}^d$, we denote its intersection with the ball B by $Q(x)$. This turns the family A to a subset of the hyperspace $Conv(B)$ of compact convex subsets of the unit ball B. Observe that the inclusion $Q(x) \subset \liminf_{y \to x, y \in \gamma} Q(y)$ in the statement of Lemma 8 holds for the cones $Q(x)$ as well.

Let S be a contraction similarity, for which $S(\gamma) \subset \gamma$. Let x_0 be its fixed point. Let O be the orthogonal part of the similarity S.

Since $Q(x_0) \subset \liminf_{x \to x_0} Q(x)$, there is an open subarc $(y, z) \ni x_0$ such that for any $x \in (y, z)$, the cone $Q(x_0)$ is contained in δ-neighborhood $N_\delta(Q(x))$ of a cone $Q(x)$.

For some sufficiently large k, the subarc $S_i^k(\gamma)$ is contained in (y, z). Then for any $\xi \in \gamma$, the point $x = S_i^k(\xi)$ lies in (y, z) and $N_\delta(Q(x)) \supset Q(x_0)$.

Since $Q(x) = O_i^k(Q(\xi))$, and $Q(x_0) = O_i^k(Q(x_0))$ and O_i^k is an isometry, $N_\delta(Q(\xi))$ must also contain $Q(x_0)$.

Thus, if $S : \gamma \to \gamma$ is a similarity and $fix(S) = x$, then for any $\delta > 0$ and any $\xi \in \gamma$, $N_\delta(Q(\xi)) \supset Q(x)$. Therefore, $Q(\xi) \supset Q(x)$ for all $\xi \in \gamma$. If we take for ξ a fixed point of some other similarity $S' : \gamma \to \gamma$, we get that $Q(\xi) = Q(x)$. Thus, the minimal cone $Q(x)$ is the same, no matter which fixed point x we choose, and we denote it by Q. If x is not a fixed point of any $S \in G(\mathcal{S})$, then $Q(x) \subset Q$. If $\sigma(x)$ is a support hyperplane to the cone $Q^+(x)$ at some fixed point x, then for any $\xi \in \gamma$ parallel hyperplane $\sigma(\xi)$ is a support hyperplane for $Q^+(\xi)$ and is thus weakly transverse to γ at the point ξ.

Suppose for some x and w in γ, $w \in \dot{\gamma}^+(x)$ and $w \in \sigma(x)$. Then $\sigma(x) = \sigma(w)$ and $V^+(\sigma, x) = V^+(\sigma, w)$. By weak transversality of σ at the points x and w, the subarc $[x, w]$ lies in $V^+(\sigma, x) \cap V^-(\sigma, w) = \sigma$. Then the whole arc γ lies in a hyperplane. The contradiction shows that the hyperplanes parallel to $\sigma(x)$ are transverse to γ at any point.■

Lemma 10. *Let γ be a self-similar Jordan arc, which has a hyperplane transverse to γ at any of its points. Then there is such transverse hyperplane σ, that for any similarity $S_i \in \mathcal{S}$, $O_i(\sigma) = \sigma$.*

Proof. Let G_O be a group generated by orthogonal parts O_i of the similarities $S_i \in \mathcal{S}$. For any $O \in G_O$, the image $O(Q)$ is either Q or $-Q$. The space $\mathbb{R}^d$ is a direct sum of two orthogonal subspaces $X_0 \oplus X_1$, where X_0 is the space of all such x that for any $O \in G_O$, $O(\{x, -x\}) = \{x, -x\}$ and $X_1 = X_0^\perp$.

Consider the intersection $X_0 \cap Q$. This intersection is a convex cone Q' in X_0. Take a support hyperplane Y to the cone Q' at the point 0 in the space X_0. Then $Y + X_1$ is a support hyperplane for Q in $\mathbb{R}^d$.

Suppose contrary. Then there is some $z \in (Y + X_1) \cap \dot{Q}$. The point z has unique representation in the form $z = x + y$, where $x \in X_1$, $x \neq 0$ and $y \in Y$. Consider the convex hull W of the orbit $G_O(x)$. It's barycenter is fixed by the group G_O, therefore it is 0. Then the barycenter of the convex hull of the orbit $G_O(z)$ is y.

Take a ball $B(z, \varepsilon) \subset Q$. The convex hull of a set $\bigcup\limits_{O \in G_O} O(B(z, \varepsilon))$ contains the ball $B(y, \varepsilon)$, therefore $y \in \dot{Q}$, which is impossible. So $\dot{Q} \cap (Y + X_1) = \varnothing$.

At the same time, for any $O \in G_O$ the transformation O sends the hyperplane $Y + X_1$ to itself.■

The proof of Theorem 2. Let γ be a self-similar Jordan arc, which is not a line segment. By Theorem 4.1 in [12], the arc γ may be represented as a component $\gamma^{(u)}$ of the invariant set of some multizipper $\mathcal{Z}$, for which the maps $S_e, e \in E$ are the elements of the semigroup $G(\mathcal{S})$. Let $z_i^{(u)}$ be the node points and $\Gamma = \langle V, E, \alpha, \omega \rangle$ be the structural graph of $\mathcal{Z}$. Passing, if necessary, to a subarc of γ, we may suppose that the graph Γ is strongly connected and the multizipper $\mathcal{Z}$ is regular.

If γ contains such dense subset $D \subset \gamma$, that for any $x \in D$ there is a hyperplane $\sigma(x)$, weakly transverse to γ, then by Lemma 7, such hyperplane $\sigma(x)$ exists for any $x \in \gamma$. By Lemma 9, there is a hyperplane σ, transverse to γ at any $x \in \gamma$.

By Lemma 10 there is a hyperplane σ, transverse to γ at any of its points, which is preserved by any of $O_i \in G_O$. Then the duplicates of σ are transverse to the components $\gamma^{(u)}, u \in V$ of the attractor of the multizipper $\mathcal{Z}$ at any of their points and are preserved by the orthogonal parts O_e of the similarities S_e.

Let $\Lambda^{(u)}$ be a line, orthogonal to σ in the copy $X^{(u)}$ of the space $\mathbb{R}^d$. Let $\gamma^{(u)}$ be the component of the invariant set of $\mathcal{Z}$ lying in $X^{(u)}$. Consider the orthogonal projection π of each arc $\gamma^{(u)}$ to the $\Lambda^{(u)}$.

Since the similarities S_e send the hyperplanes, parallel to σ, to the hyperplanes, parallel to σ, for each similarity $S_e \in \mathcal{Z}$, $S_e : \gamma^{(v)} \to \gamma^{(u)}$ there is a similarity $\hat{S}_e : \Lambda^{(v)} \to \Lambda^{(u)}$, satisfying the condition

$$\pi \circ S_e = \hat{S}_e \circ \pi.$$

Due to this condition each map $\hat{S}_e$ sends the set $\{\pi(z_0^{(v)}), \pi(z_{m_v}^{(v)})\}$ to the set $\{\pi(z_{i-1}^{(u)}), \pi(z_i^{(u)})\}$.

The system $\hat{\mathcal{Z}}$ is a linear multizipper with node points $\hat{z}_i^{(u)} = \pi(z_i^{(u)})$.

Since for any S_e, $\mathrm{Lip}(\hat{S}_e) = Lip(S_e)$ the similarity dimension of the multizipper $\mathcal{Z}$ is equal to the similarity dimension of $\hat{\mathcal{Z}}$ and therefore it is equal to 1. By Theorem 5, its invariant set is a collection of straight line segments. ∎

Acknowledgements Andrey Tetenov was supported by Russian Foundation of Basic Research project 13-01-00513.

References

1. Aseev, V.V.: On the regularity of self-similar zippers. In: The 6th Russian-Korean International Symposium on Science and Technology (KORUS-2002), 24–30 June 2002, Part 3, p. 167. Novosibirsk, Russia
2. Aseev, V.V., Tetenov, A.V., Kravchenko, A.S.: On selfsimilar Jordan curves on the plane. Siberian Math. J. **44**(3), 379–386 (2003)
3. Bandt, C., Kravchenko, A.: Differentiability of fractal curves. Nonlinearity **24**, 2717–2728 (2011)
4. Edgar, G.A.: Measure, Topology, and Fractal Geometry. Springer, New York (1990)
5. Gantmacher, F.R.: The Theory of Matrices, vol. 2. Chelsea, New York (1959)
6. Hutchinson, J.: Fractals and self-similarity. Indiana Univ. Math. J. **30**(5), 713–747 (1981)
7. Kravchenko, A.: Smooth self-affine curves (in Russian), Preprint No. 161, Sobolev Institute of Mathematics, Novosibirsk (2005)
8. Kuratowski, K.: Topology, vol. I. Academic/Polish Scientific Publishers, New York/ London/Warszawa (1966)
9. Levy, P.: Les courbes planes ou gauches et les surfaces composees de parties semblables au tout. J. Ecole Polytechn. III. Ser. **144**, 227–247, 249–291 (1938)
10. Mauldin, R.D., Williams, S.C.: Hausdorff dimension in graph directed constructions. Trans. Am. Math. Soc. **309**, 811–829 (1988)

11. Sierpinski, W.: Sur une courbe dont tout point est un point de ramification. Compt. Rendus Acad. Sci. Paris **160**, 302–305 (1915)
12. Tetenov, A.V.: Self-similar Jordan arcs and graph-directed systems of similarities. Siberian Math. J. **47**(5), 940–949 (2006)
13. von Koch, H.: Sur une courbe continue sans tangente, obtenue par une construction geometrique elementaire. Archiv for Matemat.; Astron. och Fys. **1**, 681–702 (1904)

Fractals in Product Fuzzy Metric Space

R. Uthayakumar and A. Gowrisankar

Abstract The purpose of this paper is to prove the Hutchinson–Barnsley operator on the product fuzzy metric space is fuzzy B-contraction. We also present the fuzzy B-contraction properties of HB operator in product fuzzy metric space. The notion of product fuzzy fractal is introduced in product fuzzy metric space in the sense of the fuzzy B-contraction.

Keywords Fractals • Fuzzy metric space • Fuzzy Iterated Function System • Hutchinson–Barnsley operator • Hausdorff fuzzy metric • Product fuzzy metric space

AMS Classification Codes: 26E50; 28A80; 47H10

1 Introduction

Fuzzy set theory was introduced by Zadeh in 1965 [17]. Many authors have introduced and discussed several notions of fuzzy metric space in different ways and also proved fixed point theorems with interesting consequent results in the fuzzy metric spaces [3–6, 9, 11, 13].

Fractal Analysis was introduced by Mandelbrot in 1975 [10] and popularized by various mathematicians. A fractal is an object which appears self-similar under varying degrees of magnification. Mathematically, sets with non-integral Hausdorff dimension which exceeds its topological dimension, are called Fractals

---

R. Uthayakumar • A. Gowrisankar (✉)
Department of Mathematics, The Gandhigram Rural Institute - Deemed University, Gandhigram 624 302, Dindigul, Tamil Nadu, India
e-mail: uthayagri@gmail.com; gowrisankargri@gmail.com

C. Bandt et al. (eds.), *Fractals, Wavelets, and their Applications*, Springer Proceedings in Mathematics & Statistics 92, DOI 10.1007/978-3-319-08105-2_9,

by Mandelbrot. Hutchinson [7] introduced the formal definition of Iterated Function Systems (IFS) and Barnsley [1] developed the theory of IFS called the Hutchinson–Barnsley theory (HB Theory) in order to define and construct the fractal as a compact invariant subset of a complete metric space generated by the Iterated Function System (IFS) of contractions. That is, Hutchinson introduced an operator on hyperspace of non-empty compact sets called as Hutchinson–Barnsley operator (HB operator) to define a fractal set as a unique fixed point by using the Banach Contraction Theorem in the complete metric space. Kramosil and Michalek [9] introduced the notion of fuzzy metric space. George and Veeramani [4] imposed some stronger conditions on the fuzzy metric space in order to obtain a Hausdorff topology.

In this paper, we introduce the concepts and properties of HB operator in the product fuzzy metric space and we present the fuzzy contraction properties of HB operator on the product fuzzy metric space with respect to the Hausdorff product fuzzy metric. Also we introduce the notion of product fuzzy fractal in product fuzzy metric space with respect to the fuzzy B-contraction.

2 Preliminary

2.1 Metric Fractals

In this section, we recall the Hutchinson–Barnsley theory (HB theory) to define and construct the fractals in the complete metric space.

Definition 2.1 ([1, 7]). Let (X, d) be a metric space and $\mathscr{K}(X)$ be the collection of all non-empty compact subsets of X. Define $d(x, B) := \inf_{y \in B} d(x, y)$ and $d(A, B) := \sup_{x \in A} d(x, B)$ for all $x \in X$ and $A, B \in \mathscr{K}(X)$. The Hausdorff metric or Hausdorff distance (H_d) is a function $H_d : \mathscr{K}(X) \times \mathscr{K}(X) \longrightarrow \mathbb{R}$ defined by $H_d\Big(A, B\Big) = \max\Big\{d(A, B), d(B, A)\Big\}$. Then H_d is a metric on $\mathscr{K}(X)$ and hence $(\mathscr{K}(X), H_d)$ is called a Hausdorff metric space.

Theorem 2.1 ([1, 7]). *If (X, d) is a complete metric space, then $(\mathscr{K}(X), H_d)$ is also a complete metric space.*

Definition 2.2 ([1, 7]). The function $f : X \longrightarrow X$ is said to be a contraction or Banach contraction mapping on a metric space (X, d), if there exists $k \in [0, 1)$ such that $d(f(x), f(y)) \leq kd(x, y), \quad \forall x, y \in X$. Here k is called a contractivity ratio of f.

Definition 2.3 ([1, 7]). Let (X, d) be a metric space and $f_n : X \longrightarrow X$, $n = 1, 2, 3, \ldots, N$ $(N \in \mathbb{N})$ be N-contraction mappings with the corresponding contractivity ratios k_n, $n = 1, 2, 3, \ldots, N$. Then the system $\{X; f_n, n = 1, 2, 3, \ldots, N\}$ is called an Iterated Function System (IFS) or Hyperbolic Iterated Function System with the ratio $k = \max_{n=1}^{N} k_n$.

Definition 2.4 ([1,7]). Let (X,d) be a metric space. Let $\{X; f_n,\ n = 1,2,3,\ldots, N; N \in \mathbb{N}\}$ be an IFS of Banach contractions. Then the Hutchinson–Barnsley operator (HB operator) of the IFS of Banach contractions is a function $F : \mathscr{K}(X) \longrightarrow \mathscr{K}(X)$ defined by $F(B) = \bigcup_{n=1}^{N} f_n(B)$, for all $B \in \mathscr{K}(X)$.

Theorem 2.2 ([1,7]). *Let (X,d) be a metric space. Let $\{X; f_n,\ n = 1,2,3,\ldots, N; N \in \mathbb{N}\}$ be an IFS of Banach contractions. Then, the HB operator (F) is a Banach contraction mapping on $(\mathscr{K}(X), H_d)$.*

Theorem 2.3 ([1, 7] HB Theorem for Metric IFS). *Let (X,d) be a complete metric space and $\{X; f_n,\ n = 1,2,3,\ldots, N; N \in \mathbb{N}\}$ be an IFS of Banach contractions. Then, there exists only one compact invariant set $A_\infty \in \mathscr{K}(X)$ of the HB operator (F) or, equivalently, F has a unique fixed point namely $A_\infty \in \mathscr{K}(X)$.*

Definition 2.5 ([1,7] Metric Fractal). The fixed point $A_\infty \in \mathscr{K}(X)$ of the HB operator F described in Theorem 2.3 is called the Attractor (Fractal) of the IFS. Sometimes $A_\infty \in \mathscr{K}(X)$ is called as Metric Fractal generated by the IFS of Banach contractions.

2.2 *Fuzzy Metric Space*

Definition 2.6 ([14]). A binary operation $* : [0,1]\times[0,1] \longrightarrow [0,1]$ is a continuous t-norm, if $([0,1], *)$ is a topological monoid with unit 1 such that $a * b \leq c * d$ whenever $a \leq c$, $b \leq d$ and $a,b,c,d \in [0,1]$.

George and Veeramani modified the Kramosil and Michalek [9] fuzzy metric space as follows:

Definition 2.7 ([4,5]). The 3-tuple $(X, M, *)$ is said to be a fuzzy metric space if X is an arbitrary set, $*$ is a continuous t-norm and M is a fuzzy set on $X \times X \times (0,\infty)$ satisfying the following conditions:

1. $M(x,y,t) > 0$,
2. $M(x,y,t) = 1$ if and only if $x = y$,
3. $M(x,y,t) = M(y,x,t)$,
4. $M(x,y,t) * M(y,z,s) \leq M(x,z,t+s)$,
5. $M(x,y,\cdot) : (0,\infty) \longrightarrow [0,1]$ is continuous,

$x,y,z \in X$ and $t,s > 0$.

Definition 2.8 ([4, 5]). Let (X,d) be a metric space. Define $a * b = a \cdot b$, the usual multiplication for all $a,b \in [0,1]$, and let M_d be the function defined on $X \times X \times (0,\infty)$ by

$$M_d(x,y,t) = \frac{t}{t + d(x,y)},$$

for all $x, y \in X$ and $t > 0$. Then $(X, M_d, *)$ is a fuzzy metric space called standard fuzzy metric space, and M_d is called as the standard fuzzy metric induced by the metric d.

Definition 2.9 ([6]). Let $(X, M, *)$ be a fuzzy metric space. The mapping $f : X \longrightarrow X$ is fuzzy contractive if there exists $k \in (0, 1)$ such that

$$\frac{1}{M(f(x), f(y), t)} - 1 \leq k\left(\frac{1}{M(x, y, t)} - 1\right)$$

for each $x, y \in X$ and $t > 0$. Here, k is called the fuzzy contractivity ratio of f.

Definition 2.10 ([6, 15]). A fuzzy B-contraction (Sehgal contraction) on a fuzzy metric space $(X, M, *)$ is a self-mapping f on X for which $M(f(x), f(y), kt) \geq M(x, y, t)$, for all $x, y \in X$ and $t > 0$, where k is fixed constant in $(0, 1)$.

Definition 2.11 ([13]). Let $(X, M, *)$ be a fuzzy metric space. Let $\mathscr{K}(X)$ be set of all non-empty compact subsets of X. Define, $M(x, B, t) := \sup_{y \in B} M(x, y, t)$ and $M(A, B, t) := \inf_{x \in A} M(x, B, t)$ for all $x \in X$ and $A, B \in \mathscr{K}(X)$. Then Hausdorff fuzzy metric (H_M) is a function $H_M : \mathscr{K}(X) \times \mathscr{K}(X) \times (0, \infty) \longrightarrow [0, 1]$ defined by

$$H_M(A, B, t) = \min\Big\{M(A, B, t), M(B, A, t)\Big\}.$$

Then H_M is a fuzzy metric on $\mathscr{K}(X)$, and hence $(\mathscr{K}_o(X), H_M, *)$ is called a Hausdorff fuzzy metric space.

Definition 2.12 ([2,16]). Let $(X, M, *)$ be a fuzzy metric space and $f_n : X \longrightarrow X$, $n = 1, 2, 3, \ldots, N$ $(N \in \mathbb{N})$ be N fuzzy B-contraction mappings. Then the system $\{X; f_n, \ n = 1, 2, 3, \ldots, N\}$ is called a Fuzzy Iterated Function System (FIFS) of fuzzy B-contraction on the fuzzy metric space $(X, M, *)$.

Definition 2.13 ([2,16]). Let $(X, M, *)$ be a fuzzy metric space. Let $\{X; f_n, n = 1, 2, 3, \ldots, N; N \in \mathbb{N}\}$ be a FIFS of fuzzy B-contractions. Then the Fuzzy Hutchinson–Barnsley operator (FHB operator) of the FIFS is a function $F : \mathscr{K}(X) \longrightarrow \mathscr{K}(X)$ defined by

$$F(B) = \bigcup_{n=1}^{N} f_n(B), \quad \text{for all} \quad B \in \mathscr{K}(X).$$

Theorem 2.4 ([6] Fuzzy Banach Contraction Theorem). *Let $(X, M, *)$ be a complete fuzzy metric space in which fuzzy contractive sequence are Cauchy. Let $f : X \longrightarrow X$ be a fuzzy contractive mapping with contractivity ratio k. Then f has a unique fixed point.*

3 Product Fuzzy Metric Space

Uthayakumar and Easwaramoorthy investigated the fuzzy IFS fractals in the fuzzy metric space [2, 16]. In this paper we implement their result into product fuzzy metric space.

Definition 3.1 ([8, 12]). Let $(X, M_x, *)$ and $(Y, M_y, *)$ be a fuzzy metric spaces and let $Z = X \times Y$. For $z_=(x_1, y_1)$,$z_2 = (x_2, y_2) \in Z$ and define $t > 0$, $M_z(z_1, z_2, t) = M_x(x_1, x_2, t) * M_y(y_1, y_2, t)$. Then $(M_z, *)$ is a fuzzy metric on Z, and the triple $(Z, M_z, *)$ is called the product fuzzy metric space of X and Y.

Definition 3.2. Let (Z, d) be a product space. Define $a * b = a.b$, the usual multiplication for all $a, b \in [0, 1]$, and let M_{z_d} be the function defined on $Z \times Z \times (0, \infty)$ by $M_{z_d}(z_1, z_2, t) = \frac{t}{t+d(z_1,z_2)}$ for all $z_1, z_2 \in Z$ and $t > 0$. Then $(Z, M_{z_d}, *)$ is a fuzzy metric space called standard product fuzzy metric space, and M_{z_d} is called as the standard product fuzzy metric induced by the metric d.

In the following theorem we use $Z = X \times X$

Theorem 3.1. *Suppose $(X, M_x, *)$ be a fuzzy metric space. Let f be a fuzzy B-contraction on X and $(Z, M_Z, *)$ be a fuzzy product space of X then the mapping $g : Z \longrightarrow Z$ defined by $g(z) = (f(x), f(y))$ for all $z = (x, y) \in Z$, is fuzzy B-contraction mapping on Z.*

Proof. Let $Z = X \times X$ and $(Z, M_z, *)$ be a product fuzzy metric space. $M_z(z_1, z_2, t) = M_x(x_1, x_2, t) * M_y(y_1, y_2, t)$ for all $z_1 = (x_1, y_1), z_2 = (x_2, y_2) \in Z$ and $t > 0$. For given $k \in (0, 1)$, we have

$$\begin{aligned} M_z(g(z_1), g(z_2), kt) &= M_z((f(x_1), f(y_1)), (f(x_2), f(y_2)), kt) \\ &= M_x(f(x_1), f(x_2), kt) * M_x(f(y_1), f(y_2), kt) \\ &\geq M_x(x_1, x_2, t) * M_x(f(y_1), f(y_2), t) \\ &= M_z(Z_1, z_2, t). \end{aligned}$$

Definition 3.3. Let $(X, M_x, *)$ be a fuzzy metric space. Let f be a fuzzy B-contraction on X and $(Z, M_z, *)$ be a product fuzzy metric space of X and $g_n : Z \longrightarrow Z, n = 1, 2, \ldots, N \in \mathbb{N}$ be N-fuzzy B-contraction mappings defined by $g_n(x, y) = (f_n(x), f_n(y))$. Then the system $\{Z; g_n, n = 1, 2, \ldots, N \in \mathbb{N}\}$ is called a Fuzzy Iterated Function System (FIFS) of fuzzy B-contraction on the product fuzzy metric space $(Z, M_z, *)$.

Definition 3.4. Let $(Z, M_z, *)$ be a product fuzzy metric space. Let $\{Z; g_n, n = 1, 2, 3 \ldots\ldots, N \in \mathbb{N}\}$ be a FIFS of fuzzy B-contraction on $(Z, M_z, *)$. Let $\mathscr{K}(Z)$ be set of all non-empty compact subsets of Z. Then the Fuzzy Hutchinson–Barnsley operator (FHB operator) of the FIFS of fuzzy B-contraction on $(Z, M_z, *)$ is a function $G : \mathscr{K}(Z) \longrightarrow \mathscr{K}(Z)$ defined by

$$G(B) = \bigcup_{n=1}^{N} g_n(B), \quad \text{for all} \quad B \in \mathscr{K}(X).$$

That is,

$$G(B) = \bigcup_{n=1}^{N} (f_n(x), f_n(y)), \quad \text{for all } z = (x, y) \in B \in \mathscr{K}(X)$$

Definition 3.5. Let $(Z, M_z, *)$ be a complete fuzzy metric space. Let $\{Z; g_n, n = 1, 2, 3 \ldots\ldots, N \in \mathbb{N}\}$ be a FIFS of fuzzy B-contraction on $(Z, M_z, *)$ and G be the FHB operator of the FIFS of fuzzy B-contraction on $(Z, M_z, *)$. We say that the set $A_\infty \in \mathscr{K}(Z)$ is product fuzzy Attractor (Product fuzzy fractal) of the given FIFS of fuzzy B-contractions, if A_∞ is a unique fixed point of the FHB operator G. Such $A_\infty \in \mathscr{K}(Z)$ is also called as Product Fuzzy Fractal generated by the FIFS of fuzzy B-contractions.

Theorem 3.2. *Let $(Z, M_z, *)$ be a product fuzzy metric space and let $(\mathscr{K}(Z), H_{M_z}, *)$ be the corresponding Hausdorff product fuzzy metric space. Suppose $g : Z \longrightarrow Z$ is a fuzzy B-contraction function on $(Z, M_z, *)$. Then for $k \in (0, 1)$*

$$H_{M_z}(g(A), g(B), kt) \geq H_{M_z}(A, B, t)$$

for all $A, B \in \mathscr{K}(Z)$ and $t > 0$.

Proof. Let $Z = X \times X$ and $(Z, M_z, *)$ be a product fuzzy metric space. $M_z(z_1, z_2, t) = M_x(x_1, x_2, t) * M_y(y_1, y_2, t)$ for all $z_1 = (x_1, y_1), z_2 = (x_2, y_2) \in Z$ and $t > 0$. Fix $t > 0$. Let $A, B \in \mathscr{K}(Z)$. By using Theorem 3.1 for given $k \in (0, 1)$, we get

$$\begin{aligned}
M_z(g(z_1), g(z_2), kt) &\geq M_z(z_1, z_2, t), \quad \text{for all } z_1, z_2 \in Z \\
M_z(g(z_1), g(z_2), kt) &\geq M_z(z_1, z_2, t), \\
&\qquad\qquad \text{for all } z_1 \in A \text{ and } z_2 \in B \\
\sup_{z_2 \in B} M_z(g(z_1), g(z_2), kt) &\geq \sup_{z_2 \in B} M_z(z_1, z_2, t), \quad \text{for all } z_1 \in A \\
M_z(g(z_1), g(B), kt) &\geq M_z(z_1, B, t), \quad \text{for all } z_1 \in A \\
\inf_{z_1 \in A} M_z(g(z_1), g(B), kt) &\geq \inf_{z_1 \in A} M_z(z_1, B, t), \quad \text{for all } z_1 \in A \\
M_z(g(A), g(B), kt) &\geq M_z(A, B, t) \\
M_z(g(B), g(A), kt) &\geq M_z(B, A, t) \\
\min(M_z(g(A), g(B), kt), M_z(g(B), g(A), kt)) &\geq \min(M_z(A, B, t), M_z(B, A, t)) \\
i, e., M_z(g(A), g(B), kt) &\geq M_z(A, B, t).
\end{aligned}$$

Theorem 3.3. *Let $(Z, M_z, *)$ be a product fuzzy metric space and let $(\mathscr{K}(Z), H_{M_z}, *)$ be the corresponding Hausdorff product fuzzy metric space. Suppose $g_n : Z \longrightarrow Z, n = 1, 2, \ldots, N \in \mathbb{N}$ be N fuzzy B-contractions functions on $(Z, M_z, *)$. Then the HB operator is fuzzy B-contraction on $(\mathscr{K}(Z), H_{M_z}, *)$.*

Proof. Fix $t > 0$. Let $A, B \in \mathscr{K}(Z)$, for given $k \in (0, 1)$. Using Theorem 3.2 we get

$$\begin{aligned} H_{M_z}(G(A), G(B), kt) &= H_{M_z}\left(\bigcup_{n=1}^{N}(g_n(A), \bigcup_{n=1}^{N}(g_n(B), kt\right) \\ &\geq \min_{n=1}^{N} H_{M_z}(g_n(A), g_n(B), kt) \\ &\geq H_{M_z}(A, B, t) \\ H_{M_z}(G(A), G(B), kt) &\geq H_{M_z}(A, B, t). \end{aligned}$$

Above theorems are proved in the product fuzzy metric space $Z = X \times X$. Also it can be extended up to finite case $Z = X_1 \times X_2 \times X_3 \ldots \times X_n$.

4 Conclusion

In this study, we have introduced the concepts and properties of HB operator in the product fuzzy metric space. Also we have presented the fuzzy contraction properties of HB operator on the product fuzzy metric space with respect to the Hausdorff product fuzzy metric. Besides that we have introduced the notion of product fuzzy fractal in product fuzzy metric space with respect to the fuzzy B-contraction.

Acknowledgements The research work has been supported by University Grants Commission, Government of India, New Delhi, India under the schemes of *UGC—Major Research Project* with Grant No.: F.No. 42-21/2013 (SR)/dated 12.03.2013 and *UGC—Special Assistance Programme (DRS-II)*.

References

1. Barnsley, M.F.: Fractals Everywhere, 2nd edn. Academic, Boston (1993)
2. Easwaramoorthy, D., Uthayakumar, R.: Analysis on fractals in fuzzy metric spaces. Fractals **19**(3), 379–386 (2011)
3. Farnoosh, R., Aghajani, A., Azhdari, P.: Contraction theorems in fuzzy metric space. Chaos Solitons Fractals **41**, 854–858 (2009)
4. George, A., Veeramani, P.: On some result in fuzzy metric spaces. Fuzzy Sets Syst. **64**, 395–399 (1994)

5. George, A., Veeramani, P.: On some results of analysis for fuzzy metric spaces. Fuzzy Sets Syst. **90**, 365–368 (1997)
6. Gregori, V., Sapena, A.: On fixed - point theorems in fuzzy metric spaces. Fuzzy Sets Syst. **125**, 245–252 (2002)
7. Hutchinson, J.E.: Fractals and self similarity. Indiana Univ. Math. J. **30**, 713–747 (1981)
8. Kocinac, L.D.R.: Selection properties in fuzzy metric spaces. Filomat **26**(2), 305–312 (2012)
9. Kramosil, O., Michalek, J.: Fuzzy metric and statistical metric space. Kybernetika **11**, 326–334 (1975)
10. Mandelbrot, B.B.: The Fractal Geometry of Nature. W.H. Freeman and Company, New York (1983)
11. Mihet, D.: On fuzzy contractive mappings in fuzzy metric spaces. Fuzzy Sets Syst. **158**, 915–921 (2007)
12. Rahmat, M.R.S., Noorani, M.S.M.: Product of fuzzy metric spaces and fixed point theorems. Int. J. Contemp. Math. Sci. **15**(3), 703–212 (2008)
13. Rodriguez-Lopez, J., Romaguera, S.: The Hausdorff fuzzy metric on compact sets. Fuzzy Sets Syst. **147**, 273–283 (2004)
14. Schweizer, B., Sklar, A.: Statistical metric spaces. Pac. J. Math. **10**, 314–334 (1960)
15. Sehgal, B., Bharucha-Reid, A.T.: Fixed points of contraction mapping on probabilistic metric spaces. Math. Syst. Theory **6**, 97–102 (1671)
16. Uthayakumar, R., Easwaramoorthy, D.: Hutchinson-Barnsley operator in fuzzy metric spaces. World Acad. Sci. Eng. Technol. **56**, 1372–1376 (2011)
17. Zadeh, L.A.: Fuzzy sets. Inf. Control **8**, 338–353 (1965)

Some Properties on Koch Curve

R. Uthayakumar and A. Nalayini Devi

Abstract Many physical problems on fractal domains lead to nonlinear models involving reaction–diffusion equations, problems on elastic fractal media or fluid flow through fractal regions, etc. The prevalence of fractal-like objects in nature has led both mathematicians and physicists to study various processes on fractals. In recent years there has been an increasing interest in studying nonlinear partial differential equations on fractals, also motivated and stimulated by the considerable amount of literature devoted to the definition of a Laplace-type operator for functions on fractal domains. The energy of a function defined on a post critically finite (p.c.f) self-similar fractal can be written as a sum of directional energies. A general concept of graph energy defined on a finite connected graph is given. A work about the graph energy is mainly concerned on a Koch curve. First graphs on this Koch curve are built. These graphs produced from the initial graph by iteration repeatedly. Find the energy renormalization constant. Second we find the non-normalized and Normalized Laplacian of a Koch Curve. With the help of this we examine the Laplacian Renormalization constant and forbidden eigenvalues. Finally we develop the Spectral decimation function of Koch Curve.

Keywords Energy renormalization constant • Laplacian renormalization constant • Forbidden eigenvalues • Spectral decimation function

R. Uthayakumar (✉)
Department of Mathematics, Gandhigram Rural Institute-Deemed University, Gandhigram, Tamil Nadu, India
e-mail: uthayagri@gmail.com

A.N. Devi
Department of Mathematics, NPR College of Engineering and Technology, Natham, Tamil Nadu, India
e-mail: nithy26676@gmail.com

C. Bandt et al. (eds.), *Fractals, Wavelets, and their Applications*, Springer Proceedings in Mathematics & Statistics 92, DOI 10.1007/978-3-319-08105-2_10,

1 Introduction

Laplacian on fractal was defined on the Sierpinski gasket SG as a diffusion process by Kusuoka [4] and Goldstein [2]. Kigami [3] constructed the Laplacian analytically, both as a renormalized limit of difference operators and through a weak formulation using the theory of Dirichlet forms. Later, the theory of Laplacians was extended to other fractals, including nested fractals and p.c.f. self-similar sets by Lindsrom and Kigami. Analysis on fractals has bloomed since then and many classical results of smooth analysis have found their analogues on the "rough" objects. It has been first observed by Fukushima and Shima [5] and Teplyaev [7] that the eigenvalues of the Laplacian on the Sierpinski gasket and its higher dimensional analogues exhibit the phenomenon of spectral decimation. In particular, the spectrum and eigen functions of the Laplacian on the Sierpinski gasket were studied by Fukushima[1] and Shima and using the so-called spectral decimation method, which originated in physics literature and was generalized by Shima and Teplyaev. In particular, the eigenvalues of the Laplacian on the fractal, which admits spectral decimation, can be calculated by means of a certain polynomial or rational function. Using that we obtain the spectral decimation function, normalized and non-normalized Laplacian for Koch curve at level 1.

2 Basic Definitions

Definition 1 (Spectral Decimation). [6]
The Laplace operator on a post critically finite self-similar fractal G admits spectral decimation, if there exists a rational function R, a finite set A and a constant $\lambda > 1$ such that all eigenvalues of Δ can be written in the form

$$\lambda^m \lim_{n\to\infty} \lambda^n R^{(-n)}(w), w \in A, m \in N \tag{1}$$

where the pre-images of w under n-fold iteration of R have to be chosen such that the limit exists.

Definition 2 (Laplacian on Finite Graphs). For any set S, we use l(S) to denote the set of real valued functions on S and

$$l_0(V_m) = \{f \in l(V_m) : f(p) = 0\, for\ p \in V_0\} \tag{2}$$

For two sets U and V, we define

$$L(U, V) = \{A : l(U) \to l(V) and\ A\ is\ linear\}$$

In particular, L(V) means L(V; V)

Definition 3 (Vertex Degree). Let G be a simple, finite graph with the set of vertices V (G) and the set of edges E(G).We say that two vertices x; y are neighbors if they are connected by exactly one edge, denoted by x y, in the graph. For any vertex x, deg x is called the vertex degree of x in the graph and

$$degx = \sum_{x \sim y \in V(G)} 1$$

Definition 4 (Normalized and Non-normalized Laplacian). Given a function $u \in l(V(G))$, the graph (non-normalized) Laplacian of u at a vertex is defined as

$$\Delta u(x) = \sum_{x \sim y \in V(G)} (u(x) - u(y)) \tag{3}$$

and the normalized (probabilistic) Laplacian is defined as

$$\hat{\Delta} u(x) = \frac{1}{degx} \sum_{x \sim y \in V(G)} (u(x) - u(y)) \tag{4}$$

The symmetric matrix D corresponding to Δ is called the Laplacian matrix and it has the expression

$$D_{i,j} = \begin{cases} 1 \text{ if } i \neq j, \\ -degx_i \text{ if } i = j, \\ 0 \text{ if } otherwise. \end{cases} \tag{5}$$

for x_i, x_j in V(G).

The normalized Laplacian on V_m satisfies

$$\hat{\Delta} u_m(x) = \frac{H_m\,(u(x))}{\hat{\mu}_m\,(u(x))} \; for\; u \in l\,(V_m) \tag{6}$$

Here we choose the measure factor

$$r_0^{-1} = \sum_{s \in S} r_{|s|}^{-1}$$

where $\left(r_0^{-1}, r_1^{-1}, r_2^{-1} \ldots, r_{|s|}^{-1}\right) \in l(S)$. (H_m, r) is the generalized non-normalized Laplacian with weight r on the Graph G_m. We decompose the matrix H_m into

$$H_m = \begin{pmatrix} T_m & J_m^t \\ J_m & X_m \end{pmatrix} \tag{7}$$

where $T_m \in L(V_0)$, $J_m \in L(V_m \setminus V_0)$ $X_m \in L(V_m \setminus V_0)$.

3 The Koch Curve

It is this similarity between the whole and its parts, even infinitesimal ones that makes us consider this curve of von Koch as a line truly marvelous among all. If it were gifted with life, it would not be possible to destroy it without annihilating it whole, for it would be continually reborn from the depths of its triangles, just as life in the universe is. Begin with a straight line (the blue segment in the top figure). Divide it into three equal segments and replace the middle segment by the two sides of an equilateral triangle of the same length as the segment being removed (the two red segments in the middle figure). Now repeat, taking each of the four resulting segments, dividing them into three equal parts and replacing each of the middle segments by two sides of an equilateral triangle. Continue this construction. The Koch curve is the limiting curve obtained by applying this construction an infinite number of times. For a proof that this construction does produce a "limit" that is an actual curve, i.e. the continuous image of the unit interval. The first iteration for the Koch curve consists of taking four copies of the original line segment, each scaled by $r = 1/3$. Two segments must be rotated by 60°, one counterclockwise and one clockwise (Figs. 1 and 2).

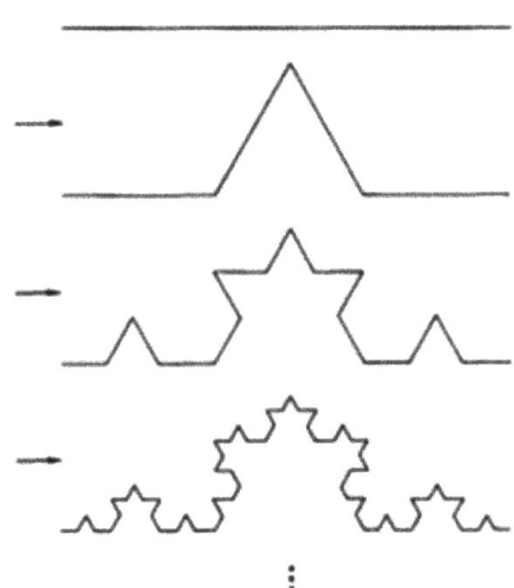

Fig. 1 Koch curve

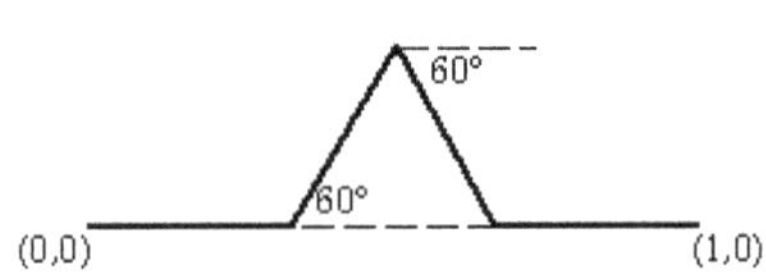

Fig. 2 Koch curve graph of Γ_1

Along with the required translations, this yields the following IFS

$$f_1(x) = \begin{pmatrix} 0.333 & 0 \\ 0 & 0.333 \end{pmatrix} x$$

scale by r

$$f_2(x) = \begin{pmatrix} 0.167 & -0.289 \\ 0.289 & 0.167 \end{pmatrix} x + \begin{pmatrix} 0.333 \\ 0 \end{pmatrix}$$

scale by r,rotation by 60°

$$f_3(x) = \begin{pmatrix} 0.167 & 0.289 \\ -0.289 & 0.167 \end{pmatrix} x + \begin{pmatrix} 0.500 \\ 0.289 \end{pmatrix}$$

scale by r,rotation by −60°

$$f_4(x) = \begin{pmatrix} 0.333 & 0 \\ 0 & 0.333 \end{pmatrix} x + \begin{pmatrix} 0.667 \\ 0 \end{pmatrix}$$

scale by r

The fixed invariant set of these IFS is the Koch curve.

4 Renormalized Energy

Renormalized Energy: We define the renormalized graph energy ε_m by

$$\varepsilon_m(u) = \alpha^{-m} E_m(u) \tag{8}$$

For ε_m and E_m are bilinear

$$\varepsilon_m(u, v) = \alpha^{-m} E_m(u, v). \tag{9}$$

Here $0 < |\alpha| < 1$, which is called renormalization factor.

5 Energy Renormalization Constant

Strong Harmonic Structure: We define the diagonal matrix $M_{i,i} = -X_{i,i}$. The generalized Laplacian is said to have a strong harmonic structure if there exists rational functions $K_D(\lambda)$ and $K_T(\lambda)$ such that $X + \lambda M$ is invertible then

$$T - J^t (X + \lambda M)^{-1} J = K_D(\lambda) D + K_T(\lambda) T \tag{10}$$

Here $K_D(0)$ is the energy renormalization constant.

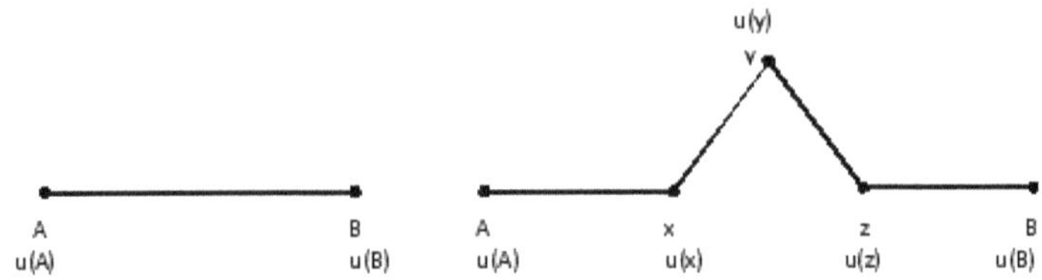

Fig. 3 Graph of Γ_0, Γ_1

Using the graphs Γ_0 and Γ_1 of Koch curve (Fig. 3) the boundary points $V_0 = \{A, B\}$. Let D be the Laplacian matrix on V_0

$$D = \begin{pmatrix} -1 & 1 \\ 1 & -1 \end{pmatrix}$$

and measure factor $r = (1, 1, 1, 1)$

Let H_1 be the standard Laplacian on V_1 is

$$H_1 = \begin{pmatrix} -1 & 0 & 1 & 0 & 0 \\ 0 & -1 & 0 & 0 & 1 \\ 1 & 0 & -2 & 1 & 0 \\ 0 & 0 & 1 & -2 & 1 \\ 0 & 1 & 0 & 1 & -2 \end{pmatrix}$$

From H_1 we get $T_1 = \begin{pmatrix} -1 & 0 \\ 0 & -1 \end{pmatrix}$, $J_1 = \begin{pmatrix} 1 & 0 & 0 \\ 0 & 0 & 1 \end{pmatrix}$, $X_1 = \begin{pmatrix} -2 & 1 & 0 \\ 1 & -2 & 1 \\ 0 & 1 & -2 \end{pmatrix}$ and $M_1 = \begin{pmatrix} 2 & 0 & 0 \\ 0 & 2 & 0 \\ 0 & 0 & 2 \end{pmatrix}$

$$(X + \lambda M)^{-1} = \frac{1}{2\lambda - 2} \begin{pmatrix} (2\lambda - 2)^2 - 1 & -(2\lambda - 2) & 1 \\ -(2\lambda - 2) & (2\lambda - 2)^2 & -(2\lambda - 2) \\ 1 & -(2\lambda - 2) & (2\lambda - 2)^2 - 1 \end{pmatrix}$$

$$T - J^t (X + \lambda M)^{-1} J = \frac{1}{(2\lambda - 2)(4\lambda^2 - 8\lambda + 2)} \times \begin{pmatrix} -4(\lambda - 1)(2\lambda^2 - 3\lambda) + 1 & -1 \\ -1 & -4(\lambda - 1)(2\lambda^2 - 3\lambda) + 1 \end{pmatrix}$$

From the above equation we get

$$K_D(\lambda) = \frac{-1}{(2\lambda - 2)(4\lambda^2 - 8\lambda + 2)}$$

and

$$K_T(\lambda) = \frac{(2\lambda)(2\lambda - 3)}{(2\lambda - 2)(4\lambda^2 - 8\lambda + 2)}$$

Then the energy renormalization factor $K_D(0) = \frac{1}{4}$
and the renormalized energy at level m $\epsilon_m(u) = (\frac{1}{4})^{-m} E_m(u)$.

In general $\epsilon_{m+1}(u) = (\frac{1}{4})^{-m+1} E_{m+1}(u) = \epsilon_0(u)$.

Hence the sequence $\{\epsilon_m(u)\}$is a non-decreasing sequence. It is constant when u is a linear function. Then if

$$\epsilon(u) = \lim_{n\to\infty} \epsilon_m(u) = \left(\frac{1}{4}\right)^{-m} E_m(u)$$

exists we say that u has a finite energy.

6 Laplacian Renormalization Factor

The spectral decimation function defined and denoted by

$$R(\lambda) = \frac{\lambda - K_T(\lambda)}{K_D(\lambda)} \tag{11}$$

that is

$$R(\lambda) = \frac{\lambda - \frac{(2\lambda)(2\lambda-3)}{(2\lambda-2)(4\lambda^2-8\lambda+2)}}{\frac{-1}{(2\lambda-2)(4\lambda^2-8\lambda+2)}} \tag{12}$$

$\Rightarrow R(\lambda) = -8\lambda\,(\lambda - 1)^2\,(\lambda - 2)$.

Proposition 5.1 (Shima). *The spectral decimation function R satisfies* $R(0) = 0$ *and* $R'(0) = \frac{r_0^{-1}}{K_D(0)}$ *where* r_0^{-1} *is a measure factor.*

Proof.

$$R(\lambda) = -8\lambda\,(\lambda - 1)^2\,(\lambda - 2) \Rightarrow R(0) = -8(0)\,(0 - 1)^2\,(0 - 2) = 0$$

and $R'(0) = 16$ and the right-hand side $\frac{r_0^{-1}}{K_D(0)} = \frac{4}{1/4} = 16$.

Hence R satisfies Proposition 5.1.

The Laplacian renormalization constant is $R'(0) = 16$.

7 Normalized Laplacian

We define the diagonal matrix $W = \begin{pmatrix} -T & 0 \\ 0 & M \end{pmatrix}$ and the normalized Laplacian $\hat{\Delta}_0$ and $\hat{\Delta}_1$ are as follows:

$$\hat{\Delta}_0 = -T^{-1}D$$

$$\hat{\Delta}_1 = -W^{-1}H_1$$

normalized Laplacian for Γ_0 and Γ_1 of Koch curve.

We have

$$W = \begin{pmatrix} 1\,0\,0\,0\,0 \\ 0\,1\,0\,0\,0 \\ 0\,0\,2\,0\,0 \\ 0\,0\,0\,2\,0 \\ 0\,0\,0\,0\,2 \end{pmatrix}$$

$$\hat{\Delta}_0 = -\begin{pmatrix} -1 & 0 \\ 0 & -1 \end{pmatrix}^{-1} \begin{pmatrix} -1 & 1 \\ 1 & -1 \end{pmatrix} = \begin{pmatrix} -1 & 1 \\ 1 & -1 \end{pmatrix}$$

$$\hat{\Delta}_1 = -\begin{pmatrix} 1\,0\,0\,0\,0 \\ 0\,1\,0\,0\,0 \\ 0\,0\,2\,0\,0 \\ 0\,0\,0\,2\,0 \\ 0\,0\,0\,0\,2 \end{pmatrix}^{-1} \begin{pmatrix} -1 & 0 & 1 & 0 & 0 \\ 0 & -1 & 0 & 0 & 1 \\ 1 & 0 & -2 & 1 & 0 \\ 0 & 0 & 1 & -2 & 1 \\ 0 & 1 & 0 & 1 & -2 \end{pmatrix} = \begin{pmatrix} -1 & 0 & 1 & 0 & 0 \\ 0 & -1 & 0 & 0 & 1 \\ 1/2 & 0 & -1 & 1/2 & 0 \\ 0 & 0 & 1/2 & -1 & 1/2 \\ 0 & 1/2 & 0 & 1/2 & -1 \end{pmatrix}$$

8 Forbidden Eigenvalues

Definition 6 (Forbidden Eigenvalues). We denote $\Psi = \{\lambda \in R : K_D(\lambda) = 0$ or $det(X + \lambda M) = 0\}$ and the elements in Ψ are the forbidden eigenvalues.

$\Psi_k = \left\{\lambda \in \Psi : \lambda \text{ is an eigenvalue of } -\hat{\Delta}_k\right\}$ and the elements in Ψ_k are the initial eigenvalues at step k or forbidden eigenvalues at step k.

To obtain the forbidden eigenvalues put $K_D(\lambda) = 0$ or $det(X + \lambda M) = 0$. But here $K_D(\lambda) = 0$ is not possible. So we consider $det(X + \lambda M) = 0$

That is $\lambda = 1, \frac{2\pm\sqrt{2}}{2}$

Therefore $\Psi = 1, \frac{2\pm\sqrt{2}}{2}$

To find the forbidden eigenvalues at step 1 or initial eigenvalues at step 1 is an eigenvalue of -$\hat{\Delta}_1$ and also in Ψ.

The eigenvalues of -$\hat{\Delta}_1$ is 0.0761, 0.6173, 1.9239, 1.3827, 1.

Therefore forbidden eigenvalues at step 1 is 1 only.

References

1. Fukushima, M., Shima, T.: On a spectral analysis for the sierpinski gasket. Potential Anal. **1**, 1–35 (1992)
2. Goldstein, S.: Random walks and diffusion on fractals. IMA Math. Appl. **8**, 121–129 (1987)
3. Kigami, J.: Analysis on Fractals. Cambridge University Press, Cambridge (2001)
4. Kusuoka, S.: A diffusion process on a fractal. In: Probabilistic Methods in Mathematical Physics, pp. 251–274. Academic, Boston (1987)
5. Shima, T.: On eigenvalue problems for Laplacians on p.c.f. self-similar sets. Jpn. J. Ind. Appl. Math. **13**, 1–23 (1996)
6. Strichartz, R.: Differential Equations on Fractals. Princeton University Press, Princeton (2006)
7. Teplyaev, A.: Spectral analysis on infinite Sierpinski gaskets. J. Funct. Anal. **159**, 537–567 (1999)

Projections of Mandelbrot Percolation in Higher Dimensions

Károly Simon and Lajos Vágó

Abstract We consider the fractal percolations which are one of the most well-studied examples of random Cantor sets. Rams and Simon studied the projections of fractal percolation sets on the plane. We extend the scope of their theorem and generalize it to higher dimensions. An extended version of this note is avaible on the arxiv [7].

Keywords Random cantor sets • Fractal percolation • Mandelbrot percolation • Projection of fractal sets

1 Introduction

Fractal percolations, or Mandelbrot percolations on the plane are defined in the following way: Fix an integer $M \geqslant 2$ and probabilities $0 < p_{i,j} < 1$, $i, j = 1, \ldots, M$. Then partition the unit square K into M^2 congruent squares of side length $1/M$, let us call these $K_{i,j}$, $i, j = 1, \ldots, M$. Then retain all small squares $K_{i,j}$ with probability $p_{i,j}$ independently from each other, or discard them otherwise. Repeat this procedure in the retained squares ad infinitum to finally get a random set E called fractal percolation.

It is well known that if all the probabilities $p_{i,j}$ are greater than $1/M$, then conditioned on non-emptiness E has Hausdorff dimension greater than 1 a.s. [3,5], and if the probabilities $p_{i,j}$ are smaller than a critical probability p_c, then it is totally disconnected [2]. However, in [6] the authors gave a rather complicated technical condition under which the orthogonal projection of E (which is a random dust) in all directions contains some interval, conditioned on E being nonempty.

K. Simon • L. Vágó (✉)
Department of Stochastics, Institute of Mathematics, Technical University of Budapest, P.O. Box 91, 1521 Budapest, Hungary
e-mail: simonk@math.bme.hu; vagolala@math.bme.hu

C. Bandt et al. (eds.), *Fractals, Wavelets, and their Applications*, Springer Proceedings in Mathematics & Statistics 92, DOI 10.1007/978-3-319-08105-2_11,

In our work, we generalize this result to higher dimensions, i.e. for fixed $d \geqslant 2$ and $d > k \geqslant 1$ we consider orthogonal projections of d-dimensional fractal percolation to all k-dimensional planes containing the origo at once, and obtain the same result as in dimension 2. We adapt the same random inverse Markov operator as in [6]. Also some geometrical issues have to be handled in the higher dimensional case.

1.1 Notations

We use the higher dimensional analogues of the notations of [6]. We denote by $E = E(\omega)$ the d-dimensional Mandelbrot percolation on the unit cube K, with retaining probabilities $p_{i^{(1)},\dots,i^{(d)}}$, $i^{(j)} = 1,\dots,M$. Fix $1 \leqslant k \leqslant d-1$. For $\alpha = \left(\mathbf{a}^{(1)},\dots,\mathbf{a}^{(k)}\right)$, $\mathbf{a}^{(1)},\dots,\mathbf{a}^{(k)}$ orthogonal unit vectors in $\mathbb{R}^d$ let P_α be the plane spanned by the vectors in α:

$$P_\alpha = span\{\mathbf{a}^{(1)},\dots,\mathbf{a}^{(k)}\}.$$

We consider the orthogonal projections $proj_\alpha$ of E to each k-dimensional planes P_α. Note that this projection is the higher dimensional pair of projection $proj_\alpha$ of [6]. Our goal is to determine the parameters $p_{i^{(1)},\dots,i^{(d)}}$ for which almost surely $int\{proj_\alpha E\} \neq \emptyset$ for all α, conditioned on E being nonempty.

In addition, we consider radial projections as well, i.e. given $t \in \mathbb{R}^d$, the radial projection with center t of set E is denoted by $Proj_t(E)$ and is defined as the set of vectors under which points of $E \setminus \{t\}$ are visible from t.

Given $\alpha = \left(\mathbf{a}^{(1)},\dots,\mathbf{a}^{(k)}\right)$ let $\gamma_\alpha = \left(\mathbf{c}^{(1)},\dots,\mathbf{c}^{(n-k)}\right)$ be collection of unit vectors such that $\alpha \cup \gamma_\alpha$ consists of pairwise orthogonal vectors in $\mathbb{R}^d$. It will be useful to handle projections parallel to some sides of the unite cube separately from other directions.

Assumption 1.1 *Let us first consider only the nonparallel planes, i.e. for which there exists a vector* $\mathbf{v} \in P_\alpha^\perp$ *such that* $\mathbf{v}_i \neq 0$ *for all* $i = 1,\dots,d$.

1.2 Results

Later we will define Condition $A(\alpha)$ on the set of probabilities $p_{i^{(1)},\dots,i^{(d)}}$, $i^{(j)} = 1,\dots,M$, which is the most important tool for the following theorems. At this point it is enough to know that if $p_{i^{(1)},\dots,i^{(d)}} > M^{-(d-k)}$ for all $i^{(j)} = 1,\dots,M$, then Condition $A(\alpha)$ holds for all α.

Theorem 1.2. *Let* d *and* $1 \leqslant k < d$ *be fixed, and suppose that Condition* $A(\alpha)$ *holds for all* α *satisfying Assumption 1.1. In addition, to control parallel directions we suppose that for all distinct* $j_1,\dots,j_k \in \{1,\dots,d\}$ *and for all* $i^{(j_1)},\dots,i^{(j_k)} \in \{0,\dots,M-1\}$

$$\sum_{i^{(j_{k+1})},\dots,i^{(j_d)}=0}^{M-1} p_{i^{(1)},\dots,i^{(d)}} > 1,$$

where $\{j_{k+1},\dots,j_d\} = \{1,\dots,d\} \setminus \{j_1,\dots,j_k\}$. *Then almost surely for* all α *orthogonal projections* $proj_\alpha(E)$ *have nonempty interior, conditioned on* E *being nonempty.*

Theorem 1.3. *Suppose that the conditions of Theorem 1.2 hold. Then almost surely for* all α *and* all t *orthogonal projections* $proj_\alpha(E)$ *and radial projections* $Proj_t(E)$ *have nonempty interior, conditioned on* E *being nonempty.*

The case of parallel directions is fully covered by the paper of Falconer and Grimmett [4], so now on we suppose that Assumption 1.1 holds.

The rest of the paper is organized as follows. In Sect. 2 we consider orthogonal projections of the fractal percolation and prove Theorem 1.2. Then in Sect. 3 we turn our attention to radial projections and using the same argument as in [6] we show that Theorem 1.3 holds.

2 Orthogonal Projections

2.1 Notations

2.1.1 Projection Π_α

It is useful to consider another linear projection called Π_α of E instead of $proj_\alpha$. To define it let $\mathbf{e}^{(i)}$ stand for the vector with all 0 entries except for the i-th, which is 1. Let $\binom{[d]}{k}$ stand for the k element subsets of $[d] = \{1,\dots,d\}$. Then, for $S = \{i_1,\dots,i_k\} \in \binom{[d]}{k}$ let P_S stand for the coordinate-plane spanned by the unit vectors corresponding to S:

$$P_S := span\left\{\mathbf{e}^{(i_1)},\dots,\mathbf{e}^{(i_k)}\right\}. \tag{1}$$

Then Π_α is the linear projection to P_S (with a suitable S defined later), in direction γ_α, that is for $\mathbf{x} \in \mathbb{R}^d$

$$\Pi_\alpha(\mathbf{x}) = \left\{span\{\mathbf{c}^{(1)},\dots,\mathbf{c}^{(d-k)}\} + \mathbf{x}\right\} \cap P_S \in \mathbb{R}^d, \tag{2}$$

where $\mathbf{c}^{(1)},\dots,\mathbf{c}^{(n-k)}$ are the vectors of γ_α. Let

$$\underline{\underline{C}} = \left[\mathbf{c}^{(1)}\middle|\dots\middle|\mathbf{c}^{(d-k)}\right].$$

One can easily describe the projection defined in (2) by matrix operations as well. Let $\mathbf{h}$ be defined by

$$\left[\begin{array}{c|c} \underline{\underline{i}} & -\underline{\underline{C_1}} \\ \hline \underline{\underline{0}} & -\underline{\underline{C_2}} \end{array}\right] \mathbf{h} = \mathbf{x}, \tag{3}$$

where $\underline{\underline{i}}$ is the $k \times k$ identity matrix; $\underline{\underline{0}}$ is a $(d-k) \times k$ matrix with all zero entries; $\underline{\underline{C_1}}$ is a $k \times (d-k)$ matrix with the rows of $\underline{\underline{C}}$ corresponding to S (in the order of $i_1, \ldots, i_k$); and $\underline{\underline{C_2}}$ is a $(d-k) \times (d-k)$ matrix formed by the rest of the rows of $\underline{\underline{C}}$ (in order, let us denote these by $i_{k+1}, \ldots, i_d$). Then

$$\Pi_\alpha(\mathbf{x}) = \sum_{j=1}^{k} \mathbf{h}_j \mathbf{e}^{(i_j)} = \sum_{j=k+1}^{d} \mathbf{h}_j \mathbf{c}^{(j-k)} + \mathbf{x} \tag{4}$$

if $\mathbf{h}$ is well defined. Using the Cauchy–Binet formula [1, Sect. 4.6, pp. 208–214] we obtain that

$$1 = \det\left(\underline{\underline{C}}^T \underline{\underline{C}}\right) = \sum_{S' \in \binom{[d]}{k}} \left(\det\left(\underline{\underline{C_2}}(S')\right)\right)^2.$$

Thus for any α there exists $S' \in \binom{[d]}{k}$ such that

$$\left|\det\left(\underline{\underline{C_2}}(S')\right)\right| \geqslant \frac{1}{\sqrt{\binom{d}{k}}}.$$

Let us choose this S' and denote the matrix in (3) by $\underline{\underline{M}}_\alpha$. Since $\det(\underline{\underline{M}}_\alpha) = -\det(\underline{\underline{C_2}}(S))$, hence $\mathbf{h}$ exists and unique.

By symmetry, without any loss of generality we fix $S = \{1, \ldots, k\}$ and restrict the set of directions to

$$A_S = \left\{\alpha \;\middle|\; \left|det\left(\underline{\underline{C_2}}(S)\right)\right| > \frac{1}{2\sqrt{\binom{d}{k}}}\right\}. \tag{5}$$

Then let Π_α be the projection to P_S for $\alpha \in A_S$. Then all α belongs to some of the possible coordinate planes $P_{S'}$, so now on we can restrict ourselves to S and A_S.

As it is argued in [6], for any α $int\{proj_\alpha E\} \neq \emptyset$ iff $int\{\Pi_\alpha E\} \neq \emptyset$. In addition, $\Pi_\alpha E$ lays in the same plane for all α, which will be useful when considering several directions at once, e.g. when considering nonlinear projections.

2.1.2 Condition A

Let us denote by Δ_α the Π_α projection of the unit cube K. Let $\varphi_{\underline{i}_n^{(1)},\dots,\underline{i}_n^{(d)}} : \mathbb{R}^d \to \mathbb{R}^d$, $\underline{i}_n^{(1)},\dots,\underline{i}_n^{(d)} \in \{0,\dots,M-1\}^n$ be defined by

$$\varphi_{\underline{i}_n^{(1)},\dots,\underline{i}_n^{(d)}}(\mathbf{x}) = \frac{1}{M^n}\mathbf{x} + \sum_{l=1}^{n} \frac{1}{M^l}\left(i_l^{(1)},\dots,i_l^{(d)}\right).$$

The $\varphi_{\underline{i}_n^{(1)},\dots,\underline{i}_n^{(d)}}$ image of the unit cube is a shifted cube of size $1/M^n$ and is called $K_{\underline{i}_n^{(1)},\dots,\underline{i}_n^{(d)}}$, i.e.

$$K_{\underline{i}_n^{(1)},\dots,\underline{i}_n^{(d)}} = \varphi_{\underline{i}_n^{(1)},\dots,\underline{i}_n^{(d)}}(K).$$

In addition we introduce the function $\psi_{\alpha,\underline{i}_n^{(1)},\dots,\underline{i}_n^{(d)}} : \Delta_\alpha \to \Delta_\alpha$ which is the inverse of $\Pi_\alpha \circ \varphi_{\underline{i}_n^{(1)},\dots,\underline{i}_n^{(d)}}$.

The following operator is defined on functions from Δ_α to nonnegative reals, vanishing on the boundary of Δ_α. This is one of the main tools of this paper, and is defined by

$$F_\alpha f(x) = \sum_{i^{(1)},\dots,i^{(d)}:x\in\Pi_\alpha\left(K_{i^{(1)},\dots,i^{(d)}}\right)} p_{i^{(1)},\dots,i^{(d)}} \cdot f \circ \psi_{\alpha,i^{(1)},\dots,i^{(d)}}(x).$$

It is easy to see that the n-th iterate of F_α equals

$$F_\alpha^n f(x) = \sum_{\underline{i}_n^{(1)},\dots,\underline{i}_n^{(d)}:x\in\Pi_\alpha\left(K_{\underline{i}_n^{(1)},\dots,\underline{i}_n^{(d)}}\right)} p_{\underline{i}_n^{(1)},\dots,\underline{i}_n^{(d)}} \cdot f \circ \psi_{\alpha,\underline{i}_n^{(1)},\dots,\underline{i}_n^{(d)}}(x),$$

where $p_{\underline{i}_n^{(1)},\dots,\underline{i}_n^{(d)}} = \prod_{j=1}^n p_{i_j^{(1)},\dots,i_j^{(d)}}$.

Definition 2.1 (Condition A). We say that the percolation satisfies condition $A(\alpha)$ if there exist sets $I_1^\alpha, I_2^\alpha \subset \Delta_\alpha$ which are similar to Δ_α and which have the same central as Δ_α, and a positive integer r such that

$$(i) I_1^\alpha \subset int I_2^\alpha,\ I_2^\alpha \subset int\Delta_\alpha,$$
$$(ii) F_\alpha^{r_\alpha}\mathbb{1}_{I_1^\alpha} \geqslant 2\mathbb{1}_{I_2^\alpha}.$$

This is the place where the geometrical complexity of the problem differs from that of the original case in [6]: If $d = 2$ and $k = 1$, then Δ_α is simply a section. However, if, for example, $d = 3$ and $k = 2$, then Δ_α is a hexagon, which carries some extra technical difficulties in the proof in the next section. The left side of Fig. 1 shows the mutual position of Δ_α, I_2^α and I_1^α for a fixed α and for $d = 3$, $k = 2$.

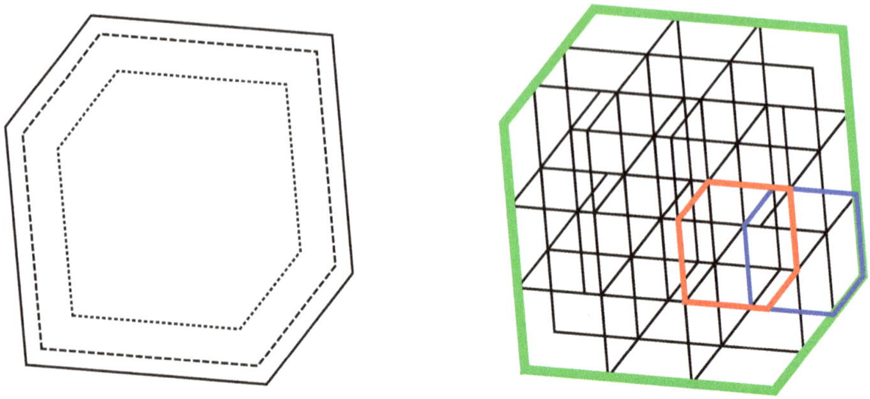

Fig. 1 We project from $d = 3$ to $k = 2$ dimension. **(a)** Δ_α(*continuous*), I_2^α(*dashed*) and I_1^α(*dotted*). **(b)** W_n with $n = 1$, $M = 3$. The big bold faced green contour is W_0. The blue and the red small contours correspond to cubes $K_{3,1,1}$ and $K_{3,3,2}$, respectively. As their coordinates show, these cubes are separated from each other, however after projection their top and bottom faces intersect. Hence we have at least two *rational* classes (see the definition in the proof of Lemma 2.4): the sides on the *top* and on the *bottom*

The following condition is stronger than Condition A, but it is easier to check it.

Definition 2.2 (Condition B). We say that the percolation satisfies condition $B(\alpha)$ if there exists a nonnegative continuous function $f : \Delta_\alpha \to \mathbb{R}$ such that f vanishes exactly on the boundaries of Δ_α and $\exists \varepsilon > 0$:

$$F_\alpha f \geqslant (1+\varepsilon) f. \tag{6}$$

In the following sections we show that condition B implies A (Sect. 2.2), which implies that $int\{\Pi_\alpha E\} \neq \emptyset$ conditioned on E being nonempty (Sect. 2.3), and for certain choice of the parameters $p_{i_1,\dots,i_d}$ we show some functions f satisfying Condition $B(\alpha)$ for all α (Sect. 2.4).

2.2 Condition B Implies Condition A

Proposition 2.3. *For all α satisfying Assumption 1.1 condition $B(\alpha)$ implies condition $A(\alpha)$.*

Lemma 2.4. *For all α satisfying Assumption 1.1 there exists an integer $n > 0$ and there exist sets $I_1^\alpha, I_2^\alpha \subset \Delta_\alpha$ with the same properties as in the definition of Condition A except for (ii), which is replaced by (ii^*):*

$$(ii^*) \forall x \in I_2^\alpha \;\; F_\alpha^n g_1(x) \geqslant (1+\varepsilon)\, g_2(x).$$

where $g_1 = f|_{I_1}$, $g_2 = f|_{I_2}$.

Proof of Proposition 2.3. Using Lemma 2.4 Condition A holds with the smallest multiple r of n satisfying

$$(1+\varepsilon)^{r/n} \geqslant 2\frac{\max_{x\in I_1} g_1(x)}{\min_{x\in I_2} g_2(x)}.$$

□

To prove Lemma 2.4 we need the following definition. Let us denote by W_0 the boundary of Δ_α and by W_n the projection of the union of the boundaries of the $1/M^n$ size (level n) cubes. The right side of Fig. 1 shows W_n with $d=3, k=2$, $n=1$, and $M=3$.

Sketch of the Proof of Lemma 2.4: Basically we would like to follow the idea of the proof of [6, Lemma 8], but since the projection onto the k-dimensional plane is geometrically more complicated, now we explain how the proof is carried out. The main difference is that while the proof of case $d=2, k=1$ uses the fact that the sets W_0 and $W_n \setminus W_0$ are separated, the same is not true if $k>1$. This difficulty is handled by splitting the sides of W_0 to *rational* and *irrational classes* and by showing that for rational classes some kind of periodicity occurs, while for irrational classes there is a separation similar to that of case $k=1$ and therefore we can use the continuity of function f.

Proof of Lemma 2.4. Fix α and suppose that Condition $B(\alpha)$ holds for some f and $\varepsilon>0$. By Assumption 1.1 the Π_α image of the $k-1$-dimensional sides of K are $k-1$-dimensional. Some of these form W_0, call these the sides contained in W_0. Let us say that each of these sides forms a *class*, so totally we have $2\binom{d}{k-1}$ classes. In addition, for any $n \geqslant 1$ the sides contained in W_n can be ordered into these classes in such a way that the sides with given relative position to the projection of their cube are ordered into the same class. Note that each side belongs to two different cubes and hence to two different classes. In this proof two types of classes are handled separately which are called rational and irrational. To define these it will be useful to use another indices to denote the level n cubes: for $l=1,\dots,d$, $i^{(l)}=1,\dots,M^n$ let

$$K_{i^{(1)},\dots,i^{(d)}} = M^{-n}[i^{(1)}-1, i^{(1)}] \times \dots \times [i^{(d)}-1, i^{(d)}].$$

If we do not specify the level of a cube given by these coordinates, then it can be any level $n \geqslant 1$.

We denote by $Q^u_{i^{(1)},\dots,i^{(d)}}$ the $k-1$-dimensional plane containing the class u side of $\Pi_\alpha(K_{i^{(1)},\dots,i^{(d)}})$. Let us call a class u *rational* if for n big enough there exists two level n cubes $K_{i^{(1)},\dots,i^{(d)}}$ and $K_{j^{(1)},\dots,j^{(d)}}$ such that

$$Q^u_{i^{(1)},\dots,i^{(d)}} = Q^u_{j^{(1)},\dots,j^{(d)}} \text{ , but } U_{i^{(1)},\dots,i^{(d)}} \neq U_{j^{(1)},\dots,j^{(d)}},$$

where $U_{i^{(1)},\dots,i^{(d)}}$ and $U_{j^{(1)},\dots,j^{(d)}}$ are the $k-1$-dimensional sides of cubes $K_{i^{(1)},\dots,i^{(d)}}$ and $K_{j^{(1)},\dots,j^{(d)}}$, respectively, corresponding to the class u sides of shadows

$\Pi_\alpha(K_{i^{(1)},\dots,i^{(d)}})$ and $\Pi_\alpha(K_{j^{(1)},\dots,j^{(d)}})$. See Fig. 1. The other classes are called irrational.

First we discuss rational classes. For any $n \geqslant 1$ we give a uniform lower bound in $x \in \Delta_\alpha$ on the ratio of the number of level n cubes $K_{i^{(1)},\dots,i^{(d)}}$ for which $x \in \Pi_\alpha\left(K_{i^{(1)},\dots,i^{(d)}}\right)$ and the distance between x and any rational side of $K_{i^{(1)},\dots,i^{(d)}}$ is greater than η_1/M^n with some $\eta_1 > 0$, compared to the number of those cubes for which $x \in \Pi_\alpha\left(K_{i^{(1)},\dots,i^{(d)}}\right)$.

Suppose that $Q^u_{\hat{i}^{(1)},\dots,\hat{i}^{(d)}} = Q^u_{\tilde{i}^{(1)},\dots,\tilde{j}^{(d)}}$. Then for any $n \geqslant 1$ and for any level n indices $i^{(1)},\dots,i^{(d)}$

$$Q^u_{i^{(1)}+k_1(\tilde{i}-\hat{i}^{(1)}),\dots,i^{(d)}+k_1(\tilde{i}^{(d)}-\hat{i}^{(d)})} = Q^u_{i^{(1)}+k_2(\tilde{i}^{(1)}-\hat{i}^{(1)}),\dots,i^{(d)}+k_2(\tilde{i}^{(d)}-\hat{i}^{(d)})}$$

for any $k_1,k_2 \in \mathbb{Z}$ as long as $0 \leqslant i^{(l)} + k_j(\tilde{i}^{(l)} - \hat{i}^{(l)}) \leqslant M^n$ for all $l = 1,\dots,d$, $j = 1,2$. Hence for all $\mathbf{x} \in \Delta_\alpha$ and for any indices $i^{(1)},\dots,i^{(d)}$

$$dist\left(\mathbf{x}\,;\ Q^u_{i^{(1)},\dots,i^{(d)}}\right) = dist\left(\mathbf{x}\,;\ Q^u_{i^{(1)}+k(\tilde{i}^{(1)}-\hat{i}^{(1)}),\dots,i^{(d)}+k(\tilde{i}^{(d)}-\hat{i}^{(d)})}\right)$$

holds since the planes are the same, where $dist(\ .\ ;\ .\ )$ is the usual distance in $\mathbb{R}^d$. We can prove that the same is true while considering all rational classes at once: There exist $\hat{j}^{(1)},\dots,\hat{j}^{(d)} \in \mathbb{Z}^d$ such that for all rational class u, for all level n, for all $i^{(1)},\dots,i^{(d)}$ and for all k

$$dist\left(\mathbf{x}\,;\ Q^u_{i^{(1)},\dots,i^{(d)}}\right) = dist\left(\mathbf{x}\,;\ Q^u_{i^{(1)}+k\,\hat{j}^{(1)},\dots,i^{(d)}+k\,\hat{j}^{(d)}}\right), \tag{7}$$

as long as $0 \leqslant i^{(l)} + k\ \hat{j}^{(l)} \leqslant M^n$ for all $l = 1,\dots,d$.

We show that there exist $\eta_1 > 0$ such that for all $\mathbf{x} \in \Delta_\alpha \setminus B_{\eta_1/M}(W_0)$ for all level n cube $K_0 = K_{i^{(1)},\dots,i^{(d)}}$ such that $\mathbf{x} \in \Pi_\alpha(K_0)$ the projection of one of the level n neighbors of K_0 or the projection of K_0 itself contains the η_1/M^n neighborhood of $\mathbf{x}$. To see this let us denote by K_i, $i = 1,\dots,2d$ the level n cubes which have a common $d-1$-dimensional side with K_0. For $i = 0,\dots,2d$ we denote by V_i the open shadows of these cubes, i.e.

$$V_i = int\ \Pi_\alpha(K_i).$$

Using Assumption 1.1 it is easy to see that the sets V_i cover $\Pi_\alpha(K_0)$. Figure 2 shows $\Pi_\alpha(K_0)$ and the covering for $d = 3$ and $k = 2$. Hence we can choose η_1 to be smaller than M^n times the Lebesgue number of the finite open covering $\{V_i\}_{i=0}^{2d}$, which is always positive [8, Theorem 0.20] and by similarity proportional to M^{-n}. For the definition of the Lebesgue number of a covering, see [8]. Note that if one of the cubes K_i, $i = 1,\dots,2d$ is out of the unit cube, then we don't need that cube for the cover, since our goal is to cover $K_0 \setminus B_{\eta_1/M}(W_0)$. Another note is that the same argument remains valid for any η such that $\eta_1 > \eta > 0$ holds.

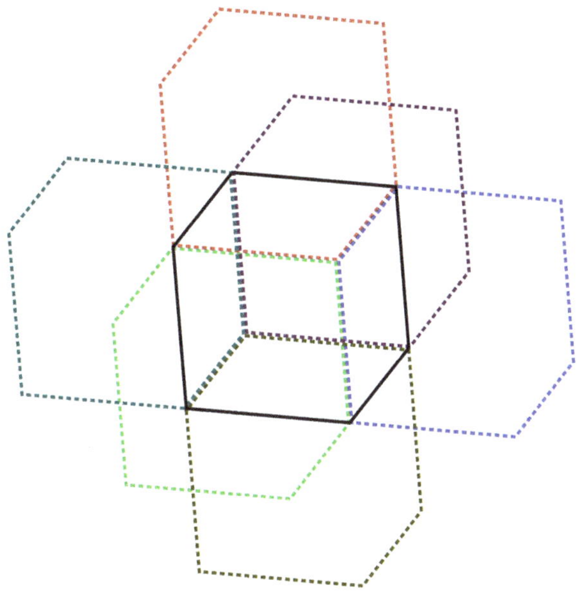

Fig. 2 We project from $d = 3$ to $k = 2$ dimension. The covering of $\Pi_\alpha(K_0)$ (*continuous*) with the open sets V_i (*dotted*), $i = 0, \ldots, 6$

Now we are able to handle rational classes. For $\epsilon > 0$ and for $\mathbf{x} \in \mathbb{R}^d$ let us denote by $B_\epsilon(\mathbf{x})$ the ϵ neighborhood of $\mathbf{x}$, i.e.

$$B_\epsilon(\mathbf{x}) = \left\{\mathbf{y} \in \mathbb{R}^d \;\middle\|\; \|\mathbf{x} - \mathbf{y}\|_2 \leqslant \epsilon\right\}.$$

By the definition of η_1 for all $\mathbf{x} \in \Delta_\alpha \setminus B_{\eta_1/M}(W_0)$ and for all n there exists a level n cube $K_{i^{(1)},\ldots,i^{(d)}}$ such that

$$B_{\eta_1/M^n}(\mathbf{x}) \subseteq \Pi_\alpha(K_{i^{(1)},\ldots,i^{(d)}}).$$

In addition, using the periodicity described in (7), cubes with such property follow each other periodically, with period independent of $\mathbf{x}$. Hence we conclude that there exists $\vartheta > 0$ such that for any $\eta_1 > \eta > 0$, for any level n and for all $\mathbf{x} \in \Delta_\alpha \setminus B_{\eta/M}(W_0)$

$$\begin{aligned}\#\left\{(i^{(1)}, \ldots, i^{(d)}) \;\middle|\; \mathbf{x} \in \Pi_\alpha\left(K_{i^{(1)},\ldots,i^{(d)}}\right) \;\&\; dist\left(\mathbf{x}\,;\; Q^u_{i^{(1)},\ldots,i^{(d)}}\right) > \eta/M^n \;\forall u\right\} \geqslant \\ \geqslant \vartheta \#\left\{(i^{(1)}, \ldots, i^{(d)}) \;\middle|\; \mathbf{x} \in \Pi_\alpha\left(K_{i^{(1)},\ldots,i^{(d)}}\right)\right\}.\end{aligned} \tag{8}$$

Note that ϑ does not depend on η_1, only the length of a period has effect on it. This equation will be sufficient to handle rational classes.

To handle irrational classes as well let us say that the number of these classes is C_2. It is clear that there exists an integer n such that

$$\vartheta(1+\varepsilon)^n - C_2 > 1+\varepsilon, \tag{9}$$

where ε was defined in Condition $B(\alpha)$. Choose $\eta_2 > 0$ to be small enough to be smaller than the half of the smallest distance between any two level n sides falling in the same irrational class and to have

$$\inf_{\mathbf{x}\in\Delta_\alpha\setminus B_{\eta_2/M}(W_0)} f(\mathbf{x}) \geqslant \sup_{\mathbf{x}\in B_{\eta_2/M^n}(W_0)} f(\mathbf{x}), \tag{10}$$

which can be easily achieved by the continuity of f. Setting $\eta = \min\{\eta_1, \eta_2\}$, we define the sets needed by

$$I_1 := \Delta_\alpha \setminus B_\eta(W_0) \text{andby} I_2 := \Delta_\alpha \setminus B_{\eta/M}(W_0).$$

When showing that the assertion of the lemma holds we distinguish two cases:

- If $\mathbf{x} \in \Delta_\alpha \setminus B_{\eta/M^n}(W_n)$, then using the definition of F_α and (6) we obtain

$$F_\alpha^n g_1(\mathbf{x}) = F_\alpha^n f(\mathbf{x}) \geqslant (1+\varepsilon)^n f(\mathbf{x}) > (1+\varepsilon)\, g_2(\mathbf{x}).$$

- The other case is when $\mathbf{x} \in B_{\eta/M^n}(W_n) \setminus B_{\eta/M}(W_0)$. Using (8) and the definition of η

$$\#\left\{(\underline{i}_n, \underline{j}_n) \,\middle|\, \mathbf{x} \in \Pi_\alpha\left(\varphi_{\underline{i}_n, \underline{j}_n}\left(B_\eta(W_0)\right)\right)\right\} \leqslant$$
$$\leqslant (1-\vartheta)\#\left\{(\underline{i}_n, \underline{j}_n) \,\middle|\, \mathbf{x} \in \Pi_\alpha\left(K_{\underline{i}_n, \underline{j}_n}\right)\right\} + C_2.$$

Putting this together with (10) and with the definition of F_α we get

$$F_\alpha^n g_1(\mathbf{x}) \geqslant F_\alpha^n f(\mathbf{x}) - (1-\vartheta) F_\alpha^n f(\mathbf{x}) - C_2 f(\mathbf{x}) = \vartheta F_\alpha^n f(\mathbf{x}) - C_2 f(\mathbf{x})\ \forall \mathbf{x} \in \Delta_\alpha. \tag{11}$$

Then by (11), (6), and (9) we have

$$F_\alpha^n g_1(\mathbf{x}) \geqslant (1+\varepsilon)\, f(\mathbf{x}) + (\vartheta(1+\varepsilon)^n - 1 - \varepsilon - C_2)\, f(\mathbf{x}) \geqslant (1+\varepsilon)\, g_2(\mathbf{x}).$$

□

2.3 Condition A Implies Nonempty Interior

2.3.1 Robustness

To handle all directions at once we show that the robustness property described in [6, Sect. 4.3] holds in the higher dimensional case as well. Suppose that condition $A(\alpha)$ holds for some $\alpha = (\mathbf{a}^{(1)}, \ldots, \mathbf{a}^{(k)})$ with I_1^α, I_2^α and r. Let δ be the Hausdorff distance between I_1^α and I_2^α and let I_1' be the $\delta/2$ neighborhood of I_1^α. Let the distance between α and $\beta = \left(\mathbf{b}^{(1)}, \ldots, \mathbf{b}^{(k)}\right)$ be

$$\|\alpha - \beta\| = \max_{i=1,\ldots,k} \left\|\mathbf{a}^{(i)} - \mathbf{b}^{(i)}\right\|_2. \tag{12}$$

Proposition 2.5 (Robustness). *Fix α, I_1^α, I_2^α, δ and I_1' as before. Then there exists a constant $0 < c(\alpha) < \infty$ such that if $\|\alpha - \beta\| < c(\alpha)$, then*

$$\Pi_\alpha \circ \varphi_{\underline{i_r}^{(1)},\ldots,\underline{i_r}^{(d)}}(I_1^\alpha) \subset \Pi_\beta \circ \varphi_{\underline{i_r}^{(1)},\ldots,\underline{i_r}^{(d)}}(I_1').$$

Proof. It is enough to show that

$$\left\|\Pi_\alpha \mathbf{x} - \Pi_\beta \mathbf{x}\right\|_2 < M^{-r}\delta/2 \quad \text{forall } \mathbf{x} \in \bigcup_{\underline{i_r}^{(1)},\ldots,\underline{i_r}^{(d)}} \varphi_{\underline{i_r}^{(1)},\ldots,\underline{i_r}^{(d)}}(I_1').$$

Note that $\Pi_\alpha \mathbf{x}$ was defined in (4), and that given α (and S') we denote the matrix in (3) by M_α. Then

$$\left\|\Pi_\alpha \mathbf{x} - \Pi_\beta \mathbf{x}\right\|_2 \leqslant \left\|\underline{\underline{M_\alpha}}^{-1} - \underline{\underline{M_\beta}}^{-1}\right\| \|\mathbf{x}\|_2 \leqslant \left\|\frac{adj\left(\underline{\underline{M_\alpha}}\right)}{det\left(\underline{\underline{M_\alpha}}\right)} - \frac{adj\left(\underline{\underline{M_\beta}}\right)}{det\left(\underline{\underline{M_\beta}}\right)}\right\| \sqrt{d}$$

$$\leqslant \frac{\sqrt{d}}{\left|det\left(\underline{\underline{M_\alpha}}\right)\right|}\left\|adj\left(\underline{\underline{M_\alpha}}\right) - adj\left(\underline{\underline{M_\beta}}\right)\right\| + \sqrt{d}\left\|adj\left(\underline{\underline{M_\beta}}\right)\right\| \left|\frac{1}{det\left(\underline{\underline{M_\alpha}}\right)} - \frac{1}{det\left(\underline{\underline{M_\beta}}\right)}\right|, \tag{13}$$

where $\|.\|$ denotes the induced norm of $\|.\|_2$. Since the determinants are uniformly separated from zero (5) and M_β has entries in $[0, 1]$, hence the expression on the right-hand side of (13) can be arbitrarily small with $c(\alpha)$ small enough. □

Corollary 2.6. *Similarly for any $\varepsilon > 0$ if*

$$\|\alpha - \beta\| < c(\alpha)\varepsilon/2,$$

then by roughly estimating $\left\|\underline{\underline{M_\alpha}}^{-1} - \underline{\underline{M_\beta}}^{-1}\right\|$, we obtain that

$$\left\| \Pi_\alpha \mathbf{x} - \Pi_\beta \mathbf{x} \right\|_2 < \varepsilon M^{-r}\delta/2 \quad \textit{for all } \mathbf{x} \in \bigcup_{\underline{i}_r^{(1)},\dots,\underline{i}_r^{(d)}} \varphi_{\underline{i}_r^{(1)},\dots,\underline{i}_r^{(d)}}(I_1').$$

Corollary 2.7. *An immediate consequence of Proposition 2.5 is that Condition A holds for all directions in*

$$J = B_{c(\alpha)}(\alpha) \cap A_S$$

with the same I_1', I_2^α and r.

2.3.2 The Proof

Hence we can restrict ourselves to one such range J. The proof follows the proof of [6, Sect. 5].

Note that it is enough to prove that the nonempty interior in Theorem 1.2 exists with positive probability. This is because almost surely conditioned on E being nonempty for any N there exists n such that there are at least N many retained level n cubes which will not vanish totally. In addition, events happening in different cubes are independent and statistically similar. Hence if the interior of all orthogonal projections is nonempty with positive probability, then the same holds almost surely.

Proof of Theorem 1.2. Let us write I_1 for I_1', I_2 for I_2^α and δ' for the new Hausdorff distance $\delta/2$. Assume $\mathbf{x}, \mathbf{y} \in \Delta_\alpha$, $\|x - y\|_2 < \delta' M^{-nr}/2$, and that for $\alpha, \beta \in J$

$$\|\alpha - \beta\| < c(\alpha) M^{-(n-1)d^3 r} 2^{-d^3}.$$

Then from Corollary 2.6 with $\varepsilon = M^{-(n-1)r}/2$, it follows that $\left|\Pi_\alpha(\mathbf{x}) - \Pi_\beta(\mathbf{y})\right|_2 < \delta' M^{-nr}/2$, and hence

$$G_\beta^r \mathbb{1}_{I_2}(\psi_{\underline{i}_{nr}^{(1)},\dots,\underline{i}_{nr}^{(d)}}(x)) \geqslant G_\alpha^r \mathbb{1}_{I_1}(\psi_{\underline{i}_{nr}^{(1)},\dots,\underline{i}_{nr}^{(d)}}(y)). \tag{14}$$

For given n let X_n be a $\delta' M^{-nr}/2$ dense subset of I_1 and Y_n be a $c(\alpha)M^{-(n-1)d^3 r}$ 2^{-d^3} dense subset of J such that

$$\#(X_n \times Y_n) \leqslant c M^{d^4 nr},$$

with some constant c. For any $(x, \theta) \in I_1' \times J$ we define a sequence of random variables

$$V_n(x,\theta) = \#\left\{(\underline{i}_{nr}^{(1)},\dots,\underline{i}_{nr}^{(d)}) \in \mathcal{E}_{nr} \;\middle|\; x \in \Pi_\theta \circ \varphi_{\underline{i}_{nr}^{(1)},\dots,\underline{i}_{nr}^{(d)}}(I_2)\right\},$$

where $\mathcal{E}_n = \mathcal{E}_n(\omega)$ stands for the set of retained level n cubes. We prove that with positive probability $V_n(\mathbf{x},\theta) \geqslant (3/2)^n$ for all $n, \mathbf{x}, \theta$. Let us use induction on n and note that the $n = 0$ case is obvious. For $(\mathbf{y}, \kappa) \in X_{n+1} \times Y_{n+1}$ let

$$Z(y,\kappa)=\left\{(\mathbf{x},\theta)\in I_1'\times J\,|\,|\mathbf{x}-\mathbf{y}|<\delta' M^{-(n+1)r}/2\ \&\ |\theta-\kappa|<M^{-nd^3r}2^{d^3}c(\alpha)\right\}.$$

Clearly the sets $Z(y,\kappa)$ cover $I_1'\times J$.

By the inductive hypothesis with positive probability $V_n(\mathbf{x},\theta)\geqslant(3/2)^n$. For each level nr cube $K_{\underline{i}^{(1)}_{nr},\ldots,\underline{i}^{(d)}_{nr}}$ in $V_n(\mathbf{x},\theta)$ the number of its sub-cubes in $V_{n+1}(\mathbf{x},\theta)$ is given by $G_\theta^r\mathbb{1}_{I_2}(\psi_{\underline{i}^{(1)}_{nr},\ldots,\underline{i}^{(d)}_{nr}}(\mathbf{x}))$, which by (14) can be bounded from below uniformly in $(\mathbf{x},\theta)\in Z(\mathbf{y},\kappa)$ by

$$G_\theta^r\mathbb{1}_{I_2}(\psi_{\underline{i}^{(1)}_{nr},\ldots,\underline{i}^{(d)}_{nr}}(\mathbf{x}))\geqslant G_\kappa^r\mathbb{1}_{I_1}(\psi_{\underline{i}^{(1)}_{nr},\ldots,\underline{i}^{(d)}_{nr}}(\mathbf{y})).$$

The expected value of this random variable is 2, and it is bounded below by 0, above by M^{dr}. Moreover, random variables coming from different level nr cubes are independent.

By Azuma–Hoeffding inequality

$$\mathbb{P}\left(\sum_{\substack{\left(\underline{i}^{(1)}_{nr},\ldots,\underline{i}^{(d)}_{nr}\right)\in\\ \in\mathcal{E}_{nr}\cap D_{nr}(\mathbf{x},I_2,\theta)}} G_\kappa^r\mathbb{1}_{I_1}(\psi_{\underline{i}^{(1)}_{nr},\ldots,\underline{i}^{(d)}_{nr}}(\mathbf{y}))\geqslant\left(\frac{3}{2}\right)^{n+1}\,\middle|\,V_n(\mathbf{y},\kappa)\geqslant\left(\frac{3}{2}\right)^n\right)\geqslant 1-\epsilon^{(3/2)^n},$$

where $0<\epsilon<1$ is fixed. Hence using the notation

$$E_n=\left\{(\forall\mathbf{x}\in X_n)(\forall\theta\in Y_n)\ V_n(\mathbf{x},\theta)\geqslant\left(\frac{3}{2}\right)^n\right\}$$

we have

$$\mathbb{P}\Big(E_{n+1}\,|\,E_n\Big)\geqslant\left(1-\epsilon^{(3/2)^n}\right)^{cM^{2knr}}.$$

Summation in n converges, hence for any $p_{i^{(1)},\ldots,i^{(d)}}$, $i^{(j)}=1,\ldots,M$ satisfying Condition B, a.s. $\Pi_\alpha(E(\omega))$ has nonempty interior for all α satisfying Assumption 1.1, conditioned on E being nonempty. □

2.4 How to Choose f?

The case of equal probabilities can be handled as in [6]: Suppose that $p_{i^{(1)},\ldots,i^{(d)}}=p>1/M^{d-k}$ for all indices $i^{(1)},\ldots,i^{(d)}$. Then let us define the function $f:\Delta_\alpha\to\mathbb{R}^+$ needed in Condition $B(\alpha)$ by

$$f(\mathbf{x}) = \sum_{i_1,\ldots,i_d=1}^{M} \left|\psi_\alpha^{-1}(\mathbf{x}) \cap K_{i_1,\ldots,i_d}\right|,$$

where $|.|$ denotes the d-dimensional Lebesgue measure. This f vanishes continuously on the borders of Δ_α, and strictly positive inside. In addition, it is obvious that if all of the $p_{i_1,\ldots,i_d}$-s are strictly greater than $1/M^{d-k}$, then

$$F_\alpha f \geqslant M^{d-k} p \cdot f.$$

Therefore the requirements of Condition $B(\alpha)$ are satisfied for all α and hence we can apply Theorem 1.3.

3 Radial Projections

In this section we consider the radial projections $Proj_t(E)$ with center $t \in \mathbb{R}^d$ of the percolation fractal. Recall that $Proj_t(E)$ is the set of vectors under which points of $E \setminus \{t\}$ are visible from t. Our goal is to prove Theorem 1.3. We do this as it is done in [6] for the two-dimensional case: We introduce a notion called *Almost linear family of projections* for which, using the robustness property, it is easy to show that almost surely for all member of the family the interior of the projection of the percolation fractal is nonempty. Then we show that radial projections can be viewed as an *Almost linear family of projections*.

3.1 Almost Linear Family of Projections

We fix the dimensions d and k and recall that $A_{\{1,\ldots,k\}}$ was defined in (5). Consider a parametrized family of projections

$$S_t(\mathbf{x}) : K \to \bigcup_{\substack{\alpha \in A_{\{1,\ldots,k\}}: \\ \text{Assumption 1.1 holds}}} \Delta_\alpha \subset span\{\mathbf{e}^{(1)}, \ldots, \mathbf{e}^{(k)}\}, \tag{15}$$

$t \in T$. For all $\mathbf{x} \in K$ we define $\alpha_t(\mathbf{x})$ by

$$S_t(\mathbf{x}) = \Pi_{\alpha_t(\mathbf{x})}(\mathbf{x}). \tag{16}$$

Note that $span\{\alpha_t(\mathbf{x})\} = S_{\alpha_t(\mathbf{x})}$ is not always well defined, since the only restriction on it is to have $(S_t(\mathbf{x}) - \mathbf{x}) \in P_{\alpha_t(\mathbf{x})}$ and $\alpha_t(\mathbf{x}) \in A_{\{1,\ldots,k\}}$. We suppose that $\alpha_t(\mathbf{x})$ is such that $S_{\alpha_t(\mathbf{x})}$ is not a coordinate plane.

Definition 3.1 (Almost Linear Family of Projections). We say that a family $\{S_t\}_{t\in T}$ (S_t satisfies (15)) is an *almost linear family of projections* if we can choose $\alpha_t(\mathbf{x})$ (defined by (16)) in such a way that the following properties are satisfied. We set J as the range of vectors for which Condition $A(\alpha)$ is satisfied with the same I_1, I_2, and r. We denote by δ the Hausdorff distance between I_1 and I_2.

i) $\alpha_t(\mathbf{x}) \in J$ for all $t \in T$ and $\mathbf{x} \in K$.
ii) $\alpha_t(\mathbf{x})$ is a Lipschitz function of $\mathbf{x}$, with the Lipschitz constant not greater than $\delta/3$. This guarantees in particular that $S_t(K_{\underline{i}_n^{(1)},\dots,\underline{i}_n^{(d)}})$ is connected.
iii) For any n we can divide T into subsets $Z_i^{(n)}$ such that whenever $t, s \in Z_i^{(n)}$ and $\mathbf{x}, \mathbf{y} \in K_{\underline{i}_n^{(1)},\dots,\underline{i}_n^{(d)}}$, we have

$$|\alpha_t(\mathbf{x}) - \alpha_s(\mathbf{y})| \leqslant c(\alpha) M^{-(n-1)d^3 r} 2^{-d^3}.$$

Moreover, we can do that in such a way that $\#\{Z_i^{(n)}\}$ grows only exponentially fast with n.

In the following we show that for an almost linear family of projections $\{S_t\}_{t\in T}$ almost surely $S_t(E)$ has nonempty interior for all t conditioned on E being nonempty. The proof follows the proof of [6, Theorem 14] and is a modified version of the proof in Sect. 2.3.2.

Proof. Let

$$V_n(x,t) = \#\left\{(\underline{i}_{nr}^{(1)},\dots,\underline{i}_{nr}^{(d)}) \in \mathcal{E}_{nr} \;\middle|\; x \in S_t \circ \varphi_{\underline{i}_{nr}^{(1)},\dots,\underline{i}_{nr}^{(d)}}(I_2)\right\}.$$

We prove inductively that with positive probability $V_n(\mathbf{x},\theta) \geqslant (3/2)^n$ for all $n, \mathbf{x}, \theta$. The $n = 0$ case is obvious, a.s. $V_0(\mathbf{x},\theta) = 1$ for all $\mathbf{x}, \theta$. For given n let X_n be a $\delta M^{-nr}/2$ dense subset of I_1. Then we can cover $I_1 \times T$ with at most exponentially many sets of the form $B_{\delta M^{-(n+1)r}/2}(\mathbf{x}_i) \times Z_j^{(n+1)r}$, $\mathbf{x}_i \in X_{n+1}$.

By the inductive hypothesis with positive probability $V_n(\mathbf{x},t) \geqslant (3/2)^n$. For each level nr cube $K_{\underline{i}_{nr}^{(1)},\dots,\underline{i}_{nr}^{(d)}}$ in $V_n(\mathbf{x},t)$ the number of its sub-cubes in $V_{n+1}(\mathbf{x},t)$ can be bounded from below by

$$G^r_{\alpha_t(\mathbf{x}_{\underline{i}_{nr}^{(1)},\dots,\underline{i}_{nr}^{(d)}})} \mathbb{1}_{I_1}(\psi_{\underline{i}_{nr}^{(1)},\dots,\underline{i}_{nr}^{(d)}}(\mathbf{x}_i)),$$

where $t \in Z_j^{(n+1)r}$ is arbitrary. Note that now $\alpha_t(\mathbf{x}_{\underline{i}_{nr}^{(1)},\dots,\underline{i}_{nr}^{(d)}})$ is fixed, i.e. we approximate by a linear projection, so we can apply Condition $A(\alpha_t(\mathbf{x}_{\underline{i}_{nr}^{(1)},\dots,\underline{i}_{nr}^{(d)}}))$.

The expected value of this random variable is 2, and it is bounded below by 0, above by M^{dr}. Moreover, random variables coming from different level nr cubes are independent. So we can apply Azuma–Hoeffding inequality as above in the proof of Theorem 1.2, thus we are done. □

3.2 Mandelbrot Umbrella

Proof of Theorem 1.3. It is easy to see that instead of radial projection $Proj_t$, it is equivalent to consider the projection R_t defined by

$$R_t(\mathbf{x}) = Line(t, \mathbf{x}) \cap P_\omega,$$

where $Line(t, \mathbf{x})$ is the line through t and $\mathbf{x}$. Moreover, as explained in [6, Sect. 3], we only need to consider radial projections with center separated from parallel directions (as in Assumption 1.1) and arbitrary big distance from K. This ensures that conditions ii) and iii) of Definition 3.1 hold. Condition i) also holds if we subdivide the family of centers to at most countably many subfamilies. Hence Theorem 1.3 holds. □

Acknowledgment Károly Simon was supported by the grant OTKA K 104745 and Lajos Vágó by the grant TÁMOP-4.2.2.B-10/1-2010-0009.

References

1. Broida, J.G., Williamson, S.G.: A Comprehensive Introduction to Linear Algebra. Addison-Wesley, Reading (1989). ISBN 0-201-50065-5
2. Chayes, J.T., Chayes, L., Durrett, R.: Connectivity properties of Mandelbrot's percolation process. Prob. Theory Rel. Fields **77**, 307–324 (1988)
3. Falconer, K.J.: Random fractals. Math. Proc. Camb. Philos. Soc. **100**, 559–582 (1986)
4. Falconer, K.J., Grimmett, G.R.: On the geometry of random Cantor sets and fractal percolation. J. Theor. Probab. **5**(3), 465–485 (1992)
5. Mauldin, R.D., Williams, S.C.: Random recursive constructions: asymptotic geometric and topological properties. Trans. Am. Math. Soc. **295**, 325–346 (1986)
6. Rams, M., Simon, K.: Michał Rams and Károly Simon. Projections of fractal percolations. Ergodic Theory Dyn. Syst. (2013), available on CJO2013. doi:10.1017/etds.2013.45. http://arxiv.org/abs/1306.3844
7. Simon, K. and Vago, L.: Projections of Mandelbrot percolation in higher dimensions, (2014). http://arxiv.org/abs/1407.2225
8. Walters, P.: An Introduction to Ergodic Theory. Springer, New York (2000)

Some Examples of Finite Type Fractals in Three-Dimensional Space

Mai The Duy

Abstract By choosing the contraction functions in the Iterated Function System we extend the construction from two-dimensional space to three-dimensional space to build self-similar sets in 3-space. We also extend the neighbor map concept to Iterated Function Systems with different contraction factors in order to identify examples with finite type. Some interesting examples of self-similar sets in three-dimensional space are given.

Keywords Self-similar set • Fractal • Finite type • Self-affine tile

1 Introduction

A fractal in general is a rough or fragmented geometric shape that can be split into parts, each of which is a reduced-size copy of the whole. This essential property is called self-similarity. A fractal usually has Hausdorff dimension which is greater than its topological dimension. Now with the aid of computer programs, fractal geometry has recently grown and is continuing to grow and we can visualize the beauty of many of the images that they have discovered.

Self-similar sets are a class of fractals which can be rigorously defined and treated by mathematical methods. In 1981 Hutchinson [14] rigorously defined self-similar sets by this equation

$$F = f_1(F) \cup f_2(F) \cup \ldots \cup f_m(F),$$

M.T. Duy (✉)
Duy Tan University, Da Nang, Vietnam
e-mail: vnenlighten@yahoo.com

C. Bandt et al. (eds.), *Fractals, Wavelets, and their Applications*, Springer Proceedings in Mathematics & Statistics 92, DOI 10.1007/978-3-319-08105-2_12,

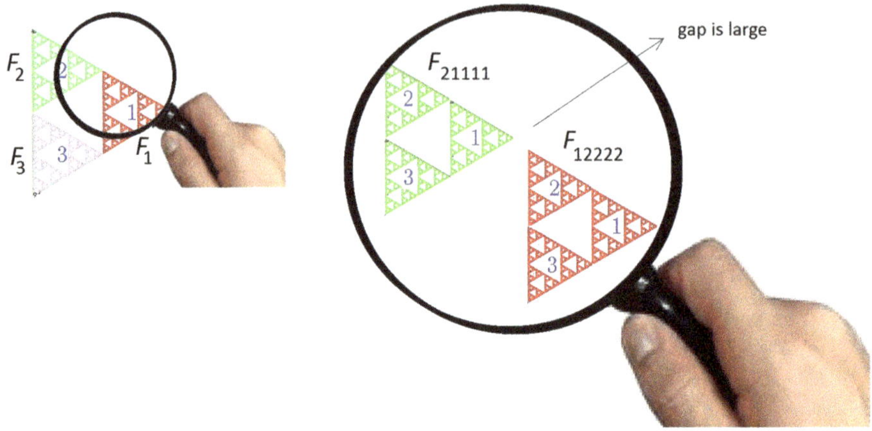

Fig. 1 Using the neighbor maps we can magnify infinitely any self-similar sets

where f_i, $i = 1, \ldots, m$ are contracting maps (IFS) on $\mathbb{R}^d$. Hutchinson proved that for given maps there is exactly one compact nonempty set, F, which fulfils the equation. This set F is called fractal or attractor of IFS. See [7, 12] for details.

Neighbor maps can be considered as a representation of relative position of pairs of nonempty intersecting sub-pieces. They were first introduced by Bandt and Graf [3]. We extend the definition of Bandt on neighbor maps and neighbor graphs. Two sub-pieces are neighbors if they intersect each other and the relation between their sizes must be the same with the relation between the sizes of the pieces on the original fractal. We also extend fractal constructions from two-dimensional space to three-dimensional space. There are many fractal shapes in nature, such as fern leafs, clouds, and mountains. We try to find the self-similar or self-affine structure of those things. From the geometric point of view, the interesting self-similar sets is the self-similar sets that have the sub-pieces just touching or have exact overlap. So we have to enlarge them to see whether they are Cantor sets or they have overlap, as indicated in Fig. 1. We use the strong conept of neighbor map to decide such questions analytically.

2 Magnify Fractals: The Neighbor Maps

Fractals are sets or entities that look the same under magnification. Small pieces of such a set are similar to the whole set. Such sets are "self-similar."

To obtain an interesting structure in the self-similar sets, it is often required that overlaps between pieces are sufficiently thin or just touching, which is expressed by the open set condition (OSC):

Definition. We say that the IFS $\{f_1, \ldots, f_m\}$ satisfies *the OSC*, if there exists an open set V such that

$$\bigcup_{i=1}^{m} f_i(V) \subset V \text{ and } f_i(V) \cap f_j(V) = \emptyset, \ \forall i \neq j \in \{1, \ldots, m\}.$$

We call such open set V a feasible open set of the f_i, or of F.

The OSC controls the overlap of the sub-pieces F_i of fractal F. If an IFS satisfies the OSC, then the Hausdorff dimension and the self-similarity dimension of the attractor coincide.

However, it is not easy to check the OSC. In 1992 Bandt and Graf introduced an algebraic equivalent for OSC [3]. We take some notations, let $f_i : \mathbb{R}^d \to \mathbb{R}^d$ contractive similarities with contraction factor r, $i \in I = \{1, \ldots, m\}$ and $F = \bigcup_{i=1}^{n} f_i(F)$, $I^n := \{(u_i)_{i=1,\ldots,n} \mid u_i \in I \ \forall i = 1, \ldots, n\}$, $I^* := \bigcup_{n=1}^{\infty} I^n$, for $u := u_1 \ldots u_n \in I^n$, define $f_u := f_{u_1} \circ \ldots \circ f_{u_n}$ and $F_u := f_u(F)$. Given an IFS $\{f_1, \ldots, f_m\}$, for each $u, v \in I^*$, $u = u_1 u_2 \ldots$ and $v = v_1 v_2 \ldots$, where $u_k, v_k \in I$, $k \in \mathbb{N}$. Let $\mathcal{N} = \{h = f_u^{-1} f_v \mid u, v \in I^*, u_1 \neq v_1\}$. The algebraic formulation of OSC reads as following theorem

Theorem ([3]). *The iterated function system $\{f_1, \ldots, f_m\}$ satisfies the OSC if and only if there exists $\delta > 0$ such that $\| h - id \| > \delta$, for all $h \in \mathcal{N}$.*

The norm in this theorem is the norm on affine maps, which can be $\| g \| := \| A \| + |b|$ if $g = Ax + b$, where

$$\| A \| = \max\{\| Ax \| \ | x \in \mathbb{R}^d \text{ with } \ \| x \| \leq 1\}.$$

In 2001, Bandt [2] described an algorithm deciding on separation, when all the contraction factors are equal to r. The algorithm is as followed: starting with identity map id, we applied the automorphism

$$h_{ij}(g) := f_i^{-1}.g.f_j, \quad i, j = 1, \cdots, m, \text{and} i \neq j$$

Repeat this process with the obtained maps belonging to a neighborhood U of id until all the maps run out of U. The reality of this algorithm is confirmed by the following proposition:

Proposition ([2, Lemma 4.1]). *Given similarities $f_i = rA_i(x + a_i)$, where $r \in (0, 1)$, and A_i are orthogonal matrices. Let U be the neighborhood of id in the space of similarities defined as*

$$U := \{sBx + b \mid |b| \leq \frac{(1+s)c}{1-r}\} \text{ where } c = \max_{i \in \{1, \cdots, m\}} |a_i|.$$

Then the complement of U is mapped into itself by each h_{ij}.

Neighbor maps can be considered as a representation of relative position of pairs of nonempty intersecting sub-pieces. They were first introduced by Bandt and Graf [2]. We take the Sierpinski gasket to explain the neighbor maps. Suppose that our

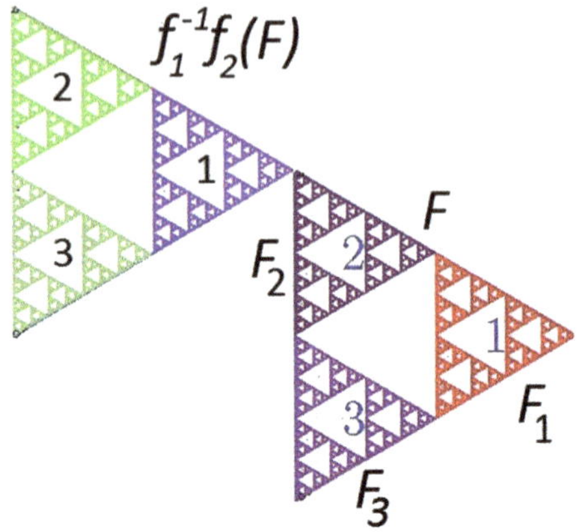

Fig. 2 The relative position of sub-pieces

Sierpinski gasket has three sub-pieces in the origin (see Fig. 2) and the IFS has three contraction functions $\{f_1, f_2, f_3\}$. The relative position of sub-pieces F_1 and F_2 was represented by the map $g = f_1^{-1} f_2(F)$. The relative position between sub-pieces F_1 and F_2 is the same with the relative position between the fractal F and its image $f_1^{-1} f_2(F)$. Suppose that $F_1 \bigcap F_2$ have very small gap (they do not touching when the contraction factors r are smaller than 0.5, let us take $r = 0.499$). Because the relative position between sub-pieces F_1 and F_2 is the same with the relative position between the fractal F and its image $f_1^{-1} f_2(F)$ then $F \bigcap f_1^{-1} f_2(F)$ also has a gap. Because F is larger than F_1 then the gap of $F \bigcap f_1^{-1} f_2(F)$ is bigger than the gap of $F_1 \bigcap F_2$. When we go to the next level the relative position between F_{12} and F_{21} is the same with the relative position between the fractal F and $f_{12}^{-1} f_{21}(F)$. On a deeper level, the relative position between F_{12222} and F_{21111} coincides with the relative position between the fractal F and $f_{12222}^{-1} f_{21111}(F)$. So we can imagine that we enlarge F_{12222} equal to F so the gap between F_{12222} and F_{21111} also to be enlarged that we can see in Fig. 1. It leads us to understand that when we apply the Bandt's algorithm this example then the algorithm will never stop because the gap goes bigger and bigger out of the neighborhood U and never come back to h_{ij}. Bandt and Graf have given the definition of neighbor map for the maps which have the same contraction factors. Given $f_1, \ldots, f_n$ a *neighbor map* is an element of the set $\{f_u^{-1} f_v \mid F_u \bigcap F_v \neq \emptyset, u \neq v \in I^n\}$.

We extend the definition of Bandt on neighbor maps and neighbor graphs. Two subpieces are neighbors if they intersect each other and the relation between their sizes must be the same with the relation between the sizes of the pieces on the original fractal. A type is a standardized relative position of two intersecting pieces of the fractal.

Definition. We say two pieces F_u and F_v are *neighbors* if

$$F_u \cap F_v \neq \emptyset, u, v \in I^*, u_1 \neq v_1, \min_{k,l \in I} \frac{r_k}{r_l} \leq r_{u,v} \leq \max_{k,l \in I} \frac{r_k}{r_l}$$

where $u = u_1 u_2 \ldots, v = v_1 v_2 \ldots$, and $r_{u,v}$ is the contraction factor of the contracting map $f_u^{-1} f_v$, $r_{u,v} = \frac{r_v}{r_u}$.

With above definition two pieces F_u and F_v are neighbors if they intersect each other and their sizes can be comparable that they are not big different.

Definition. A neighbor map is an element of the set

$$V = \{f_u^{-1} f_v \mid F_u \cap F_v \neq \emptyset, F_u \text{ and} F_v \text{ are neighbors}\}$$

3 The Neighbor Graph-Finite Type Fractals

The two pieces f_u and F_v are neighbors, and the type of that relation between f_u and F_v is presented by the neighbor map $h = f_u^{-1} f_v$. Every fractal that has a finite type neighbor map is called a finite type fractal. A type is a standardized relative position of two intersecting pieces of the fractal and the neighbor graph will show the relation between types. More detail we have

Definition. The neighbor graph $G = (V, E)$ of an IFS $\{f_i\}_{i=1}^m$ is given by the sets

$$V = \{f_u^{-1} f_v \mid F_u \cap F_v \neq \emptyset, F_u \text{ and } F_v \text{ are neighbors}\},$$
$$E = \{(g, h, ij) \in V \times V \times I^2 \mid h = f_i^{-1} g f_j \quad \min_{k,l \in I} \frac{r_k}{r_l} \leq r_h \leq \max_{k,l \in I} \frac{r_k}{r_l}\}$$

where $i, j \in I$, $r_h = \frac{r_j}{r_i} r_{u,v}$.

The fractals that have finite type neighbors are finite fractals. If a self-similar set is a finite type fractal, then the Bandt's algorithm will stop after sometime and we get the number of types. In the following part of this paper we give many new examples of finite type fractals in three-dimensional space.

4 Some Three-Dimensional Examples: The Choice of IFS

Until now there are just a few examples of fractals in $\mathbb{R}^d$ with $d \geq 3$. We have two examples that are well known: the Menger sponge and the fractal tetrahedron (http://www.mathpaint.blogspot.com) and recently, the three-dimensional twindragon [5] and the fractal octahedron, the three-dimensional modification of Sierpinski's triangle [6, 16]. Rendering the pictures in this paper we use software packages from France [9] and Russia [13, 15] for three-dimensional fractals. These new visualization tools enhance the study of three-dimensional fractals.

Fig. 3 The rectangular cuboid in Example 2

In this paper, we follow [11] to turn fractals in plane to fractals in space. Sometimes we change the number of functions in the IFS (Example 6) but in most cases the number of functions is the same in the 2D and 3D version (Examples 1–5).

A similarity map in the plane is specified by a rotation matrix. When we go to space we have to combine many rotation. In this paper we use three rotation matrices M, M', M'' where M is the rotation by $\frac{3\pi}{2}$ in the X-axis combined with the rotation by $\frac{3\pi}{2}$ in the Y-axis and M' is the rotation by $\frac{3\pi}{2}$ in the X-axis combined with the rotation by $\frac{\pi}{2}$ in the Y-axis, more exactly:

$$M = \begin{bmatrix} 0 & 1 & 0 \\ 0 & 0 & 1 \\ 1 & 0 & 0 \end{bmatrix}, M' = \begin{bmatrix} 0 & -1 & 0 \\ 0 & 0 & 1 \\ -1 & 0 & 0 \end{bmatrix}, \text{and } M'' = \begin{bmatrix} 0 & 0 & 1 \\ -1 & 0 & 0 \\ 0 & -1 & 0 \end{bmatrix}$$

Example 1. The simplest 3D fractal is the rectangular cuboid

$$F = [0, n] \times [0, \sqrt[3]{n}] \times [0, \sqrt[3]{n^2}]$$

with the IFS: $\{f_k(x) = rMx + (k-1)\}$, $k = 1, 2, \ldots, n$ and $r = 1/\sqrt[3]{n}$.

Example 2. The rectangular cuboid can be made by different sizes of sub-pieces (Fig. 3):

$$F = [0, 2] \times [0\sqrt[3]{2}] \times [0, \sqrt[3]{4}]$$

The IFS has three functions:

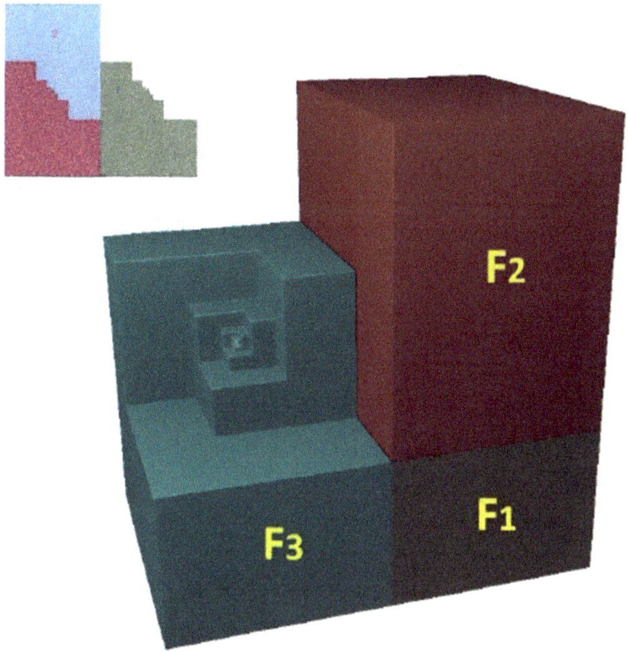

Fig. 4 The spiral fractal

$f_1(x) = rMx$, $f_2(x) = r^2M^2x + (1, 0, 0)'$ and $f_3(x) = r^2M''x + (1, \sqrt[3]{2}, \sqrt[3]{4})'$, $r = 1/\sqrt[3]{2}$

Example 3 (See Fig. 4). The spiral fractal is made of three sub-pieces of equal size. The IFS: $f_1(x) = rMx - v, f_2(x) = -rMx - v$, $f_3(x) = rMx$ where $r = 1/\sqrt[3]{3}$, $v = (1, 0, 0)'$. To decide whether the spiral fractal is of finite type, we investigate Fig. 5 and count the numbers of sub-pieces around F. When neighboring pieces meeting in a single point or a line are neglected we can see eight-face neighbor types. Other versions of spiral fractals with different sizes of sub-pieces can be seen in Fig. 6.

Example 4 (See Fig. 6). The spiral fractals made of three different sizes of sub-pieces. The IFS: $f_1(x) = rMx + v, f_2(x) = r^2M^2x$, $f_3(x) = -r^2M^2x + v$ where $r = 1/\sqrt[3]{2}$, $v = (-1/3, \sqrt[3]{2}/3, \sqrt[3]{4}/3)'$. It follows from [1] that all these examples are tiles, that is, the whole space can be tiled by congruent copies of F. See also [10].

Example 5 (See Fig. 7). The new Menger sponge uses only four contraction functions in the IFS: $f_1(x) = rMx + v, f_2(x) = rMx + (2, 0, 0)'$, $f_3(x) = r^2M^2x + (1, 0, 0)'$, $f_4(x) = r^2M^2x + (1, 0, 2/\sqrt[3]{3})'$ where $r = 1/\sqrt[3]{3}$.

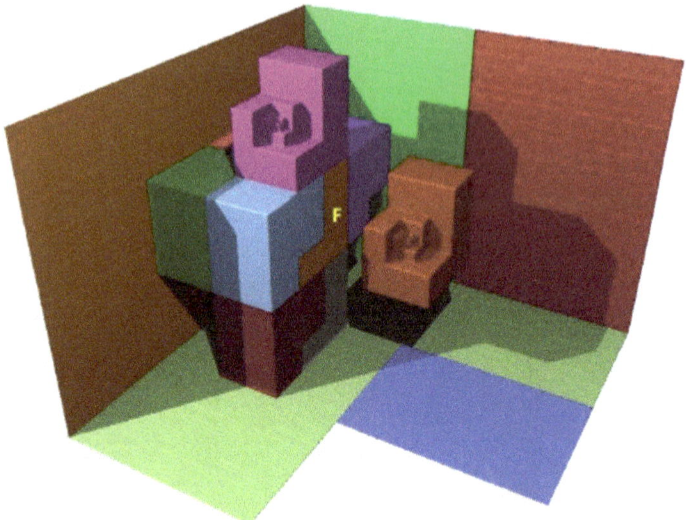

Fig. 5 Eight-face neighbor types of the spiral fractal

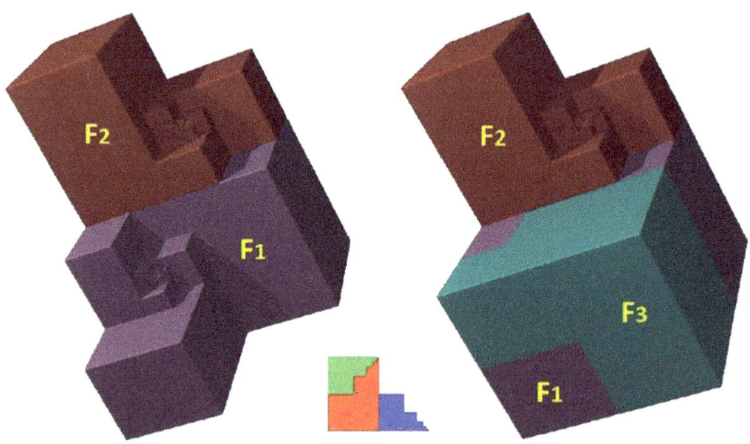

Fig. 6 The spiral fractal with three different sub-pieces

Example 6 (See Fig. 8). The two-dimensional 'golden pentagasket' is a combination of the overlapping 'golden gasket' presented in [8] and the fractal n-gons in [4]. Its 3-dimensional counterpart will be called 'golden dodecahedron'. It is produced from an IFS with 50 similarity maps $f_k(x) = \delta(x - a_k) + a_k$ if $k = 1, \ldots, 20$, and $f_k(x) = \delta^2(x - a_k) + a_k$, if $k = 21, \ldots, 50$, where δ is the golden ratio and a_k are the vertex points on the dodecahedron and the midpoints of the lines which connect that vertex points (see Fig. 9).

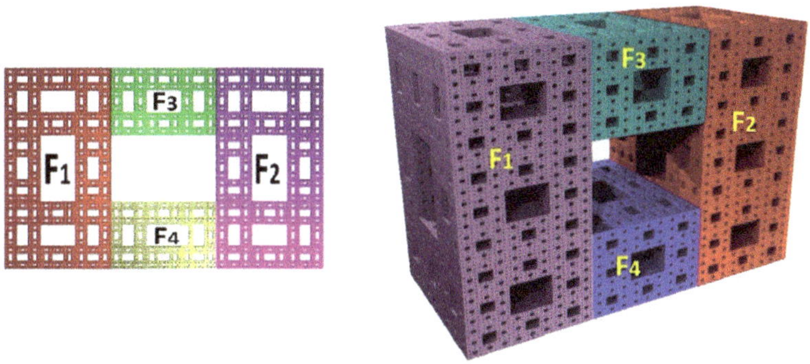

Fig. 7 The new Menger sponge with four contraction functions in IFS

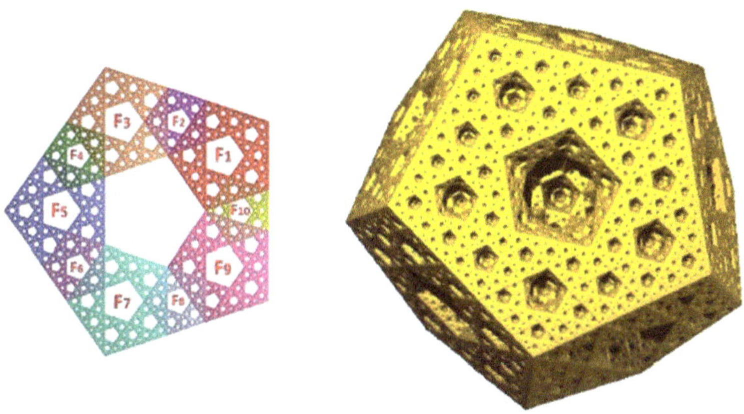

Fig. 8 The golden dodecahedron fractal

When cutting the golden dodecahedron fractal we have slices as we can see in Figs. 10 and 11. My friend Ruediger Zeller told me that the golden dodecahedron fractal contains plane segments and in the center there is a hole. We can imply that holes exist almost everywhere, but it need to be proved so there is much left to explore about this golden dodecahedron fractal.

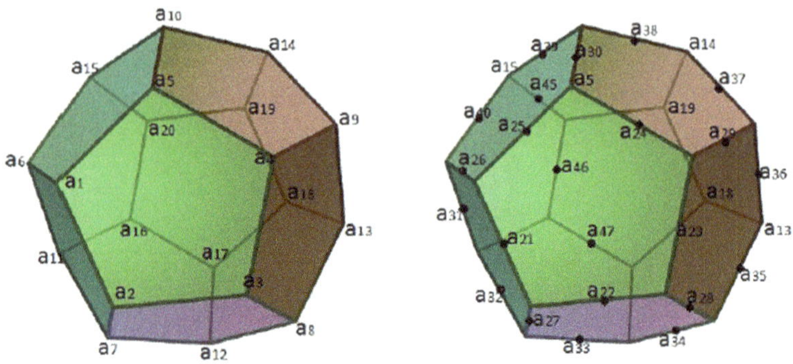

Fig. 9 The 50 fixed points of the functions in IFS of the golden dodecahedron

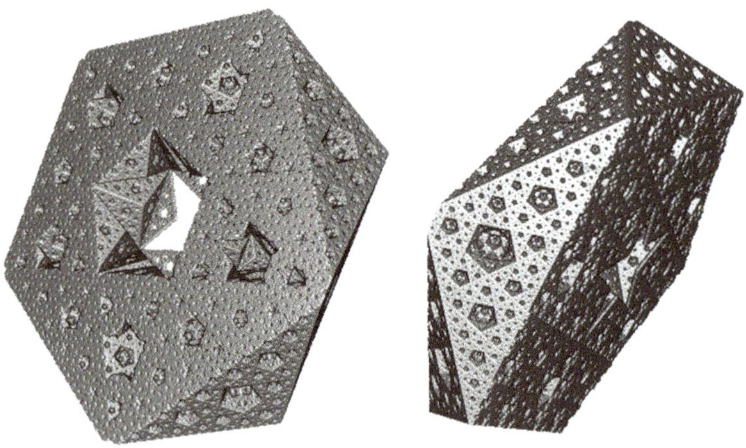

Fig. 10 Two sides of cutting slides of the golden dodecahedron

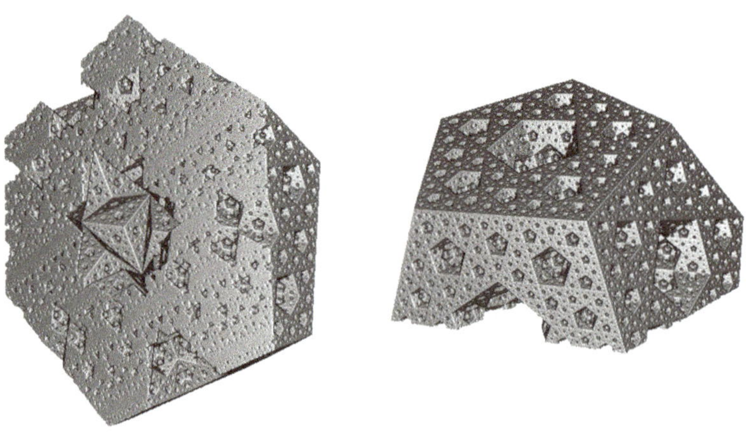

Fig. 11 There is a hole in the center of the golden dodecahedron fractal

References

1. Bandt, C.: Self-similar sets 5. Integer matrices and fractal tilings of Rn. Proc. Am. Math. Soc. **112**, 549–562 (1991)
2. Bandt, C.: Self-similar measures. In: Fiedler, B. (ed.) Ergodic Theory, Analysis and Efficient Simulation of Dynamical Systems, pp. 31–46. Springer, Berlin (2001)
3. Bandt, C., Graf, S.: Self-similar sets 7. A characterization of selfsimilar fractals with positive Hausdorff measure. Proc. Am. Math. Soc. **114**, 995–1001 (1992)
4. Bandt, C., Hung, N.V.: Fractal n-gons and their Mandelbrot sets. Nonlinearity **21**, 2653–2670 (2008)
5. Bandt, C., Mesing, M.: Fractals of finite type. Banach Center Publ. **84**, 131–148 (2009)
6. Bandt, C., Duy, M., Mesing, M.: Three-dimensional fractals. Math. Intell. **32**(3), 12–18 (2010) (web sites: http://www.springerlink.com/content/w200578285042037/)
7. Barnsley, M.F.: Fractals Everywhere, 2nd edn. Academic, Boston (1993)
8. Broomhead, D., Montaldi, J., Sidorov, N.: Golden gaskets: variations on the Sierpinski sieve. Nonlinearity **17**, 1455–1480 (2004)
9. Chaoscope software. Chaoscope software developed by N. Desprez (2003) http://www.chaoscope.org/
10. Darst, R., Palagallo, J.A., Price, T.E.: Fractal tiling in the plane. Math. Mag. **71**, 22–23 (1988)
11. Duy, M.: Some self-similar constructions in two and three dimensions and their neighbor geometry. Dissertation, Greifswald University (2011) (web sites: http://ub-ed.ub.uni-greifswald.de/opus/volltexte/2011/993/)
12. Falconer, K.J.: Fractal Geometry. Mathematical Foundations and Applications. Wiley, New York (1990)
13. Fractracer software, created by Mekhontsev Dmitriy (2010) http://fractracer.com/
14. Hutchinson, J.E.: Fractals and Self-similarity. Indiana Univ. Math. J. **30**, 713–747 (1981)
15. IFS Builder 3d software, created by Kravchenko Alexei and Mekhontsev Dmitriy, graduates of Novosibirsk State University (NSU) (2011) http://fractal.nsu.ru/builder3d_en.htm
16. Jones, H., Campa, A.: Fractals based on regular polygons and polyhedra. In: Patrikalakis, N.M. (ed.) Scientific Visualization of Physical Phenomena, pp. 299–314. Springer, New York (1991)

Fractals in Partial Metric Spaces

S. Minirani and Sunil Mathew

Abstract Partial metric space is a generalisation of metric space due to non zero self-distance. In this paper, we discuss the nature of fractals in a partial metric space.

Keywords Partial metric space • Complete space • Fractals • Iterated function system

2000 AMS Subject Classification: Primary ; 28A80, 54E50, 54E99

1 Introduction

The notion of Partial Metric Space was introduced by Steve G. Mathews in 1992 [4] as a part of the study of denotational semantics of dataflow networks. It was originated from a program in Computer Science known as Gilles Kahn's Model of parallel computation. Partial Metric Space is a generalisation of metric space with non-zero self-distances. Whenever the self-distance becomes zero, it is a metric space. S.G. Mathews established the precise relationship between partial metric space and the weightable quasi-metric spaces, and proved a partial metric generalisation of Banach's contraction mapping theorem. S.J. O'Neil proposed one significant change to Mathews' definition of the partial metrics, and that was to extend their range from $\mathbb{R}^+$ to $\mathbb{R}$ [3].

S. Minirani (✉) • S. Mathew
Department of Mathematics, National Institute of Technology, Calicut, Kerala, India
e-mail: miniranis@yahoo.com; sm@nitc.ac.in

C. Bandt et al. (eds.), *Fractals, Wavelets, and their Applications*, Springer Proceedings in Mathematics & Statistics 92, DOI 10.1007/978-3-319-08105-2_13,

2 Partial Metric Space

This section discusses some of the basic definitions in the area which is by Mathews.

Definition 2.1. A partial metric space is a pair $(X, p : X \times X \to \mathbb{R})$ such that

P_1: $p(x,x) \leq p(x,y)$ (nonnegativity and small self-distances)
P_1: $If p(x,x) = p(x,y) = p(y,y)$, then $x = y$ (indistance implies equality)
P_2: $p(x,y) = p(y,x)$ (symmetry), and
P_3: $p(x,z) \leq p(x,y) + p(y,z) - p(y,y)$ (triangularity).

If $p(x,y) = 0$, then P_0 and P_1 imply that $x = y$, but the converse does not hold always.

A trivial example of a partial metric space is the pair $(\mathbb{R}^+, p)$, where $p : \mathbb{R}^+ \times \mathbb{R}^+ \to \mathbb{R}^+$ defined as $p(x,y) = \max\{x,y\}$

2.1 *Contraction Fixed Point Theorem*

We now consider how a familiar theorem from the theory of metric spaces can be carried over to partial metric spaces. Complete Spaces, Cauchy sequences, and the contraction fixed point theorem can be generalised to partial metric spaces as follows.

Definition 2.2. A sequence $x = (x_n)$ of points in a partial metric space (X, p) is Cauchy, if there exists $a \geq 0$ such that for each $\epsilon > 0$ there exists k such that for all $n, m > k$, $\mid p(x_n, x_m) - a \mid < \epsilon$.

In other words, x is Cauchy if the numbers $p(x_n, x_m)$ converge to some a as n and m approach ∞, that is, if $\lim_{n,m\to\infty} p(x_n, x_m) = a$. Then $\lim_{n,m\to\infty} p(x_n, x_n) = a$, and so if (X, p) is a metric space $a = 0$.

Definition 2.3. A sequence $x = (x_n)$ of points in a partial metric space (X, p) converges to y in X if

$$\lim_{n,m\to\infty} p(x_n, y) = \lim_{n,m\to\infty} p(x_n, x_n) = p(y,y)$$

Thus if a sequence converges to a point, then the self-distances converge to the self-distance of that point.

Definition 2.4. A partial metric space (X, p) is complete if every Cauchy sequence in X converges in X.

Definition 2.5. A set A is closed in (X, p) if $A = \overline{A}$, where $\overline{A}$ is the closure of A.

Definition 2.6. A set A is a bounded subset in (X, p) if there exist an $x \in X$ and $M \geq 0$ such that for all $a \in A$, we have $p(x,a) < p(a,a) + M$.

Definition 2.7. For a partial metric space (X, p), the open ball with center x and radius $\epsilon > 0$ is defined as $B_p(x, \epsilon) = \{y \in X : p(x, y) < p(x, x) + \epsilon\}$.

Definition 2.8. A subset A of (X, p) is said to be totally bounded if for every $\epsilon > 0$, there are $x_1, \ldots, x_n$ in X such that $A \subseteq \bigcup_{i=1}^{n} B_p(x_i, \epsilon)$.

Definition 2.9. For a partial metric space (X, p), a contraction is a function $f : X \rightarrow X$ for which there exists a $c \in [0, 1)$ such that for all $x, y \in X$, $p(f(x), f(y)) \leq c.p(x, y)$.

Theorem 2.10. *For each contraction f over a complete partial metric space (X, p) there exists a unique $x \in X$ such that $x = f(x)$. Also, $p(x, x) = 0$.*

Thus the contraction fixed point theorem is extended to partial metric spaces. It says that the fixed point has self-distance 0, which is trivial in the case of metric spaces.

If p is a partial metric on X, then the function $d : X \times X \rightarrow \mathbb{R}^+$ defined by

$$d_p(x, y) = 2p(x, y) - p(x, x) - p(y, y)$$

is a metric on X. Also, a sequence $x = (x_n)$ converges to a point x in (X, d_p) if and only if

$$\lim_{n,m \to \infty} p(x_n, x_m) = \lim_{n \to \infty} p(x_n, x) = p(x, x).$$

Lemma 2.11. *(i) A sequence (x_n) in X is a Cauchy sequence in (X, p) if and only if it is a Cauchy sequence in metric space (X, d_p).*
(ii) A partial metric space (X, p) is complete if and only if the metric space (X, d_p) is complete.

3 Fractals in Partial Metric Spaces

We work in a partial metric space (X, p) and $H_p(X)$ denote the set of all nonempty compact subsets of the partial metric space (X, p). The compact subsets are the closed and bounded subsets of X induced by the partial metric p. Now we discuss the nature of fractals in a partial metric space.

Definition 3.1. Let (X, p) be a partial metric space. For $A, B \in H_p(X)$ and $x \in X$, let us define

$$r_p(x, A) = inf\{p(x, a) : a \in A\}$$

and

$$\rho_p(A, B) = sup\{r_p(a, B) : a \in A\}$$

Similarly we define

$$\rho_p(B, A) = sup\{p(b, A) : b \in B\}$$

Then the *Hausdorff partial metric* h_p on $H_p(X)$ is defined by

$$h_p(A, B) = max\{\rho_p(A, B), \rho_p(B, A)\}$$

Remark. For nonempty set A in (X, p), an element $a \in \overline{A}$ if and only if $p(a, A) = p(a, a)$. Before that we will prove some important results required for later proofs.

Result 1: Let (x_n) and (y_n) be sequences in a partial metric space (X, p). If (x_n) converges to x and (y_n) converges to y, then the sequence $(p(x_n, y_n))$ converges to $p(x, y)$.

Proof. Let $\epsilon > 0$. Since sequence (x_n) converges to x, by definition, there exists an N_1 such that for all $n \geq N_1$, $p(x_n, x) - p(x, x) < \epsilon/2$. Similarly, there exists an N_2 such that for all $n \geq N_2$, $p(y_n, y) - p(y, y) < \epsilon/2$.

Thus

$$\begin{aligned} p(x_n, y_n) &\leq p(x_n, x) + p(x, y) + p(y, y_n) - p(x, x) - p(y, y) \\ &< \epsilon/2 + \epsilon/2 + p(x, y) \end{aligned}$$

$\Rightarrow p(x_n, y_n) - p(x, y) < \epsilon$. Therefore by definition, the sequence $(p(x_n, y_n))$ converges to $p(x, y)$. □

Result 2: If (z_k) is a sequence in a partial metric space (X, p) with the property that $p(z_k, z_{k+1}) - p(z_k, z_k) < 1/2^k$ for all k and for some $a \geq 0$, then the sequence (z_k) is a Cauchy sequence.

Proof. Let $\epsilon > 0$ and choose a positive integer N such that $\frac{1}{2^{N-1}} < \epsilon$. Then for all $n > m \geq N$ we find that

$$\begin{aligned} p(z_m, z_n) &\leq p(z_m, z_{m+1}) + p(z_{m+1}, z_{m+2}) + \cdots + p(z_{n-1}, z_n) \\ &\quad - p(z_{m+1}, z_{m+1}) - p(z_{m+2}, z_{m+2}) - \cdots - p(z_{n-1}, z_{n-1}) \\ &< \frac{1}{2^m} + \frac{1}{2^{m+1}} + \cdots + \frac{1}{2^{n-1}} + p(z_m, z_m) \\ &< \Sigma_{k=m}^{\infty} \frac{1}{2^k} + p(z_m, z_m) \end{aligned}$$

$$= \frac{1}{2^{m-1}} + p(z_m, z_m)$$
$$\leq \frac{1}{2^{N-1}} + p(z_m, z_m)$$
$$< \epsilon + b$$

where $b = p(z_m, z_m) \geq 0$.

Thus it follows that (z_k) is a Cauchy sequence in (X, p). □

Lemma 1. *Let (X, p) be the metric space and let A be a closed subset of X. If a sequence (a_n) converges to x and $a_n \in A$ for all n, then $x \in A$.*

Proof. Suppose (a_n) is a sequence that converges to x and $a_n \in A$ for all n. If there exists a positive integer n such that $a_n = x$, then it is clear that $x \in A$. If there does not exist a positive integer n such that $a_n = x$, then x is a limit point of A and since A is closed, $x \in A$. □

Theorem 3.2. *Let $x \in X$ and $A, B, C \in H_p(X)$.*

(1) $r_p(x, A) = p(x, x)$ if and only if $x \in A$.
(2) $\rho_p(A, B) = h_p(A, A)$ if and only if $A \subseteq B$.
(3) There exists $a_x \in A$ such that $r_p(x, A) = p(x, a_x)$.
(4) There exist $a^* \in A$ and $b^* \in B$ such that $\rho(A, B) = p(a^*, b^*)$.
(5) If $A \subseteq B$, then $r_p(x, B) \leq r_p(x, A)$.
(6) If $B \subseteq C$, then $\rho_p(A, C) \leq \rho_p(A, B)$.
(7) $\rho_p(A \cup B, C) = \max\{\rho_p(A, C), \rho_p(B, C)\}$.
(8) $\rho_p(A, B) \leq \rho_p(A, C) + \rho_p(C, B) - \inf_{c \in C} p(c, c)$.

Proof. (1) Let $x \in A$, then $\inf\{p(x, a) : a \in A\} = p(x, x)$, since $p(x, x) \leq p(x, y)$ for all $x \in X$. This implies $r_p(x, A) = p(x, x)$. Conversely, let $r_p(x, A) = p(x, x)$. Then for each positive integer n, there exists $a_n \in A$: $p(x, a_n) - p(x, x) < \frac{1}{n}$. Then by definition $(a_n) \to x$. Since A is compact, it is closed. So by Lemma 1, $x \in A$.

(2) Let $A \subseteq B$ and $a \in A$. Then $a \in B$. Then by Property 1, $r_p(a, B) = p(a, a)$. So $\rho_p(A, B) = \sup\{p(a, a) : a \in A\} = h_p(A, A)$. Conversely, let $a \in A$. We have

$$\rho_p(A, B) = h_p(A, A)$$

i.e.

$$\sup\{r_p(a, B) : a \in A\} = sup\{p(a, a) : a \in A\}$$

$$\Rightarrow r_p(a, B) = p(a, a) \Rightarrow a \in B \Rightarrow A \subseteq B.$$

(3) By definition of infimum, let (a_n) be a sequence in A such that $p(x, a_n)$ converges to $r_p(x, A)$. Since A is compact, there exists a subsequence (a_{n_k}) of (a_n) that converges to some $a_x \in A$. Then the sequence $(p(x, a_{n_k}))$ converges to $p(x, a_x)$. Since the limit is unique, we have $r_p(x, A) = p(x, a_x)$.

(4) By definition of supremum, let $\{a_n\}$ be a sequence in A such that $r_p(a_n, B)$ converges to $\rho_p(A, B)$. Then by Property 3, there exists a sequence $\{b_n\}$ in B such that $r_p(a_n, B) = p(a_n, b_n)$. Since A is compact, there exists a subsequence (a_{n_k}) of (a_n) that converges to some $a^* \in A$. Since B is compact, there exists a subsequence (b_{n_k}) of (b_n) that converges to some $b^* \in B$. Then $(p(a_{n_k}, b_{n_k}))$ converges to $p(a^*, b^*)$ which implies $p(a^*, b^*) = \rho_p(A, B)$.

(5) Suppose $A \subseteq B$ and $x \in X$. Let $a \in A$, then $a \in B$. Hence it follows that $p(x, a) \geq \inf\{p(x, b) : b \in B\} = r_p(x, B)$. This is true for all $a \in A$. So we have $r_p(x, A) = \inf\{p(x, a) : a \in A\} \geq r_p(x, B)$

(6) Suppose $B \subseteq C$. Then by Property 5, $r_p(a, C) \leq r_p(a, B)$ for all $a \in A$. Taking supremum over A on both the sides gives $\rho_p(A, C) \leq \rho_p(A, B)$.

(7) We have

$$\begin{aligned}\rho_p(A \cup B, C) &= \sup\{r_p(x, C) : x \in A \cup B\} \\ &= \max\{\sup\{r_p(x, C) : x \in A\}, \sup\{r_p(x, C) : x \in B\}\} \\ &= \max\{\rho_p(A, C), \rho_p(B, C)\}\end{aligned}$$

(8) We know that

$$p(a, b) \leq p(a, c) + p(c, b) - p(c, c)$$

which implies

$$r_p(a, B) \leq p(a, c) + r_p(c, B) - p(c, c),$$

by taking infimum over B. i.e.,

$$r_p(a, B) + p(c, c) \leq p(a, c) + p(c, B)$$

So $r_p(a, B) + p(c, c) \leq p(a, c) + \rho_p(C, B)$
$\Rightarrow r_p(a, B) + \inf_{c \in C} p(c, c) \leq r_p(a, C) + \rho_p(C, B)$, by taking infimum over C.
Since a is an arbitrary element in A, the result follows.

□

Proposition 3.3. *h_p is a partial metric on $H_p(X)$.*

Proof. We have to prove the following axioms:

P_0: $h_p(A, A) \leq h_p(A, B)$
P_1: $h_p(A, B) = h_p(B, A)$
P_2: If $h_p(A, A) = h_p(A, B) = h_p(B, B)$, then $A = B$
P_3: $h_p(A, C) \leq h_p(A, B) + h_p(B, A) - \inf_{b \in B} p(b, b)$

P_0 : By definition, $\rho_p(A,A) = sup\{p(a,A) : a \in A\}$ which implies $\rho_p(A,A) = sup\{p(a,a) : a \in A\}$. Since $p(a,a) \leq p(a,b)$ for all $a \in A$ and $b \in B$

$$\rho_p(A,A) \leq \rho_p(A,B)$$

Similarly

$$\rho_p(A,A) \leq \rho_p(B,A)$$

Hence by definition

$$h_p(A,A) \leq h_p(A,B)$$

P_1 : By definition of $h_p(A,B)$, it is symmetric, since $p(a,b) = p(b,a)$.
P_2 : $h_p(A,B) = h_p(A,A) = h_p(B,B)$

i.e. $max\{\rho_p(A,B), \rho_p(A,B)\} = max\{\rho_p(A,A)\} = max\{\rho_p(B,B)\}$

$\Rightarrow$ $\rho_p(A,B) = \rho_p(A,A) = \rho_p(B,B)$
$\Rightarrow$ $r_p(a,B) = r_p(a,A) = r_p(b,B)$ for all $a \in A$ and $b \in B$
$\Rightarrow$ $p(a,b) = p(a,a) = p(b,b) \Rightarrow a = b$ for all $a \in A$ and $b \in B$
$\Rightarrow$ $A = B$.

P_3 : We will prove this using Property 8 of Theorem 3.2

$$\begin{aligned} h_p(A,B) &= max\{\rho_p(A,B), \rho_p(B,A)\} \\ &\leq max\{\rho_p(A,C)+\rho_p(C,B)-\inf p(c,c)+\rho_p(B,C)+\rho_p(C,A)-\inf p(c,c)\} \\ &= max\{\rho_p(A,C)+\rho_p(C,B), \rho_p(B,C)+\rho_p(C,A)\} - \inf p(c,c) \\ &\leq max\{\rho_p(A,C), \rho_p(C,A)\} + max\{\rho_p(C,B), \rho_p(B,C)\} - \inf p(c,c) \\ &= h_p(A,C) + h_p(C,B) - \inf p(c,c) \end{aligned}$$

Hence the proof. □

Proposition 3.4. $h_p(A,B) = 0 \Rightarrow A = B$

Proof.

$$h_p(A,B) = 0 \Rightarrow \rho_p(A,B) = \rho_p(B,A) = 0$$

$$\begin{aligned} \rho_p(A,B) = 0 &\Rightarrow r_p(a,B) = 0, \forall a \in A \\ &\Rightarrow p(a,a) \leq \rho_p(A,B) = 0, \forall a \in A \\ &\Rightarrow p(a,a) = 0, \forall a \in A \end{aligned}$$

$$\Rightarrow r_p(a, B) = p(a, a), \forall a \in A$$

$$\Rightarrow a \in \overline{B} = B$$

$$\Rightarrow A \subseteq B$$

Similarly, starting with $\rho_p(B, A) = 0$ gives $B \subseteq A$.

Thus we have $A = B$ □

Proposition 3.5. *If* $A = B$, *then* $H_p(A, B) = \sup_{a \in A} p(a, a)$

Proof. If $A = B$, $H_p(A, B) = H_p(A, A) = \rho_p(A, A) = \sup\{p(a, A) : a \in A\} = \sup\{p(a, a) : a \in A\}$ □

To be a complete partial metric space, every Cauchy sequence in $(H_p(X), h_p)$ must converge to a point in $(H_p(X)$. Therefore, in order to prove that the partial metric space $(H_p(X), h_p)$ is complete, we will choose an arbitrary Cauchy sequence (A_n) in $(H_p(X)$ and show that it converges to some A in $(H_p(X)$.

Definition 3.6. Given a set $A \in (H_p(X)$ and a positive number ϵ, the set $A + \epsilon$ is defined as $\{x \in X : r_p(x, A) \leq h_p(A, A) + \epsilon\}$.

Now we need to show that this set is closed for all possible choices of A and ϵ.

Proposition 3.7. $A + \epsilon$ *is closed for all possible choices of* $A \in (H_p(X)$ *and* $\epsilon > 0$.

Proof. Let $A \in (H_p(X))$ and $\epsilon > 0$. Let us assume that x is a limit point of $A + \epsilon$. Then there exists a sequence (x_n) of points in $A + \epsilon$ that converges to x. Since $x_n \in A + \epsilon$ for all n, by definition $r_p(x_n, A) \leq h_p(A, A) + \epsilon$ for all n. By Property 3 of Theorem 3.2, there exists $a_n \in A$ such that $r_p(x_n, A) = p(x_n, a_n)$. Thus $p(x_n, a_n) \leq h_p(A, A) + \epsilon$ for all n. Since A is compact, each sequence (a_n) has a subsequence (x_{n_k}) which converges to some $a \in A$. Since (x_n) converges to x, any subsequence (x_{n_k}) of (x_n) also converges to x. Then, the sequence $(p(x_{n_k}, a_{n_k}))$ converges to $p(x, a)$. Since (x_{n_k}) and (a_{n_k}) are the subsequences of (x_n) and (a_n), respectively, $p(x_{n_k}, a_{n_k}) \leq h_p(A, A) + \epsilon$ for all k. Therefore, $p(x, a) \leq h_p(A, A) + \epsilon$. By definition, $r_p(x, A) \leq h_p(A, A) + \epsilon$, so $x \in A + \epsilon$. Since x is arbitrary, $A + \epsilon$ is closed. □

To show $(H_p(X), h_p)$ is complete, we will have to show that there exist an N such that for all $n > N$, $h_p(A_n, A) - h_p(A, A) < \epsilon$. The following theorem gives an alternate way of proving convergence.

Theorem 3.8. *For* $A, B \in H_p(X)$ *and* $\epsilon > 0$. *Then* $h_p(A, B) - h_p(A, A) \leq \epsilon$ *if and only if* $A \subseteq B + \epsilon$ *and* $B \subseteq A + \epsilon$

Proof. By symmetry, it is enough to prove that $\rho_p(B, A) \leq h_p(A, A) + \epsilon$ if and only if $B \subseteq A + \epsilon$.

Let $B \subseteq A + \epsilon$. By definition, for every $b \in B$, $r_p(b, A) \leq h_p(A, A) + \epsilon$ which implies $\rho_p(B, A) \leq h_p(A, A) + \epsilon$.

Now let $\rho_p(B, A) \leq h_p(A, A)+\epsilon$. Then for every $b \in B$, $r_p(b, A) \leq h_p(A, A)+\epsilon$. By the definition of the set $A + \epsilon$ it follows that $B \subseteq A + \epsilon$. □

Before proving the Extension Lemma we will see some results and a lemma which will be used later in the proofs.

Lemma 3.9 (Extension Lemma). *Let (A_n) be a Cauchy sequence in $H_p(X)$ and let (n_k) be an increasing sequence of positive integers. If (x_{n_k}) is a Cauchy sequence in (X, p) for which $x_{n_k} \in A_{n_k}$ for all k, then there exists a Cauchy sequence (y_n) in X such that $y_n \in A_n$ for all n and $y_{n_k} = x_{n_k}$ for all k.*

Proof. Let (x_{n_k}) be a Cauchy sequence in (X, p) for which $x_{n_k} \in A_{n_k}$ for all k. For each n such that $n_{k-1} < n \leq n_k$, use Property 3 of Theorem 3.2 to choose $y_n \in A_n$ such that $r_p(x_{n_k}, A_n) = p(x_{n_k}, y_n)$. Then, by definition of r_p and ρ_p we can see that

$$p(x_{n_k}, y_n) = r_p(x_{n_k}, A_n) \leq \rho_p(A_{n_k}, A_n) \leq h_p(A_{n_k}, A_n).$$

Now, since $x_{n_k} \in A_{n_k}$, then $p(x_{n_k}, y_{n_k}) = r_p(x_{n_k}, A_{n_k}) = p(x_{n_k}, x_{n_k})$, by Property 1 of Theorem 3.2. Hence it follows that $x_{n_k} = y_{n_k}$, for all k.

Let $\epsilon > 0$. Since (x_{n_k}) is a Cauchy sequence in (X, p), there exists a positive integer K and $a \geq 0$ such that $p(x_{n_k}, x_{n_j}) - a < \epsilon/3$ for all $k, j \geq K$. Since (A_n) is a Cauchy sequence in $H_p(X)$, by definition there exist a positive integer $N > n_K$ and $b \geq 0$ such that $h_p(A_n, A_m) - b < \epsilon/3$ for all $m, n \geq N$. Then there exists integers $j, k \geq K$ such that $n_{k-1} < n \leq n_k$ and $n_{j-1} < m \leq n_j$. Then

$$\begin{aligned}
p(y_n, y_m) &\leq p(y_n, x_{n_k}) + p(x_{n_k}, x_{n_j}) + p(x_{n_j}, y_m) - p(x_{n_k}, x_{n_k}) - p(x_{n_j}, x_{n_j}) \\
&= r_p(x_{n_k}, A_n) + p(x_{n_k}, x_{n_j}) + r_p(x_{n_j}, A_m) - p(x_{n_k}, x_{n_k}) - p(x_{n_j}, x_{n_j}) \\
&\leq \rho_p(A_{n_k}, A_n) + p(x_{n_k}, x_{n_j}) + \rho_p(A_{n_j}, A_m) - p(x_{n_k}, x_{n_k}) - p(x_{n_j}, x_{n_j}) \\
&\leq h_p(A_{n_k}, A_n) + p(x_{n_k}, x_{n_j}) + h_p(A_{n_j}, A_m) - p(x_{n_k}, x_{n_k}) - p(x_{n_j}, x_{n_j}) \\
&< b + \epsilon/3 + a + \epsilon/3 + b + \epsilon/3 - p(x_{n_k}, x_{n_k}) - p(x_{n_j}, x_{n_j}) \\
&< \epsilon + c
\end{aligned}$$

where $c = 2b + a - p(x_{n_k}, x_{n_k}) - p(x_{n_j}, x_{n_j})$.

Since $p(x_{n_k}, x_{n_k}) \leq h_p(A_{n_k}, h_p(A_{n_k}))$ for all k and $p(x_{n_k}, x_{n_k}) \leq p(x_{n_k}, x_{n_j})$, we have $c \geq 0$. Hence by definition (y_n) is a Cauchy sequence in (X, p) such that $y_n \in A_n$ for all n and $y_{n_k} = x_{n_k}$ for all k. This completes the proof. □

The following lemma makes use of the Extension Lemma to guarantee that A is closed and nonempty.

Lemma 3.10. *Let (A_n) be a sequence in $H_p(X)$ and let A be the set of all points $x \in X$ such that there is a sequence (x_n) that converges to x and satisfies $x_n \in A_n$ for all n. If A_n is a Cauchy sequence, then the set A is closed and nonempty.*

Proof. First we will prove that A is nonempty. Since (A_n) is a Cauchy sequence, we can find a sequence of integers n_k and $b \geq 0$ such that $h_p(A_m, A_n) - b < \frac{1}{2^k}$ for all $m, n \geq n_k$. Let x_{n_1} be a fixed point in A_{n_1}. By Property 2 of Theorem 3.2, we can choose $x_{n_2} \in A_{n_2}$ such that $p(x_{n_1}, x_{n_2}) = r_p(x_{n_1}, A_{n_2})$. Then by definition of r_p, ρ_p and h_p, we find that

$$p(x_{n_1}, x_{n_2}) = r_p(x_{n_1}, A_{n_2}) \leq \rho_p(A_{n_1}, A_{n_2}) \leq h_p(A_{n_1}, A_{n_2}) < b + \frac{1}{2}.$$

Similarly, we can choose $x_{n_3} \in A_{n_3}$ such that

$$p(x_{n_2}, x_{n_3}) = r_p(x_{n_2}, A_{n_3}) \leq \rho_p(A_{n_2}, A_{n_3}) \leq h_p(A_{n_2}, A_{n_3}) < b + \frac{1}{2^2}.$$

Continuing this process we can construct a sequence (x_{n_k}) where $x_{n_k} \in A_{n_k}$ for all k. Then by the Extension Lemma, there exists a Cauchy sequence (y_n) in X such that $y_n \in A_n$ for all n and $y_{n_k} = x_{n_k}$ for all k. Since X is complete, the Cauchy sequence (y_n) converges to a point $y \in X$. Since $y_n \in A_n$ for all n, by the definition of the set, $y \in A$. Therefore A is nonempty.

Now we will prove that A is closed. Suppose a is a limit point of A. There exists a sequence (a_k) in A which converges to a. Since each $a_k \in A$, there exists a sequence (y_n) which converges to a_k and for each n, $y_n \in A_n$. Consequently, we can choose an increasing sequence (n_k) of integers such that $x_{n_k} \in A_{n_k}$ $p(x_{n_k}, a_k) - p(a_k, a_k) < \frac{1}{k}$ for all k. Then

$$p(x_{n_k}, a) \leq p(x_{n_k}, a_k) + p(a_k, a) - p(a_k, a_k).$$

Thus as $n \to \infty$, $p(a_k, a_k) \to p(a, a)$ and hence by definition, (x_{n_k}) converges to a. Every convergent sequence is Cauchy, so (x_{n_k}) is a Cauchy sequence for which $x_{n_k} \in A_{n_k}$ for all k. By the extension lemma, (x_{n_k}) can be extended to a convergent Cauchy sequence (y_n) in X such that $y_n \in A_n$ for all n and $y_{n_k} = x_{n_k}$, and $a \in A$, so A is closed. □

Since closed and totally bounded sets are compact, it remains to show that A is totally bounded. In the next lemma we will prove this.

Lemma 3.11. *Let $\{D_n\}$ be a sequence of totally bounded sets in X and let A be any subset of X. If for each $\epsilon > 0$, there exists a positive integer N such that $A \subseteq D_N + \epsilon$, then A is totally bounded.*

Proof. Let $\epsilon > 0$. Choose a positive integer N so that $A \subseteq D_N + \frac{\epsilon}{4}$. Since D_N is totally bounded, by definition we can choose a finite set $\{x_i : 1 \leq i \leq q\}$ where $x_{i \in D_N}$ such that $D_N \subseteq \bigcup_{i=1}^{q} B_p(x_i, \frac{\epsilon}{4})$. By reordering the $x_i's$, we may assume that $B_p(x_i, \frac{\epsilon}{4}) \cap A \neq \phi$ for $1 \leq i \leq k$, and $B_p(x_i, \frac{\epsilon}{4}) \cap A = \phi$ for $i > k$. Then for each $1 \leq i \leq k$, let $y_i \in B_p(x_i, \frac{\epsilon}{4}) \cap A$. We claim that $A \subseteq \bigcup_{i=1}^{k} B_p(y_i, \epsilon$.

Let $a \in A$. Then $a \in D_N + \frac{\epsilon}{4}$, so $r_p(a, D_N) < h_p(D_N, D_N) + \frac{\epsilon}{4}$. By Property 3 of Theorem 3.2, there exists $x \in D_N$ such that $p(a, x) = r_p(a, D_N)$. Then

$$p(a, x_i) \leq p(a, x) + p(x, x_i) - p(x, x)$$

$$p(a, x_i) - p(x_i, x_i) \leq p(a, x) + p(x, x_i) - p(x_i, x_i) - p(x, x)$$

$$p(a, x_i) - p(x_i, x_i) \leq p(a, x) + \frac{\epsilon}{4} - p(x, x)$$

$$p(a, x_i) - p(x_i, x_i) \leq \frac{\epsilon}{4} + \frac{\epsilon}{4} = \frac{\epsilon}{2}$$

Since $p(x, x_i) - p(x_i, x_i) \leq \frac{\epsilon}{4}$ and $p(a, x) - p(x, x) \leq \frac{\epsilon}{4}$.

So $a \in B_p(x_i, \frac{\epsilon}{2})$ for some $1 \leq i \leq k$. Thus we have $y_i \in B_p(x_i, \frac{\epsilon}{2}) \cap A$ such that $p(x_i, y(i)) - p(x_i, x_i) < \frac{\epsilon}{2}$. It follows that

$$p(a, y_i) \leq p(a, x_i) + p(x_i, y_i) - p(x_i, x_i)$$

$$p(a, y_i) \leq p(x_i, y_i) + \frac{\epsilon}{2}$$

$$p(a, y_i) - p(y_i, y_i) \leq p(x_i, y_i) - p(y_i, y_i) + \frac{\epsilon}{2}$$

$$p(a, y_i) - p(y_i, y_i) \leq \frac{\epsilon}{2} + \frac{\epsilon}{2} = \epsilon.$$

Thus for each $a \in A$ we found y_i for $1 \leq i \leq p$ such that $a \in B_p(y_i, \epsilon)$, then it follows that $A \subseteq \bigcup_{i=1}^{k} B_p(y_i, \epsilon)$. Thus by definition, A is totally bounded. This completes the proof. □

After proving these important results, we are now to prove our main result.

Theorem 3.12. *The space $H_p(X)$ is complete in the partial metric h_p*

Proof. Let (A_n) be a Cauchy sequence in $H_p(X)$, and A be the set of all points $x \in X$ such that there is a sequence (x_n) that converges to x and $x_n \in A_n$ for all n. We have to prove that $A \in H_p(X)$ and (A_n) converges to A.

By Lemma 3.10, the set A is closed and nonempty. Let $\epsilon > 0$. Since (A_n) is Cauchy, there exists a positive integer N such that $h_p(A_n, A_m) - h_p(A, A) < \epsilon$ for all $m, n \geq N$. By Theorem 3.8, then $A_m \subseteq A_n + \epsilon$ for all $m > n \geq N$. Let $a \in A$. Then we have to show that $a \in A_n + \epsilon$.

Fix $n \geq N$. By definition of the set A, there exists a sequence (x_i) such that $x_i \in A_i$ for all i and (x_i) converges to a. By Proposition 3.7, $A_n + \epsilon$ is closed. Since $x_i \in A_n + \epsilon$ for each i, then it follows that $a \in A_n + \epsilon$. This shows that $A \subseteq A_n + \epsilon$. By Lemma 3.11, A is totally bounded. Also A is complete since it is the closed subset of a complete partial metric space. Since A is nonempty, complete, and totally bounded, then A is compact and thus $A \in H_p(X)$.

To show that (A_n) converges to A, we need to show that there exists a positive integer N such that $h_p(A_n, A) - h_p(A, A) < \epsilon$, for all $n \geq N$. By Theorem 3.8, we need to show $A \subseteq A_n + \epsilon$ and $A_n \subseteq A + \epsilon$.

We have already proved that there exists N such that $A \subseteq A_n + \epsilon$ for all $n \geq N$. To prove $A_n \subseteq A + \epsilon$, choose an N such that $h_p(A_n, A_m) - h_p(A, A) < \frac{\epsilon}{2}$ for all $m, n \geq N$. Since (A_n) is Cauchy in $H_p(X)$, there exist a strictly increasing sequence n_i of positive integers such that $n_1 > N$ and $h_p(A_m, A_n) - h_p(A, A) < \frac{\epsilon}{2^{i+1}}$ for all $m, n > n_i$.

Using the Property 3 of Theorem 3.2, we have

since $A_n \subseteq A_{n_1 + \frac{\epsilon}{2}}$, there exists $x_{n_1} \in A_{n_1}$ such that $p(y, x_{n_1}) - p(x_{n_1}, x_{n_1}) \leq \frac{\epsilon}{2}$

since $A_{n_1} \subseteq A_{n_1 + \frac{\epsilon}{2}}$, there exists $x_{n_2} \in A_{n_2}$ such that $p(x_{n_1}, x_{n_2}) - p(x_{n_2}, x_{n_2}) \leq \frac{\epsilon}{4}$

Continuing this process we obtain a sequence (x_{n_i}) such that for all positive integers i, we have $x_{n_i} \in A_{n_i}$ and $p(x_{n_i}, x_{n_{i+1}}) - p(x_{n_{i+1}}, x_{n_{i+1}}) < \frac{\epsilon}{2^{i+1}}$. Thus by Result 2 we see that (x_{n_i}) is a Cauchy sequence and so by Extension Lemma the limit of the sequence a is in A.

Also,

$$\begin{aligned} p(y, x_{n_i}) &\leq p(y, x_{n_1}) + p(x_{n_1}, x_{n_2}) + \cdots + p(x_{n_{i-1}}, x_{n_i}) \\ &- [p(x_{n_1}, x_{n_1}) + p(x_{n_2}, x_{n_2}) \cdots + p(xn_{i-1}, xn_{i-1})] \\ &\leq \frac{\epsilon}{2} + p(x_{n_1}, x_{n_1}) + \frac{\epsilon}{4} + p(x_{n_2}, x_{n_2}) + \cdots + \frac{\epsilon}{2^i} + p(x_{n_i}, x_{n_i}) \\ &- [p(x_{n_1}, x_{n_1}) + p(x_{n_2}, x_{n_2}) \cdots + p(xn_{i-1}, xn_{i-1})] \\ &< \epsilon + p(x_{n_i}, x_{n_i}) \end{aligned}$$

Thus $p(y, x_{n_i}) - p(x_{n_i}, x_{n_i}) < \epsilon$ for all i. Hence it follows that $p(y, a) - p(a, a) \leq \epsilon$ and so $y \in A + \epsilon$.

Thus there exists an N such that for all $n \geq N$, $h_p(A_n, A) - h_p(A, A) < \epsilon$ which implies (A_n) converges to $A \in H_p(X)$. Therefore, if (X, p) is complete, then $(H_p(X), h_p)$ is complete. □

Thus the generalisation of Banach's fixed point theorem to the partial metric spaces is applicable to the space $(H_p(X), h_p)$, thereby ensuring a unique fixed point for every contraction mapping in $H_p(X)$. This unique fixed point is a deterministic fractal in the partial metric space which can be defined as an attractor of an IFS_p.

Definition 3.13. A (hyperbolic) *Iterated Function System* (IFS_p) consists of a complete partial metric space (X, p) together with a finite set of contraction mappings $w_j : X \longrightarrow X$ with respect to the contraction factors s_j, for $j = 1, 2, \ldots, N$. The notation of the IFS_p just defined is $\{X; w_j, j = 1, 2, \ldots, N\}$ and its contraction factor is $s = max\{s_j : j = 1, 2, \ldots, N\}$.

Definition 3.14. An IFS_p is said to be *homogeneous* if the contraction factor $s_j = s$ for all $j = 1, 2, \ldots, N$.

Theorem 3.15. *Let* $\{X; w_j, j = 1, 2, \ldots, N\}$ *be a hyperbolic* IFS_p *with contraction factor s. Then the transformation* $W : H_p(X) \longrightarrow H_p(X)$ *defined by* $W(B) = \bigcup_{i=1}^{N} w_j(B)$ *for all* $B \in H_p(X)$ *is a contraction mapping on the complete metric space* $(H_p(X), h_p)$ *with contraction factor s. That is,* $h_p(W(B), W(C)) \leq s.h_p(B, C)$ *for all* $B, C \in H_p(X)$*. Its unique fixed point,* $A \in H_p(X)$*, obeys* $A = W(A) = \bigcup_{j=1}^{N} w_j(A)$ *and is given by* $A = \lim_{n\to\infty} W^{on}(B)$ *for any* $B \in H_p(X)$.

Definition 3.16. The fixed point $A \in H_p(X)$ described in the Theorem 3.15 is called the *attractor* of the IFS_p.

4 Conclusion

The nature of fractals in a partial metric space is studied and all the related results are discussed. The definition of fractals in partial metric space is done analogous to that done by Barnsley [2].

References

1. Aydi, H., et al.: Partial Hausdorff metric and Nadler's fixed point theorem on partial metric spaces. Topol. Appl. **159**, 3234–3242 (2012)
2. Barnsley, M.: Fractals Everywhere. Academic, Boston (1988)
3. Heckmann, R.: Approximation of metric spaces by partial metric spaces. Appl. Categorical Struct. **7**, 71–83 (1999)
4. Mathews, S.G.: Partial metric topology. In: Proceedings of 8th Summer Conference on Topology and its Applications, New York (1992)

Part II
Wavelet Theory

Frames and Extension Problems I

Ole Christensen

Abstract In this article we present a short survey of frame theory in Hilbert spaces. We discuss Gabor frames and wavelet frames and set the stage for a discussion of various extension principles; this will be presented in the article *Frames and extension problems II* (joint with H.O. Kim and R.Y. Kim).

Keywords Frames • Gabor systems • Wavelet systems • Extension problems

1 Introduction

Frames provide us with a convenient tool to obtain expansions in Hilbert spaces of a similar type as the one that arise via orthonormal bases. However, the frame conditions are significantly weaker, which makes frames much more flexible. For this reason frame theory has attracted much attention in recent years, especially in connection with its concrete manifestations within Gabor analysis and wavelet analysis.

In this article we give a short overview of the general theory for frames in Hilbert spaces, as well as its concrete realizations in Gabor analysis and wavelet analysis. We set the stage for a discussion of various extension principles to be presented in the article *Frames and extension problems II* (joint paper with H.O. Kim and R.Y. Kim).

O. Christensen (✉)
Department of Applied Mathematics and Computer Science, Technical University of Denmark, Building 303, 2800 Lyngby, Denmark
e-mail: ochr@dtu.dk

C. Bandt et al. (eds.), *Fractals, Wavelets, and their Applications*, Springer Proceedings in Mathematics & Statistics 92, DOI 10.1007/978-3-319-08105-2_14,

2 A Survey on Frames and Operators

General frames were introduced already in the paper [18] by Duffin and Schaeffer in 1952. Apparently it did not find much use at that time, until it got re-introduced by Young in his book [31] from 1982. After that, Daubechies, Grossmann and Morlet took the key step of connecting frames with wavelets and Gabor systems in the paper [15].

2.1 General Frame Theory

Let $\mathcal{H}$ be a separable Hilbert space with the inner product $\langle \cdot, \cdot \rangle$ linear in the first entry. A countable family of elements $\{f_k\}_{k \in I}$ in $\mathcal{H}$ is a

1. *Bessel sequence* if there exists a constant $B > 0$ such that

$$\sum_{k \in I} |\langle f, f_k \rangle|^2 \leq B||f||^2, \ \forall f \in \mathcal{H};$$

2. *frame* for $\mathcal{H}$ if there exist constants $A, B > 0$ such that

$$A||f||^2 \leq \sum_{k \in I} |\langle f, f_k \rangle|^2 \leq B||f||^2, \ \forall f \in \mathcal{H}; \tag{1}$$

   The numbers A, B in (1) are called *frame bounds.*
3. *Riesz basis* for $\mathcal{H}$ if $\overline{\text{span}}\{f_k\}_{k \in I} = \mathcal{H}$ and there exist constants $A, B > 0$ such that

$$A \sum |c_k|^2 \leq \left|\left|\sum c_k f_k\right|\right|^2 \leq B \sum |c_k|^2. \tag{2}$$

   for all finite sequences $\{c_k\}$.

Every orthonormal basis is a Riesz basis, and every Riesz basis is a frame [the bounds A, B in (2) are frame bounds]; a frame which is not a Riesz basis is said to be *overcomplete* or *redundant.* Riesz bases and frames are natural tools to gain more flexibility than possible with an orthonormal basis. For an overview of the general theory for frames and Riesz bases we refer to [2] and [3]; a deeper treatment is given in the books [4, 6]. Here, we just mention that the difference between a Riesz basis and a frame is that the elements in a frame might be dependent. More precisely, a frame $\{f_k\}_{k \in I}$ is a Riesz basis if and only if

$$\sum_{k \in I} c_k f_k = 0, \ \{c_k\} \in \ell^2(I) \Rightarrow c_k = 0, \ \forall k \in I.$$

Associated with a Bessel sequence $\{f_k\}_{k=1}^\infty$, the *pre-frame operator* or *synthesis operator* is

$$T : \ell^2(\mathbb{N}) \to \mathcal{H},\ T\{c_k\}_{k=1}^\infty = \sum_{k=1}^\infty c_k f_k.$$

The operator T is bounded for any Bessel sequence $\{f_k\}_{k=1}^\infty$. The adjoint operator of T is called the *analysis operator* and is given by

$$T^* : \mathcal{H} \to \ell^2(\mathbb{N}),\ \ T^* f = \{\langle f, f_k\rangle\}_{k=1}^\infty.$$

Finally, the *frame operator* is defined by

$$S : \mathcal{H} \to \mathcal{H},\ \ Sf = TT^* f = \sum_{k=1}^\infty \langle f, f_k\rangle f_k.$$

The following classical result shows that any frame leads to an expansion of the elements in $\mathcal{H}$ as a (infinite) linear combinations of the frame elements. It also shows that the general expansion simplifies considerably for tight frames. Finally, the last part of the result shows that for frames that are not Riesz bases, the coefficients in the series expansion of an element $f \in \mathcal{H}$ are not unique:

Theorem 2.1. *Let $\{f_k\}_{k=1}^\infty$ be a frame with frame operator S. Then the following holds:*

(i) Each $f \in \mathcal{H}$ has the decompositions

$$f = \sum_{k=1}^\infty \langle f, S^{-1} f_k\rangle f_k = \sum_{k=1}^\infty \langle f, f_k\rangle S^{-1} f_k.$$

(ii) If $\{f_k\}_{k=1}^\infty$ is a tight frame with frame bound A, then $S = AI$, and

$$f = \frac{1}{A}\sum_{k=1}^\infty \langle f, f_k\rangle f_k,\ \forall f \in \mathcal{H}. \tag{3}$$

(iii) If $\{f_k\}_{k=1}^\infty$ is an overcomplete frame, there exist frames $\{g_k\}_{k=1}^\infty \neq \{S^{-1} f_k\}_{k=1}^\infty$ for which

$$f = \sum_{k=1}^\infty \langle f, g_k\rangle f_k,\ \forall f \in \mathcal{H}. \tag{4}$$

Any Bessel sequence $\{g_k\}_{k=1}^\infty$ satisfying (4) for a given frame $\{f_k\}_{k=1}^\infty$ is called a *dual frame* of $\{f_k\}_{k=1}^\infty$. The special choice $\{g_k\}_{k=1}^\infty = \{S^{-1} f_k\}_{k=1}^\infty$ is called the *canonical dual frame.* In order to avoid confusion we note that if (4) holds for two Bessel sequences $\{f_k\}_{k=1}^{\infty}$ and $\{g_k\}_{k=1}^{\infty}$, they are automatically frames:

Lemma 2.2. *If $\{f_k\}_{k=1}^{\infty}$ and $\{g_k\}_{k=1}^{\infty}$ are Bessel sequences and* (4) *holds, then $\{f_k\}_{k=1}^{\infty}$ and $\{g_k\}_{k=1}^{\infty}$ are dual frames.*

Note that duality between Bessel sequences $\{f_k\}_{k=1}^{\infty}$ and $\{g_k\}_{k=1}^{\infty}$ can be expressed entirely in terms of operators. In fact, if T, U denote the pre-frame operators for $\{f_k\}_{k=1}^{\infty}$, respectively, $\{g_k\}_{k=1}^{\infty}$, the sequences are dual frames if and only if

$$TU^* = I.$$

2.2 Operators on $L^2(\mathbb{R})$

In order to construct concrete frames in the Hilbert space $L^2(\mathbb{R})$, we need to consider some important classes of operators.

Definition 2.3 (Translation, Modulation, Dilation). Consider the following classes of linear operators on $L^2(\mathbb{R})$:

(i) For $a \in \mathbb{R}$, the operator T_a, called translation by a, is defined by

$$(T_a f)(x) := f(x-a), \; x \in \mathbb{R}. \tag{5}$$

(ii) For $b \in \mathbb{R}$, the operator E_b, called modulation by b, is defined by

$$(E_b f)(x) := e^{2\pi i b x} f(x), \; x \in \mathbb{R}. \tag{6}$$

(iii) For $c > 0$, the operator D_c, called dilation by c, is defined by

$$(D_c f)(x) := \frac{1}{\sqrt{c}} f(\frac{x}{c}), \; x \in \mathbb{R}. \tag{7}$$

(iv) The dyadic dilation operator is defined by

$$(Df)(x) := 2^{1/2} f(2x), \; x \in \mathbb{R}.$$

All the above operators are linear, bounded, and unitary. We will also need the Fourier transform, for $f \in L^1(\mathbb{R})$ defined by

$$\hat{f}(\gamma) := \int_{-\infty}^{\infty} f(x) e^{-2\pi i \gamma x} dx.$$

The Fourier transform is extended to a unitary operator on $L^2(\mathbb{R})$ in the usual way.

The operators T_a, E_b, D, and $\mathcal{F}$ are related by the following commutator relations:

$$T_a E_b = e^{-2\pi iba} E_b T_a, \quad T_b D = D T_{b/a}, \quad DE_b = E_{b/a} D$$
$$\mathcal{F} T_a = E_{-a} \mathcal{F}, \quad \mathcal{F} E_a = T_a \mathcal{F}, \quad \mathcal{F} D = D^{-1} \mathcal{F}.$$

3 Gabor Systems

Gabor systems in $L^2(\mathbb{R})$ have the form

$$\{e^{2\pi imbx} g(x - na)\}_{m,n\in\mathbb{Z}}$$

for some $g \in L^2(\mathbb{R})$, $a, b > 0$. Using operator notation, we can write a Gabor system as $\{E_{mb}T_{na}g\}_{m,n\in\mathbb{Z}}$.

We will not go into a general description of Gabor analysis and its role in time–frequency analysis, but just refer to the books [19–21].

Letting $\chi_{[0,1]}$ denote the characteristic function for the interval [0, 1], it is easy to show that $\{E_m T_n \chi_{[0,1]}\}_{m,n\in\mathbb{Z}}$ is an orthonormal basis for $L^2(\mathbb{R})$. But the function $\chi_{[0,1]}$ is discontinuous and has very slow decay in the Fourier domain, so this function is not suitable for time–frequency analysis. For the sake of time–frequency analysis we want the frame generator g to be a continuous function with compact support. The following classical result shows that this more or less forces us to work with frames.

Lemma 3.1. *If g is be a continuous function with compact support, then*

(i) $\{E_{mb}T_{na}g\}_{m,n\in\mathbb{Z}}$ cannot be an ONB.
(ii) $\{E_{mb}T_{na}g\}_{m,n\in\mathbb{Z}}$ cannot be a Riesz basis.
(iii) $\{E_{mb}T_{na}g\}_{m,n\in\mathbb{Z}}$ can be a frame if $0 < ab < 1$;

In addition to (iii), if $0 < ab < 1$, it is always possible to find a function $g \in C_c(\mathbb{R})$ such that $\{E_{mb}T_{na}g\}_{m,n\in\mathbb{Z}}$ is a Gabor frame. We also note that no matter whether g is continuous or not, Gabor frames $\{E_{mb}T_{na}g\}_{m,n\in\mathbb{Z}}$ for $L^2(\mathbb{R})$ only exist if $ab \leq 1$.

Bessel sequences of the form $\{E_{mb}T_{na}g\}_{m,n\in\mathbb{Z}}$ will play a central role in some of the open problems to be considered in this article, so let us state a classical sufficient condition that is easy to verify.

Lemma 3.2. *Let g be a bounded function with compact support. Then $\{E_{mb}T_{na}g\}_{m,n\in\mathbb{Z}}$ is a Bessel sequence for any $a, b > 0$.*

For a Gabor system $\{E_{mb}T_{na}g\}_{m,n\in\mathbb{Z}}$, the frame operator commutes with the operators E_{mb}, T_{na}, $m, n \in \mathbb{Z}$. We will need the result below, which is almost identical to Lemma 9.3.1 in [6].

Lemma 3.3. *Let $g, h \in L^2(\mathbb{R})$ and $a, b > 0$ be given, and assume that $\{E_{mb}T_{na}g\}_{m,n\in\mathbb{Z}}$ and $\{E_{mb}T_{na}h\}_{m,n\in\mathbb{Z}}$ are Bessel sequences. Then the following holds:*

(i) Letting T and U denote the preframe operators for $\{E_{mb}T_{na}g\}_{m,n\in\mathbb{Z}}$ and $\{E_{mb}T_{na}h\}_{m,n\in\mathbb{Z}}$,

$$TUE_{mb}T_{na} = E_{mb}T_{na}TU, \ \forall m, n \in \mathbb{Z}.$$

(ii) If $\{E_{mb}T_{na}g\}_{m,n\in\mathbb{Z}}$ is a frame with frame operator $S = TT^*$, then

$$S^{-1}E_{mb}T_{na} = E_{mb}T_{na}S^{-1}, \ \forall m, n \in \mathbb{Z}.$$

Lemma 3.3 (ii) implies that for a Gabor frame $\{E_{mb}T_{na}g\}_{m,n\in\mathbb{Z}}$ with associated frame operator S, the canonical dual frame also has Gabor structure, in contrast with the situation we encountered for wavelet frames. However, even for a nice frame $\{E_{mb}T_{na}g\}_{m,n\in\mathbb{Z}}$ it is nontrivial to control the properties of the canonical dual frame $\{E_{mb}T_{na}S^{-1}g\}_{m,n\in\mathbb{Z}}$, so often it is a better strategy to construct dual pairs $\{E_{mb}T_{na}g\}_{m,n\in\mathbb{Z}}$,$\{E_{mb}T_{na}h\}_{m,n\in\mathbb{Z}}$ such that g and h have required properties. Dual pairs of Gabor frames have been characterized by Ron and Shen [26] and Janssen [23]:

Theorem 3.4. *Two Bessel sequences $\{E_{mb}T_{na}g\}_{m,n\in\mathbb{Z}}$ and $\{E_{mb}T_{na}h\}_{m,n\in\mathbb{Z}}$ form dual frames for $L^2(\mathbb{R})$ if and only if*

$$\sum_{k\in\mathbb{Z}} \overline{g(x - n/b - ka)}h(x - ka) = b\delta_{n,0}, \ a.e. \ x \in [0, a].$$

One of the most important results in Gabor analysis is the so-called *duality principle.* It was discovered almost simultaneously by three groups of researchers, namely Daubechies et al. [16], Janssen [22], and Ron and Shen [26]. It concerns the relationship between frame properties for a function g with respect to the *lattice* $\{(na, mb)\}_{m,n\in\mathbb{Z}}$ and with respect to the so-called *dual lattice* $\{(n/b, m/a)\}_{m,n\in\mathbb{Z}}$:

Theorem 3.5. *Given $g \in L^2(\mathbb{R})$ and $a, b > 0$, the following are equivalent:*

(i) $\{E_{mb}T_{na}g\}_{m,n\in\mathbb{Z}}$ is a frame for $L^2(\mathbb{R})$ with bounds A, B;
(ii) $\{\frac{1}{\sqrt{ab}} E_{m/a}T_{n/b}g\}_{m,n\in\mathbb{Z}}$ is a Riesz sequence with bounds A, B.

The intuition behind the duality principle is that if $\{E_{mb}T_{na}g\}_{m,n\in\mathbb{Z}}$ is a frame for $L^2(\mathbb{R})$, then $ab \leq 1$, i.e., the sampling points $\{(na, mb)\}_{m,n\in\mathbb{Z}}$ are "sufficiently dense." Therefore the points $\{(n/b, m/a)\}_{m,n\in\mathbb{Z}}$ are "sparse," in the sense that $\frac{1}{ab} \geq 1$. Technically, this implies that the functions $\{\frac{1}{\sqrt{ab}} E_{m/a}T_{n/b}g\}_{m,n\in\mathbb{Z}}$ are linearly independent and only span a subspace of $L^2(\mathbb{R})$. The reason for the importance of the duality principle is that in general it is much easier to check that a system of vectors is a Riesz sequence than to check that it is a frame. The duality principle is clearly related with the *Wexler–Raz theorem* stated next, which was discovered in 1994.

Theorem 3.6. *If the Gabor systems* $\{E_{mb}T_{na}g\}_{m,n\in\mathbb{Z}}$ *and* $\{E_{mb}T_{na}h\}_{m,n\in\mathbb{Z}}$ *are Bessel sequences, then the following are equivalent:*

(i) The Gabor systems $\{E_{mb}T_{na}g\}_{m,n\in\mathbb{Z}}$ *and* $\{E_{mb}T_{na}h\}_{m,n\in\mathbb{Z}}$ *are dual frames;*
(ii) The Gabor systems $\{\frac{1}{\sqrt{ab}}E_{m/a}T_{n/b}g\}_{m,n\in\mathbb{Z}}$ *and* $\{\frac{1}{\sqrt{ab}}E_{m/a}T_{n/b}h\}_{m,n\in\mathbb{Z}}$ *are biorthogonal, i.e.,*

$$\langle\frac{1}{\sqrt{ab}}E_{m/a}T_{n/b}g, \frac{1}{\sqrt{ab}}E_{m'/a}T_{n'/b}h\rangle = \delta_{m,m'}\delta_{n,n'}.$$

Theorem 3.4 characterizes pairs of dual Gabor frames, but it does not show how to construct convenient pairs of Gabor frames. A class of convenient dual pairs of frames are constructed in [5, 8]:

Theorem 3.7. *Let* $N \in \mathbb{N}$. *Let* $g \in L^2(\mathbb{R})$ *be a real-valued bounded function for which* $supp\ g \subseteq [0, N]$ *and*

$$\sum_{n\in\mathbb{Z}} g(x-n) = 1. \tag{8}$$

Let $b \in]0, \frac{1}{2N-1}]$. *Define* $\tilde{g} \in L^2(\mathbb{R})$ *by*

$$h(x) = \sum_{n=-N+1}^{N-1} a_n g(x+n),$$

where

$$a_0 = b,\ \ a_n + a_{-n} = 2b,\ n = 1, 2, \cdots, N-1.$$

Then g and h generate dual frames $\{E_{mb}T_n g\}_{m,n\in\mathbb{Z}}$ *and* $\{E_{mb}T_n\tilde{g}\}_{m,n\in\mathbb{Z}}$ *for* $L^2(\mathbb{R})$.

Let us apply Theorem 3.7 to the classical B-splines $B_N,\ N \in \mathbb{N}$, given inductively by

$$B_1(x) := \chi_{[0,1]}(x),\ \ B_{N+1}(x) := B_N * B_1(x) = \int_0^1 B_N(x-t)\,dt. \tag{9}$$

Example 3.8. The conditions in Theorem 3.7 are satisfied for any B-spline $B_N,\ N \in \mathbb{N}$. Some choices of the coefficients a_n are given by (Fig. 1):

1) Take

$$a_0 = b, a_n = 0 \text{ for } n = -N+1, \ldots, -1, a_n = 2b, n = 1, \ldots N-1.$$

This choice gives the dual frame generated by the function with shortest support.

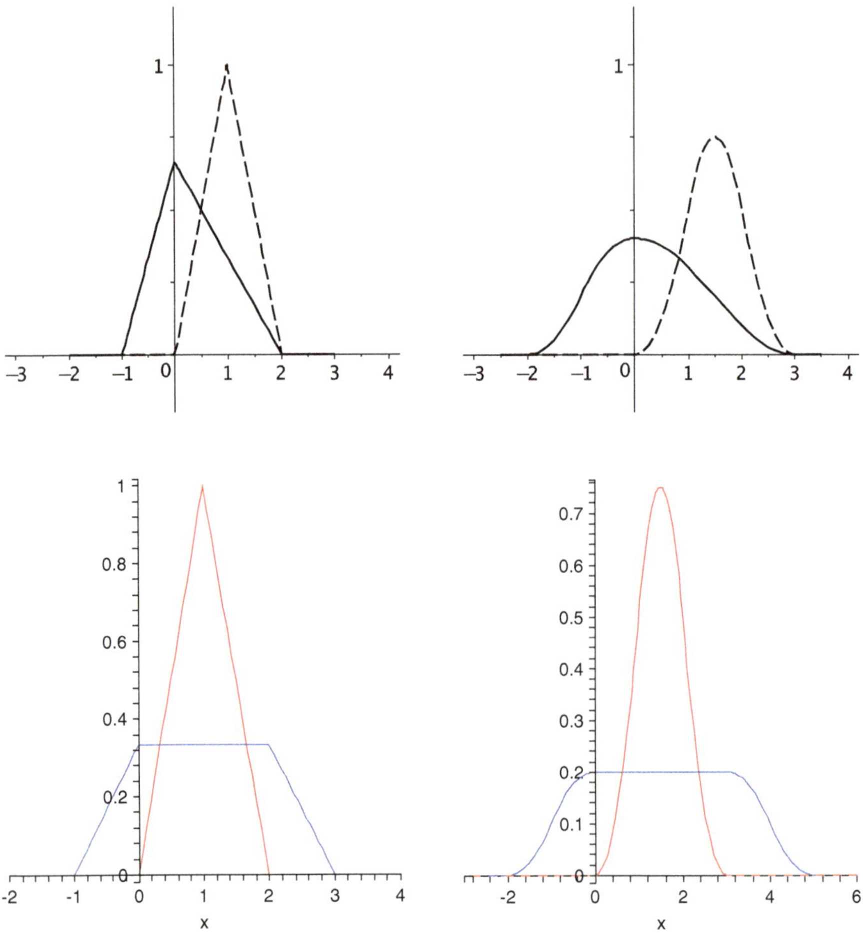

Fig. 1 The generators B_2 and B_3 and some dual generators

2) Take

$$a_{-N+1} = a_{-N+2} = \cdots = a_{N-1} = b:$$

if g is symmetric, this leads to a symmetric dual generator

$$\tilde{g}(x) = b \sum_{n=-N+1}^{N-1} g(x+n).$$

□

4 Wavelet Systems in $L^2(\mathbb{R})$

A *wavelet system* in $L^2(\mathbb{R})$ has the form $\{a^{j/2}\psi(a^j x - kb)\}_{j,k\in\mathbb{Z}}$ for some parameters $a > 1, b > 0$ and a given function $\psi \in L^2(\mathbb{R})$. Introducing the scaling operators and the translation operators, the wavelet system can be written as $\{D_{a^j}T_{kb}\psi\}_{j,k\in\mathbb{Z}}$.

There are also characterizing equations for dual wavelet frames; see [11]. They are formulated in terms of the Fourier transform:

Theorem 4.1. *Given $a > 1$, $b > 0$, two Bessel sequences $\{D_{a^j}T_{kb}\psi\}_{j,k\in\mathbb{Z}}$ and $\{D_{a^j}T_{kb}\tilde{\psi}\}_{j,k\in\mathbb{Z}}$, where $\psi, \tilde{\psi} \in L^2(\mathbb{R})$, form dual wavelet frames for $L^2(\mathbb{R})$ if and only if the following two conditions hold:*

(i) $\sum_{j\in\mathbb{Z}} \overline{\hat{\psi}(a^j\gamma)}\hat{\tilde{\psi}}(a^j\gamma) = b$ *for a.e.* $\gamma \in \mathbb{R}$.
(ii) For any number $\alpha \neq 0$ of the form $\alpha = m/a^j$, $m, j \in \mathbb{Z}$,

$$\sum_{(j,m)\in I_\alpha} \overline{\hat{\psi}(a^j\gamma)}\hat{\tilde{\psi}}(a^j\gamma + m/b) = 0, \; a.e. \; \gamma \in \mathbb{R},$$

where $I_\alpha := \{(j,m) \in \mathbb{Z}^2 \mid \alpha = m/a^j\}$.

We will present a few aspects of wavelet theory, beginning with the dyadic wavelet systems and classical multiresolution analysis.

4.1 Dyadic Wavelet Systems

A systems of functions of the form $\{D^j T_k \psi\}_{j,k\in\mathbb{Z}}$, where $\psi \in L^2(\mathbb{R})$ is a fixed function, is called a *dyadic wavelet system.* Note that $D^j T_k\psi(x) = 2^{j/2}\psi(2^j x - k)$, $x \in \mathbb{R}$. Given a frame $\{D^j T_k\psi\}_{j,k\in\mathbb{Z}}$ for $L^2(\mathbb{R})$, the associated frame operator is

$$S : L^2(\mathbb{R}) \to L^2(\mathbb{R}), \; Sf = \sum_{j,k\in\mathbb{Z}} \langle f, D^j T_k\psi\rangle D^j T_k\psi,$$

and the frame decomposition reads

$$f = \sum_{j,k\in\mathbb{Z}} \langle f, S^{-1}D^j T_k\psi\rangle D^j T_k\psi, \; f \in L^2(\mathbb{R}).$$

In order to use the frame decomposition we need to calculate the numbers $\langle f, S^{-1}D^j T_k\psi\rangle$ for all $j,k \in \mathbb{Z}$, i.e., a double-infinite sequence of numbers. One can show that

$$S^{-1}D^j T_k\psi = D^j S^{-1}T_k\psi,$$

so in practice it is enough to calculate the action of S^{-1} on the functions $T_k\psi$, and then apply the scaling D^j. Unfortunately, in general

$$D^j S^{-1} T_k \psi \neq D^j T_k S^{-1} \psi.$$

Thus, we cannot expect the canonical dual frame of a wavelet frame to have wavelet structure. As a concrete example (taken from [10, 13]), let $\{D^j T_k \psi\}_{j,k\in\mathbb{Z}}$ be a wavelet orthonormal basis for $L^2(\mathbb{R})$. Given $\epsilon \in]0,1[$, let $\theta = \psi + \epsilon D\psi$. Then $\{D^j T_k \theta\}_{j,k\in\mathbb{Z}}$ is a Riesz basis, but the canonical dual frame of $\{D^j T_k \theta\}_{j,k\in\mathbb{Z}}$ does *not* have the wavelet structure. Since the dual is unique for a Riesz basis, this example demonstrates that there are wavelet frames where no dual with wavelet structure exists. On the other hand, Bownik and Weber [1] have given an interesting example of a wavelet frame $\{D^j T_k \psi\}_{j,k\in\mathbb{Z}}$ for which the canonical dual does not have the wavelet structure, but other dual frames with wavelet structure exist.

4.2 *Classical Multiresolution Analysis*

Multiresolution analysis is a tool to construct orthonormal bases for $L^2(\mathbb{R})$ of the form $\{D^j T_k \psi\}_{j,k\in\mathbb{Z}}$ for a suitably chosen function $\psi \in L^2(\mathbb{R})$. Such a function ψ is called a *wavelet.* Its original definition of a multiresolution analysis was given by Mallat and Meyer [24, 25] is as follows:

Definition 4.2. A multiresolution analysis for $L^2(\mathbb{R})$ consists of a sequence of closed subspaces $\{V_j\}_{j\in\mathbb{Z}}$ of $L^2(\mathbb{R})$ and a function $\phi \in V_0$ such that

(i) $\cdots V_{-1} \subset V_0 \subset V_1 \cdots$
(ii) $\cap_j V_j = \{0\}$ and $\overline{\cup_j V_j} = L^2(\mathbb{R})$
(iii) $f \in V_j \Leftrightarrow Df \in V_{j+1}$.
(iv) $f \in V_0 \Rightarrow T_k f \in V_0, \ \forall k \in \mathbb{Z}$.
(v) $\{T_k\phi\}_{k\in\mathbb{Z}}$ is an orthonormal basis for V_0.

A multiresolution analysis is in fact generated just by a suitable choice of the function ϕ : if the conditions in Definition 4.2 are satisfied, then necessarily

$$V_j = \overline{\text{span}}\{D^j T_k \phi\}_{k\in\mathbb{Z}}, \ \forall j \in \mathbb{Z}.$$

The following result, due to Mallat and Meyer [24, 25], shows how to construct a wavelet based on a multiresolution analysis. Other proofs can be found in [7, 14, 30].

Theorem 4.3. *Assume that the function $\phi \in L^2(\mathbb{R})$ generates a multiresolution analysis. Then the following holds:*

(i) There exists a 1-periodic function $H_0 \in L^2(0,1)$ such that

$$\hat{\phi}(2\gamma) = H_0(\gamma)\hat{\phi}(\gamma), \ \gamma \in \mathbb{R}. \tag{10}$$

(ii) Define the 1-periodic function H_1 by

$$H_1(\gamma) := \overline{H_0(\gamma + \frac{1}{2})}\, e^{-2\pi i \gamma}. \tag{11}$$

Also, define the function ψ via

$$\hat{\psi}(2\gamma) := H_1(\gamma)\hat{\phi}(\gamma). \tag{12}$$

Then ψ is a wavelet.

The definition in (12) is quite indirect: it defines the function ψ in terms of its Fourier transform, so we have to apply the inverse Fourier transform in order to obtain an expression for ψ. This actually leads to an explicit expression of the function ψ in terms of the given function ϕ:

Proposition 4.4. *Assume that (12) holds for a 1-periodic function $H_1 \in L^2(0,1)$,*

$$H_1(\gamma) = \sum_{k\in\mathbb{Z}} d_k e^{2\pi i k\gamma}. \tag{13}$$

Then

$$\psi(x) = \sqrt{2}\sum_{k\in\mathbb{Z}} d_k D T_{-k}\phi(x) = 2\sum_{k\in\mathbb{Z}} d_k \phi(2x+k),\ x \in \mathbb{R}. \tag{14}$$

The classical example of a wavelet generated by a multiresolution analysis is the *Haar wavelet*,

$$\psi(x) = \begin{cases} 1 & \text{if } x \in [0, \frac{1}{2}[\\ -1 & \text{if } x \in [\frac{1}{2}, 1[\\ 0 & \text{otherwise} \end{cases}$$

It is generated by the function $\phi = \chi_{[0,1]}$. In 1989 Daubechies managed to construct an important class of compactly supported wavelets with very good approximation properties. We will not go into a detailed discussion of these, but just refer to, e.g., [14, 30].

4.3 *The Unitary Extension Principle*

In this section we present results by Ron and Shen, which enables us to construct tight wavelet frames generated by a collection of functions $\psi_1, \ldots, \psi_n$. Our presentation is based on the papers [27–29]. Note also that a more flexible tool, the oblique extension principle, has later been introduced by two groups of researchers, see [12, 17].

The generators $\psi_1, \ldots, \psi_n$ will be constructed on the basis of a function which satisfy a refinement equation, and since we will work with all those functions simultaneously it is convenient to change our previous notation slightly and denote the refinable function by ψ_0.

General Setup: Let $\psi_0 \in L^2(\mathbb{R})$. Assume that $\lim_{\gamma \to 0} \hat{\psi}_0(\gamma) = 1$ and that there exists a function $H_0 \in L^\infty(\mathbb{T})$ such that

$$\hat{\psi}_0(2\gamma) = H_0(\gamma)\hat{\psi}_0(\gamma). \tag{15}$$

Let $H_1, \ldots, H_n \in L^\infty(\mathbb{T})$, and define $\psi_1, \ldots, \psi_n \in L^2(\mathbb{R})$ by

$$\hat{\psi}_\ell(2\gamma) = H_\ell(\gamma)\hat{\psi}_0(\gamma), \ \ell = 1, \ldots, n. \tag{16}$$

Finally, let H denote the $(n+1) \times 2$ matrix-valued function defined by

$$H(\gamma) = \begin{pmatrix} H_0(\gamma) & T_{1/2}H_0(\gamma) \\ H_1(\gamma) & T_{1/2}H_1(\gamma) \\ \cdot & \cdot \\ \cdot & \cdot \\ H_n(\gamma) & T_{1/2}H_n(\gamma) \end{pmatrix}. \tag{17}$$

We will frequently suppress the dependence on γ and simply speak about the matrix H. The purpose is to find $H_1, \ldots, H_n$ such that

$$\{D^j T_k \psi_1\}_{j,k \in \mathbb{Z}} \cup \{D^j T_k \psi_2\}_{j,k \in \mathbb{Z}} \cup \cdots \cup \{D^j T_k \psi_n\}_{j,k \in \mathbb{Z}} \tag{18}$$

constitute a tight frame. The *unitary extension principle* by Ron and Shen shows that a condition on the matrix H will imply this:

Theorem 4.5. *Let $\{\psi_\ell, H_\ell\}_{\ell=0,\ldots,n}$ be as in the general setup, and assume that the 2×2 matrix $H(\gamma)^* H(\gamma)$ is the identity for a.e. γ. Then the multi-wavelet system $\{D^j T_k \psi_\ell\}_{j,k \in \mathbb{Z}, \ell=1,\ldots,n}$ constitutes a tight frame for $L^2(\mathbb{R})$ with frame bound equal to one.*

As an application, we show how one can construct compactly supported tight spline frames.

Example 4.6. Fix any $m = 1, 2, \ldots,$ and consider the function

$$\psi_0 = \chi_{[-\frac{1}{2},\frac{1}{2}]} * \chi_{[-\frac{1}{2},\frac{1}{2}]} * \cdots * \chi_{[-\frac{1}{2},\frac{1}{2}]} \quad (2m \text{ factors}).$$

The function ψ_0 is known as a B-spline of order $2m$, although it is defined using the function $\chi_{[-\frac{1}{2},\frac{1}{2}]}$ rather than $\chi_{[0,1]}$ as we did in (9). Note that

$$\hat{\psi}_0(\gamma) = \frac{\sin^{2m}(\pi\gamma)}{(\pi\gamma)^{2m}}.$$

It is clear that $\lim_{\gamma\to 0}\hat{\psi}_0(\gamma) = 1$, and by direct calculation,

$$\hat{\psi}_0(2\pi\gamma) = \cos^{2m}(\pi\gamma)\hat{\psi}_0(\gamma).$$

Thus ψ_0 satisfies the refinement equation with

$$H_0(\gamma) = \cos^{2m}(\pi\gamma).$$

Let $\binom{2m}{\ell}$ denote the binomial coefficients $\frac{(2m)!}{(2m-\ell)!\ell!}$ and define the 1-periodic bounded functions $H_1, H_2, \ldots, H_{2m}$ by

$$H_\ell(\gamma) = \sqrt{\binom{2m}{\ell}}\sin^\ell(\pi\gamma)\cos^{2m-\ell}(\pi\gamma).$$

Then

$$H(\gamma) = \begin{pmatrix} H_0(\gamma) & T_{1/2}H_0(\gamma) \\ H_1(\gamma) & T_{1/2}H_1(\gamma) \\ \cdot & \cdot \\ \cdot & \cdot \\ H_n(\gamma) & T_{1/2}H_n(\gamma) \end{pmatrix}$$

$$= \begin{pmatrix} \cos^{2m}(\pi\gamma) & \sin^{2m}(\pi\gamma) \\ \sqrt{\binom{2m}{1}}\sin(\pi\gamma)\cos^{2m-1}(\pi\gamma) & -\sqrt{\binom{2m}{1}}\cos(\pi\gamma)\sin^{2m-1}(\pi\gamma) \\ \sqrt{\binom{2m}{2}}\sin^2(\pi\gamma)\cos^{2m-2}(\pi\gamma) & \sqrt{\binom{2m}{2}}\cos^2(\pi\gamma)\sin^{2m-2}(\pi\gamma) \\ \cdot & \cdot \\ \cdot & \cdot \\ \sqrt{\binom{2m}{2m}}\sin^{2m}(\pi\gamma) & \sqrt{\binom{2m}{2m}}\cos^{2m}(\pi\gamma) \end{pmatrix}.$$

Now consider the 2×2 matrix $M := H(\gamma)^* H(\gamma)$. Using the binomial formula

$$(x+y)^{2m} = \sum_{\ell=0}^{2m} \binom{2m}{\ell} x^\ell y^{2m-\ell}$$

we see that the first entry in the first row of M is

$$M_{1,1} = \sum_{\ell=0}^{2m} \binom{2m}{\ell} \sin^{2\ell}(\pi\gamma)\cos^{2(2m-\ell)}(\pi\gamma) = 1.$$

A similar argument gives that $M_{2,2} = 1$. Also,

$$\begin{aligned} M_{1,2} &= \sin^{2m}(\pi\gamma)\cos^{2m}(\pi\gamma)\left(1 - \binom{2m}{1} + \binom{2m}{2} - \cdots + \binom{2m}{2m}\right) \\ &= \sin^{2m}(\pi\gamma)\cos^{2m}(\pi\gamma)(1-1)^{2m} = 0. \end{aligned}$$

Thus M is the identity on $\mathbb{C}^2$ for all γ; by Theorem 4.5 this implies that the $2m$ functions $\psi_1, \dots, \psi_{2m}$ defined by

$$\begin{aligned} \hat{\psi}_\ell(\gamma) &= H_\ell(\gamma/2)\hat{\psi}_0(\gamma/2) \\ &= \sqrt{\binom{2m}{\ell}} \frac{\sin^{2m+\ell}(\pi\gamma/2)\cos^{2m-\ell}(\pi\gamma/2)}{(\pi\gamma/2)^{2m}} \end{aligned}$$

generate a multiwavelet frame for $L^2(\mathbb{R})$.

Frequently one takes a slightly different choice of H_ℓ, namely,

$$H_\ell(\gamma) = i^\ell \sqrt{\binom{2m}{\ell}} \sin^\ell(\pi\gamma)\cos^{2m-\ell}(\pi\gamma).$$

Inserting this expression in $\hat{\psi}_\ell(\gamma) = H_\ell(\gamma/2)\hat{\psi}_\ell(\gamma/2)$ and using the commutator relations for the operators $\mathcal{F}, D, T_k$ shows that ψ_ℓ is a finite linear combination with real coefficients of the functions

$$DT_k\psi_0,\ k = -m, \dots, m.$$

It follows that ψ_ℓ is a real-valued spline with support in $[-m, m]$, degree $2m-1$, smoothness class C^{2m-2}, and knots at $\mathbb{Z}/2$. Note in particular that we obtain smoother generators by starting with higher order splines, but that the price to pay is that the number of generators increases as well. □

Note that the unitary extension principle has a more convenient (but mathematically equivalent) formulation in the *oblique extension principle,* which was discovered independently and simultaneously by Daubechies et al. [17] and Chui et al. [12]. We will not go into a discussion of this, but just refer to the original articles, as well as the compressed presentation in [4, 6] for a quick overview.

References

1. Bownik, M., Weber, E.: Affine frames, GMRA's, and the canonical dual. Stud. Math. **159**, 453–479 (2003)
2. Casazza, P.G.: The art of frame theory. Taiwan. J. Math. **4**(2), 129–201 (2000)
3. Christensen, O.: Frames, bases, and discrete Gabor/wavelet expansions. Bull. Am. Math. Soc. **38**(3), 273–291 (2001)
4. Christensen, O.: An Introduction to Frames and Riesz Bases. Birkhäuser, Boston (2003)
5. Christensen, O.: Pairs of dual Gabor frames with compact support and desired frequency localization. Appl. Comput. Harmon. Anal. **20**, 403–410 (2006)
6. Christensen, O.: Frames and Bases. An Introductory Course. Birkhäuser, Boston (2007)
7. Christensen, O.: Functions, Spaces, and Expansions. Birkhäuser, Boston (2010)
8. Christensen, O., Kim, R.Y.: On dual Gabor frame pairs generated by polynomials. J. Fourier Anal. Appl. **16**, 1–16 (2010)
9. Christensen, O., Kim, H.O., Kim, R.Y.: Frames and extension problems II (2013, to appear) Note that this is the first paper in a series of two. The continuation "Frames and extension problems II" is published in the same issue
10. Chui, C.: Wavelets - A Tutorial in Theory and Practice. Academic, San Diego (1992)
11. Chui, C.K., Shi, X.: Orthonormal wavelets and tight frames with arbitrary real dilations. Appl. Comput. Harmon. Anal. **9**, 243–264 (2000)
12. Chui, C., He, W., Stöckler, J.: Compactly supported tight and sibling frames with maximum vanishing moments. Appl. Comput. Harmon. Anal. **13**, 224–262 (2002)
13. Daubechies, I.: The wavelet transformation, time-frequency localization and signal analysis. IEEE Trans. Inform. Theory **36**, 961–1005 (1990)
14. Daubechies, I.: Ten Lectures on Wavelets. SIAM, Philadelphia (1992)
15. Daubechies, I., Grossmann, A., Meyer, Y.: Painless nonorthogonal expansions. J. Math. Phys. **27**, 1271–1283 (1986)
16. Daubechies, I., Landau, H.J., Landau, Z.: Gabor time-frequency lattices and the Wexler-Raz identity. J. Fourier Anal. Appl. **1**, 437–478 (1995)
17. Daubechies, I., Han, B., Ron, A., Shen, Z.: Framelets: MRA-based constructions of wavelet frames. Appl. Comput. Harmon. Anal. **14**(1), 1–46 (2003)
18. Duffin, R.S., Schaeffer, A.C.: A class of nonharmonic Fourier series. Trans. Am. Math. Soc. **72**, 341–366 (1952)
19. Feichtinger, H.G., Strohmer, T. (eds.): Gabor Analysis and Algorithms: Theory and Applications. Birkhäuser, Boston (1998)
20. Feichtinger, H.G., Strohmer, T. (eds.): Advances in Gabor Analysis. Birkhäuser, Boston (2002)
21. Gröchenig, K.: Foundations of Time-Frequency Analysis. Birkhäuser, Boston (2000)
22. Janssen, A.J.E.M.: Duality and biorthogonality for Weyl-Heisenberg frames. J. Fourier Anal. Appl. **1**(4), 403–436 (1995)
23. Janssen, A.J.E.M.: The duality condition for Weyl-Heisenberg frames. In: Feichtinger, H.G., Strohmer, T. (eds.) Gabor Analysis: Theory and Applications. Birkhäuser, Boston (1998)
24. Mallat, S.: Multiresolution approximation and wavelets. Trans. Am. Math. Soc. **315**, 69–88 (1989)
25. Meyer, Y.: Wavelets and Operators. Hermann, Paris (1990)

26. Ron, A., Shen, Z.: Weyl-Heisenberg systems and Riesz bases in $L^2(\mathbb{R}^d)$. Duke Math. J. **89**, 237–282 (1997)
27. Ron, A., Shen, Z.: Affine systems in $L_2(\mathbb{R}^d)$: the analysis of the analysis operator. J. Funct. Anal. **148**, 408–447 (1997)
28. Ron, A., Shen, Z.: Affine systems in $L_2(R^d)$ II: dual systems. J. Fourier Anal. Appl. **3**, 617–637 (1997)
29. Ron, A., Shen, Z., Compactly supported tight affine spline frames in $L_2(R^d)$. Math. Comput. **67**, 191–207 (1998)
30. Walnut, D.: An Introduction to Wavelet Analysis. Birkhäuser, Boston (2001)
31. Young, R.: An Introduction to Nonharmonic Fourier Series. Academic, New York (1980) (revised first edition 2001)

Frames and Extension Problems II

Ole Christensen, Hong Oh Kim, and Rae Young Kim

Abstract This article is a follow-up on the article *Frames and Extension Problems I.* Here we will go into more recent progress on the topic and also present some open problems.

Keywords Frames • Gabor systems • Wavelet systems • Extension problems

1 Introduction

Based on the article *Frames and Extension Problems I,* see [3], we discuss recent progress and open problems concerning extension of Bessel sequences to frames and dual pairs of frames. We first consider the extension problem in general Hilbert spaces in Sect. 2. The special case of Gabor frames is discussed in Sect. 3. In Sect. 4 the similar (but much more complicated) problem for wavelet systems is considered, without use of any assumption of multiresolution structure. Finally, in Sect. 5, we present a few recent results about extension of wavelet Bessel systems to frames with two or three generators. These results use the multiresolution structure.

O. Christensen
Department of Mathematics and Computer Science, Technical University of Denmark, Building 303, 2800 Lyngby, Denmark
e-mail: ochr@dtu.dk

H.O. Kim (✉)
Division of General Studies, UNIST-gil 50, Ulsan 689-798, Republic of Korea
e-mail: kimhong@kaist.edu

R.Y. Kim
Department of Mathematics, Yeungnam University, 214-1, Dae-dong, Gyeongsan-si, Gyeongsangbuk-do 712-749, Republic of Korea
e-mail: rykim@ynu.ac.kr

C. Bandt et al. (eds.), *Fractals, Wavelets, and their Applications*, Springer Proceedings in Mathematics & Statistics 92, DOI 10.1007/978-3-319-08105-2_15,

2 The Extension Problem in Hilbert Spaces

Extension problems have a long history in frame theory. It has been shown by several authors (see, e.g., [1,14]) that for any Bessel sequence $\{f_k\}_{k=1}^{\infty}$ in a separable Hilbert space $\mathcal{H}$, there exists a sequence $\{g_k\}_{k=1}^{\infty}$ such that $\{f_k\}_{k=1}^{\infty} \cup \{g_k\}_{k=1}^{\infty}$ is a tight frame for $\mathcal{H}$. A natural generalization to construction of dual frame pairs appeared in [4]; we need to refer to the proof later, so we include it here as well.

Theorem 2.1. *Let $\{f_i\}_{i\in I}$ and $\{g_i\}_{i\in I}$ be Bessel sequences in $\mathcal{H}$. Then there exist Bessel sequences $\{p_j\}_{i\in J}$ and $\{q_j\}_{i\in J}$ in $\mathcal{H}$ such that $\{f_i\}_{i\in I} \cup \{p_j\}_{i\in J}$ and $\{g_i\}_{i\in I} \cup \{q_j\}_{i\in J}$ form a pair of dual frames for $\mathcal{H}$.*

Proof. Let T and U denote the preframe operators for $\{f_i\}_{i\in I}$ and $\{g_i\}_{i\in I}$, respectively, i.e.,

$$T, U : \ell^2(I) \to \mathcal{H},\ T\{c_i\}_{i\in I} = \sum_{i\in I} c_i f_i,\ \ U\{c_i\}_{i\in I} = \sum_{i\in I} c_i g_i.$$

Let $\{a_j\}_{i\in J}, \{b_j\}_{i\in J}$ denote any pair of dual frames for $\mathcal{H}$. Then

$$\begin{aligned} f = UT^*f + (I - UT^*)f &= \sum_{i\in I}\langle f, f_i\rangle g_i + \sum_{j\in I}\langle (I - UT^*)f, a_j\rangle b_j \\ &= \sum_{i\in I}\langle f, f_i\rangle g_i + \sum_{j\in I}\langle f, (I - UT^*)^* a_j\rangle b_j \end{aligned}$$

The sequences $\{f_i\}_{i\in I}, \{g_i\}_{i\in I}$, and $\{b_j\}_{i\in J}$ are Bessel sequences by definition, and one can verify that $\{(I - UT^*)^* a_j\}_{j\in J}$ is a Bessel sequence as well. The result now follows from Lemma 2.2 in [3]. □

The reason for the interest in this more general version of the frame extension is that it often is possible to construct dual pairs of frames with properties that are impossible for tight frames. This is illustrated in the next section.

3 The Extension Problem for Gabor Frames

Li and Sun showed in [14] that if $ab \le 1$ and $\{E_{mb}T_{na}g_1\}_{m,n\in\mathbb{Z}}$ is a Bessel sequences in $L^2(\mathbb{R})$, then there exists a Gabor systems $\{E_{mb}T_{na}g_2\}_{m,n\in\mathbb{Z}}$ such that $\{E_{mb}T_{na}g_1\}_{m,n\in\mathbb{Z}} \cup \{E_{mb}T_{na}g_2\}_{m,n\in\mathbb{Z}}$ is a tight frame for $L^2(\mathbb{R})$. However, if we ask for extra properties of the functions g_1 and g_2 such an extension might be impossible. For example, if the given function g_1 has compact support, it is natural to ask for the function g_2 having compact support as well, but by Li and Sun [14] the existence of such a function is only guaranteed if $|\text{supp} g_1| \le b^{-1}$. On the other hand, such an extension can always be obtained in the setting of dual frame pairs [4]:

Theorem 3.1. *Let* $\{E_{mb}T_{na}g_1\}_{m,n\in\mathbb{Z}}$ *and* $\{E_{mb}T_{na}h_1\}_{m,n\in\mathbb{Z}}$ *be Bessel sequences in* $L^2(\mathbb{R})$, *and assume that* $ab \le 1$. *Then the following hold:*

(i) There exist Gabor systems $\{E_{mb}T_{na}g_2\}_{m,n\in\mathbb{Z}}$ *and* $\{E_{mb}T_{na}h_2\}_{m,n\in\mathbb{Z}}$ *in* $L^2(\mathbb{R})$ *such that*

$$\{E_{mb}T_{na}g_1\}_{m,n\in\mathbb{Z}} \cup \{E_{mb}T_{na}g_2\}_{m,n\in\mathbb{Z}} \text{ and } \{E_{mb}T_{na}h_1\}_{m,n\in\mathbb{Z}} \cup \{E_{mb}T_{na}h_2\}_{m,n\in\mathbb{Z}}$$

form a pair of dual frames for $L^2(\mathbb{R})$.

(ii) If g_1 *and* h_1 *have compact support, the functions* g_2 *and* h_2 *can be chosen to have compact support.*

Proof. Let us give the proof of (i). Let T and U denote the preframe operators for $\{E_{mb}T_{na}g_1\}_{m,n\in\mathbb{Z}}$ and $\{E_{mb}T_{na}h_1\}_{m,n\in\mathbb{Z}}$, respectively. Then

$$UT^*f = \sum_{m,n\in\mathbb{Z}} \langle f, E_{mb}T_{na}g_1\rangle E_{mb}T_{na}h_1.$$

Consider the operator $\Phi := I - UT^*$, and let $\{E_{mb}T_{na}r_1\}_{m,n\in\mathbb{Z}}$, $\{E_{mb}T_{na}r_2\}_{m,n\in\mathbb{Z}}$ denote any pair of dual frames for $L^2(\mathbb{R})$. By the proof of Theorem 2.1, $\{E_{mb}T_{na}g_1\}_{m,n\in\mathbb{Z}} \cup \{\Phi^* E_{mb}T_{na}r_1\}_{m,n\in\mathbb{Z}}$ and $\{E_{mb}T_{na}h_1\}_{m,n\in\mathbb{Z}} \cup \{E_{mb}T_{na}r_2\}_{m,n\in\mathbb{Z}}$ are dual frames for $L^2(\mathbb{R})$. By Lemma 2.6 in [3] we know that Φ^* commutes with the time-frequency shift operators $E_{mb}T_{na}$. This concludes the proof. □

4 An Extension Problem for Wavelet Frames

It turns out that the extension problem for wavelet systems is considerably more involved than for Gabor systems. In order to explain this, consider the proof of Theorem 2.1 and assume that $\{f_i\}_{i\in I}$ and $\{g_i\}_{i\in I}$ have wavelet structure, i.e., $\{f_i\}_{i\in I} = \{D^jT_k\psi_1\}_{j,k\in\mathbb{Z}}$ and $\{g_i\}_{i\in I} = \{D^jT_k\widetilde{\psi_1}\}_{j,k\in\mathbb{Z}}$ for some $\psi_1, \widetilde{\psi_1} \in L^2(\mathbb{R})$. Assume further that these sequences are Bessel sequences, with preframe operators T, U, respectively. Then, still referring to the proof of Theorem 2.1, $(I - UT^*)^*a_j = (I - TU^*)a_j$. Unfortunately the operator TU^* in general does not commute with D^jT_k, so even if we choose $\{a_j\}_{i\in J}$ to have wavelet structure, the system $\{(I - TU^*)a_j\}_{j\in J}$ might not be a wavelet system. Thus, we cannot apply the proof technique from the Gabor case. The following partial result was obtained in [4].

Theorem 4.1. *Let* $\{D^jT_k\psi_1\}_{j,k\in\mathbb{Z}}$ *and* $\{D^jT_k\widetilde{\psi_1}\}_{j,k\in\mathbb{Z}}$ *be Bessel sequences in* $L^2(\mathbb{R})$. *Assume that the Fourier transform of* $\widetilde{\psi_1}$ *satisfies*

$$supp\, \widehat{\widetilde{\psi_1}} \subseteq [-1, 1]. \tag{1}$$

Then there exist wavelet systems $\{D^j T_k \psi_2\}_{j,k\in\mathbb{Z}}$ *and* $\{D^j T_k \widetilde{\psi_2}\}_{j,k\in\mathbb{Z}}$ *such that*

$$\{D^j T_k \psi_1\}_{j,k\in\mathbb{Z}} \cup \{D^j T_k \psi_2\}_{j,k\in\mathbb{Z}} \text{ and } \{D^j T_k \widetilde{\psi_1}\}_{j,k\in\mathbb{Z}} \cup \{D^j T_k \widetilde{\psi_2}\}_{j,k\in\mathbb{Z}}$$

form dual frames for $L^2(\mathbb{R})$*. If we further assume that* $\widehat{\psi_1}$ *is compactly supported and that*

$$supp\ \widehat{\widetilde{\psi_1}} \subseteq [-1,1] \setminus [-\epsilon, \epsilon]$$

for some $\epsilon > 0$*, the functions* ψ_2 *and* $\widetilde{\psi_2}$ *can be chosen to have compactly supported Fourier transforms as well.*

In the Gabor case, no assumption of compact support was necessary, neither for the given functions nor their Fourier transform. From this point of view it is natural to ask whether the assumption (1) is necessary in Theorem 4.1.

Question: Let $\{D^j T_k \psi_1\}_{j,k\in\mathbb{Z}}$ and $\{D^j T_k \widetilde{\psi_1}\}_{j,k\in\mathbb{Z}}$ be Bessel sequences in $L^2(\mathbb{R})$.

(i) Do there exist functions $\psi_2, \widetilde{\psi_2} \in L^2(\mathbb{R})$ such that

$$\{D^j T_k \psi_1\}_{j,k\in\mathbb{Z}} \cup \{D^j T_k \psi_2\}_{j,k\in\mathbb{Z}} \text{ and } \{D^j T_k \widetilde{\psi_1}\}_{j,k\in\mathbb{Z}} \cup \{D^j T_k \widetilde{\psi_2}\}_{j,k\in\mathbb{Z}} \quad (2)$$

form dual frames for $L^2(\mathbb{R})$?

(ii) If $\widehat{\psi_1}$ and $\widehat{\widetilde{\psi_1}}$ are compactly supported, can we find compactly supported functions ψ_2 and $\widetilde{\psi_2} \in L^2(\mathbb{R})$ such that the functions in (2) form dual frames?

The problem (i) can also be formulated in the negative way: can we find just one example of a pair of Bessel sequences $\{D^j T_k \psi_1\}_{j,k\in\mathbb{Z}}$ and $\{D^j T_k \widetilde{\psi_1}\}_{j,k\in\mathbb{Z}}$ that cannot be extended to a pair of dual wavelet frames, each with two generators? The open question is strongly connected to the following conjecture by Han [8]:

Conjecture by Deguang Han. *Let* $\{D^j T_k \psi_1\}_{j,k\in\mathbb{Z}}$ *be a wavelet frame with upper frame bound B. Then there exists* $D > B$ *such that for each* $K \geq D$*, there exists* $\widetilde{\psi_1} \in L^2(\mathbb{R})$ *such that* $\{D^j T_k \psi_1\}_{j,k\in\mathbb{Z}} \cup \{D^j T_k \widetilde{\psi_1}\}_{j,k\in\mathbb{Z}}$ *is a tight frame for* $L^2(\mathbb{R})$ *with bound K.*

The paper [8] contains an example showing that (again in contrast with the Gabor setting) it might not be possible to extend the Bessel system $\{D^j T_k \psi_1\}_{j,k\in\mathbb{Z}}$ to a tight frame without enlarging the upper bound; hence it is essential that the conjecture includes the option that the extended wavelet system has a strictly larger frame bound than the upper frame bound B for $\{D^j T_k \psi_1\}_{j,k\in\mathbb{Z}}$. We also note that Han's conjecture is based on an example where supp $\widehat{\psi_1} \subseteq [-1,1]$, i.e., a case that is covered by Theorem 4.1.

Observe that a pair of wavelet Bessel sequences always can be extended to dual wavelet frame pairs by adding *two* pairs of wavelet systems. In fact, we can always add one pair of wavelet systems that cancels the action of the given wavelet system, and another one that yields a dual pair of wavelet frames by itself. Thus, the issue is really whether it is enough to add one pair of wavelet systems, as stated in the formulation of the open problem.

Note that extension problems have a long history in frame theory. Most of the results deal with the unitary extension principle (UEP) [16, 17] and its variants, and are thus based on the assumption of an underlying refinable function. The open problems formulated in this section are not based on such an assumption.

5 Extension Problems via the UEP

In this section we present recent results from [5]; more information and examples can be found there. We will consider the extension problem for wavelet systems in $L^2(\mathbb{R})$ that are generated from the UEP by Ron and Shen. That is, we consider wavelet system $\{D^j T_k \psi_1\}_{j,k\in\mathbb{Z}}$ generated from a given scaling function and characterize the existence of a UEP-type wavelet system $\{D^j T_k \psi_2\}_{j,k\in\mathbb{Z}}$ generated by the same scaling function, such that the system $\{D^j T_k \psi_1\}_{j,k\in\mathbb{Z}} \cup \{D^j T_k \psi_2\}_{j,k\in\mathbb{Z}}$ forms a Parseval frame for $L^2(\mathbb{R})$, i.e., a tight frame with frame bound 1. In the process of doing so, we identify two conditions on the filters associated with the scaling function and with ψ_1, which are necessary for any extension of $\{D^j T_k \psi_1\}_{j,k\in\mathbb{Z}}$ to a tight UEP-type frame with any number of generators. Interestingly, we are able to show that these conditions imply that we can always construct a Parseval frame by adding *at most two* wavelet systems.

Let $\mathbb{T}$ denote the unit circle which will be identified with $[-1/2, 1/2]$. Also, for $f \in L^1(\mathbb{R}) \cap L^2(\mathbb{R})$ we denote the Fourier transform by $\mathcal{F}f(\gamma) = \hat{f}(\gamma) = \int_{-\infty}^{\infty} f(x)e^{-2\pi i x\gamma}dx$. As usual, the Fourier transform is extended to a unitary operator on $L^2(\mathbb{R})$.

In the rest of the paper we will use the following setup.

General Setup. Consider a *scaling function* $\varphi \in L^2(\mathbb{R})$, i.e., a function such that $\hat{\varphi}$ is continuous at the origin and $\hat{\varphi}(0) = 1$, and there exists a function $m_0 \in L^\infty(\mathbb{T})$ (called a *refinement mask*) such that $\hat{\varphi}(2\gamma) = m_0(\gamma)\hat{\varphi}(\gamma)$, a.e.$\gamma \in \mathbb{R}$. Given functions $m_1, m_2, \dots, m_n \in L^\infty(\mathbb{T})$, consider the functions $\psi_\ell \in L^2(\mathbb{R})$ defined by

$$\widehat{\psi_\ell}(2\gamma) = m_\ell(\gamma)\hat{\varphi}(\gamma), \; \ell = 1, \dots, n. \tag{3}$$

In the classical UEP-setup by Ron and Shen, one search for functions $m_1, m_2, \dots, m_n \in L^\infty(\mathbb{T})$ such that

$$\{D^j T_k \psi_1\}_{j,k\in\mathbb{Z}} \cup \cdots \cup \{D^j T_k \psi_n\}_{j,k\in\mathbb{Z}}$$

is a Parseval frame. We will modify this slightly. In fact, we will consider a given refinement mask m_0 and a given filter $m_1 \in L^\infty(\mathbb{T})$, and derive equivalent conditions for the existence of appropriate functions $m_2, \dots, m_n \in L^\infty(\mathbb{T})$ for the cases $n = 2$ and $n = 3$.

We will base the analysis on the *UEP*, which is formulated in terms of the $(n + 1) \times 2$ matrix-valued function M defined by

$$M(\gamma) = \begin{pmatrix} m_0(\gamma) & m_0\left(\gamma + \frac{1}{2}\right) \\ m_1(\gamma) & m_1\left(\gamma + \frac{1}{2}\right) \\ \vdots & \vdots \\ m_n(\gamma) & m_n\left(\gamma + \frac{1}{2}\right) \end{pmatrix}. \tag{4}$$

Proposition 5.1 (UEP by Ron and Shen [16]). *Let $\varphi \in L^2(\mathbb{R})$ be a scaling function and $m_0 \in L^\infty(\mathbb{T})$ the corresponding refinement mask. For each $\ell = 1, \cdots, n$, let $m_\ell \in L^\infty(\mathbb{T})$, and define $\psi_\ell \in L^2(\mathbb{R})$ by (3). If the corresponding matrix-valued function M satisfies*

$$M(\gamma)^* M(\gamma) = I, \ a.e. \ \gamma \in \mathbb{T}, \tag{5}$$

then $\{D^j T_k \psi_i : j \in \mathbb{Z}, k \in \mathbb{Z}, 1 \leq i \leq n\}$ is a Parseval frame for $L^2(\mathbb{R})$.

With the additional constraint that the generating functions should be symmetric, the issue of constructing Parseval wavelet frames with two or three generators has attracted quite some attention in the literature, see, e.g., the papers [15] by Petukhov, [13] by Jiang, [18] by Selesnick and Abdelnour, and the papers [11, 12] by Han and Mo. For example, in the paper [11] B-splines were used as scaling functions, while a more general approach, valid for real-valued, compactly supported, and symmetric scaling functions, was provided in [12]. Other cases where a UEP-based construction with n generators can be modified to a Parseval frame with two or three generators have been considered in [6, 7]. These papers are based on the so-called oblique extension principle, which is known to be equivalent to the UEP. However, a characterization of the conditions that ensure the possibility of extension with two or three generators, as provided in the current paper, has not been available before.

Note that the analysis in the current paper is complementary to the one in Sect. 4, where the key condition for obtaining an extension of a (general) wavelet system $\{D^j T_k \psi_1\}_{j,k \in \mathbb{Z}}$ to a tight frame of the same form is that $\widehat{\psi_1}$ is compactly supported. The extension principle applied in the current paper usually involves functions that are compactly supported in time (even though this is not strictly necessary).

In the current paper we have restricted our attention to wavelet systems in $L^2(\mathbb{R})$. An interesting discussion of the complexity of the extension problem for wavelet systems in higher dimensions, together with several deep results, recently appeared in [2].

In the rest of the paper we assume that we have given functions $m_0, m_1 \in L^\infty(\mathbb{R})$ as described in the general setup. Associated with functions $m_2, \cdots, m_n \in L^\infty(\mathbb{T})$, we consider the $(n - 1) \times 2$ matrix-valued function $M_{2,n}$ defined by

$$M_{2,n}(\gamma) = \begin{pmatrix} m_2(\gamma) & m_2\left(\gamma + \frac{1}{2}\right) \\ \vdots & \vdots \\ m_n(\gamma) & m_n\left(\gamma + \frac{1}{2}\right) \end{pmatrix}.$$

Note that

$$\begin{aligned}
&M_{2,n}(\gamma)^* M_{2,n}(\gamma) \\
&= \begin{pmatrix} \overline{m}_2(\gamma) & \cdots & \overline{m}_n(\gamma) \\ \overline{m}_2(\gamma+1/2) & \cdots & \overline{m}_n(\gamma+1/2) \end{pmatrix} \begin{pmatrix} m_2(\gamma) & m_2\left(\gamma + \frac{1}{2}\right) \\ \vdots & \vdots \\ m_n(\gamma) & m_n\left(\gamma + \frac{1}{2}\right) \end{pmatrix} \\
&= M(\gamma)^* M(\gamma) - \begin{pmatrix} \overline{m}_0(\gamma) & \overline{m}_1(\gamma) \\ \overline{m}_0(\gamma+1/2) & \overline{m}_1(\gamma+1/2) \end{pmatrix} \begin{pmatrix} m_0(\gamma) & m_0(\gamma+1/2) \\ m_1(\gamma) & m_1(\gamma+1/2) \end{pmatrix} \qquad (6) \\
&= M(\gamma)^* M(\gamma) - \\
&\begin{pmatrix} |m_0(\gamma)|^2 + |m_1(\gamma)|^2 & \overline{m_0(\gamma)}m_0(\gamma+1/2) + \overline{m_1(\gamma)}m_1(\gamma+1/2) \\ \overline{m_0(\gamma+1/2)}m_0(\gamma) + \overline{m_1(\gamma+1/2)}m_1(\gamma) & |m_0(\gamma+1/2)|^2 + |m_1(\gamma+1/2)|^2 \end{pmatrix}
\end{aligned}$$

We define

$$M^{\alpha,\beta}(\gamma) := \begin{pmatrix} M_\alpha(\gamma) & \overline{M}_\beta(\gamma) \\ M_\beta(\gamma) & M_\alpha(\gamma+1/2) \end{pmatrix}, \qquad (7)$$

where

$$\begin{aligned}
M_\alpha(\gamma) &:= 1 - |m_0(\gamma)|^2 - |m_1(\gamma)|^2; \\
M_\beta(\gamma) &:= -m_0(\gamma)\overline{m}_0(\gamma+1/2) - m_1(\gamma)\overline{m}_1(\gamma+1/2).
\end{aligned}$$

Then the above calculation shows that

$$M(\gamma)^* M(\gamma) = I \Leftrightarrow M_{2,n}(\gamma)^* M_{2,n}(\gamma) = M^{\alpha,\beta}(\gamma). \qquad (8)$$

The following lemma gives two necessary conditions for the existence of $m_2, \cdots, m_n$ such that the equivalent conditions in (8) hold.

Lemma 5.2. *Suppose that $m_0, m_1, \cdots, m_n \in L^\infty(\mathbb{T})$ satisfy that $M(\gamma)^* M(\gamma) = I$ for a.e. $\gamma \in \mathbb{T}$, then the Hermitian matrix $M^{\alpha,\beta}(\gamma)$ is positive semidefinite and*

(a) $|m_0(\gamma)|^2 + |m_1(\gamma)|^2 \le 1,\ a.e.\ \gamma \in \mathbb{T}$;
(b) $M_\alpha(\gamma) M_\alpha(\gamma+1/2) \ge \left|M_\beta(\gamma)\right|^2,\ a.e.\ \gamma \in \mathbb{T}$.

On the other hand, if (a) and (b) are satisfied then $M^{\alpha,\beta}(\gamma)$ is positive semidefinite.

We are now ready to state the condition for extension to a UEP-type wavelet system $\{D^j T_k \psi_1\}_{j,k\in\mathbb{Z}}$ to a Parseval frame by adding just one UEP-type wavelet system.

Theorem 5.3. *Let $\varphi \in L^2(\mathbb{R})$ be a scaling function and $m_0 \in L^\infty(\mathbb{T})$ the corresponding refinement mask. Let $m_1 \in L^\infty(\mathbb{T})$, and define $\psi_1 \in L^2(\mathbb{R})$ by* (3)*. Assume that condition (a) in Lemma 5.2 is satisfied. Then the following are equivalent:*

(a) *There exists a 1-periodic function m_2 such that the matrix-valued function M in* (4) *with $n = 2$ satisfies that*

$$M(\gamma)^* M(\gamma) = I, \ a.e. \ \gamma \in \mathbb{T}; \tag{9}$$

(b) $M_\alpha(\gamma) M_\alpha(\gamma + 1/2) = M_\beta(\gamma) M_\beta(\gamma + 1/2)$.

In the affirmative case, the multi-wavelet system $\{D^j T_k \psi_l\}_{l=1,2;j,k\in\mathbb{Z}}$, with ψ_2 defined by (3)*, forms a Parseval frame for $L^2(\mathbb{R})$.*

If the necessary conditions in Lemma 5.2 are satisfied, then we can always extend $\{D^j T_k \psi_l\}_{j,k\in\mathbb{Z}}$ to a Parseval wavelet frame by adding *two* wavelet systems:

Theorem 5.4. *Let $\varphi \in L^2(\mathbb{R})$ be a scaling function and $m_0 \in L^\infty(\mathbb{T})$ the corresponding refinement mask. Let $m_1 \in L^\infty(\mathbb{T})$, and define $\psi_1 \in L^2(\mathbb{R})$ by* (3)*. Assume that the functions m_0, m_1 satisfy (a) and (b) in Lemma 5.2. Then there exist $m_2, m_3 \in L^\infty(\mathbb{T})$ such that $\{D^j T_k \psi_l\}_{l=1,2,3;j,k\in\mathbb{Z}}$, with ψ_2, ψ_3 defined by* (3)*, forms a Parseval frame.*

Note that Theorem 5.4 is related with Theorem 1.2 in [12], where it is shown that certain conditions on a scaling function imply the existence of three functions that generate a Parseval wavelet frame. However, the spirit of these two results is different: while the goal of Theorem 1.2 in [12] is to provide sufficient conditions for wavelet constructions that have attractive properties from the point of applications (i.e., symmetry properties and a high number of vanishing moments), the purpose of our result is to guarantee the existence of three functions generating a Parseval frame under the weakest possible conditions. We also note that for the case where the refinement mask m_0 is a trigonometric polynomial, the problem of characterizing associated Parseval frames generated by two or three symmetric functions has been solved in [9, 10].

Acknowledgements This research was supported by Basic Science Research Program through the National Research Foundation of Korea (NRF) funded by the Ministry of Education (2013R1A1A2A10011922).

References

1. Casazza, P., Leonhard, N.: Classes of finite equal norm Parseval frames. Contemp. Math. **451**, 11–31 (2008)
2. Charina, M., Putinar, M., Scheiderer, C., Stöckler, J.: A real algebra perspective on multivariate tight wavelet frames. Preprint (2012)
3. Christensen, O.: Frames and extension problems I (2013, to appear)
4. Christensen, O., Kim, H.O., Kim, R.Y.: Extensions of Bessel sequences to dual pairs of frames. Appl. Comput. Harmon. Anal. **34**(2), 224–233 (2013)
5. Christensen, O., Kim, H.O., Kim, R.Y.: On Parseval wavelet frame with two or three generators via the unitary extension principle. Can. Math. Bull. **57**, 254–263 (2014)
6. Chui, C., He, W., Stöckler, J.: Compactly supported tight and sibling frames with maximum vanishing moments. Appl. Comput. Harmon. Anal. **13**(3), 226–262 (2002)
7. Daubechies, I., Han, B., Ron, A., Shen, Z.: Framelets: MRA-based constructions of wavelet frames. Appl. Comput. Harmon. Anal. **14**(1), 1–46 (2003)
8. Han, D.: Dilations and completions for Gabor systems. J. Fourier Anal. Appl. **15**, 201–217 (2009)
9. Han, B.: Matrix splitting with symmetry and symmetric tight framelet filter banks with two high-pass filters. Appl. Comput. Harmon. Anal. (2013). http://dx.doi.org/10.1016/j.acha.2012.08.007
10. Han, B.: Symmetric tight framelet filter banks with three high-pass filters (2013)
11. Han, B., Mo, Q.: Tight wavelet frames generated by three symmetric B-spline functions with high vanishing moments. Proc. Am. Math. Soc. **132**(1), 77–86 (2003)
12. Han, B., Mo, Q.: Symmetric MRA tight wavelet frames with three generators and high vanishing moments. Appl. Comput. Harmon. Anal. **18**, 67–93 (2005)
13. Jiang, Q.T.: Parametrizations of masks for tight affine frames with two symmetric/antisymmetric generators. Adv. Comput. Math. **18**, 247–268 (2003)
14. Li, D.F., Sun, W.: Expansion of frames to tight frames. Acta Math. Sin. Engl. Ser. **25**, 287–292 (2009)
15. Petukhov, A.: Symmetric framelets. Constr. Approx. **19**, 309–328 (2003)
16. Ron, A., Shen, Z.: Affine systems in $L_2(\mathbb{R}^d)$: the analysis of the analysis operator. J. Funct. Anal. **148**, 408–447 (1997)
17. Ron, A., Shen, Z.: Affine systems in $L_2(R^d)$ II: dual systems. J. Fourier Anal. Appl. **3**, 617–637 (1997)
18. Selesnick, I.W., Abdelnour, A.F.: Symmetric wavelet tight frames with two generators. Appl. Comput. Harmon. Anal. **17**, 211–225 (2004)

Local Fractal Functions and Function Spaces

Peter R. Massopust

Abstract We introduce local iterated function systems (IFSs) and present some of their basic properties. A new class of local attractors of local IFSs, namely local fractal functions, is constructed. We derive formulas so that these local fractal functions become elements of various function spaces, such as the Lebesgue spaces L^p, the smoothness spaces C^n, the homogeneous Hölder spaces $\dot{C}^s$, and the Sobolev spaces $W^{m,p}$.

Keywords Iterated function system (IFS) • Local iterated function system • Attractor • Fractal interpolation • Read-Bajraktarević operator • Fractal function • Lebesgue space • Hölder space • Sobolev space

AMS Subject Classification (2010): 28A80, 37C70, 41A05, 41A30, 42B35

1 Introduction

Iterated function systems (IFSs), for short IFSs, are a powerful means for describing fractal sets and for modeling or approximating natural objects. IFSs were first introduced in [4, 12] and subsequently investigated by numerous authors. Within the fractal image compression community a generalization of IFSs was proposed in [5] whose main purpose was to obtain efficient algorithms for image coding.

P.R. Massopust (✉)
Centre of Mathematics, Research Unit M6, Technische Universität München, Boltzmannstrasse 3, 85747, Garching b. München, Germany

Helmholtz Zentrum München, Ingolstädter Landstrasse 1, 85764 Neuherberg, Germany
e-mail: peter.massopust@helmholtz-muenchen.de; massopust@ma.tum.de

C. Bandt et al. (eds.), *Fractals, Wavelets, and their Applications*, Springer Proceedings in Mathematics & Statistics 92, DOI 10.1007/978-3-319-08105-2_16,

In [8], this generalization of a traditional IFS, called a local IFS, was reconsidered but now from the viewpoint of approximation theory and from the standpoint of developing computationally efficient numerical methods based on fractal methodologies. In the current paper, we continue this former exploration of local IFSs and consider a special class of attractors, namely those that are the graphs of functions. We will derive conditions under which such local fractal functions are elements of certain function spaces which are important in harmonic analysis and numerical mathematics.

The structure of this paper is as follows. We present the traditional IFSs in Sect. 2 in a more general and modern setting and state some of their properties. Section 3 introduces local IFSs and discusses some characteristics of this newly rediscovered concept. Local fractal functions and their connection to local IFSs are investigated in Sect. 4. In Sect. 5 we briefly consider tensor products of local fractal functions. Local fractal functions in Lebesgue spaces are presented in Sect. 6, in smoothness and Hölder spaces in Sect. 7, and in Sobolev spaces in Sect. 8.

2 Iterated Function Systems

In this section, we introduce the traditional IFS and highlight some of its fundamental properties. For more details and proofs, we refer the reader to [3, 4, 6, 12] and the references stated therein.

Throughout this paper, we use the following notation. The set of positive integers is denoted by $\mathbb{N} := \{1, 2, 3, \ldots\}$, the set of nonnegative integers by $\mathbb{N}_0 = \mathbb{N} \cup \{0\}$, and the ring of integers by $\mathbb{Z}$. We denote the closure of a set S by $\overline{S}$ and its interior by $\overset{\circ}{S}$. In the following, $(\mathbb{X}, d_X)$ always denotes a complete metric space with metric $d_{\mathbb{X}}$.

Definition 1. Let $N \in \mathbb{N}$. If $f_n : \mathbb{X} \to \mathbb{X}$, $n = 1, 2, \ldots, N$, are continuous mappings, then $\mathcal{F} := (\mathbb{X}; f_1, f_2, \ldots, f_N)$ is called an *IFS*.

By a slight abuse of notation and terminology, we use the same symbol $\mathcal{F}$ for the IFS, the set of functions in the IFS, and for the following set-valued mapping defined on the class of all subsets $2^{\mathbb{X}}$ of $\mathbb{X}$. Define $\mathcal{F} : 2^{\mathbb{X}} \to 2^{\mathbb{X}}$ by

$$\mathcal{F}(B) := \bigcup_{f \in \mathcal{F}} f(B), \quad B \in 2^{\mathbb{X}}.$$

Denote by $\mathbb{H} = \mathbb{H}(\mathbb{X})$ the hyperspace of all nonempty compact subsets of $\mathbb{X}$. The hyperspace $(\mathbb{H}, d_{\mathbb{H}})$ becomes a complete metric space when endowed with the Hausdorff metric $d_{\mathbb{H}}$ (cf. [10])

$$d_{\mathbb{H}}(A, B) := \max\{\max_{a \in A} \min_{b \in B} d_X(a, b), \max_{b \in B} \min_{a \in A} d_X(a, b)\}.$$

Since $\mathcal{F}(\mathbb{H}) \subset \mathbb{H}$, we can also treat $\mathcal{F}$ as a mapping $\mathcal{F} : \mathbb{H} \to \mathbb{H}$. When $U \subset \mathbb{X}$ is nonempty, we may write $\mathbb{H}(U) = \mathbb{H}(\mathbb{X}) \cap 2^U$. We denote by $|\mathcal{F}|$ the number of distinct mappings in $\mathcal{F}$.

A metric space $\mathbb{X}$ is termed *locally compact* if every point of $\mathbb{X}$ has a neighborhood that contains a compact neighborhood. The following information, a proof of which can be found in [6], is foundational.

Theorem 1. *(i) If* $(\mathbb{X}, d_{\mathbb{X}})$ *is compact then* $(\mathbb{H}, d_{\mathbb{H}})$ *is compact.*
(ii) If $(\mathbb{X}, d_{\mathbb{X}})$ *is locally compact then* $(\mathbb{H}, d_{\mathbb{H}})$ *is locally compact.*
(iii) If $\mathbb{X}$ *is locally compact, or if each* $f \in \mathcal{F}$ *is uniformly continuous, then* $\mathcal{F} : \mathbb{H} \to \mathbb{H}$ *is continuous.*
(iv) If $f : \mathbb{X} \to \mathbb{X}$ *is a contraction mapping for each* $f \in \mathcal{F}$*, then* $\mathcal{F} : \mathbb{H} \to \mathbb{H}$ *is a contraction mapping.*

For $B \subset \mathbb{X}$, let $\mathcal{F}^k(B)$ denote the k-fold composition of $\mathcal{F}$, i.e., the union of $f_{i_1} \circ f_{i_2} \circ \cdots \circ f_{i_k}(B)$ over all finite words $i_1 i_2 \cdots i_k$ of length k. Define $\mathcal{F}^0(B) := B$.

Definition 2. A nonempty compact set $A \subset \mathbb{X}$ is said to be an *attractor* of the IFS $\mathcal{F}$ if

(i) $\mathcal{F}(A) = A$, and if
(ii) there exists an open set $U \subset \mathbb{X}$ such that $A \subset U$ and $\lim_{k\to\infty} \mathcal{F}^k(B) = A$, for all $B \in \mathbb{H}(U)$, where the limit is in the Hausdorff metric.

The largest open set U such that (ii) is true is called the *basin of attraction* (for the attractor A of the IFS $\mathcal{F}$).

Note that if U_1 and U_2 satisfy condition (ii) in Definition 2 for the same attractor A then so does $U_1 \cup U_2$. We also remark that the invariance condition (i) is not needed; it follows from (ii) for $B := A$.

We will use the following observation [14, Proposition 3(vii)], [9, p. 68, Proposition 2.4.7].

Lemma 1. *Let* $\{B_k\}_{k=1}^{\infty}$ *be a sequence of nonempty compact sets such that* $B_{k+1} \subset B_k$*, for all* $k \in \mathbb{N}$*. Then* $\bigcap_{k\geq 1} B_k = \lim_{k\to\infty} B_k$ *where convergence is with respect to the Haudorff metric* $d_{\mathbb{H}}$.

The next result shows how one may obtain the attractor A of an IFS. For the proof, we refer the reader to [6]. Note that we do not assume that the functions in the IFS $\mathcal{F}$ are contractive.

Theorem 2. *Let* $\mathcal{F}$ *be an IFS with attractor* A *and basin of attraction* U*. If the map* $\mathcal{F} : \mathbb{H}(U) \to \mathbb{H}(U)$ *is continuous then*

$$A = \bigcap_{K\geq 1} \overline{\bigcup_{k\geq K} \mathcal{F}^k(B)}, \qquad \text{for all } B \subset U \text{ such that } \overline{B} \in \mathbb{H}(U).$$

The quantity on the right-hand side here is sometimes called the *topological upper limit* of the sequence $\{\mathcal{F}^k(B) \mid k \in \mathbb{N}\}$. (See, for instance, [10].)

A subclass of IFSs is obtained by imposing additional conditions on the functions that comprise the IFS. The definition below introduces this subclass.

Definition 3. An IFS $\mathcal{F} = (\mathbb{X}; f_1, f_2, \dots, f_N)$ is called *contractive* if there exists a metric d^* on $\mathbb{X}$, which is equivalent to d, such that each $f \in \mathcal{F}$ is a contraction with respect to the metric d^*, i.e., there is a constant $c \in [0, 1)$ such that

$$d^*(f(x_1), f(x_2)) \leq c\, d(x_1, x_2),$$

for all $x_1, x_2 \in \mathbb{X}$.

By item (iv) in Theorem 1, the mapping $\mathcal{F} : \mathbb{H} \to \mathbb{H}$ is then also contractive on the complete metric space $(\mathbb{H}, d_{\mathbb{H}})$, and thus possesses a unique attractor A. This attractor satisfies the *self-referential equation*

$$A = \mathcal{F}(A) = \bigcup_{f \in \mathcal{F}} f(A). \tag{1}$$

In the case of a contractive IFS, the basin of attraction for A is $\mathbb{X}$ and the attractor can be computed via the following procedure: Let K_0 be any set in $\mathbb{H}(\mathbb{X})$ and consider the sequence of iterates

$$K_m := \mathcal{F}(K_{m-1}) = \mathcal{F}^m(K_0), \quad m \in \mathbb{N}.$$

Then K_m converges in the Hausdorff metric to the attractor A as $m \to \infty$, i.e., $d_{\mathbb{H}}(K_m, A) \to 0$ as $m \to \infty$.

For the remainder of this paper, the emphasis will be on contractive IFSs, respectively, contractive local IFSs. We will see that the self-referential equation (1) plays a fundamental role in the construction of fractal sets and in the determination of their geometric and analytic properties.

3 From IFS to Local IFS

The concept of *local* IFS is a generalization of an IFS as defined above and was first introduced in [5] and reconsidered in [8]. In what follows, $N \in \mathbb{N}$ always denotes a positive integer and $\mathbb{N}_N := \{1, \dots, N\}$.

Definition 4. Suppose that $\{\mathbb{X}_i \,|\, i \in \mathbb{N}_N\}$ is a family of nonempty subsets of a metric space $\mathbb{X}$. Further assume that for each $\mathbb{X}_i$ there exists a continuous mapping $f_i : \mathbb{X}_i \to \mathbb{X}$, $i \in \mathbb{N}_N$. Then $\mathcal{F}_{\text{loc}} := \{\mathbb{X}; (\mathbb{X}_i, f_i) \,|\, i \in \mathbb{N}_N\}$ is called a *local IFS*.

Note that if each $\mathbb{X}_i = \mathbb{X}$, then Definition 4 coincides with the usual definition of a standard (global) IFS on a complete metric space. However, the possibility of choosing the domain for each continuous mapping f_i different from the entire space X adds additional flexibility as will be recognized in the sequel.

Definition 5. A local IFS $\mathcal{F}_{\text{loc}}$ is called *contractive* if there exists a metric d^* equivalent to d with respect to which all functions $f \in \mathcal{F}_{\text{loc}}$ are contractive (on their respective domains).

With a local IFS we associate a set-valued operator $\mathcal{F}_{\text{loc}} : 2^{\mathbb{X}} \to 2^{\mathbb{X}}$ by setting

$$\mathcal{F}_{\text{loc}}(S) := \bigcup_{i=1}^{N} f_i(S \cap \mathbb{X}_i). \tag{2}$$

By a slight abuse of notation, we use the same symbol for a local IFS and its associated operator.

Definition 6. A subset $A \in 2^{\mathbb{X}}$ is called a *local attractor* for the local IFS $\{\mathbb{X}; (\mathbb{X}_i, f_i) \mid i \in \mathbb{N}_N\}$ if

$$A = \mathcal{F}_{\text{loc}}(A) = \bigcup_{i=1}^{N} f_i(A \cap \mathbb{X}_i). \tag{3}$$

In (3) we allow for $A \cap \mathbb{X}_i$ to be the empty set. Thus, every local IFS has at least one local attractor, namely $A = \emptyset$. However, it may also have many distinct ones. In the latter case, if A_1 and A_2 are distinct local attractors, then $A_1 \cup A_2$ is also a local attractor. Hence, there exists a largest local attractor for $\mathcal{F}_{\text{loc}}$, namely the union of all distinct local attractors. We refer to this largest local attractor as *the* local attractor of a local IFS $\mathcal{F}_{\text{loc}}$.

Remark 1. There exists an alternative definition for (2). We could consider the mappings f_i as defined on all of $\mathbb{X}$ in the following sense: For any $S \in 2^{\mathbb{X}}$, let

$$f_i(S) := \begin{cases} f_i(S \cap \mathbb{X}_i), & S \cap \mathbb{X}_i \neq \emptyset; \\ \emptyset, & S \cap \mathbb{X}_i = \emptyset, \end{cases} \qquad i \in \mathbb{N}_N.$$

Now suppose that $\mathbb{X}$ is compact and the $\mathbb{X}_i$, $i \in \mathbb{N}_N$, are closed, i.e., compact in $\mathbb{X}$. If in addition the local IFS $\{\mathbb{X}; (\mathbb{X}_i, f_i) \mid i \in \mathbb{N}_N\}$ is contractive then the local attractor can be computed as follows. Let $K_0 := \mathbb{X}$ and set

$$K_n := \mathcal{F}_{\text{loc}}(K_{n-1}) = \bigcup_{i \in \mathbb{N}_N} f_i(K_{n-1} \cap \mathbb{X}_i), \quad n \in \mathbb{N}.$$

Then $\{K_n \mid n \in \mathbb{N}_0\}$ is a decreasing nested sequence of compact sets. *If* each K_n is nonempty then by the Cantor Intersection Theorem,

$$K := \bigcap_{n \in \mathbb{N}_0} K_n \neq \emptyset.$$

Using [14, Proposition 3(vii)], we see that

$$K = \lim_{n\to\infty} K_n,$$

where the limit is taken with respect to the Hausdorff metric on $\mathbb{H}$. This implies that

$$K = \lim_{n\to\infty} K_n = \lim_{n\to\infty} \bigcup_{i\in\mathbb{N}_N} f_i(K_{n-1}\cap \mathbb{X}_i) = \bigcup_{i\in\mathbb{N}_N} f_i(K\cap \mathbb{X}_i) = \mathcal{F}_{\text{loc}}(K).$$

Thus, $K = A_{\text{loc}}$. A condition which guarantees that each K_n is nonempty is that $f_i(\mathbb{X}_i) \subset \mathbb{X}_i$, $i \in \mathbb{N}_N$. (See also [5].)

In the above setting, one can derive a relation between the local attractor A_{loc} of a contractive local IFS $\{\mathbb{X}; (\mathbb{X}_i, f_i) \,|\, i \in \mathbb{N}_N\}$ and the (global) attractor A of the associated (global) IFS $\{\mathbb{X}; f_i \,|\, i \in \mathbb{N}_N\}$. To this end, let the sequence $\{\mathbb{K}_n \,|\, n \in \mathbb{N}_0\}$ be defined as above. The unique attractor A of the IFS $\mathcal{F} := \{\mathbb{X}; f_i \,|\, i \in \mathbb{N}_N\}$ is obtained as the fixed point of the set-valued map $\mathcal{F} : \mathbb{H} \to \mathbb{H}$,

$$\mathcal{F}(B) = \bigcup_{i\in\mathbb{N}_N} f_i(B), \tag{4}$$

where $B \in \mathbb{H}$. If the IFS $\mathcal{F}$ is contractive, then the set-valued mapping (4) is contractive on $\mathbb{H}$ and its unique fixed point is obtained as the limit of the sequence of sets $\{A_n \,|\, n \in \mathbb{N}_0\}$ with $A_0 := \mathbb{X}$ and

$$A_n := \mathcal{F}(A_{n-1}), \quad n \in \mathbb{N}.$$

Note that $K_0 = A_0 = \mathbb{X}$ and, assuming that $K_{n-1} \subseteq A_{n-1}$, $n \in \mathbb{N}$, it follows by induction that

$$K_n = \bigcup_{i\in\mathbb{N}_N} f_i(K_{n-1}\cap X_i) \subseteq \bigcup_{i\in\mathbb{N}_N} f_i(K_{n-1}) \subseteq \bigcup_{i\in\mathbb{N}_N} f_i(A_{n-1}) = A_n.$$

Hence, upon taking the limit with respect to the Hausdorff metric as $n \to \infty$, we obtain $A_{\text{loc}} \subseteq A$. This proves the next result.

Proposition 1. *Let $\mathbb{X}$ be a compact metric space and let $\mathbb{X}_i$, $i \in \mathbb{N}_N$, be closed, i.e., compact in $\mathbb{X}$. Suppose that the local IFS $\mathcal{F}_{\text{loc}} := \{\mathbb{X}; (\mathbb{X}_i, f_i) \,|\, i \in \mathbb{N}_N\}$ and the IFS $\mathcal{F} := \{\mathbb{X}; f_i \,|\, i \in \mathbb{N}_N\}$ are both contractive. Then the local attractor A_{loc} of $\mathcal{F}_{\text{loc}}$ is a subset of the attractor A of $\mathcal{F}$.*

Contractive local IFSs are point-fibered provided $\mathbb{X}$ is compact and the subsets $\mathbb{X}_i$, $i \in \mathbb{N}_N$, are closed. To show this, define the code space of a local IFS by $\Omega := \prod_{n\in\mathbb{N}} \mathbb{N}_N$ and endowed it with the product topology $\mathfrak{T}$. It is known that Ω is metrizable and that $\mathfrak{T}$ is induced by the metric $d_F : \Omega \times \Omega \to \mathbb{R}$,

$$d_F(\sigma, \tau) := \sum_{n\in\mathbb{N}} \frac{|\sigma_n - \tau_n|}{(N+1)^n},$$

where $\sigma = (\sigma_1 \ldots \sigma_n \ldots)$ and $\tau = (\tau_1 \ldots \tau_n \ldots)$. (As a reference, see for instance [10], Theorem 4.2.2.) The elements of Ω are called codes.

Define a set-valued mapping $\gamma : \Omega \to \mathbb{K}(\mathbb{X})$, where $\mathbb{K}(\mathbb{X})$ denotes the hyperspace of all compact subsets of X, by

$$\gamma(\sigma) := \bigcap_{n=1}^{\infty} f_{\sigma_1} \circ \cdots \circ f_{\sigma_n}(\mathbb{X}),$$

where $\sigma = (\sigma_1 \ldots \sigma_n \ldots)$. Then $\gamma(\sigma)$ is point-fibered, i.e., a singleton. Moreover, in this case, the local attractor A equals $\gamma(\Omega)$. (For details about point-fibered IFSs and attractors, we refer the interested reader to [13], Chaps. 3–5.)

Example 1. Let $\mathbb{X} := [0,1] \times [0,1]$ and suppose that $0 < x_2 < x_1 < 1$ and $0 < y_2 < y_1 < 1$. Define

$$\mathbb{X}_1 := [0, x_1] \times [0, y_1] \qquad \text{and} \qquad \mathbb{X}_2 := [x_2, 1] \times [y_2, 1].$$

Furthermore, let $f_i : \mathbb{X}_i \to \mathbb{X}$, $i = 1, 2$, be given by

$$f_1(x, y) := (s_1 x, s_1 y) \quad \text{and} \quad f_2(x, y) := (s_2 x + (1 - s_2)x_2, s_2 y + (1 - s_2)y_2),$$

respectively, where $s_1, s_2 \in [0, 1)$.

The (global) IFS $\{\mathbb{X}; f_1, f_2\}$ has the line segment $A = \{(x, \frac{y_2}{x_2} x) \,|\, 0 \le x \le x_2\}$ as its unique attractor. The local attractor of the local IFS $\{\mathbb{X}; (\mathbb{X}_1, f_1), (\mathbb{X}_2, f_2)\}$ is given by $A_{\mathrm{loc}} = \{(0,0)\} \cup \{(x_2, y_2)\}$, the union of the fixed point for f_1 and f_2, respectively.

4 Local Fractal Functions

In this section, we introduce bounded local fractal functions as the fixed points of operators acting on the complete metric space of bounded functions. We will see that the graph of a local fractal functions is the local attractor of an associated local IFS and that the set of discontinuities of a bounded local fractal function is at most countably infinite. We follow the exhibition presented in [8]. For the theory of (global) fractal functions, the reader may consult [2, 15, 17].

To this end, let X be a nonempty connected set and $\{X_i \,|\, i \in \mathbb{N}_N\}$ a family of nonempty connected subsets of X. Suppose $\{u_i : X_i \to X \,|\, i \in \mathbb{N}_N\}$ is a family of bijective mappings with the property that

(P) $\{u_i(X_i) \,|\, i \in \mathbb{N}_N\}$ forms a (set-theoretic) partition of X: $X = \bigcup_{i=1}^N u_i(X_i)$ and $u_i(X_i) \cap u_j(X_j) = \emptyset$, for all $i \neq j \in \mathbb{N}_N$.

Now suppose that $(\mathbb{Y}, d_{\mathbb{Y}})$ is a complete metric space with metric $d_{\mathbb{Y}}$. A mapping $f : X \to \mathbb{Y}$ is called *bounded* (with respect to the metric $d_{\mathbb{Y}}$) if there exists an $M > 0$ so that for all $x_1, x_2 \in X$, $d_{\mathbb{Y}}(f(x_1), f(x_2)) < M$.

Denote by $B(X, \mathbb{Y})$ the set

$$B(X,\mathbb{Y}) := \{f : X \to \mathbb{Y}\,|\,f \text{ is bounded}\}.$$

Endowed with the metric

$$d(f,g) := \sup_{x\in X} d_{\mathbb{Y}}(f(x), g(x)),$$

$(B(X,\mathbb{Y}), d)$ becomes a complete metric space. In a similar fashion, we define $B(X_i,\mathbb{Y}), i \in \mathbb{N}_N$.

Under the usual addition and scalar multiplication of functions, the spaces $B(X_i,\mathbb{Y})$ and $B(X,\mathbb{Y})$ become metric linear spaces [18]. Recall that a *metric linear space* is a vector space endowed with a metric under which the operations of vector addition and scalar multiplication become continuous.

For $i \in \mathbb{N}_N$, let $v_i : X_i \times \mathbb{Y} \to \mathbb{Y}$ be a mapping that is uniformly contractive in the second variable, i.e., there exists an $\ell \in [0,1)$ so that for all $y_1, y_2 \in \mathbb{Y}$

$$d_{\mathbb{Y}}(v_i(x,y_1), v_i(x,y_2)) \le \ell\, d_{\mathbb{Y}}(y_1,y_2), \quad \forall x \in X. \tag{5}$$

Define a *Read–Bajactarević (RB) operator* $\Phi : B(X,\mathbb{Y}) \to \mathbb{Y}^X$ by

$$\Phi f(x) := \sum_{i=1}^N v_i(u_i^{-1}(x), f_i \circ u_i^{-1}(x))\, \chi_{u_i(X_i)}(x), \tag{6}$$

where $f_i := f|_{X_i}$ and

$$\chi_M(x) := \begin{cases} 1, & x \in M \\ 0, & x \notin M \end{cases},$$

denotes the characteristic function of a set M. Note that Φ is well-defined, and since f is bounded and each v_i contractive in its second variable, $\Phi f \in B(X,\mathbb{Y})$.

Moreover, by (5), we obtain for all $f, g \in B(X,\mathbb{Y})$ the following inequality:

$$\begin{aligned} d(\Phi f, \Phi g) &= \sup_{x\in X} d_{\mathbb{Y}}(\Phi f(x), \Phi g(x)) \\ &= \sup_{x\in X} d_{\mathbb{Y}}(v(u_i^{-1}(x), f_i(u_i^{-1}(x))), v(u_i^{-1}(x), g_i(u_i^{-1}(x)))) \\ &\le \ell \sup_{x\in X} d_{\mathbb{Y}}(f_i \circ u_i^{-1}(x), g_i \circ u_i^{-1}(x)) \le \ell\, d_{\mathbb{Y}}(f,g). \end{aligned} \tag{7}$$

To simplify notation, we had set $v(x,y) := \sum_{i=1}^N v_i(x,y)\,\chi_{X_i}(x)$ in the above equation. In other words, Φ is a contraction on the complete metric space $B(X,\mathbb{Y})$ and, by the Banach Fixed Point Theorem, has therefore a unique fixed point $\mathfrak{f}$ in $B(X,\mathbb{Y})$. This unique fixed point will be called a *local fractal function* $\mathfrak{f} = \mathfrak{f}_\Phi$ (generated by Φ).

Next, we would like to consider a special choice of mappings v_i. To this end, we require the concept of an F-space. We recall that a metric $d : \mathbb{Y} \times \mathbb{Y} \to \mathbb{R}$ is called *complete* if every Cauchy sequence in $\mathbb{Y}$ converges with respect to d to a point of $\mathbb{Y}$, and *translation-invariant* if $d(x+a, y+a) = d(x, y)$, for all $x, y, a \in \mathbb{Y}$.

Definition 2. A topological vector space $\mathbb{Y}$ is called an ***F*** *-space* [18] if its topology is induced by a complete translation-invariant metric d.

Now suppose that $\mathbb{Y}$ is an F-space. Denote its metric by $d_{\mathbb{Y}}$. We define mappings $v_i : X_i \times \mathbb{Y} \to \mathbb{Y}$ by

$$v_i(x, y) := \lambda_i(x) + S_i(x)\, y, \quad i \in \mathbb{N}_N, \tag{8}$$

where $\lambda_i \in B(X_i, \mathbb{Y})$ and $S_i : X_i \to \mathbb{R}$ is a function.

If in addition we require that the metric $d_{\mathbb{Y}}$ is homogeneous, that is,

$$d_{\mathbb{Y}}(\alpha y_1, \alpha y_2) = |\alpha| d_{\mathbb{Y}}(y_1, y_2), \quad \forall \alpha \in \mathbb{R}\ \forall y_1.y_2 \in \mathbb{Y},$$

then v_i given by (8) satisfies condition (5) provided that the functions S_i are bounded on X_i with bounds in $[0, 1)$. For then

$$\begin{aligned} d_{\mathbb{Y}}(\lambda_i(x) + S_i(x)\, y_1, \lambda_i(x) + S_i(x)\, y_2) &= d_{\mathbb{Y}}(S_i(x)\, y_1, S_i(x)\, y_2) \\ &= |S_i(x)| d_{\mathbb{Y}}(y_1, y_2) \\ &\le \|S_i\|_{\infty, X_i}\, d_{\mathbb{Y}}(y_1, y_2) \\ &\le s\, d_{\mathbb{Y}}(y_1, y_2). \end{aligned}$$

Here, we denoted the supremum norm with respect to X_i by $\| \bullet \|_{\infty, X_i}$, and set $s := \max\{\|S_i\|_{\infty, X_i} \mid i \in \mathbb{N}_N\}$.

Thus, for a *fixed* set of functions $\{\lambda_1, \ldots, \lambda_N\}$ and $\{S_1, \ldots, S_N\}$, the associated RB operator (6) has now the form

$$\Phi f = \sum_{i=1}^{N} \lambda_i \circ u_i^{-1}\, \chi_{u_i(X_i)} + \sum_{i=1}^{N} (S_i \circ u_i^{-1}) \cdot (f_i \circ u_i^{-1})\, \chi_{u_i(X_i)},$$

or, equivalently,

$$\Phi f_i \circ u_i = \lambda_i + S_i \cdot f_i, \quad \text{on } \mathrm{X_i}, \forall\ \mathrm{i} \in \mathrm{N_N},$$

with $f_i = f|_{X_i}$.

Theorem 3. *Let $\mathbb{Y}$ be an F-space with homogeneous metric $d_{\mathbb{Y}}$. Let X be a nonempty connected set and $\{X_i \mid i \in \mathbb{N}_N\}$ a collection of nonempty connected subsets of X. Suppose that $\{u_i : X_i \to X \mid i \in \mathbb{N}_N\}$ is a family of bijective mappings satisfying property* (P).

Let $\boldsymbol{\lambda} := (\lambda_1, \ldots, \lambda_N) \in \bigtimes_{i=1}^{N} B(X_i, \mathbb{Y})$ *and* $\boldsymbol{S} := (S_1, \ldots, S_N) \in \bigtimes_{i=1}^{N} B(X_i, \mathbb{R})$.

Define a mapping $\Phi : \left(\bigtimes_{i=1}^{N} B(X_i, \mathbb{Y})\right) \times \left(\bigtimes_{i=1}^{N} B(X_i, \mathbb{R})\right) \times B(X, \mathbb{Y}) \to B(X, \mathbb{Y})$ *by*

$$\Phi(\boldsymbol{\lambda})(\boldsymbol{S})f = \sum_{i=1}^{N} \lambda_i \circ u_i^{-1}\, \chi_{u_i(X_i)} + \sum_{i=1}^{N} (S_i \circ u_i^{-1}) \cdot (f_i \circ u_i^{-1})\, \chi_{u_i(X_i)}. \tag{9}$$

If $\max\{\|S_i\|_{\infty,X_i} \mid i \in \mathbb{N}_N\} < 1$ *then the operator* $\Phi(\boldsymbol{\lambda})(\boldsymbol{S})$ *is contractive on the complete metric space* $B(X, \mathbb{Y})$ *and its unique fixed point* $\mathfrak{f}$ *satisfies the self-referential equation*

$$\mathfrak{f} = \sum_{i=1}^{N} \lambda_i \circ u_i^{-1}\, \chi_{u_i(X_i)} + \sum_{i=1}^{N} (S_i \circ u_i^{-1}) \cdot (\mathfrak{f}_i \circ u_i^{-1})\, \chi_{u_i(X_i)}, \tag{10}$$

or, equivalently

$$\mathfrak{f} \circ u_i = \lambda_i + S_i \cdot \mathfrak{f}_i, \quad on\ \mathrm{X_i}, \forall\, \mathrm{i} \in \mathbb{N}_\mathrm{N}, \tag{11}$$

where $\mathfrak{f}_i = \mathfrak{f}|_{X_i}$.

Proof. The statements follow directly from the considerations preceding the theorem. □

The fixed point $\mathfrak{f}$ in (10) is called a *bounded local fractal function* or, for short, *local fractal function.*

Remark 2. Note that the local fractal function $\mathfrak{f}$ generated by the operator Φ defined by (9) does not only depend on the family of subsets $\{X_i \mid i \in \mathbb{N}_N\}$ but also on the two N-tuples of bounded functions $\boldsymbol{\lambda} \in \underset{i=1}{\overset{N}{\times}} B(X_i, \mathbb{Y})$ and $\boldsymbol{S} \in \underset{i=1}{\overset{N}{\times}} B(X_i, \mathbb{R})$. The fixed point $\mathfrak{f}$ should therefore be written more precisely as $\mathfrak{f}(\boldsymbol{\lambda})(\boldsymbol{S})$. However, for the sake of notational simplicity, we usually suppress this dependence for both $\mathfrak{f}$ and Φ.

Example 2. Suppose $X := [0, 1]$ and $\mathbb{Y} := \mathbb{R}$. In Fig. 1, we display the graph of a randomly generated local fractal function where the λ_i's and the S_i's were chosen to have random constant values.

The following result found in [11] and, in more general form, in [16] is the extension to the setting of local fractal functions.

Theorem 4. *The mapping* $\boldsymbol{\lambda} \mapsto \mathfrak{f}(\boldsymbol{\lambda})$ *defines a linear isomorphism from* $\underset{i=1}{\overset{N}{\times}} B(X_i, \mathbb{Y})$ *to* $B(X, \mathbb{Y})$.

Proof. Let $\alpha, \beta \in \mathbb{R}$ and let $\boldsymbol{\lambda}, \boldsymbol{\mu} \in \underset{i=1}{\overset{N}{\times}} B(X_i, \mathbb{Y})$. Injectivity follows immediately from the fixed point equation (10) and the uniqueness of the fixed point: $\boldsymbol{\lambda} = \boldsymbol{\mu} \iff \mathfrak{f}(\boldsymbol{\lambda}) = \mathfrak{f}(\boldsymbol{\mu}), .$

Linearity follows from (10), the uniqueness of the fixed point and injectivity:

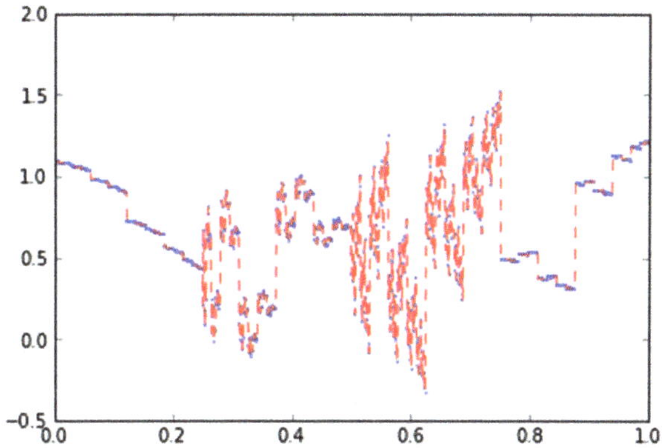

Fig. 1 A randomly generated local fractal function

$$\mathfrak{f}(\alpha\boldsymbol{\lambda} + \beta\boldsymbol{\mu}) = \sum_{i=1}^{N} (\alpha\lambda_i + \beta\mu_i) \circ u_i^{-1}\, \chi_{u_i(X_i)} + \sum_{i=1}^{N} (S_i \circ u_i^{-1}) \cdot (f_i^*(\alpha\boldsymbol{\lambda} + \beta\boldsymbol{\mu}) \circ u_i^{-1})\, \chi_{u_i(X_i)}$$

and

$$\alpha\mathfrak{f}(\boldsymbol{\lambda}) + \beta\mathfrak{f}(\boldsymbol{\mu}) = \sum_{i=1}^{N} (\alpha\lambda_i + \beta\mu_i) \circ u_i^{-1}\, \chi_{u_i(X_i)} + \sum_{i=1}^{N} (S_i \circ u_i^{-1}) \cdot (\alpha f_i^*(\boldsymbol{\lambda}) + \beta f_i^*(\boldsymbol{\mu})) \circ u_i^{-1})\, \chi_{u_i(X_i)}.$$

Hence, $\mathfrak{f}(\alpha\boldsymbol{\lambda} + \beta\boldsymbol{\mu}) = \alpha\mathfrak{f}(\boldsymbol{\lambda}) + \beta\mathfrak{f}(\boldsymbol{\mu})$.

For surjectivity, we define $\lambda_i := \mathfrak{f} \circ u_i - S_i \cdot \mathfrak{f}$, $i \in \mathbb{N}_N$. Since $\mathfrak{f} \in B(X, \mathbb{Y})$, we have $\boldsymbol{\lambda} \in \bigtimes_{i=1}^{N} B(X_i, \mathbb{Y})$. Thus, $\mathfrak{f}(\boldsymbol{\lambda}) = \mathfrak{f}$. □

The next results give information about the set of discontinuities of a bounded local fractal function $\mathfrak{f}$. The proof can be found in [8].

Theorem 5. *Let Φ be given as in* (9). *Assume that for all $i \in \mathbb{N}_N$ the u_i are contractive and the λ_i are continuous on $\overline{X_i}$. Further assume that*

$$\max\{\|S_i\|_{\infty,X_i} \mid i \in \mathbb{N}_N\} < 1,$$

and that the fixed point $\mathfrak{f}$ is bounded everywhere. Then the set of discontinuities of $\mathfrak{f}$ is at most countably infinite.

Next, we exhibit the relation between the graph G of the fixed point $\mathfrak{f}$ of the operator Φ given by (9) and the local attractor of an associated contractive local IFS. To this end, we need to require that $\mathbb{X}$ is a closed subset of a complete metric space, hence complete itself. Consider the complete metric space $\mathbb{X} \times \mathbb{Y}$ and define mappings $w_i : \mathbb{X}_i \times \mathbb{Y} \to \mathbb{X} \times \mathbb{Y}$ by

$$w_i(x, y) := (u_i(x), v_i(x, y)), \quad i \in \mathbb{N}_N.$$

Assume that the mappings $v_i : \mathbb{X}_i \times \mathbb{Y} \to \mathbb{Y}$ in addition to being uniformly contractive in the second variable are also uniformly Lipschitz continuous in the first variable, i.e., that there exists a constant $L > 0$ so that for all $y \in \mathbb{Y}$,

$$d_{\mathbb{Y}}(v_i(x_1, y), v_i(x_2, y)) \leq L\, d_{\mathbb{X}}(x_1, x_2), \quad \forall x_1, x_2 \in \mathbb{X}_i, \quad \forall i \in \mathbb{N}_N.$$

Denote by $a := \max\{a_i \,|\, i \in \mathbb{N}_N\}$ the largest of the Lipschitz constants of the mappings $u_i : \mathbb{X}_i \to \mathbb{X}$ and let $\theta := \frac{1-a}{2L}$. It is straight-forward to show that the mapping $d_\theta : (\mathbb{X} \times \mathbb{Y}) \times (\mathbb{X} \times \mathbb{Y}) \to \mathbb{R}$ given by

$$d_\theta := d_{\mathbb{X}} + \theta\, d_{\mathbb{Y}}$$

is a metric for $\mathbb{X} \times \mathbb{Y}$ compatible with the product topology on $\mathbb{X} \times \mathbb{Y}$.

Theorem 6. *The family $\mathcal{W}_{\text{loc}} := \{\mathbb{X} \times \mathbb{Y}; (\mathbb{X}_i \times \mathbb{Y}, w_i) \,|\, i \in \mathbb{N}_N\}$ is a contractive local IFS in the metric d_θ and the graph $G(\mathfrak{f})$ of the local fractal function $\mathfrak{f}$ associated with the operator Φ given by* (9) *is an attractor of $\mathcal{W}_{\text{loc}}$. Moreover,*

$$G(\Phi\mathfrak{f}) = \mathcal{W}_{\text{loc}}(G(\mathfrak{f})), \tag{12}$$

where $\mathcal{W}_{\text{loc}}$ denotes the set-valued operator (2) *associated with the local IFS $\mathcal{W}_{\text{loc}}$.*

Proof. We first show that $\{\mathbb{X} \times \mathbb{Y}; (X_i \times \mathbb{Y}, w_i) \,|\, i \in \mathbb{N}_N\}$ is a contractive local IFS. For this purpose, let $(x_1, y_1), (x_2, y_2) \in \mathbb{X}_i \times \mathbb{Y}$, $i \in \mathbb{N}_N$, and note that

$$\begin{aligned} d_\theta(w_i(x_1, y_1), w_i(x_2, y_2)) &= d_{\mathbb{X}}(u_i(x_1), u_i(x_2)) + \theta d_{\mathbb{Y}}(v_i(x_1, y_1), v_i(x_2, y_2)) \\ &\leq a\, d_{\mathbb{X}}(x_1, x_2) + \theta d_{\mathbb{Y}}(v_i(x_1, y_1), v_i(x_2, y_1)) \\ &\qquad + \theta d_{\mathbb{Y}}(v_i(x_2, y_1), v_i(x_2, y_2)) \\ &\leq (a + \theta L) d_{\mathbb{X}}(x_1, x_2) + \theta\, s\, d_{\mathbb{Y}}(y_1, y_2) \\ &\leq q\, d_\theta((x_1, y_1), (x_2, y_2)). \end{aligned}$$

Here we used (5) and set $q := \max\{a + \theta L, s\} < 1$.

The graph $G(\mathfrak{f})$ of $\mathfrak{f}$ is an attractor for the contractive local IFS $\mathcal{W}_{\text{loc}}$, for

$$\begin{aligned}\mathcal{W}_{\text{loc}}(G(\mathfrak{f})) &= \bigcup_{i=1}^{N} w_i(G(\mathfrak{f}) \cap \mathbb{X}_i) = \bigcup_{i=1}^{N} w_i(\{(x, \mathfrak{f}(x) \mid x \in \mathbb{X}_i\} \\ &= \bigcup_{i=1}^{N} \{(u_i(x), v_i(x, \mathfrak{f}(x))) \mid x \in \mathbb{X}_i\} = \bigcup_{i=1}^{N} \{(u_i(x), \mathfrak{f}(u_i(x))) \mid x \in \mathbb{X}_i\} \\ &= \bigcup_{i=1}^{N} \{(x, \mathfrak{f}(x)) \mid x \in u_i(\mathbb{X}_i)\} = G(\mathfrak{f}).\end{aligned}$$

That (12) holds follows from the above computation and the fixed point equation for $\mathfrak{f}$ written in the form

$$\mathfrak{f} \circ u_i(x) = v_i(x, \mathfrak{f}(x)), \quad x \in \mathbb{X}_i, \quad i \in \mathbb{N}_N.$$

□

5 Tensor Products of Local Fractal Functions

In this section, we define the tensor product of local fractal functions thus extending the previous construction to higher dimensions.

For this purpose, we follow the notation and of the previous section, and assume that X and $\overline{X}$ are nonempty connected sets, and $\{X_i \mid i \in \mathbb{N}_N\}$ and $\{\overline{X}_i \mid i \in \mathbb{N}_N\}$ are families of nonempty connected subsets of X and $\overline{X}$, respectively. Analogously, we define finite families of bijections $\{u_i : X_i \to X \mid i \in \mathbb{N}_N\}$ and $\{\overline{u}_i : \overline{X}_i \to X \mid i \in \mathbb{N}_N\}$ requiring both to satisfy condition (P).

Furthermore, we assume that $(\mathbb{Y}, \|\bullet\|_{\mathbb{Y}})$ is a *Banach algebra*, i.e., a Banach space that is also an associate algebra for which multiplication is continuous:

$$\|y_1 y_2\|_{\mathbb{Y}} \leq \|y_1\|_{\mathbb{Y}} \, \|y_2\|_{\mathbb{Y}}, \quad \forall\, y_1, y_2 \in \mathbb{Y}.$$

Let $f \in B(X, \mathbb{Y})$ and $\overline{f} \in B(\overline{X}, \mathbb{Y})$. The tensor product of f with $\overline{f}$, written $f \otimes \overline{f} : X \times \overline{X} \to \mathbb{Y}$, with values in $\mathbb{Y}$ is defined by

$$(f \otimes \overline{f})(x, \overline{x}) := f(x)\overline{f}(\overline{x}), \quad \forall\, (x, \overline{x}) \in X \times \overline{X}.$$

As f and $\overline{f}$ are bounded, the inequality

$$\|(f \otimes \overline{f})(x, \overline{x})\|_{\mathbb{Y}} = \|f(x)\overline{f}(\overline{x}\|_{\mathbb{Y}} \leq \|f(x)\|_{\mathbb{Y}} \, \|\overline{f}(\overline{x})\|_{\mathbb{Y}},$$

implies that $f \otimes \overline{f}$ is bounded. Under the usual addition and scalar multiplication of functions, the set

$$B(X \times \overline{X}, \mathbb{Y}) := \{f \otimes \overline{f} : X \times \overline{X} \to \mathbb{Y} \,|\, f \otimes \overline{f} \text{ is bounded}\}$$

becomes a complete metric space when endowed with the metric

$$d(f \otimes \overline{f}, g \otimes \overline{g}) := \sup_{x \in X} \|f(x) - g(x)\|_{\mathbb{Y}} + \sup_{\overline{x} \in \overline{X}} \|\overline{f}(\overline{x}) - \overline{g}(\overline{x})\|_{\mathbb{Y}}.$$

Now let $\Phi : B(X, \mathbb{Y}) \to B(X, \mathbb{Y})$ and $\overline{\Phi} : B(\overline{X}, \mathbb{Y}) \to B(\overline{X}, \mathbb{Y})$ be contractive RB-operators of the form (6). We define the tensor product of Φ with $\overline{\Phi}$ to be the RB-operator $\Phi \otimes \overline{\Phi} : B(X \times \overline{X}, \mathbb{Y}) \to B(X \times \overline{X}, \mathbb{Y})$ given by

$$(\Phi \otimes \overline{\Phi})(f \otimes \overline{f}) := (\Phi f) \otimes (\overline{\Phi f}).$$

It follows that $\Phi \otimes \overline{\Phi}$ maps bounded functions to bounded functions. Furthermore, $\Phi \otimes \overline{\Phi}$ is contractive on the complete metric space $(B(X \times \overline{X}, \mathbb{Y}), d)$. To see this, note that

$$\begin{aligned} \sup_{x \in X} \|(\Phi f)(x) - (\Phi g)(x)\|_{\mathbb{Y}} &+ \sup_{\overline{x} \in \overline{X}} \|(\Phi \overline{f})(\overline{x}) - (\Phi \overline{g})(\overline{x})\|_{\mathbb{Y}} \\ &\leq \ell \sup_{x \in X} \|f(x) - g(x)\|_{\mathbb{Y}} + \overline{\ell} \sup_{\overline{x} \in \overline{X}} \|\overline{f}(\overline{x}) - \overline{g}(\overline{x})\|_{\mathbb{Y}} \\ &\leq \max\{\ell, \overline{\ell}\}\, d(f \otimes \overline{f}, g \otimes \overline{g}), \end{aligned}$$

where we used (7) and denoted the uniform contractivity constant of $\overline{\Phi}$ by $\overline{\ell}$.

The unique fixed point of the RB-operator $\Phi \otimes \overline{\Phi}$ will be called a *tensor product local fractal function* and its graph a *tensor product local fractal surface*.

6 Lebesgue Spaces $L^p(\mathbb{R})$

We may construct local fractal functions on spaces other than $B(X, \mathbb{Y})$. (See also [8].) In this section, we derive conditions under which local fractal functions are elements of the Lebesgue spaces L^p for $p > 0$. To this end, we assume again that the functions v_i are given by (8) and that $\mathbb{X} := [0, 1]$ and $\mathbb{Y} := \mathbb{R}$. We consider the metric on $\mathbb{R}$ and $\mathbb{X} = [0, 1]$ as being induced by the L^1-norm. Note that endowed with this norm $B(\mathbb{X}, \mathbb{R})$ becomes a Banach space.

Recall that the Lebesgue spaces $L^p[0, 1]$, $1 \leq p \leq \infty$, are obtained as the completion of the space $C[0, 1]$ of real-valued continuous functions on $[0, 1]$ with respect to the L^p-norm

$$\|f\|_{L^p} := \left(\int_{[0,1]} |f(x)|^p \, dx \right)^{1/p}.$$

For $0 < p < 1$, the spaces $L^p(\mathbb{R})$ are defined as above but instead of a norm, a metric is used to obtain completeness. More precisely, define

$$d_p(f,g) := \|f - g\|_{L^p}^p,$$

where $\|\bullet\|_{L^p}$ is the norm introduced above. Then $(L^p(\mathbb{R}), d_p)$ is an F-space. (Note that the inequality $(a+b)^p \le a^p + b^p$ holds for all $a, b \ge 0$.) For more details, we refer to [19].

We have the following result for RB-operators defined on the Lebesgue spaces $L^p[0,1]$, $0 < p \le \infty$. The case $p \in [1,\infty]$ was already considered in [8], but for the sake of completeness we reproduce the proof.

Theorem 7. *Suppose that $\{X_i \mid i \in \mathbb{N}_N\}$ is a family of half-open intervals of $[0,1]$. Further suppose that $\{x_0 := 0 < x_1 < \cdots < x_N := 1\}$ is a partition of $[0,1]$ and that $\{u_i \mid i \in \mathbb{N}_N\}$ is a family of affine mappings from X_i onto $[x_{i-1}, x_i)$, $i = 1,\ldots,N-1$, and from $X_N^+ := X_N \cup u_N^{-1}(1-)$ onto $[x_{N-1}, x_N]$, where u_N maps X_N onto $[x_{N-1}, x_N)$.*

The operator $\Phi : L^p[0,1] \to \mathbb{R}^{[0,1]}$, $p \in (0,\infty]$, defined by

$$\Phi g := \sum_{i=1}^N (\lambda_i \circ u_i^{-1})\, \chi_{u_i(X_i)} + \sum_{i=1}^N (S_i \circ u_i^{-1}) \cdot (g_i \circ u_i^{-1})\, \chi_{u_i(X_i)}, \tag{13}$$

where $g_i = g|_{X_i}$, $\lambda_i \in L^p(X_i,[0,1])$, and $S_i \in L^\infty(X_i,\mathbb{R})$, $i \in \mathbb{N}_N$, maps $L^p[0,1]$ into itself. Moreover, if

$$\begin{cases} \displaystyle\sum_{i=1}^N a_i\, \|S_i\|_{\infty,X_i}^p < 1, & p \in (0,1); \\[2ex] \displaystyle\left(\sum_{i=1}^N a_i\, \|S_i\|_{\infty,X_i}^p\right)^{1/p} < 1, & p \in [1,\infty); \\[2ex] \max\{\|S_i\|_{\infty,X_i} \mid i \in \mathbb{N}_N\} < 1, & p = \infty, \end{cases} \tag{14}$$

where a_i denotes the Lipschitz constant of u_i, then Φ is contractive on $L^p[0,1]$, and its unique fixed point $\mathfrak{f}$ is an element of $L^p[0,1]$.

Proof. Note that under the hypotheses on the functions λ_i and S_i as well as the mappings u_i, Φf is well-defined and an element of $L^p[0,1]$. It remains to be shown that under conditions (14), Φ is contractive on $L^p[0,1]$.

We start with $1 \le p < \infty$. If $g, h \in L^p[0,1]$ then

$$\|\Phi g - \Phi h\|_p^{L^p} = \int_{[0,1]} |\Phi g(x) - \Phi h(x)|^p dx$$

$$
\begin{aligned}
&= \int\limits_{[0,1]} \left|\sum_{i=1}^{N}(S_i \circ u_i^{-1})(x)[(g_i \circ u_i^{-1})(x) - (h_i \circ u_i^{-1})(x)]\, \chi_{u_i(X_i)}(x)\right|^p dx \\
&= \sum_{i=1}^{N} \int\limits_{[x_{i-1},x_i]} \left|(S_i \circ u_i^{-1})(x)[(g_i \circ u_i^{-1})(x) - (h_i \circ u_i^{-1})(x)]\right|^p dx \\
&= \sum_{i=1}^{N} a_i \int\limits_{X_i} |S_i(x)[g_i(x) - h_i(x)]|^p\, dx \\
&\le \sum_{i=1}^{N} a_i\, \|S_i\|_{\infty,X_i}^p \int\limits_{X_i} |g_i(x) - h_i(x)|^p\, dx = \sum_{i=1}^{N} a_i\, \|S_i\|_{\infty,X_i}^p\, \|g_i - h_i\|_{L^p,X_i}^p \\
&= \sum_{i=1}^{N} a_i\, \|S_i\|_{\infty,X_i}^p\, \|g_i - h_i\|_{L^p}^p \le \left(\sum_{i=1}^{N} a_i\, \|S_i\|_{\infty,X_i}^p\right) \|g - h\|_{L^p}^p.
\end{aligned}
$$

The case $0 < p < 1$ now follows in similar fashion. We again have after substitution and rearrangement

$$
\begin{aligned}
d_p(\Phi g, \Phi h) &= \sum_{i=1}^{N} a_i \int\limits_{X_i} |S_i(x)[g_i(x) - h_i(x)]|^p\, dx \\
&= \sum_{i=1}^{N} a_i\, \|S_i\|_{\infty,X_i}^p\, \|g_i - h_i\|_{L^p}^p \le \left(\sum_{i=1}^{N} a_i\, \|S_i\|_{\infty,X_i}^p\right) \|g - h\|_{L^p}^p \\
&= \left(\sum_{i=1}^{N} a_i\, \|S_i\|_{\infty,X_i}^p\right) d_p(g, h).
\end{aligned}
$$

Now let $p = \infty$. Then

$$
\begin{aligned}
\|\Phi g - \Phi h\|_\infty &= \left\|\sum_{i=1}^{N}(S_i \circ u_i^{-1})(x)[(g_i \circ u_i^{-1})(x) - (h_i \circ u_i^{-1})(x)]\, \chi_{u_i(X_i)}(x)\right\|_\infty \\
&\le \max_{i\in\mathbb{N}_N} \left\|(S_i \circ u_i^{-1})(x)[(g_i \circ u_i^{-1})(x) - (h_i \circ u_i^{-1})(x)]\right\|_{\infty,X_i} \\
&\le \max_{i\in\mathbb{N}_N} \|S_i\|_{\infty,X_i}\, \|g_i - h_i]\|_{\infty,X_i} = \max_{i\in\mathbb{N}_N} \|S_i\|_{\infty,X_i}\, \|g_i - h_i]\|_\infty \\
&\le \left(\max_{i\in\mathbb{N}_N} \|S_i\|_{\infty,X_i}\right) \|g - h]\|_\infty
\end{aligned}
$$

These calculations prove the claims. □

Remark 3. The proof of the theorem shows that the conclusions also hold under the assumption that the family of mappings $\{u_i : X_i \to \mathbb{X} \,|\, i \in \mathbb{N}_N\}$ is generated by the following functions.

(i) Each u_i is a bounded diffeomorphism of class C^k, $k \in \mathbb{N} \cup \{\infty\}$, from X_i to $[x_{i-1}, x_i)$ (obvious modification for $i = N$). In this case, the a_i's are given by $a_i = \sup\{\left|\frac{du_i}{dx}(x)\right| \,|\, x \in X_i\}, i \in \mathbb{N}_N$.

(ii) Each u_i is a bounded invertible function in C^ω, the class of real-analytic functions from X_i to $[x_{i-1}, x_i)$ and its inverse is also in C^ω. (Obvious modification for $i = N$.) The a_i's are given as above in item (i).

7 Smoothness Spaces C^n and Hölder Spaces $\dot{C}^s$

Our next objective is to derive conditions on the partition $\{X_i \,|\, i \in \mathbb{N}_N\}$ of $\mathbb{X} := [0, 1]$ and the function tuples $\boldsymbol{\lambda}$ and $\boldsymbol{S}$ so that we obtain a continuous or even differentiable local fractal function $\mathfrak{f} : [0, 1] \to \mathbb{R}$. To this end, consider the complete metric linear space $C := C^0(\mathbb{X}) := \{f : [0, 1] \to \mathbb{R} \,|\, f \text{ continuous}\}$ endowed with the supremum norm $\| \bullet \|_\infty$.

7.1 *Binary Partition of* $\mathbb{X}$

We introduce the following subsets of $\mathbb{X} = [0, 1]$ which play an important role in fractal-based numerical analysis as they give discretizations for efficient computations. For more details, we refer to [8] and partly to [7].

Assume that $N \in 2\mathbb{N}$ and let

$$\mathbb{X}_{2j-1} := \mathbb{X}_{2j} := \left[\frac{2(j-1)}{N}, \frac{2j}{N}\right], \quad j = 1, \ldots, \frac{2}{N}. \tag{15}$$

Define affine mappings $u_i : \mathbb{X}_i \to [0, 1]$ so that

$$u_i(\mathbb{X}_i) := \left[\frac{i-1}{N}, \frac{i}{N}\right], \quad i = 1, \ldots, N. \tag{16}$$

In explicit form, the u_i's are given by

$$u_{2j-1}(x) = \frac{x}{2} + \frac{j-1}{N} \quad \text{and} \quad u_{2j}(x) = \frac{x}{2} + \frac{j}{N}, \quad x \in \mathbb{X}_{2j-1} = \mathbb{X}_{2j}.$$

Note that here $u_i(\mathbb{X}_i) \subsetneq \mathbb{X}_i$, $\forall\, i \in \mathbb{N}_N$. Clearly, $\{u_i(\mathbb{X}_i) \,|\, i \in \mathbb{N}_N\}$ is a partition of $[0, 1]$. We denote the distinct endpoints of the partitioning intervals $\{u_i(\mathbb{X}_i)\}$ by

$\{x_0 < x_1 < \ldots < x_N\}$ where $x_0 = 0$ and $x_N = 1$, and refer to them as *knot points* or simply as *knots*.

Furthermore, we assume that we are given interpolation values at the endpoints of the intervals $\mathbb{X}_{2j-1} = \mathbb{X}_{2j}$:

$$\mathscr{I} := \left\{(x_{2j}, y_j) \mid j = 0, 1, \ldots, N/2\right\}. \tag{17}$$

Let

$$C_{\mathscr{I}} := \{f \in C \mid f(x_{2j}) = y_j,\ \forall\, j = 0, 1, \ldots, N/2\}.$$

Then $C_{\mathscr{I}}$ is a closed metric subspace of C. We consider an RB operator Φ of the form (9) acting on $C_{\mathscr{I}}$.

In order for Φ to map $C_{\mathscr{I}}$ into itself one needs to require that $\lambda_i, S_i \in C(\mathbb{X}_i) := C(\mathbb{X}_i, \mathbb{R}) := \{f : \mathbb{X}_i \to \mathbb{R} \mid f \text{ continuous}\}$ and that

$$y_{j-1} = \Phi f(x_{2(j-1)}) \quad \wedge \quad y_j = \Phi f(x_{2j}), \quad j = 1, \ldots, N/2, \tag{18}$$

where $x_{2j} := (2j)/N$. Note that the preimages of the knots $x_{2(j-1)}$ and x_{2j} are the endpoints of $\mathbb{X}_{2j-1} = \mathbb{X}_{2j}$. Substituting the expression for Φ into (18) and collecting terms yields

$$\begin{aligned} \lambda_{2j-1}(x_{2(j-1)}) + \big(S_{2j-1}(x_{2(j-1)}) - 1\big)\, y_{j-1} &= 0, \\ \lambda_{2j}(x_{2j}) + \big(S_{2j}(x_{2j}) - 1\big)\, y_j &= 0, \end{aligned} \tag{19}$$

for all $j = 1, \ldots, N/2$.

To ensure continuity of Φf across [0, 1], the following join-up conditions at the oddly indexed knots need to be imposed. (They are the images of the midpoints of the intervals $\mathbb{X}_{2j-1} = \mathbb{X}_{2j}$.)

$$\Phi f(x_{2j-1}-) = \Phi f(x_{2j-1}+), \quad j = 1, \ldots, N/2. \tag{20}$$

A simple calculation gives

$$\lambda_{2j}(x_{2(j-1)}) + S_{2j}(x_{2(j-1)}) y_{j-1} = \lambda_{2j-1}(x_{2j}) + S_{2j-1}(x_{2j}) y_j, \tag{21}$$

for all $j = 1, \ldots, N/2$. In case all functions λ_i and S_i are constant, (21) reduces to the condition given in [7, Example 2]. Two tuples of functions $\boldsymbol{\lambda}, \boldsymbol{S} \in \displaystyle\mathop{\times}_{i=1}^{N} C(\mathbb{X}_i)$ are said to have property (J) if they satisfy (19) and (21).

We summarize these results in the next theorem.

Theorem 8. *Let* $\mathbb{X} := [0, 1]$ *and let* $N \in 2\mathbb{N}$. *Suppose that subsets of* $\mathbb{X}$ *are given by* (15) *and the associated mappings* u_i *by* (16). *Further suppose that*

$\mathscr{I}$ is as in (17) *and that $\boldsymbol{\lambda}, \boldsymbol{S} \in \mathop{\times}\limits_{i=1}^{N} C(\mathbb{X}_i)$ have property (J). Then the RB operator Φ as given in* (9) *maps $C_{\mathscr{I}}$ into itself and is well-defined. If, in addition,* $\max\{\|S_i\|_{\infty,\mathbb{X}_i} \mid i \in \mathbb{N}_N\} < 1$, *then Φ is a contraction and thus possesses a unique fixed point $\mathfrak{f} : [0,1] \to \mathbb{R}$ satisfying $\mathfrak{f}(x_{2j}) = y_j$, $\forall\, j = 0, 1, \ldots, N/2$.*

We call this unique fixed point a *continuous local fractal interpolation function.*

Proof. It remains to be shown that under the condition $\max\{\|S_i\|_{\infty,\mathbb{X}_i} \mid i \in \mathbb{N}_N\} < 1$, Φ is contractive on $C_{\mathscr{I}}$. This, however, follows immediately from the case $p = \infty$ in the proof of Theorem 7. □

Theorem 8 can be adapted to the setting of Hölder spaces. For this purpose, we introduce the *homogeneous Hölder space* $\dot{C}^s(\Omega)$, $0 < s < 1$, as the family of all functions $f \in C(\Omega)$, $\Omega \subseteq \mathbb{R}$, for which

$$|f|_{\dot{C}^s(\Omega)} := \sup_{x \neq x' \in \Omega} \frac{|f(x) - f(x')|}{|x - x'|^s} < \infty.$$

$|\bullet|_{\dot{C}^s(\Omega)}$ is a homogeneous semi-norm making $\dot{C}^s$ into a complete locally convex topological vector space, i.e., a Fréchet space.

Theorem 9. *Let $\mathbb{X} := [0,1]$ and let $N \in 2\mathbb{N}$. Assume that subsets of $\mathbb{X}$ are given by* (15)*, associated mappings u_i by* (16)*, and that $\mathscr{I}$ is as in* (17)*. Assume further that $\boldsymbol{\lambda} \in \mathop{\times}\limits_{i=1}^{N} \dot{C}^s(\mathbb{X}_i)$, $\boldsymbol{S} \in \mathop{\times}\limits_{i=1}^{N} C(\mathbb{X}_i)$, and that have property condition (J). Then the RB operator* (9) *maps $\dot{C}^s := \dot{C}^s(\mathbb{X})$ into itself and is well defined. Furthermore, if*

$$2^s \max\{\|S_i\|_{\infty,\mathbb{X}_i} \mid i \in \mathbb{N}_N\} < 1$$

then Φ is contractive on $\dot{C}^s$ and has a unique fixed point $\mathfrak{f} \in \dot{C}^s$.

In case the last conclusion of the above theorem holds, we say that the fixed point $\mathfrak{f}$ is a *local fractal function of class $\dot{C}^s$.*

Proof. First we show that $\Phi f \in \dot{C}^s$. For $x, x' \in [0,1]$, note that there exist $i, i' \in \mathbb{N}_N$ so that $x \in u_i(\mathbb{X}_i)$ and $x' \in u_{i'}(\mathbb{X}_{i'})$. Therefore,

$$\begin{aligned}
|\Phi f(x) - \Phi f(x')| &\le \left|\lambda_i(u_i^{-1}(x)) - \lambda_{i'}(u_{i'}^{-1}(x'))\right| \\
&\quad + \left|(S_i(u_i^{-1}(x)) \cdot (f_i(u_i^{-1}(x)) - (S_{i'}(u_{i'}^{-1}(x')) \cdot (f_{i'}(u_{i'}^{-1}(x'))\right| \\
&\le \left|\lambda_i(u_i^{-1}(x)) - \lambda_{i'}(u_{i'}^{-1}(x'))\right| \\
&\quad + \max\{\|S_i\|_{\infty,\mathbb{X}_i}\} \left|f_i(u_i^{-1}(x)) - f_{i'}(u_{i'}^{-1}(x'))\right|.
\end{aligned}$$

Using the fact that $|x - x'|^s = 2^{-s}|u_i^{-1}(x) - u_{i'}^{-1}(x')|$ and employing the properties of the supremum, we thus obtain

$$|\Phi f|_{\dot{C}^s} \leq 2^s \left(\sum_{i\in\mathbb{N}_N} |\lambda_i|_{\dot{C}^s(X_i)} + \max\{\|S_i\|_{\infty,\mathbb{X}_i}\} |f|_{\dot{C}^s} \right) < \infty.$$

To establish the contractivity of Φ, note that

$$\begin{aligned} &|(\Phi f - \Phi g)(x) - (\Phi f - \Phi g)(x')| = \\ &\qquad |S_i(u_i^{-1}(x)) \cdot (f_i - g_i)(u_i^{-1}(x)) - S_{i'}(u_{i'}^{-1}(x')) \cdot (f_{i'} - g_{i'})(u_{i'}^{-1}(x'))| \\ &\qquad \leq \max\{\|S_i\|_{\infty,\mathbb{X}_i}\} |(f_i - g_i)(u_i^{-1}(x)) - (f_{i'} - g_{i'})(u_{i'}^{-1}(x'))| \end{aligned}$$

As above, using again $|x - x'|^s = 2^{-s}|u_i^{-1}(x) - u_{i'}^{-1}(x')|$ and that f is defined on all of $[0, 1]$, this yields

$$|\Phi f - \Phi g|_{\dot{C}^s} \leq 2^s \max\{\|S_i\|_{\infty,\mathbb{X}_i}\} |f - g|_{\dot{C}^s}.$$

□

Just as in the case of splines, we can impose join-up conditions and choose the function tuples $\boldsymbol{\lambda}$ and $\boldsymbol{S}$ so that the RB operator (9) maps the space of continuously differentiable functions into itself. More precisely, suppose that $\Omega \subseteq \mathbb{R}$. Let $C^n(\Omega) := C^n(\Omega, \mathbb{R}) := \{f : \Omega \to \mathbb{R} \,|\, D^k f \in C, \forall k = 1, \ldots, n\}$, where D denotes the ordinary differential operator. The linear space $C^n(\Omega)$ is a Banach space under the norm

$$\|f\|_{C^n(\Omega)} := \sum_{k=0}^{n} \|D^k f\|_{\infty,\Omega}.$$

We write C^n for $C^n(\mathbb{X})$ and will delete the Ω from the norm notation when $\Omega := \mathbb{X} = [0, 1]$.

As we require C^n-differentiability across $\mathbb{X} = [0, 1]$, we impose C^n-interpolation values at the endpoints of the intervals $\mathbb{X}_{2j-1} = \mathbb{X}_{2j}$:

$$\mathscr{I}^{(n)} := \left\{(x_{2j}, \boldsymbol{y}_j^{(n)}) \,|\, j = 0, 1, \ldots, N/2\right\}, \tag{22}$$

where $\boldsymbol{y}_j^{(n)} := (y_j^{(0)}, y_j^{(1)}, \ldots, y_j^{(n)})^T \in \mathbb{R}^{n+1}$ is a given interpolation vector. Let

$$C^n_{\mathscr{I}^{(n)}} := \{f \in C^n \,|\, D^k f(x_{2j}) = y_j^{(k)}, \,\forall\, k = 0, 1, \ldots, n; \,\forall\, j = 0, 1, \ldots, N/2\}.$$

Then $C^n_{\mathscr{I}^{(n)}}$ is a closed metric subspace of C^n.

In order for Φ to map $C^n_{\mathscr{I}^{(n)}}$ into itself, choose $\lambda_i, S_i \in C^n(\mathbb{X}_i)$, $i \in \mathbb{N}_N$, so that

$$y_{j-1}^{(k)} = D^k \Phi f(x_{2(j-1)}) \quad \wedge \quad y_j^{(k)} = D^k \Phi f(x_{2j}), \tag{23}$$

for all $k = 0, 1, \ldots, n$ and for all $j = 1, \ldots, N/2$.

At the midpoints of the intervals $\mathbb{X}_{2j-1} = \mathbb{X}_{2j}$, the function tuples $\boldsymbol{\lambda}$ and $\boldsymbol{S}$ need to additionally satisfy the C^n-join-up conditions

$$D^k \Phi f(x_{2j-1}-) = D^k \Phi f(x_{2j-1}+), \quad \forall k = 0, 1, \ldots, n; \; \forall j = 1, \ldots, N/2. \tag{24}$$

Theorem 10. *Let* $\mathbb{X} := [0, 1]$ *and let* $N \in 2\mathbb{N}$. *Assume that subsets of* $\mathbb{X}$ *are given by* (15)*, associated mappings* u_i *by* (16)*, and that* $\mathscr{I}^{(n)}$ *is as in* (22). *Assume further that* $\boldsymbol{\lambda}, \boldsymbol{S} \in \mathop{\times}\limits_{i=1}^{N} C^n(\mathbb{X}_i)$*, and that they satisfy conditions* (23) *and* (24). *Then the RB operator* (9) *maps* $C^n_{\mathscr{I}^{(n)}}$ *into itself and is well defined. Furthermore, if*

$$2^n \max_{i \in \mathbb{N}_N} \max_{k=0,1,\ldots,n} \left\{ \sum_{l=0}^{k} \binom{n-k+l}{l} \| D^l S_i \|_{\infty,\mathbb{X}_i} \right\} < 1 \tag{25}$$

then Φ *is contractive on* $C^n_{\mathscr{I}^{(n)}}$ *and has a unique fixed point* $\mathfrak{f} \in \mathbb{C}^n_{\mathscr{I}^{(n)}}$.

We refer to this fixed point $\mathfrak{f}$ as a *local fractal function of class* $C^n_{\mathscr{I}^{(n)}}$.

Proof. The statements that Φ is well defined and maps $C^n_{\mathscr{I}^{(n)}}$ into itself is implied by the conditions imposed on $\boldsymbol{\lambda}$ and $\boldsymbol{S}$. It remains to be shown that under condition (25) the RB operator Φ is contractive. To this end, consider $f, g \in C^n_{\mathscr{I}^{(n)}}$. Then

$$\begin{aligned} D^k \Phi f(x) - D^k \Phi g(x) &= \sum_{i \in \mathbb{N}_N} D^k \left[S_i(u_i^{-1}(x)) \cdot (f_i(u_i^{-1}(x)) - g_i(u_i^{-1}(x))) \right] \chi_{u_i(\mathbb{X}_i)} \\ &= \sum_{i \in \mathbb{N}_N} \sum_{l=0}^{k} \binom{k}{l} 2^k \left[(D^{k-l}(f_i - g_i))(u_i^{-1}(x)) \cdot (D^l S_i)(u_i^{-1}(x) \right] \chi_{u_i(\mathbb{X}_i)}, \end{aligned}$$

where we applied the Leibnitz Differentiation Rule. Therefore,

$$\| D^k \Phi f - D^k \Phi g \|_\infty \le 2^k \sum_{i \in \mathbb{N}_N} \sum_{l=0}^{k} \binom{k}{l} \| D^l S_i \|_{\infty,\mathbb{X}_i} \| D^{k-l}(f - g) \|_\infty.$$

Hence,

$$\begin{aligned} \| \Phi f - \Phi g \|_{C^n} &= \sum_{k=0}^{n} \| D^k \Phi f - D^k \Phi g \|_\infty \\ &\le 2^n \sum_{i \in \mathbb{N}_N} \sum_{k=0}^{n} \sum_{l=0}^{k} \binom{k}{l} \| D^l S_i \|_{\infty,\mathbb{X}_i} \| D^{k-l}(f - g) \|_\infty \\ &= 2^n \sum_{i \in \mathbb{N}_N} \sum_{k=0}^{n} \sum_{l=0}^{k} \binom{n-k+l}{l} \| D^l S_i \|_{\infty,\mathbb{X}_i} \| D^{n-k}(f - g) \|_\infty \end{aligned}$$

The last equality is proven directly by computation or mathematical induction. Thus,

$$\|\Phi f - \Phi g\|_{C^n} \leq \left(2^n \max_{i\in\mathbb{N}_N} \max_{k=0,1,\dots,n} \left\{\sum_{l=0}^{k} \binom{n-k+l}{l} \|D^l S_i\|_{\infty,\mathbb{X}_i}\right\}\right) \|f-g\|_{C^n},$$

and the statement follows. □

7.2 Vanishing Endpoint Conditions for S_i

Here, we consider a more general set-up than in the previous subsection. We assume again that $\mathbb{X} := [0,1]$ and let $\mathbb{X}_i := [a_i, b_i]$, for $i \in \mathbb{N}_N$, be N different subintervals of positive length. We further assume that $\{0 =: x_0 < x_1 < \ldots < x_{N-1} < x_N := 1\}$ is a partition of $\mathbb{X}$ and that we have chosen an enumeration in such a way that the mappings $u_i : \mathbb{X}_i \to \mathbb{X}$ satisfy

$$u_i([a_i, b_i]) := [x_{i-1}, x_i], \quad \forall\, i \in \mathbb{N}_N.$$

In particular, note that $a_1 = x_0$, $b_N = x_N$, and $u_i(b_i) = x_i = u_{i+1}(a_{i+1})$, for all interior knots $x_1, \ldots, x_{N-1}$. We assume that the u_i are affine functions but that they are not necessarily contractive.

Let

$$\mathscr{I} := \{(x_j, y_j) \mid j = 0, 1, \ldots, N\}. \tag{26}$$

be a given set of interpolation points and let

$$C_{\mathscr{I}} := \{f \in C \mid f(x_j) = y_j, \forall\, j = 0, 1, \ldots, N\}. \tag{27}$$

Our objective in this subsection is to construct a local fractal function that belongs to $C_{\mathscr{I}}$ and which is generated by an RB operator of the form (9). For this purpose, we need to impose continuity conditions at the interpolation points. More precisely, we require that for an $f \in C_{\mathscr{I}}$,

$$\begin{aligned} \Phi f(x_0) = y_0, \quad \Phi f(x_N) = y_N, \\ \Phi f(x_i-) = y_i = \Phi f(x_i+), \quad i = 1, \ldots, N-1. \end{aligned} \tag{28}$$

Substituting the expression for Φ into these equations and simplifying yields

$$\lambda_1(x_0) + S_1(x_0)y_0 = y_0, \quad \lambda_N(x_N) + S_N(x_N)y_N = y_N$$
$$\lambda_i(b_i) + S_i(b_i)f(b_i) = y_i = \lambda_{i+1}(a_{i+1}) + S_{i+1}(a_{i+1})f(a_{i+1}), \quad i = 1, \ldots, N-1.$$

Since these equations require unavailable knowledge of f at the points a_i and b_i, we impose the following vanishing endpoint conditions on the functions S_i:

$$S_i(a_i) = 0 = S_i(b_i), \ \forall i = 1, \ldots, N. \tag{29}$$

Thus the requirements on the functions λ_i reduce to

$$\lambda_1(x_0) = y_0, \quad \lambda_N(x_N) = y_N$$
$$\lambda_i(b_i) = y_i = \lambda_{i+1}(a_{i+1}), \quad i = 1, \ldots, N-1.$$

Function tuples $\boldsymbol{\lambda}$ and $\boldsymbol{S}$ satisfying (28) and (29) are said to have property (S).

A class of functions S_i for which conditions (29) hold is, for instance, the class of polynomial B-splines B_n of order $2 < n \in \mathbb{N}$ centered at the midpoint of the interval $[a_i, b_i]$. Polynomial B-splines B_n have even the property that all derivatives up to order $n-2$ vanish at the endpoints: $D^k B_n(a_i) = 0 = D^k B_n(b_i)$, for all $k = 0, 1 \ldots, n-2$.

The above considerations now entail the next theorem.

Theorem 11. *Let $\mathbb{X}$ and $\mathbb{X}_i$, $i \in \mathbb{N}_N$, be as defined above. Let $\mathscr{I}$ be as in* (26). *Suppose that $\boldsymbol{\lambda}, \boldsymbol{S} \in \mathop{\times}\limits_{i=1}^{N} C(\mathbb{X}_i)$ and that they have property (S). The RB operator* (9) *maps $C_{\mathscr{I}}$ as given by* (27) *into itself and is well defined. If in addition*

$$\max\{\|S_i\|_{\infty,\mathbb{X}_i} \mid i \in \mathbb{N}_N\} < 1,$$

then Φ is contractive on $C_{\mathscr{I}}$.

The fixed point $\mathfrak{f}$ of Φ is called again a *continuous local fractal interpolation function*.

Proof. The assumptions on $\boldsymbol{\lambda}$ and $\boldsymbol{S}$ guarantee that Φ is well defined and maps $C_{\mathscr{I}}$ into itself. The contractivity of Φ under the given condition follows immediately from the proof of Theorem 7. □

For the particular setting at hand, one may, of course, also construct fractal functions of class $\dot{C}^s$ and C^n by imposing the relevant conditions on the function tuples $\boldsymbol{\lambda}$ and $\boldsymbol{S}$ and choose the appropriate interpolation sets. We rely on the diligent reader to provide these conditions and prove the corresponding results.

8 Sobolev Spaces $W^{m,p}$

The final type of function space we consider is the Sobolev spaces $W^{m,p}$ with $m \in \mathbb{N}_0$ and $1 \leq p \leq \infty$. To this end, let $\Omega \subset \mathbb{R}$ be open and

$$C^{m,p}(\Omega) := \{f \in C^\infty(\Omega) \mid D^k f \in L^p(\Omega),\ \forall\, k = 0, 1, \ldots, m\}.$$

Define functionals $\|\bullet\|_{m,p}$, $m \in \mathbb{N}_0$ and $1 \leq p \leq \infty$, as follows:

$$\|f\|_{m,p} := \begin{cases} \left(\sum_{k=0}^{m} \|D^k f\|_{L^p}^p\right)^{1/p}, & 1 \le p < \infty; \\ \max_{k\in\{0,1,\ldots,m\}} \{\|D^k f\|_\infty\}, & p = \infty. \end{cases}$$

The closure of $C^{m,p}(\Omega)$ in the norm $\| \bullet \|_{m,p}$ produces the Sobolev space $W^{m,p}(\Omega)$. The ordinary derivatives D^k in $C^{m,p}(\Omega)$ have a continuous extension to $W^m(L^p)(\Omega)$. These extensions are then the weak derivatives $D^{(k)}$. The Sobolev space $W^{m,p}(\Omega)$ is a Banach space when endowed with the norm $\| \bullet \|_{m,p}$. For more details, we refer the reader to [1].

Now suppose $X := (0,1)$ and $\{X_i \mid i \in \mathbb{N}_N\}$ is a collection of nonempty open intervals of X. Further suppose that $\{x_1 < \cdots < x_{N-1}\}$ is a partition of X and that $\{u_i : X_i \to X\}$ is a family of affine mappings with the property that $u_i(X_i) = (x_{i-1}, x_i)$, for all $i \in \mathbb{N}_N$, where we set $x_0 := 0$ and $x_N := 1$. We write $W^{m,p}$ for $W^{m,p}(X)$.

Theorem 12. *Under the assumptions stated above, let* $\boldsymbol{\lambda} \in \mathop{\times}\limits_{i=1}^{N} W^{m,p}(X_i)$ *and let* $\boldsymbol{S} := (s_1, \ldots, s_N) \in \mathbb{R}^N$. *Then the RB operator* $\Phi : W^{m,p} \to \mathbb{R}^{(0,1)}$, $m \in \mathbb{N}_0$ *and* $1 \le p \le \infty$, *defined by*

$$\Phi g := \sum_{i=1}^{N} (\lambda_i \circ u_i^{-1})\, \chi_{u_i(X_i)} + \sum_{i=1}^{N} s_i (g_i \circ u_i^{-1})\, \chi_{u_i(X_i)},$$

has range contained in $W^{m,p}$ *and is well defined. Moreover, if*

$$\begin{cases} \left(\max_{k\in\{0,1,\ldots,m\}} \sum_{i\in\mathbb{N}_N} \frac{|s_i|^p}{a_i^{kp-1}}\right)^{1/p} < 1, & 1 \le p < \infty; \\ \sum_{i\in\mathbb{N}_N} \frac{|s_i|}{a_i^k} < 1, & p = \infty, \end{cases} \tag{30}$$

then Φ *is contractive on* $W^{m,p}$.

The unique fixed point $\mathfrak{f}$ of Φ is called a *local fractal function of class* $W^{m,p}$.

Proof. That Φ is well defined and has range contained in $W^{m,p}$ follows from the assumption on the function tuple $\boldsymbol{\lambda}$ and the fact that if the weak derivative of a function f exits and u_i is a diffeomorphism, then the weak derivative of $f \circ u_i^{-1}$ exists and equals $(D^{(1)} f)(u_i^{-1}) \cdot Du_i^{-1}$.

To prove contractivity on $W^{m,p}$, suppose that $g, h \in W^{m,p}$, $k \in \{0, 1, \ldots, m\}$. Denote the ordinary derivative of u_i by a_i. Note that $a_i > 0$ but may be larger than one. Then, for $1 \le p < \infty$, we obtain the following estimates.

$$
\begin{aligned}
\|D^{(k)}\Phi g - D^{(k)}\Phi h\|_{L^p}^p &= \int_X \left| D^{(k)} \sum_{i\in\mathbb{N}_N} s_i (g_i - h_i)(u_i^{-1})(x) \right|^p \chi_{u_i(X_i)} dx \\
&\le \sum_{i\in\mathbb{N}_N} |s_i|^p \int_{u_i(X_i)} \left| D^{(k)}(g_i - h_i)(u_i^{-1}(x)) \right|^p \left(\frac{1}{a_i}\right)^{kp} dx \\
&\le \sum_{i\in\mathbb{N}_N} |s_i|^p \left(\frac{1}{a_i}\right)^{kp-1} \int_{X_i} \left| D^{(k)}(g_i - h_i)(x) \right|^p dx \\
&\le \left(\sum_{i\in\mathbb{N}_N} |s_i|^p \left(\frac{1}{a_i}\right)^{kp-1} \right) \|D^{(k)} g - D^{(k)} h\|_{L^p}^p .
\end{aligned}
$$

Summing over $k = 0, 1, \ldots, m$, and factoring out the maximum value of the expression in parentheses, proves the statement.

Similarly, for $p = \infty$, we get

$$
\begin{aligned}
\left| D^{(k)} g(x) - D^{(k)} h(x) \right| &= \left| \sum_{i\in\mathbb{N}_N} s_i D^{(k)}(g_i - h_i)(u_i^{-1})(x) \left(\frac{1}{a_i^k}\right) \chi_{u_i(X_i)}(x) \right| \\
&\le \sum_{i\in\mathbb{N}_N} \frac{|s_i|}{a_i^k} \left| D^{(k)}(g_i - h_i)(u_i^{-1})(x) \chi_{u_i(X_i)}(x) \right| \\
&\le \sum_{i\in\mathbb{N}_N} \frac{|s_i|}{a_i^k} \left\| D^{(k)} g - D^{(k)} h \right\|_\infty ,
\end{aligned}
$$

verifying the assertion. □

Acknowledgements The author wishes to thank the Mathematical Sciences Institute of The Australian National University for its kind hospitality and support during his research visit in May 2013 which initiated the investigation into local IFSs.

References

1. Adams, R., Fourier, J.: Sobolev Spaces, 2nd edn. Academic, New York (2003)
2. Barnsley, M.F.: Fractal functions and interpolation. Constr. Approx. **2**, 303–329 (1986)
3. Barnsley, M.F.: Fractals Everywhere. Dover, New York (2012)
4. Barnsley, M.F., Demko, S.: Iterated function systems and the global construction of fractals. Proc. R. Soc. Lond. A **399**, 243–275 (1985)
5. Barnsley, M.F., Hurd, L.P.: Fractal Image Compression. AK Peters Ltd., Wellesly (1993)
6. Barnsley, M.F., Massopust, P.: Bilinear Fractal Interpolation and Box Dimension (2012) [arXiv:1209.3139]
7. Barnsely, M.F., Hegland, M., Massopust, P.: Self-referential descriptions of analytic functions. Preprint (2013)
8. Barnsely, M.F., Hegland, M., Massopust, P.: Numerics and Fractals (2013) [arXiv:1309:0972]
9. Edgar, G.A.: Measure, Topology, and Fractal Geometry. Springer, New York (1990)

10. Engelking, R.: General Topology. Helderman Verlag, Berlin (1989)
11. Geronimo, J., Hardin, D., Massopust, P.: Fractal functions and wavelets expansions based on several scaling functions. J. Approx. Theory **78**(3), 373–401 (1994)
12. Hutchinson, J.E.: Fractals and self similarity. Indiana Univ. J. Math. **30**, 713–747 (1981)
13. Kieninger, B.: Iterated Function Systems on Compact Hausdorff Spaces. Ph.D. Thesis, Augsburg University, Berichte aus der Mathematik. Shaker-Verlag, Aachen (2002)
14. Leśniak, K.: Stability and invariance of multivalued iterated function systems. Math. Slovaca **53**, 393–405 (2003)
15. Massopust, P.R.: Fractal Functions, Fractal Surfaces, and Wavelets. Academic, San Diego (1994)
16. Massopust, P.R.: Fractal functions and their applications. Chaos Solitons Fractals **8**(2), 171–190 (1997)
17. Massopust, P.R.: Interpolation with Splines and Fractals. Oxford University Press, New York (2012)
18. Rolewicz, S.: Metric Linear Spaces. Kluwer Academic, Poland (1985)
19. Rudin, W.: Functional Analysis. McGraw–Hill, New York (1991)

Some Historical Precedents of the Fractal Functions

M.A. Navascués and M.V. Sebastián

Abstract With this short text, we want to pay tribute to the scientists of older generations who, through their research, led to the current state of the knowledge of the fractal functions. We review the fundamental milestones of the origin and evolution of self-similar curves that, in some cases, agree with continuous and nowhere differentiable functions, but they are not exhausted by them. Our main interest is to emphasize the lesser known examples, due to a deficient or late publication (Bolzano's map for instance). We describe different ways of defining self-similar curves. We recall the first functions without tangent, but also some fractal functions having derivative almost everywhere. Most of the models studied may seem quite paradoxical ("monsters" in the words of poincaré) as, for instance, curves with a fractal dimension of two and having tangent at every point. These instances suggest that the classification and even the defintion of fractal functions are far from being established.

Keywords Fractal functions • Non-differentiability • History of Mathematics Fractals

M.A. Navascués (✉)
Departamento de Matemática Aplicada, Escuela de Ingeniería y Arquitectura, Universidad de Zaragoza, Zaragoza, Spain
e-mail: manavas@unizar.es

M.V. Sebastián
Centro Universitario de la Defensa, Academia General Militar de Zaragoza, Zaragoza, Spain
e-mail: msebasti@unizar.es

C. Bandt et al. (eds.), *Fractals, Wavelets, and their Applications*, Springer Proceedings in Mathematics & Statistics 92, DOI 10.1007/978-3-319-08105-2_17,

1 Introduction

In this paper we inquire about the origin and historical evolution of the fractal functions. We also wish to pay tribute to the men of older generations who, through their discoveries, led to the current state of the fractal maths. In particular, we want to describe how the definition of self-similar curves has evolved over time.

The nineteenth century has been called the century of the science due to the important discoveries made in those days. In the mathematical field we may speak about a golden century, not only for the numerous contributions made, but also for their quality. Great efforts were made in order to lay the foundations of the modern mathematics: logic, good definitions, axiomatic systems, etc. because the intuition is sometimes misleading. The rigor is fully established. Some creations in the math field are:

- New geometries: Hyperbolic (Gauss, Lobachevsky, Bolyai), elliptic and Riemannian (Riemann)
- Systems of numbers: reals (Dedekind), naturals (Peano)
- Set theory (Cantor) and logic (Hilbert, Peano, Boole)
- Celestial mechanics (Bessel, Poincaré)
- New mechanics (Lie, Hamilton), etc.

Huge impetus was given to modern algebra (Abel, Galois, Cayley, etc.) and analysis (Fourier, Cauchy, Weierstrass, Riemann, Lebesgue, Darboux, etc.).

Paradoxically many mathematicians had a very naïve concept of the curves. Almost all the scientists believed that all the continuous functions had tangent, except possibly at an isolated set of points. Contradicting this hypothesis some "monsters" (in the words of Poincaré) began to appear. It is worth emphasizing the Weierstrass map, followed by those of Riemann and Darboux, as non-differentiable functions. When these so-called "irregular continua" appeared, some scientists reacted with irritation. For instance, there is a letter from Hermite to Stieltjes (1893) saying [2]: "I turn away with fear and horror from the lamentable plague of continuous functions which do not have derivatives." Well into the twentieth century Dieudonnée (one of the founders of Bourbaki) comments (1975): "Some mathematical objects, like Peano curve, are totally non-intuitive....extravagant."

Of course there were defenders of the new continua. Perrin in 1906 says [2]: "One might encounter instances where using a function without a derivative would be simpler than using one that can be differentiated. When this happens, the mathematical study of the irregular continua will prove its practical value. However this hope is nothing but a daydream, as yet" and Levy (1970) defends the non-differentiable curves: "I have always been surprised to hear it said that geometric intuition inevitably leads one to think that all continuous functions are differentiable. From my first encounter with the notion of derivative, my experience proved that the contrary is true."

2 Nineteenth Century

2.1 Bolzano's Function (1830)

The map proposed by Bolzano is the first of a long series of curves without derivative but, being the oldest, is the most modern among the classics. It is defined as the limit of a sequence of functions. The first element is a line segment. In the second term the segment is replaced by a polygonal of four edges. In the third item, every edge is replaced by another piecewise linear function, etc. We observe then a collection of elements currently in use (self-affine geometry, iterative construction, limit set).

The manuscript of Bolzano, where this function is studied, remained lost until its discovery in the National Library of Vienna in 1920 by M. Jasek. It was then entitled and published. The work does not only contain the map, but it is a complete treaty of one variable calculus.

Bolzano proved that his function is continuous and non-differentiable in a dense set of points of the interval where is defined. Later on K. Rychlik (1921) and V. Jarnik (1922) proved the continuity and the nowhere differentiability. Moreover:

- The local extremes form a countable set.
- At the local maxima the left derivative is $+\infty$ and the right derivative is $-\infty$. At the local minima the left derivative is $-\infty$ and the right derivative is $+\infty$.
- The side derivatives exist simultaneously only on a countable set, where they are infinite and with opposite signs.

2.2 Strategy[1] *#1: Definition of a Function as Sum of a Series*

Fourier published in 1822 his "Théorie Analytique de la Chaleur" where he exposed the method of trigonometric series (Chap. III) in the context of an application to the solution of the heat equation. It is easy to imagine that the mathematicians of the time would be fascinated by this theory. In this context, many examples of *continuous and non-differentiable functions* began to appear, related to Fourier series.

These are uniformly convergent functional series whose differentiated series does not converge. We can see that the fractality is closely linked to non-differentiability from the beginning! Some famous curves of this type are the functions by Riemann, Weiertrass and Darboux.

[1] We call "strategies" the different ways of definition of self-similar curves.

The *Riemann's function (1861)* is defined as

$$R(x) = \sum_{n=1}^{\infty} \frac{\sin(n^2 x)}{n^2}.$$

Riemann said that he had a proof for the non-differentiability of $R(x)$ but it was never found. Hardy (1916) and Gerver (1970) completed the proof and generalized the model to

$$h(x) = \sum_{n=1}^{+\infty} \frac{sin(n^2 x)}{n^\alpha}.$$

If $\alpha < \frac{5}{2}$ $h(x)$ has only derivative at the points

$$x_{pq} = \pi \frac{2p+1}{2q+1}; \quad p, q \in \mathbb{Q}$$

whose value is $-\frac{1}{2}$. Riemann never published this work, although it is known that Christoffel knew the results. Neuenschwander found a reference of an application of the function in the Casorati's diary.

Weierstrass's function (1872) is expressed as

$$W(x) = \sum_{n=0}^{+\infty} b^n \cos(a^n \pi x),$$

$$0 < b < 1, \quad ab > 1 + 3\pi/2, \quad a > 1 \quad \text{odd integer.}$$

It was presented in a lecture given by the author in the Prussian Sciences Academy (July 18, 1872) and published by du Bois-Reymond (1875) and the translated version in a collection of works of this time (1895). It was the first model published (with the consequent impact sometimes negative) and extensively studied later on, with generalizations reaching our days (Mandelbrot–Berry) so that it is sometimes called the Weierstrass–Mandelbrot function.

The series converges uniformly, according to the criterion of the author (M) and thus it is a continuous function. Weierstrass proved that, at every point, the upper limit of the incremental quotient is $+\infty$ and the lower limit is $-\infty$. It was thoroughly studied by G.H. Hardy (1916). He proved that the map is not differentiable at every point if $0 < b < 1, a > 1, ab \geq 1$. The fractal (box-counting) dimension is

$$D = 2 + \frac{\ln a}{\ln b}.$$

Mandelbrot proposed a variant

$$f(x) = \sum_{n=-\infty}^{+\infty} b^n(1 - \cos(a^n \pi x))$$

satisfying the self-affine equation: $f(x) = bf(ax)$. It can be randomized introducing the phases

$$W_H(x) = \sum_{n=0}^{+\infty} w^{-nH} \cos(w^n x + \varphi_n),$$

where φ_n is any sequence. The fractal dimension is always $2 - H$, independently of φ_n. It can be used to model random motions.

The original of the article by Weierstrass can be found in the collection of G.A. Edgar "Classics in Fractals" [1].

3 Turn of the Century (19–20th)

3.1 Strategy #2: Definition of Functions by Means of the N-Adic Representation of the Variable

In this way of definition, the map is calculated by means of the representation of the independent variable in a certain base (usually 2, 3 or 4). The fact of using the digits complicates the differentiability but it does not prevent it. As particular cases we have the singular functions, with the property that they do not agree with the integral of their derivative. The best known are the Cantor and Minkowski maps.

The *Cantor's function (1884)* [1] is called sometimes the devil's staircase. It is defined using a ternary representation of x:

Let $x \in [0, 1]$ and $x = \sum_{n=1}^{+\infty} \frac{a_{nx}}{3^n}$, $a_{nx} \in \{0, 1, 2\}$. Let N_x be the smallest n with $a_{nx} = 1$ if it exists and $N_x = \infty$ otherwise, then

$$C(x) = \frac{1}{2^{N_x}} + \frac{1}{2} \sum_{n=1}^{N_x - 1} \frac{a_{nx}}{2^n}.$$

It was discovered by Ludwig Sheffer (student of Cantor). It is monotone (increasing) and uniformly continuous. According to Cantor's definition: "Its geometric image consists of a set of line segments each parallel to the x axis and some interpolated points making this curve continuous." The "interpolated points" correspond with the Cantor set of the line. Its length is equal to 2 and the fractal dimension is 1. (A fractal curve with $FD = 1$!!). Its derivative is null except in the Cantor set (almost everywhere) and is self-affine [3].

The *Minkowski's (question mark) function (1911)* presents a variant of the "arithmetic" strategy, since it uses the representation of the variable as a continued fraction:

If $x \in [0, 1]$ is irrational, $x = [0; a_1, a_2, \ldots]$

$$?(x) = 2\sum_{n=1}^{+\infty} \frac{(-1)^{n+1}}{2^{a_1+a_2+\ldots+a_n}}.$$

If $x \in [0, 1]$ is rational, $x = [0; a_1, a_2, \ldots, a_m]$

$$?(x) = 2\sum_{n=1}^{m} \frac{(-1)^{n+1}}{2^{a_1+a_2+\ldots+a_n}}.$$

It is another singular function known as "question mark" due to the notation used in its definition. The question mark function is increasing and Lipschitz continuous. The derivative only takes the value 0 and $+\infty$, but it is null almost everywhere. It satisfies the functional equations:

$$?\left(\frac{x}{x+1}\right) = \frac{?(x)}{2},$$

$$?(1-x) = 1-?(x).$$

Minkowski contributed to the fractal theory in many ways. One of them is the Minkowski–Bouligand (box-counting or fractal) dimension of a set (D_M).

Let (X, d) be a metric space and A compact, $A \subseteq X$. For any $\varepsilon > 0$ let $\mathcal{N}(A, \varepsilon)$ denote the smallest number of closed balls of radius $\varepsilon > 0$ needed to cover A. Then the fractal dimension D_M is

$$D_M = \lim_{\varepsilon \to 0+} \frac{\ln(\mathcal{N}(A, \varepsilon))}{\ln(1/\varepsilon)}.$$

This number agrees with the Hausdorff dimension D on many sets, otherwise $D_M > D$.

3.2 *Strategy #3: Definition of a Function in a Self-similar Geometric Way*

The turn of the century brought a block of continuous functions given by a geometric construction of fractal type (space filling curves by Peano, Hilbert, Lebesgue and the snowflake (Von Koch) curve).

The *Hilbert's function (1891)* is a space filling curve first published in Mathematische Annalen (38). Its properties are very similar to Peano's curve (continuous non-differentiable passing through every point of a square). The definition is geometrical. The paper has two pages and three figures explaining the first steps of the geometric construction. It is considered the first visualization of a fractal geometric definition.

The *Peano's curve* was presented on January *1890* in Mathematische Annalen: "Sur una courbe qui remplit toute une aire plane." Although included in the section of geometric constructions it was defined analytically by the author. The paper contains no figures. Peano proved that it is a parametric continuous curve passing through every point of the unit square.

$$\begin{aligned} P : [0,1] &\rightarrow [0,1] \times [0,1] \\ t &\rightarrow (x(t), y(t)) \end{aligned}$$

is defined

$$\text{if} \qquad t = 0.a_1a_2 \ldots a_n \ldots (3) \qquad \text{as} \begin{cases} x(t) = 0.b_1b_2 \ldots b_n \ldots (3) \\ y(t) = 0.c_1c_2 \ldots c_n \ldots (3) \end{cases}$$

$$b_n = \begin{cases} a_{2n-1} & \text{if } a_2 + a_4 + \ldots + a_{2n-2} \text{ is even} \\ 2 - a_{2n-1} & \text{otherwise.} \end{cases}$$

$$c_n = \begin{cases} a_{2n} & \text{if } a_1 + a_3 + \ldots + a_{2n-1} \text{ is even} \\ 2 - a_{2n} & \text{otherwise.} \end{cases}$$

The curve does not admit derivative at any point, but Peano did not prove it. $P(t)$ is independent of the ternary representation of t. It was the first space filling curve discovered. Peano thought that all the space filling curves were not differentiable but this is not true!!! The Hausdorff dimension is 2.

The *Lebesgue's curve (1904)* is another space filling curve, defined for

$$t = 0.a_1a_2a_3 \ldots a_n \ldots (3) \quad \text{as} \begin{cases} x(t) = C(0.a_1a_3 \ldots a_{2n-1} \ldots), \\ y(t) = C(0.a_2a_4 \ldots a_{2n} \ldots), \end{cases}$$

where C is the Cantor function. Its peculiarity is that it is differentiable ALMOST EVERYWHERE!!!

The *Koch's curve (1904)* is a continuous, self-similar, non-differentiable curve [1] consisting of four parts, each one similar to the whole curve. Its length is infinite. It was defined by his author as: "courbe continue sans tangent, obtenue par une construction géométrique élémentaire." The definition is illustrated by means of figures. The fractal dimension is

$$D = \frac{2 \ln 2}{\ln 3} = 1.2619 \ldots .$$

Koch was a great defender of the visual mathematics. Referring to Weierstrass's function he said: "... it seems to me that this example is not satisfactory from the geometrical point of view since the function is defined by an analytic expression that hides the geometrical nature of the corresponding curve and so from this point of view one does not see why the curve has no tangent." It was studied later by Levy (1938): "Plane or space curves and surfaces consisting of parts similar to the whole" who generalized the Koch's method [1].

4 First Half of the Twentieth Century

At the beginning of the twentieth century there is a return to origins: the series are taken up in order to provide continuous curves. The difference is in the general term:

$$\sum_{n=0}^{+\infty} a_n g(w_n t),$$

where $a_n \to 0$, g is continuous, periodic and non-smooth and $|a_n| w_n \to \infty$. For the uniform convergence is sufficient that $\sum |a_n| < +\infty$ and g bounded. Some authors working in this line are

- Takagi (1903).
- Knopp (1918).
- Schoenberg (1938).
- Orlicz (1947).
- McCarthy (1953).

4.1 *Julia's Function (1931)*

Gaston Julia is the main creator of nonlinear dynamics, chaos theory and fractals. This author proposed a non-smooth model:

$$J(z) = \sum_{n=0}^{+\infty} a_n R_n(z)$$

where $a_n \in \mathbb{C}$, $R_n = R \circ R \circ \overset{n)}{\ldots} \circ R$, and $R(z)$ rational. For some a_n and R, $J(z)$ is continuous and non-differentiable on the circle. He uses arguments of nonlinear dynamics (points attractifs et répulsifs, etc.). The Weierstrass model can be deduced from it.

4.2 Fractal Dimension

In the nineteenth century there were many proposals to the concept of dimension. Concerning the fractal dimension of graphs of continuous functions, it is relevant to the work by Besicovitch and Ursell (1937): "Sets of fractional dimensions (V). On dimensional numbers of some continuous functions" [1]:

Let f be a Hölder continuous function, that is to say, satisfying the inequality:

$$|f(x) - f(y)| \leq K|x - y|^{\delta},$$

where $K \in \mathbb{R}$ and $0 < \delta \leq 1$. The Hausdorff dimension D of the graph of f satisfies the relation:

$$1 \leq D \leq 2 - \delta.$$

The authors resume the Knopp model as well. Due to this fact the Knopp function is called sometimes Besicovitch–Ursell map.

4.3 *Strategy #4: Definition of a Curve as a Random Function*

The most important model of this type is the *Brownian motion*. The name is due to the botanist *Robert Brown* who observed in the microscope (1827) that the movement of the pollen bobbles of a plant was a continuous motion with constant changes of direction. *Louis Bachelier* studied mathematically the Brownian motion in his thesis: "Théorie de la spéculation." This date (March 29, 1900) is considered the day of birth of the financial mathematics as the author proposed the Brown function as a model associated to the stock prices.

It was described and employed furtherly by *Einstein, Perrin, Wiener*, etc. Mandelbrot generalized the concept to a fractional Brownian motion.

The time increments of a motion of this type are Gaussian variables. Almost all the sample paths are continuous (when t varies in a compact interval I). Almost all the trajectories are in fact Hölder continuous and, with probability one, the graph has both Hausdorff and box dimension equal to $2 - H$, where H is the Hurst exponent of the process (in the ordinary case $H = 0.5$).

5 Second Half of the Twentieth Century

5.1 *Strategy #5: Definition of a Curve by Means of a Functional Equation*

In this case, the maps are determined as solutions of a functional equation. The fact that, in many cases, they do not have an explicit (closed form) expression makes the differentiability difficult. An important contribution is the work by *Georges de Rham* [1]: "Sur quelques courbes définies par des équations fonctionnelles" (*1957*):

Let F_0, F_1 be two transformations of the plane. One looks for a mapping $M : I = [0,1] \to \mathbb{R}^2$ satisfying the functional equations:

$$M(t/2) = F_0 M(t); \qquad M((1+t)/2) = F_1 M(t).$$

The author proves that if F_0 and F_1 satisfy certain conditions (contractivity for instance), the problem has a bounded solution, being a unique and continuous curve. As a particular case of this problem, one obtains the Koch's curve, among others. This work has as precedents the articles:

- E. Césaro: "Fonctions continues sans dérivées", Archiv der Mathematik und Physik, 10, 57–63, 1906
- G. Faber: "Ueber stetige Functionen II", Mathematische Annalen 69, 372–443, 1910.

A variant of this strategy is the definition of a fractal function as fixed point of an operator:

If (E, d) is complete metric space of continuous functions and $T : E \to E$ is a contraction then there exists $f \in E$

$$f = Tf.$$

An important result concerning this equation is the Collage Theorem: If the contractivity factor is $c \in [0,1]$ and $g \in E$

$$d(g, f) \leq \frac{d(g, Tg)}{1-c}.$$

This inequality is important for Approximation of Functions.

5.2 *Strategy #6: Definition of a Curve by Means of an Iterated Function System*

Let K be a complete metric space with respect to the distance $d(x, y)$, for $x, y \in K$, and a set of continuous mappings:

$$w_n : K \to K \quad \text{for} \quad n = 1, 2, \ldots, N.$$

Then

$$\{K, w_n \,|\, n = 1, \ldots, N\}$$

is an *Iterated Function System* (IFS) on K. If w_n are contractions the System is Hyperbolic.

The set $\mathcal{H}$ of all compact not empty subsets of K is a complete metric space. Let us define the transformation of sets $W : \mathcal{H} \to \mathcal{H}$ as

$$W(A) = \bigcup_{n=1}^{N} w_n(A) \quad \forall A \in \mathcal{H}$$

Any set $G \in \mathcal{H}$ such that

$$G = W(G)$$

is an invariant set of the IFS (G is a fixed point of W). Furthermore, G is an attractor if

$$G = \lim_{m \to \infty} W^{(m)}(A) \qquad \forall A \in \mathcal{H}.$$

A typical mapping of this strategy is the *Kiesswetter's curve (1966)*, that can be generated as attractor of four affinities (IFS)

$$w\begin{pmatrix} x \\ y \end{pmatrix} = \begin{pmatrix} a & b \\ c & d \end{pmatrix}\begin{pmatrix} x \\ y \end{pmatrix} + \begin{pmatrix} e \\ f \end{pmatrix}$$

with the coefficients given in the table.

a	b	c	d	e	f
0.25	0	0	−0.5	0	0
0.25	0	0	0.5	0.25	−0.50
0.25	0	0	0.5	0.50	0
0.25	0	0	0.5	0.75	0

It is a continuous non-differentiable function very similar to Bolzano's map. The author did not use an IFS to define the curve. His definition is "arithmetic":

If $x = \sum x_n/4^n$ where $x_n = 0, 1, 2, 3$, then

$$f(x) = \sum (-1)^{N_n} X_n/2^n,$$

where

$$X_n = \begin{cases} x_n - 2 & \text{for} \quad x_n > 0 \\ 0 & \text{for} \quad x_n = 0 \end{cases}$$

and N_n is the number of x_k such that $x_k = 0$ and $k < n$.

The function is Hölder continuous with exponent $1/2$ and its fractal dimension reaches the upper bound provided by this exponent: $3/2$.

References

1. Bailland, B., Bourget, H.: Correspondance d'Hermite et de Stieltjes, Gauthier-Villars, Paris (1905)
2. Besicovitch, A.S., Ursell, H.: Sets of fractional dimensions (V): On dimensional numbers of some continuous curves, J. London Math. Soc. **12**, 18–25 (1937)
3. Dieudonné, J.: L'abstraction et l'intuituion mathématique, Dialectica **29**, 39–54 (1975)
4. duBois-Reymond, P.: Versuch einer Klassification der willkürlichen Functionen reeller Argumente nach ihren Änderungen in den kleinsten Intervallen. J. Reine Angew. Math. **79**, 21–37 (1875)
5. Edgar, G.A. (ed.): Classics on Fractals, Addison Wesley, Reading (1993)
6. Gerver, J.: The differentiability of the Riemann function at certain rational multiples of π, Amer. J. Math. **92**, 33–55 (1970)
7. Hardy, G.H.: Weierstrass's non-differentiable function, Trans. AMS **17**, 301–325 (1916)
8. Jarnik, V.: O funkci Bolzanove, Casopis Pest. Mat. **51**, 248–262 (1922)
9. Knopp, K.: Ein einfaches Verfahren zur Bildung stetiger nirgends differenzierbaren Funktionen, Math. Z. **2**, 1–26 (1918)
10. Lévy, P.: Quelques aspects de la pensée d'un mathématicien, Albert Blanchard, Paris (1970)
11. Lévy, P.: Plane or space curves and surfaces consisting of parts similar to the whole. In: Classics on Fractals, Edgar, G.A. (ed.), Addison-Wesley, Reading (1993)
12. McCarthy, J.: An everywhere continuous nowhere differentiable function, Amer. Math. Month. **LX(10)**, 709 (1953)
13. Orlicz, W.: Sur les fonctions continues non dérivables, Fund. Math. **34**, 45–60 (1947)
14. Peitgen, H.O., Jurgens, H., Saupe, D.: Chaos and Fractals, Springer, Berlin (1992)
15. Perrin, J.: La discontinuité de la matiere. Revue du mois **1**, 323–344 (1906)
16. Rychlik, K.: Über eine Funktion aus Bolzanos handschriftlichem Nachlasse, Sitzber. d. Böhm. Ges. d. Wiss. 1–20 (1921)
17. Schoenberg, I.J.: Metric spaces and completely monotone functions. Ann. Math. **39(4)**, 811–834 (1938)
18. Takagi, T.: A simple example of the continuous function without derivative, Tokyo sugaku butsurigaku kwai kiji **1**, 176–177 (1903)

A New Class of Rational Quadratic Fractal Functions with Positive Shape Preservation

A.K.B. Chand, P. Viswanathan, and M.A. Navascués

Abstract Fractal interpolation functions (FIFs) developed through iterated function systems prove more general than their classical counterparts. However, the theory of fractal interpolation functions in the domain of *shape preserving interpolation* is not fully explored. In this paper, we introduce a new kind of iterated function system (IFS) involving rational functions of the form $\frac{p_n(x)}{q_n(x)}$, where $p_n(x)$ are quadratic polynomials determined through the interpolation conditions of the corresponding FIF and $q_n(x)$ are preassigned quadratic polynomials involving one free shape parameter. The presence of the scaling factors in our rational FIF adds a layer of flexibility to its classical counterpart and provides fractality in the derivative of the interpolant. The uniform convergence of the rational quadratic FIF to the original data generating function is established. Suitable conditions on the rational IFS parameters are developed so that the corresponding rational quadratic fractal interpolant inherits the positivity property of the given data.

Keywords Fractals • Iterated function systems • Fractal interpolation functions • Rational quadratic FIF • Uniform convergence • Positivity

---

A.K.B. Chand (✉) • P. Viswanathan
Department of Mathematics, IIT Madras, Chennai, India
e-mail: chand@iitm.ac.in; amritaviswa@gmail.com

M.A. Navascués
Departmento de Matemática Aplicada, Escuela de Ingeniería y Arquitectura, Universidad de Zaragoza, Zaragoza, Spain
e-mail: manavas@unizar.es

C. Bandt et al. (eds.), *Fractals, Wavelets, and their Applications*, Springer Proceedings in Mathematics & Statistics 92, DOI 10.1007/978-3-319-08105-2_18,

1 Introduction

Mandelbrot [16] coined the term fractal for a new class of mathematical objects that does not easily fit into the classical Euclidean geometric settings. Fractal curves have been applied successfully in various problems of natural sciences and engineering for last 35 years. These curves are constructed deterministically [1, 2] by using fractal interpolation functions (FIFs) from suitable choices of iterated function systems (IFSs). Fractal interpolation defined through a functional equation provides a constructive way to model a prescribed data set, in contrast to the descriptive ways employed in the traditional interpolation techniques. The functional equation involved in the definition of a FIF gives self-similarity on small scales. Thus, FIFs are generally self-affine in nature, and their Hausdorff–Besicovitch dimensions are non-integers. These features enable FIFs to provide non-smooth approximants suitable for data that define dynamic relations corresponding to real processes.

Barnsley and Harrington [3] observed that a FIF can be indefinitely integrated any number of times to yield a hierarchy of smoother functions, and consequently developed differentiability of a FIF. The above mentioned observation has a paramount importance in the construction of smooth fractal interpolants, and initiated a striking relationship between the classical splines and fractal functions. By allowing the admissibility of various types of boundary conditions, Chand and Kapoor [4] generalized the above construction of spline FIFs, and developed cubic spline FIFs through moments. Various classical interpolation schemes emerge as special cases of smooth fractal interpolants [5–8, 17], and therefore FIFs provide satisfactory generalization for the classical non-recursive smooth interpolants. Fractal splines provide a single specification method to obtain a very large class of interpolants with significant differences in their visual properties, which can be effectively utilized in the geometric modelling and design environment. Moreover, if an experimental data set is approximated by a spline FIF, then the fractal dimension of the graph of a certain derivative of this FIF can be used as an index for the analysis of the complexity in given data. For a brief review of fractal interpolation and related developments, the reader is invited to refer a recent short survey article by the authors [18].

Standard interpolation techniques available in the classical numerical analysis/fractal theory indeed explore a few characteristics of the data, for instance, scalar/vector nature of the data, scattered/ordered distribution of the data, and degree of smoothness. However, they often violate many additional qualitative characteristics hidden in the given data. To obtain a valid physical interpretation of the underlying process, it is important to develop interpolation schemes that honor such properties present in the data, particularly when the data are produced by some scientific phenomena. Examples of prevalent shape features of the data are positivity, monotonicity, and convexity.

Since an interpolation method which accurately represents physical reality is a demand in design and manufacturing of products such as car bodies, aircrafts, and ship hulls, in modelling of paths of particles, in abstract and physical processes, in economics, in social and physical sciences, in the description of geological and medical phenomena, etc., the development of interpolation and approximation techniques that preserve these shape characteristics is inevitable. Preserving positivity is most common and fundamental requirement especially for the visualization of data representing physical quantities that cannot be negative. In computer aided geometric design, modification of the interpolating curve demanded by engineering applications is a major issue. But, uniqueness of the interpolant makes it impossible to modify the shape under the condition that the given interpolating data is unchanged. This gives the importance of functions with shape parameters. A vast literature is available in the area of shape preserving interpolation, where the interpolants involved are the classical polynomial and rational splines with or without shape parameters [10–14, 19, 20]. Rational interpolation is well known for its better interpolating properties with less oscillatory nature, easiness with which it fits into shape modification problems and its excellent asymptotic properties compared to the polynomial models.

In this paper, we develop a new kind of rational quadratic FIF, where interpolant involves rational functions of the form $\frac{p_n(x)}{q_n(x)}$, where $p_n(x)$ are quadratic polynomials determined via interpolation conditions of the FIF and $q_n(x)$ are preassigned quadratic polynomials containing one free shape parameters. This rational FIF has zeroth order continuity. The order of continuity is increased up to first order by appropriate choice of the shape parameters. Despite its implicit nature, some simple conditions on the scaling factors and shape parameters are developed to enable automatic generation of positivity preserving rational FIFs. The presence of the scaling factors provides an additional pliability in the shape modification. It is observed that the computational complexity for the implementation of the positivity preserving fractal interpolation scheme is not too high when compared with that of its classical counterpart.

The rest of the paper is organized as follows: In Sect. 2, the notion of FIF is briefly reviewed. In Sect. 3, we introduce the rational quadratic spline FIFs with one family of shape parameters. An upper bound of the interpolation error for the rational quadratic FIF is obtained and as a consequence the uniform convergence to the original function as the interpolation step tends to zero is proved in Sect. 4. In Sect. 5, suitable values of the IFS parameters are identified so that the FIF generates positive curves for the given positive data. Further, the positivity preserving rational quadratic fractal interpolation scheme is demonstrated with some numerical examples.

2 Fractal Interpolation Function

In this section, an overview of FIF is given. For further details, the reader is referred to the well-known treatise [1, 2].

Consider a data set $\{(x_n, y_n) \in I \times \mathbb{R} : n = 1, 2, \ldots, N\}$, where $x_1 < x_2 < \cdots < x_N$, $N > 2$ and $I = [x_1, x_N] \subseteq \mathbb{R}$ is a closed interval in $\mathbb{R}$. For $n \in J = \{1, 2, \ldots, N-1\}$, let $I_n = [x_n, x_{n+1}]$ and consider the contractive homeomorphisms $L_n : I \longrightarrow I_n$ with subinterval end point conditions

$$L_n(x_1) = x_n, \quad L_n(x_N) = x_{n+1}. \tag{1}$$

Let $F_n : I \times \mathcal{R} \to \mathcal{R}$ be functions which are continuous in the first variable, contraction in the second variable with contractive factors $0 \leq |\alpha_n| < 1$, and satisfy

$$F_n(x_1, y_1) = y_n, \quad F_n(x_N, y_N) = y_{n+1} \ \forall \ n \in J. \tag{2}$$

Let $w_n : I \times \mathbb{R} \to I \times \mathbb{R}$ be defined by $w_n(x, y) = \big(L_n(x), F_n(x, y)\big)$.

By Barnsley's theorem [1], the IFS $\{I \times \mathbb{R}; w_n, \ n \in J\}$ defined above admits a unique attractor G and G is the graph of a continuous function $f : I \to \mathbb{R}$ which satisfies $f(x_n) = y_n$, $n = 1, 2, \ldots, N$. This function f is called a FIF corresponding to the IFS $\{I \times \mathbb{R}; w_n(x, y) = \big(L_n(x), F_n(x, y)\big), \ n \in J\}$. Let $\mathcal{F} = \{g : I \to \mathbb{R} |\ g$ is continuous, $g(x_1) = y_1$ and $g(x_N) = y_N\}$ and $\alpha = (\alpha_1, \alpha_2, \ldots, \alpha_{N-1})$. Then $\mathcal{F}$ endowed with the uniform metric $d_\infty(g, g^*) := \max\{|g(x) - g^*(x)| : x \in I\}$ is a complete metric space. The FIF f is the unique fixed point of the Read–Bajraktarević operator T_α on $(\mathcal{F}, d_\infty)$ defined by

$$T_\alpha g(x) = F_n\big(L_n^{-1}(x), g \circ F_n^{-1}(x)\big), \ x \in I_n, \ \ n \in J. \tag{3}$$

Therefore the FIF f satisfies the functional equation:

$$f(x) = F_n\big(L_n^{-1}(x), f \circ L_n^{-1}(x)\big), \ \ x \in I_n, \ \ n \in J. \tag{4}$$

The IFS $\{I \times \mathbb{R}; w_n(x, y) = \big(L_n(x), F_n(x, y)\big), n \in J\}$ arising from the following special type of mappings L_n and F_n is widely studied in the literature.

$$L_n(x) = a_n x + b_n, \ \ F_n(x, y) = \alpha_n y + R_n(x), \tag{5}$$

where $|\alpha_n| < 1$ and $R_n : I \to \mathbb{R}$ are suitable continuous functions such that conditions prescribed in (2) are satisfied. The free parameters α_n are called the scaling factors. The existence of differentiable polynomial spline FIFs is given in [3]. This result can be extended in the same lines to the rational functions and is described in the following:

Theorem 2.1. *Let* $\{(x_n, y_n) : n = 1, 2, \ldots, N\}$, $N > 2$ *be a given data set. Suppose that* $L_n(x) = a_n x + b_n, F_n(x, y) = \alpha_n y + R_n(x), R_n(x) = \dfrac{U_n(x)}{V_n(x)}$, $U_n(x), V_n(x)$ *are suitably chosen polynomials, and* $V_n(x) \neq 0$ for every $x \in I$. *Suppose that for some integer* $p \geq 0, |\alpha_n| < a_n^p, n \in J$. *Let* $F_{n,m}(x, y) = \frac{\alpha_n y + R_n^{(m)}(x)}{a_n^m}$, $y_{1,m} = \frac{R_1^{(m)}(x_1)}{a_1^m - \alpha_1}$, $y_{N,m} = \frac{R_{N-1}^{(m)}(x_N)}{a_{N-1}^m - \alpha_{N-1}}$, $m = 1, 2, \ldots, p$. *If the join-up conditions* $F_{n-1,m}(x_N, y_{N,m}) = F_{n,m}(x_1, y_{1,m}), n = 2, 3, \ldots, N-1$, $m = 1, 2, \ldots, p$, *are satisfied, then* $\{I \times \mathbb{R}; w_n(x, y) = \big(L_n(x), F_n(x, y)\big), \ n \in J\}$ *determines a rational FIF* $\Phi \in \mathcal{C}^p(I)$, *and* $\Phi^{(m)}$ *is the rational FIF determined by* $\{I \times \mathbb{R}; w_{n,m}(x, y) = \big(L_n(x), F_{n,m}(x, y)\big), \ n \in J\}$ *for* $m = 1, 2, \ldots, p$.

3 Rational Quadratic FIF with Shape Parameters

3.1 Construction

Let $\{(x_n, y_n) : n = 1, 2, \ldots, N\}$, $N > 2$ be a given set of data points, where $x_1 < x_2 < \cdots < x_N$. To construct the rational quadratic FIF with one family of shape parameters, let us proceed as follows:

Let $\mathcal{F} = \{f \in \mathcal{C}(I) | \ f(x_1) = y_1 \text{ and } f(x_N) = y_N\}$. Then, $\mathcal{F}$ endowed with the uniform metric is complete. Let $\alpha = (\alpha_1, \alpha_2, \ldots \alpha_{N-1}) \in (-1, 1) \times (-1, 1) \times \ldots \times (-1, 1) \subset \mathbb{R}^{N-1}$. Suppose the Read–Bajraktarević operator $T_\alpha : \mathcal{F} \longrightarrow \mathcal{F}$ is defined as

$$T_\alpha f\big(L_n(x)\big) = \alpha_n f(x) + R_n(x), \ x \in I, \ n \in J, \tag{6}$$

where $L_n(x) = a_n x + b_n$ satisfies (1) for all $n \in J$. Then we have $a_n = \frac{x_{n+1}-x_n}{x_N-x_1}$ and $b_n = \frac{x_N x_n - x_1 x_{n+1}}{x_N-x_1}$. For suitably chosen R_n, T_α is a contraction map on $\mathcal{F}$. The fixed point S of T_α is a FIF, and it satisfies the functional equation:

$$S\big(L_n(x)\big) = \alpha_n S(x) + R_n(x), \ x \in I, \ n \in J. \tag{7}$$

For the desired new kind of rational FIF, we choose R_n in the following format:

$$R_n(x) = \frac{A_n(1-\theta)^2 + B_n\theta(1-\theta) + C_n\theta^2}{1 + (r_n - 2)\theta(1-\theta)}, \ \theta = \frac{x - x_1}{x_N - x_1}. \tag{8}$$

In order to get differentiability of S, we assume that $|\alpha_n| < a_n \ \forall \ n \in J$ (see Theorem 2.1). The parameters $r_n > -2$ called shape parameters, ensure a strict positive denominator in the rational expression R_n. The constants A_n, B_n, and C_n appearing in the numerator of R_n are evaluated based on the interpolatory conditions: $S(x_n) = y_n, S(x_{n+1}) = y_{n+1}$ and $S^{(1)}(x_n^+) = d_n$. These conditions determine the constants uniquely as follows. Let $h_n = x_{n+1} - x_n$. Substituting $x = x_1$ in (7) and using (1),

$$S\big(L_n(x_1)\big) = \alpha_n S(x_1) + A_n \Longrightarrow y_n = \alpha_n y_1 + A_n \Longrightarrow A_n = y_n - \alpha_n y_1.$$

Similarly, substituting $x = x_N$ in (7) and using (1), we obtain: $C_n = y_{n+1} - \alpha_n y_N$. From (7),

$$a_n S^{(1)}\big(L_n(x)^+\big) = \alpha_n S^{(1)}(x^+) + R_n^{(1)}(x^+).$$

Taking $x = x_1$ and simplifying the above expression using (8), we obtain $B_n = r_n y_n + h_n d_n - \alpha_n[r_n y_1 + (x_N - x_1)d_1]$. With these values of A_n, B_n, and C_n, Eqs. (7)–(8) produce the desired rational quadratic FIF:

$$S\big(L_n(x)\big) = \alpha_n S(x) + \frac{U_n(x)}{V_n(x)}, \tag{9}$$

$$U_n(x) \equiv u_n(\theta) = \{y_n - \alpha_n y_1\}(1-\theta)^2 + \{r_n y_n + h_n d_n - \alpha_n [r_n y_1 + (x_N - x_1) d_1]\}$$

$$\theta\,(1-\theta) + \{y_{n+1} - \alpha_n y_N\}\,\theta^2,$$

$$V_n(x) \equiv v_n(\theta) = 1 + (r_n - 2)\theta(1-\theta),\ \theta = \frac{x - x_1}{x_N - x_1}.$$

Remark 3.1. When $\alpha_n = 0\ \forall\ n \in J$, the above rational quadratic FIF reduces to the classical rational quadratic interpolant C (see [15]), which takes the following form on $[x_n, x_{n+1}]$:

$$C(x) = \frac{P_n^*(\phi)}{Q_n^*(\phi)}, \quad \phi = \theta\big(L_n^{-1}(x)\big) = \frac{L_n^{-1}(x) - x_1}{x_N - x_1} = \frac{x - x_n}{h_n}, \tag{10}$$

$$P_n^*(\phi) = y_n(1-\phi)^2 + (r_n y_n + h_n d_n)\phi(1-\phi) + y_{n+1}\phi^2,$$

$$Q_n^*(\phi) = 1 + (r_n - 2)\phi(1-\phi).$$

When $\alpha_n = 0$ and $r_n = 2\ \forall\ n \in J$, we obtain the classical quadratic polynomial interpolant.

To implement our rational fractal interpolation scheme, we need derivative parameters $d_n, n \in J$. If d_n are not given, then we estimate them by some numerical approximation (see, for instance, [9]). In this article, we use arithmetic mean method (amm) based on the three-point difference approximation. With the notation $\Delta_n = \frac{y_{n+1} - y_n}{h_n}, n \in J$, the amm is described as follows.

At the interior point x_n $(n = 2, 3, \ldots, N-1)$,

$$d_n = \begin{cases} 0, & \text{if } \Delta_n = 0 \text{ or } \Delta_{n-1} = 0, \\ \frac{h_n \Delta_{n-1} + h_{n-1}\Delta_n}{h_n + h_{n-1}}, & \text{otherwise.} \end{cases}$$

At the end points x_1 and x_n,

$$d_1 = \begin{cases} 0, & \text{if } \Delta_1 = 0 \text{ or } \operatorname{sgn}(d_1^*) \neq \operatorname{sgn}(\Delta_1), \\ d_1^* = \Delta_1 + \frac{(\Delta_1 - \Delta_2)h_1}{h_1 + h_2}, & \text{otherwise.} \end{cases}$$

$$d_n = \begin{cases} 0, & \text{if } \Delta_{N-1} = 0 \text{ or } \operatorname{sgn}(d_N^*) \neq \operatorname{sgn}(\Delta_{N-1}), \\ d_N^* = \Delta_{N-1} + \frac{(\Delta_{N-1} - \Delta_{N-2})h_{N-1}}{h_{N-1} + h_{N-2}}, & \text{otherwise.} \end{cases}$$

For a prescribed bounded data, the above derivative approximations are bounded. Hence, if the values of the scaling factors and shape parameters are fixed, then the rational quadratic FIF defined in (9) exists, and is unique. The above discussion leads to the following existence and uniqueness theorem.

Theorem 3.1. *For bounded values of the shape parameters $r_n > -2$, the scaling factors $|\alpha_n| < a_n$ $(n \in J)$, and with the bounded derivative approximations, the rational quadratic FIF (9) with one family of shape parameter exists, and for a fixed set of parameters, the interpolant is unique.*

3.2 C^1-Smooth Rational FIF

The above rational quadratic FIF preserves only the zeroth order continuity. The quadratic polynomial involved in the definition of rational FIF S does not provide sufficient degrees of freedom to make $S^{(1)}$ to be continuous (with arbitrary values of the shape parameters) by demanding $S^{(1)}(x_n) = d_n$, $S^{(1)}(x_{n+1}) = d_{n+1}$, $n \in J$. We have the following theorem for the fractal interpolant S to be in $C^1(I)$.

Theorem 3.2. *Let $\{(x_n, y_n) : n = 1, 2, \ldots, N\}$ be given data and S be the corresponding rational quadratic FIF defined in (9). Then S preserves C^1-continuity if the scaling factors and shape parameters satisfy the following conditions:*

$$|\alpha_n| < a_n, \; r_n = \frac{h_n(d_n + d_{n+1}) - \alpha_n(x_N - x_1)(d_1 + d_N)}{y_{n+1} - y_n - \alpha_n(y_N - y_1)} \quad \forall\, n \in J. \qquad (11)$$

Here we assume that the shape parameters r_n given by the above equation satisfy $r_n > -2$.

Proof. The condition $|\alpha_n| < a_n$ is in accordance with the principle of construction of C^1-FIF (see Theorem 2.1). To get the first order continuity for S, we need to impose the following interpolatory conditions: For $n \in J$,

$$S(x_n) = y_n, \; S(x_{n+1}) = y_{n+1}, \; S^{(1)}(x_n^+) = d_n, \; S^{(1)}(x_{n+1}^-) = d_{n+1}. \qquad (12)$$

First three constraints of Eq. (12) are satisfied by the rational quadratic FIF given in (9). As $x \longrightarrow x_N^-$ we have $L_n(x) \longrightarrow x_{n+1}^-$, and consequently from (9)

$$a_n S^{(1)}(x_{n+1}^-) = \alpha_n\, S^{(1)}(x_N^-) + \frac{1}{x_N - x_1}\{-(r_n y_n + h_n d_n) + \alpha_n[r_n y_1 + d_1(x_N - x_1)] + r_n(y_{n+1} - \alpha_n y_N)\}. \qquad (13)$$

Now we impose the additional condition prescribed in (12) to achieve desired smoothness.

With some basic algebraic manipulations, we deduce the condition on r_n as

$$r_n = \frac{h_n d_{n+1} + h_n d_n - \alpha_n(x_N - x_1)(d_1 + d_N)}{y_{n+1} - y_n - \alpha_n(y_N - y_1)}, \qquad (14)$$

where α_n is selected such that $y_{n+1} - y_n - \alpha_n(y_N - y_1) \neq 0$. □

As a special case of the preceding theorem, the following corollary is worth singling out.

Corollary 3.1. *Let $\{(x_n, y_n) : n = 1, 2, \ldots, N\}$ be a data set with $\Delta_n \neq 0\ \forall\ n \in J$. Then the corresponding classical rational quadratic function C defined in (10) preserves $\mathcal{C}^1$-continuity if and only if $r_n = \frac{d_{n+1}+d_n}{\Delta_n}\ \forall\ n \in J$.*

Proof. Sufficiency part follows from the Theorem 3.2 by taking $\alpha_n = 0$ for all $n \in J$. For the necessary part observe that, if r_n is not chosen as specified in the corollary, then $C^{(1)}(x_n^-) \neq C^{(1)}(x_n^+)$, and hence C fails to be differentiable at the interior knot points. □

Examples. Consider an equally spaced data set $\{(0.5, 0.4804), (0.6, 0.5669), (0.7, 0.7262), (0.8, 0.1)\}$. The estimates for the derivative parameters obtained by the amm are: $d_1 = 0.5010$, $d_2 = 1.2290$, $d_3 = -2.3345$, and $d_4 = -10.1895$. The scaling factors α_n satisfying $|\alpha_n| < a_n$ are selected as follows: $\alpha_1 = 0.32, \alpha_2 = 0.1, \alpha_3 = 0.25$. The IFS associated with the shape parameters $r_1 = 0.5$, $r_2 = 0$, and $r_3 = -1$ is given by the mappings:

$$L_1(x) = 0.333x + 0.333, \quad q_1(x) = \frac{0.327(1-\theta)^2 + 0.165\theta(1-\theta) + 0.535\theta^2}{1 - 1.5\theta(1-\theta)},$$

$$L_2(x) = 0.333x + 0.433, \quad q_2(x) = \frac{0.519(1-\theta)^2 + 0.108\theta(1-\theta) + 0.716\theta^2}{1 - 2\theta(1-\theta)},$$

$$L_3(x) = 0.333x + 0.533, \quad q_3(x) = \frac{0.606(1-\theta)^2 - 0.877\theta(1-\theta) + 0.075\theta^2}{1 - 3\theta(1-\theta)},$$

where $\theta = \frac{10x-5}{3}$. The iterations of the IFS code generate $\mathcal{C}^0$-rational quadratic FIF in Fig. 1a. Taking $\alpha_n = 0$ for $n = 1, 2, 3$, and the shape parameters as in Fig. 1a we obtain the classical $\mathcal{C}^0$-rational quadratic interpolant in Fig. 1b. With the same scale vector as in Fig. 1a, and the shape parameters as per the prescription in (14) $\big($i.e., $r_1 = 5.2975, r_2 = 0.9127, r_3 = 0.99\big)$ the functional equation (9) is iterated to generate the $\mathcal{C}^1$-rational quadratic FIF in Fig. 1c. With $\alpha_1 = \alpha_2 = \alpha_3 = 0$ and the shape parameters chosen according to Corollary 3.1 $\big($ i.e., $r_1 = 2$, $r_2 = -0.6940$, $r_3 = 2\big)$, we generate the corresponding classical $\mathcal{C}^1$-rational quadratic spline in Fig. 1d. Therefore the fractal methodology generalizes the classical interpolant and produces different types of functions depending on the nature of the scaling factors and shape parameters. The following differences between rational quadratic FIF S and its classical counterpart C are worth mentioning. Even with arbitrary choice of the shape parameters satisfying $r_n > -2$, $C \in \mathcal{C}^0(I)$ fails to be differentiable only at the interior knot points whereas $S \in \mathcal{C}^0(I)$ may fail to be differentiable at an infinite number of points. Similarly, with special choices of the shape parameters as described in Theorem 3.2, the derivative $S^{(1)}$ of the rational quadratic FIF $S \in \mathcal{C}^1(I)$ may be nondifferentiable in a dense subset of I. On the other hand, the derivative $C^{(1)}$ of the classical rational quadratic interpolant $C \in \mathcal{C}^1(I)$ is infinitely differentiable in each of the open subintervals determined by the knot points. In fact, as $|\alpha_n|$, $n \in J$, increase from zero the irregularity (fractality) of the derivative $S^{(1)}$ of $S \in \mathcal{C}^1(I)$ quantified by means of fractal dimension increases. This can be

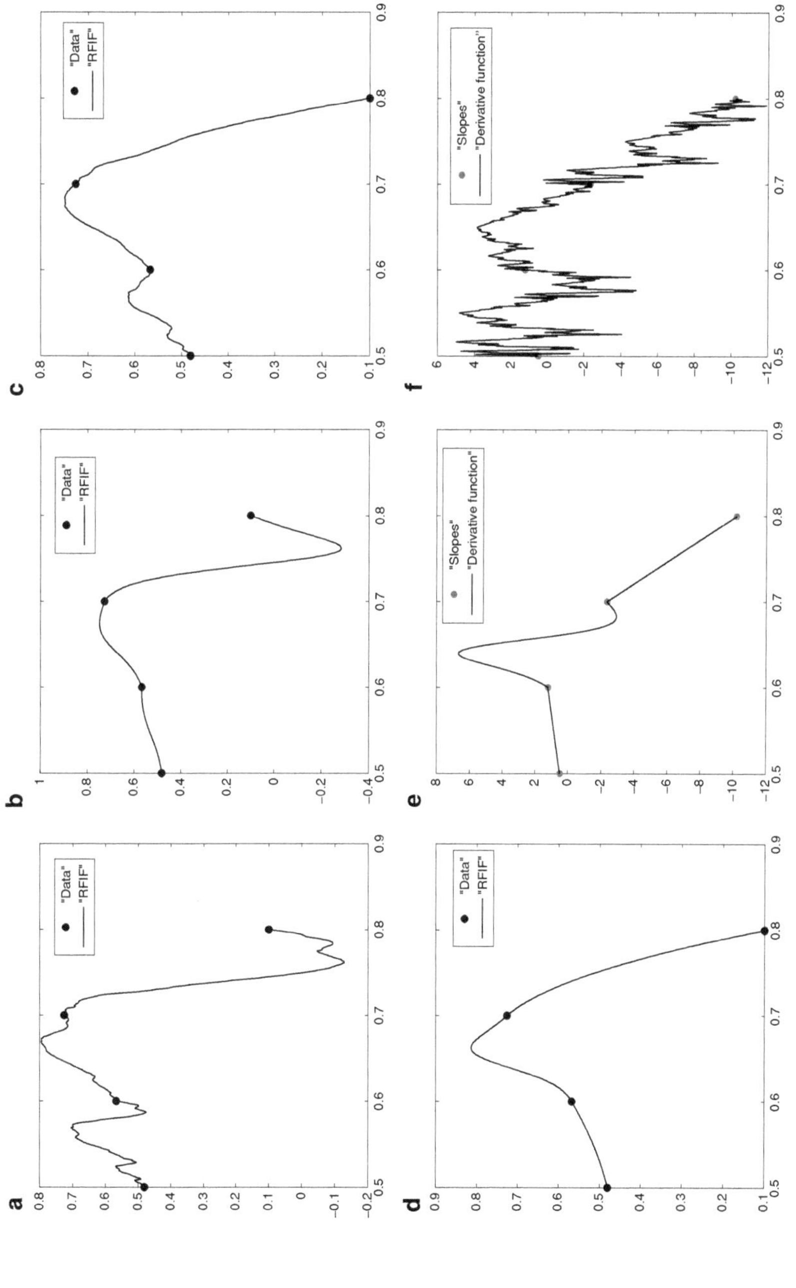

Fig. 1 Rational quadratic FIFs and derivatives. **(a)** C^0-rational quadratic FIF S_1. **(b)** Classical C^0-rational quadratic interpolant C_1. **(c)** C^1-rational quadratic FIF S_2. **(d)** Classical C^1-rational quadratic interpolant C_2. **(e)** Derivative of C_2. **(f)** Derivative of S_2

observed from Fig. 1e and f which represent the derivatives of the FIFs in Fig. 1d and c respectively. Thus our procedure expands the models of $\mathcal{C}^1$-smooth interpolation including the interpolants whose derivatives can even be non-differentiable in a dense set of points of the domain.

4 Convergence Analysis

Suppose the interpolation data are generated from the original function $\Phi \in \mathcal{C}^1(I)$. In this section, we shall prove the uniform convergence of the rational quadratic FIF S to Φ. First we estimate the pointwise error between the classical quadratic interpolant and Φ in the following:

Proposition 4.1. *Let $\Phi \in \mathcal{C}^1(I)$ and let C be the classical rational quadratic interpolant as described in (10), where d_n denotes the exact or approximate derivative of Φ at the knot x_n. Then for $x \in [x_n, x_{n+1}]$, $n \in J$,*

$$|\Phi(x) - C(x)| \leq \frac{1}{4\beta_n}\{(|r_n| + 8)\omega(f, h) + h_n|d_n|\},$$

where $\omega(f, h)$ represents the modulus of continuity of the function f, defined as $\omega(f, h) = \sup\{|f(x) - f(x')| : |x - x'| \leq h\}$, $h = \max\{h_n : n \in J\}$ and

$$\beta_n = \begin{cases} \frac{r_n+2}{4}, & \text{if } -2 < r_n < 2, \\ 1, & \text{if } r_n \geq 2. \end{cases}$$

Proof. For $x \in [x_n, x_{n+1}]$, $x = x_n + \phi h_n$. Then

$$\begin{aligned} \Phi(x) - C(x) &= \Phi(x) - \frac{P_n^*(\phi)}{Q_n^*(\phi)}, \\ &= \frac{[1 + (r_n - 2)\phi(1-\phi)]\Phi(x) - y_n(1-\phi)^2 - (r_n y_n + h_n d_n)\phi(1-\phi) - y_{n+1}\phi^2}{1 + (r_n - 2)\phi(1-\phi)}, \\ &= \frac{[(1-\phi)^2 + \phi^2 + r_n\phi(1-\phi)]\Phi(x) - y_n(1-\phi)^2 - (r_n y_n + h_n d_n)\phi(1-\phi) - y_{n+1}\phi^2}{1 + (r_n - 2)\phi(1-\phi)}, \\ &= \frac{(1-\phi)^2(\Phi(x) - y_n) + \phi^2(\Phi(x) - y_{n+1}) + r_n\phi(1-\phi)(\Phi(x) - y_n) - h_n d_n\phi(1-\phi)}{1 + (r_n - 2)\phi(1-\phi)}. \end{aligned}$$

Now we calculate a lower bound for $Q_n^*(\phi)$ as follows:

$$|Q_n^*(\phi)| = Q_n^*(\phi) = 1 + (r_n - 2)\phi(1-\phi) \geq \begin{cases} \frac{2+r_n}{4}, & \text{if } -2 < r_n < 2, \\ 1, & \text{if } r_n \geq 2. \end{cases} \tag{15}$$

Using (15) in the expression of $\Phi(x) - C(x)$, we have

$$\begin{aligned}|\Phi(x) - C(x)| \leq & \frac{1}{\beta_n}\{(1-\phi)^2|\Phi(x) - y_n| \\ & + \phi^2|\Phi(x) - y_{n+1}| + |r_n|\phi(1-\phi)|\Phi(x) - y_n| \\ & + h_n|d_n|\phi(1-\phi)\} \\ \leq & \frac{1}{\beta_n}\{2\omega(f,h) + \frac{|r_n|}{4}\omega(f,h) + \frac{1}{4}h_n|d_n|\}.\end{aligned}$$

Hence, we have the desired estimate. □

Theorem 4.1. *Let $\Phi \in \mathcal{C}^1(I)$ be the original data generating function. Let S and C, respectively, be the rational quadratic FIF and classical rational quadratic interpolant for Φ with respect to the interpolation data $\{(x_n, y_n) : n = 1, 2, \ldots, N\}$. Suppose that the rational quadratic function $R_n(\alpha_n, r_n, x)$ associated with the IFS generating the FIF S satisfies $\left|\frac{\partial R_n(\tau_n, r_n, x)}{\partial \alpha_n}\right| \leq K$ for $|\tau_n| \in (0, a_n)$ and $n \in J$. Then,*

$$\|\Phi - S\|_\infty \leq \frac{1}{4\beta}\{(|r|_\infty + 8)\omega(f,h) + h|d|_\infty\} + \frac{|\alpha|_\infty}{1-|\alpha|_\infty}(\|C\|_\infty + K),$$

where $\beta = \min\{\beta_n : n \in J\}$, $|r|_\infty = \max\{|r_n| : n \in J\}$, and $|d|_\infty = \max\{|d_n| : 1 \leq n \leq N\}$. In particular, S converges to Φ uniformly as the interpolation step tends to zero, if r_n is chosen such that $r_n \geq -1$ and if $|r|_\infty$, $|d|_\infty$ are bounded.

Proof. The rational quadratic FIF $S \in \mathcal{C}(I)$ is the fixed point of the Read–Bajraktarević operator T_α defined on the space $\mathcal{F} = \{f \in \mathcal{C}(I) | \ f(x_1) = y_1$ and $f(x_N) = y_N\}$ such that

$$T_\alpha f(x) = \alpha_n f(L_n^{-1}(x)) + R_n\big(\alpha_n, r_n, L_n^{-1}(x)\big), \tag{16}$$

where $R_n\big(\alpha_n, r_n, L_n^{-1}(x)\big) = \frac{u_n(\alpha_n, r_n, \phi)}{v_n(r_n, \phi)}$, $\phi = \frac{x - x_n}{h_n}$, $x \in I_n$ with u_n and v_n described as in (9). Let $\Pi = [-\kappa a_1, \kappa a_1] \times [-\kappa a_2, \kappa a_2] \times \ldots \times [-\kappa a_{N-1}, \kappa a_{N-1}], 0 \leq \kappa < 1$. For a given $\alpha = (\alpha_1, \alpha_2, \ldots, \alpha_{N-1}) \in \Pi \subseteq \mathbb{R}^{N-1}$ with at least one $\alpha_n \neq 0$, the rational quadratic FIF S is the fixed point of T_α, and for $\alpha^* = (0, 0, \ldots, 0) \in \Pi$, the classical rational quadratic C is the fixed point of T_{α^*}. From (9) and (16), for $x \in \mathcal{D}_n$, we have

$$\frac{\partial R_n}{\partial \alpha_n} = \frac{\bar{U}_n(\phi)}{V_n(\phi)}, \quad \bar{U}_n(\phi) = -\{y_1(1-\phi)^2 + [r_n y_1 + (x_N - x_1)d_1]\phi(1-\phi) + y_N\phi^2\}.$$

Now,

$$
\begin{aligned}
|\bar{U}_n(\phi)| &\le |y_1|(1-\phi)^2 + [|r_n||y_1| + |d_1|(x_N - x_1)]\phi(1-\phi) + |y_N|\phi^2,\\
&\le \max\{|y_1|, |y_N|\} + \frac{1}{4}[|r_n||y_1| + |d_1|(x_N - x_1)],\\
\Longrightarrow \max_{n\in J}|\bar{U}_n(\phi)| &\le \max\{|y_1|, |y_N|\} + \frac{1}{4}[|r|_\infty|y_1| + |d_1|(x_N - x_1)].
\end{aligned}
$$

Consequently, $\left|\frac{\partial R_n(\tau_n, r_n, \phi)}{\partial \alpha_n}\right| \le \dfrac{\max\{|y_1|, |y_N|\} + \frac{1}{4}[|r|_\infty|y_1| + |d_1|(x_N - x_1)]}{\beta} := K.$

For $x \in I_n$, using (16) and the mean value theorem, we have the following estimation:

$$
\begin{aligned}
|T_\alpha C(x) - T_{\alpha^*}C(x)| &= \left|\alpha_n C(L_n^{-1}(x)) + R_n(\alpha_n, r_n, \phi) - R_n(0, r_n, \phi)\right|,\\
&\le |\alpha_n|\|C\|_\infty + |\alpha_n|\left|\frac{\partial R_n(\tau_n, r_n, \phi)}{\partial \alpha_n}\right|,\\
&\le |\alpha_n|(\|C\|_\infty + K).
\end{aligned}
$$

From the above inequality we obtain

$$\|T_\alpha C - T_{\alpha^*}C\|_\infty \le |\alpha|_\infty(\|C\|_\infty + K). \tag{17}$$

Since T_α is a contractive map with contractive factor $|\alpha|_\infty$, for a fixed choice of scale vector $\alpha \ne \alpha^* \in \Pi$, we have

$$\|T_\alpha S - T_\alpha C\|_\infty \le |\alpha|_\infty\|S - C\|_\infty. \tag{18}$$

Using (17) and (18),

$$
\begin{aligned}
\|S - C\|_\infty = \|T_\alpha S - T_{\alpha^*}C\|_\infty &\le \|T_\alpha S - T_\alpha C\|_\infty + \|T_\alpha C - T_{\alpha^*}C\|_\infty,\\
&\le |\alpha|_\infty\|S - C\|_\infty + |\alpha|_\infty(\|C\|_\infty + K),
\end{aligned}
$$

which on simplification yields

$$\|S - C\|_\infty \le \frac{|\alpha|_\infty(\|C\|_\infty + K)}{1 - |\alpha|_\infty}. \tag{19}$$

From Proposition 4.1, we have

$$\|\Phi - C\|_\infty \le \frac{1}{4\beta}\{(|r|_\infty + 8)\omega(f, h) + h|d|_\infty\}. \tag{20}$$

Using (19)–(20) in $\|\Phi - S\|_\infty \leq \|\Phi - C\|_\infty + \|C - S\|_\infty$, we obtain the desired estimation. Note that $r_n \geq -1$ gives $\beta_n \geq \frac{1}{4}$, preventing that β goes to zero as partition tends to zero. Consequently, by using the boundedness of $|r|_\infty$, $|d|_\infty$, and K in the obtained error bound, it follows that S converges to Φ as the interpolation step tends to zero. □

Remark 4.1. Using a similar extremum computations as in the calculations of K, we obtain the uniform bound for C as

$$\|C\|_\infty \leq \frac{|y|_\infty + \frac{1}{4}(|r|_\infty |y|_\infty + h|d|_\infty)}{\beta}, \tag{21}$$

where $|y|_\infty = \max\{|y_n| : 1 \leq n \leq N\}$. Thus for a given data and bounded values for the derivative parameters, the rational quadratic FIF S is always bounded. An explicit upper bound for uniform norm of S can be obtained from the triangle inequality $\|S\|_\infty \leq \|S - C\|_\infty + \|C\|_\infty$, (19) and (21).

5 Positivity Preserving Rational Quadratic FIF

5.1 *Sufficient Conditions for Positivity*

The rational FIF described in Sect. 3 has deficiency as far as shape preserving issue is concerned. For example, the quadratic rational FIF generated in Fig. 1a for the given positive data fails to be positive. So, we derive sufficient conditions for shape preserving interpolant for positive data in the following theorem:

Theorem 5.1. *Let $\{(x_n, y_n) : n = 1, 2, \ldots, N\}$ be given set of strictly positive data (i.e., $y_n > 0$ for $n = 1, 2, \ldots, N$) and S be the corresponding rational quadratic FIF defined as in (9). Then the following conditions on the scaling factors and shape parameters on each subinterval $I_n, n \in J$ are sufficient for the rational FIF S to be positive on I:*

$$0 \leq \alpha_n < \min\left\{a_n, \frac{y_n}{y_1}, \frac{y_{n+1}}{y_N}\right\}, \tag{22}$$

$$r_n > \max\left\{-2, \frac{-h_n d_n + \alpha_n d_1(x_N - x_1)}{y_n - \alpha_n y_1}\right\}. \tag{23}$$

Proof. For a typical knot point x_m, we can compute $S\big(L_n(x_m)\big) = \alpha_n S(x_m) + \dfrac{U_n(\theta)}{V_n(\theta)}$ where $\theta = \frac{x_m - x_1}{x_N - x_1}$. Recall that the condition $|\alpha_n| < a_n$ is imposed in the construction of S to assure desired differentiability, and the condition $r_n > -2$ ensures a strict positive denominator $V_n(\theta)$. Hence, if $\alpha_n \geq 0$, then sufficient conditions for $S\big(L_n(x_m)\big) > 0$ reduces to $A_n > 0$, $B_n > 0$, $C_n > 0$, where A_n, B_n, and C_n are given in Sect. 3.1.

$$A_n > 0 \Longleftrightarrow y_n - \alpha_n y_1 > 0, \tag{24}$$

$$B_n > 0 \Longleftrightarrow r_n(y_n - \alpha_n y_1) + h_n d_n - \alpha_n d_1(x_N - x_1) > 0, \tag{25}$$

$$C_n > 0 \Longleftrightarrow y_{n+1} - \alpha_n y_N > 0. \tag{26}$$

If $\alpha_n < \min\{\frac{y_n}{y_1}, \frac{y_{n+1}}{y_N}\}$, then in view of Eqs. (24) and (26), we get $A_n > 0$ and $C_n > 0$. From (25) we infer that $r_n > \frac{-h_n d_n + \alpha_n d_1(x_N - x_1)}{y_n - \alpha_n y_1}$ implies $B_n > 0$. Thus conditions prescribed on the shape parameters and scaling factors by (22), (23) yield positivity of $S\big(L_n(x_m)\big)$. Note that $[x_1, x_N]$ is the attractor of the IFS $\{I, L_n : n \in J\}$. Since the graph of S is the attractor of an IFS, by property of the attractor it follows that $S\big(L_n(x_m)\big) > 0$ for all knot points x_m, $n \in J$, and $S(x_m) = y_m > 0$ implies the positivity of the rational FIF S. □

The sufficient condition for the positivity of the classical rational quadratic function given in [15] can be obtained as a corollary in the following.

Corollary 5.1. *Let $\{(x_n, y_n) : n = 1, 2, \ldots, N\}$ be given set of positive data and C be the rational quadratic function defined in (10). Then C preserves positivity if $r_n > \max\{-2, \frac{-h_n d_n}{y_n}\}$.*

Proof. Proof follows from the above theorem by taking all the scaling factors α_n to be zero. □

Remark 5.1. For a given positive data, we can estimate the derivative parameters by the amm, and then construct a $\mathcal{C}^1$- rational quadratic FIF via (11) and (9). However, for a fixed set of the scaling factors and derivative parameters, the shape parameters computed from (14) may not meet the conditions of Theorem 5.1, and hence in general the resulting $\mathcal{C}^1$- rational quadratic spline FIF may not be positive. To get a positive $\mathcal{C}^1$- rational quadratic spline FIF, one should concern with the solvability of the system governed by (14), (22), and (23).

Remark 5.2. In order to avoid the difficulty of preserving positivity in the subsequent iterations and to obtain a viable condition for the positivity preservation, we have imposed the nonnegativity condition on the scaling factors. However, the negative scaling factors can also be considered for the construction of positive rational quadratic FIFs. This can be achieved by identifying the scaling factors and shape parameters that constrain the graph of the rational quadratic FIF within an axis-aligned rectangle lying above x-axis (see, for instance, [8]).

5.2 Implementation

To implement our positivity preserving rational quadratic fractal interpolation scheme, we consider the positive data taken at random displayed in Table 1. Application of the three-point difference scheme yields the following values for the derivative parameters: $d_1 = -6.5$, $d_2 = -5.5$, $d_3 = -4.24$, $d_4 = -0.2328$,

Table 1 A positive data set

x	1	2	3	8	10	11	12	14
$f(x)$	14	8	3	0.8	0.5	0.45	0.4	0.37

Table 2 Scaling factors and shape parameters for positive rational quadratic FIFs

Figure no.	Choice of parameters							
Figure 2a	α_n	0.076	0.076	0.214	0.057	0.035	0.032	0.028
	r_n	2	3	780	0.75	1	0.5	2
Figure 2b	α_n	0.076	0.076	0.15	0.057	0.035	0.032	0.028
	r_n	2	3	780	0.75	1	0.5	2
Figure 2c	α_n	0.02	0.076	0.214	0.057	0.035	0.032	0.028
	r_n	2	3	780	0.75	1	0.5	2
Figure 2d	α_n	0.076	0.076	0.214	0.057	0.035	0.032	0.01
	r_n	2	3	780	0.75	1	0.5	2
Figure 2e	α_n	0.076	0.076	0.214	0.057	0.035	0.032	0.028
	r_n	2	3	5,000	0.75	1	0.5	2
Figure 2f	α_n	0.076	0.076	0.214	0.057	0.035	0.032	0.028
	r_n	2	3	780	0.75	1	0.5	50

$d_5 = -0.0833$, $d_6 = -0.05$, and $d_7 = -0.0383$. Theorem 5.1 bounds the scaling factors as: $\alpha_1 \in [0, 0.0769), \alpha_2 \in [0, 0.0769), \alpha_3 \in [0, 0.2142), \alpha_4 \in [0, 0.0571), \alpha_5 \in [0, 0.0357), \alpha_6 \in [0, 0.3214)$ and $\alpha_7 \in [0, 0.0285)$. With suitable choices of the scaling parameters α_n, the lower bounds for the shape parameters r_n are computed using (23). For the simplicity of presentation, the parameters involved in the IFSs generating the positive rational quadratic FIFs are displayed in Table 2.

Owing to the implicit and recursive nature of the rational quadratic FIF, each curve segment between two node points will have global properties inherited from the entire set of interpolation points, and consequently a change in a particular scaling factor may influence the entire FIF. We explore this further, by studying the nature of change in the interpolant with change in a particular scaling factor, keeping all other parameters at some fixed values. Figure 2a represents a rational quadratic spline FIF corresponding to a choice of the parameters as described in Theorem 5.1. We take this as our reference curve. Now we modify only the scaling factor α_3 (see Table 2) with respect to the parameters of Fig. 2a, and the corresponding rational FIF is generated in Fig. 2b. It is observed that the change in α_3 affects the rational fractal interpolant considerably in the interval $[x_3, x_4]$ and changes in the other subintervals are negligible. Next we modify only the scaling factor α_1, and the corresponding rational quadratic fractal interpolant is generated in Fig. 2c. In comparison with Fig. 2a, we observe that there are variations in the shape of the rational FIF in $[x_1, x_2]$ and $[x_3, x_4]$, variation in $[x_1, x_2]$ being comparatively larger. Similarly, a modification in the scaling factor α_7 (see Table 2) generates the rational quadratic FIF in Fig. 2d. In this case, there is a significant change only in $[x_7, x_8]$. Thus, even though, theoretically perturbations in a particular α_n may be propagated in the entire curve, in practice change is prominent only in the corresponding interval.

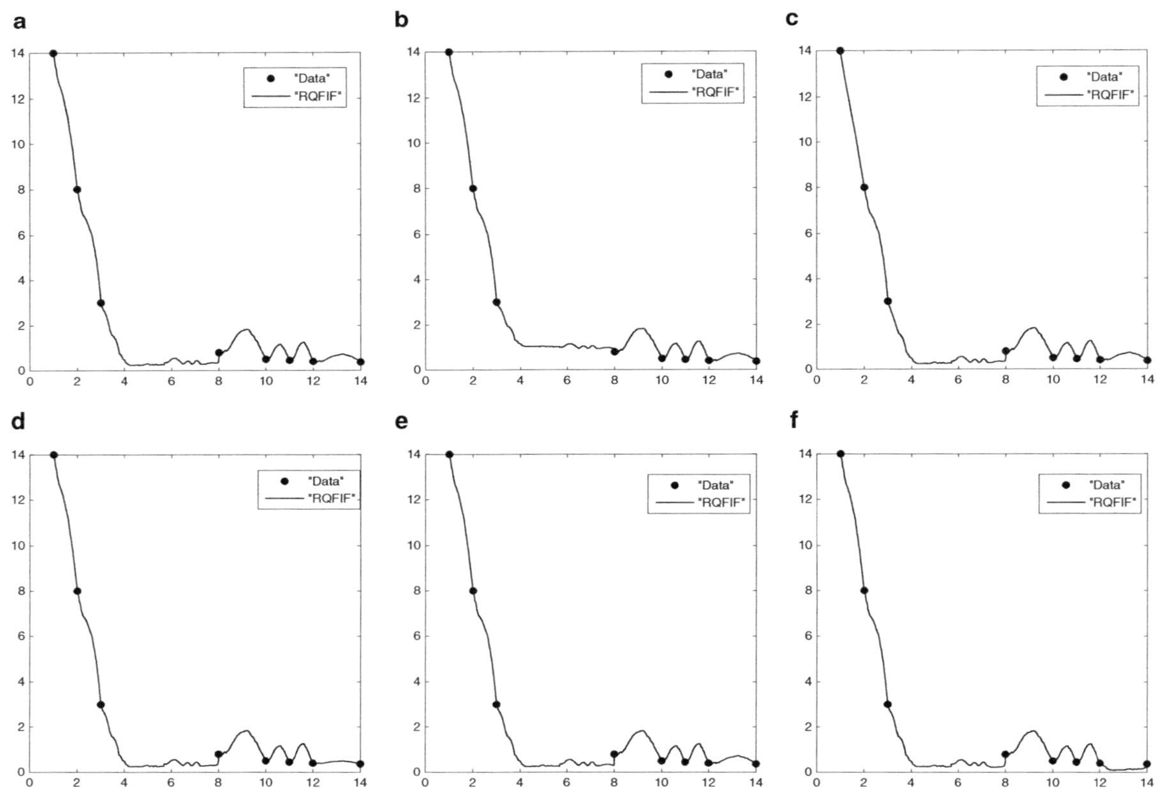

Fig. 2 Positivity preserving rational quadratic FIF for data in Table 1 with choice of parameters as indicated in Table 2. (**a**) A positive rational quadratic FIF. (**b**) Effect of α_3. (**c**) Effect of α_1. (**d**) Effect of α_7. (**e**) Effect of r_3. (**f**) Effect of r_7

Now to study the effect of shape parameters, we modify r_3 and fix all other scaling factors and shape parameters as in Fig. 2a. We observe that a change in r_3 in the allowable range $[779.25, \infty)$ does not produce considerable shape modification. Figure 2e shows a rational fractal interpolant with a specific value of r_3 (see Table 2). Now a modification of r_7 with respect to parameters of Fig. 2a generates the rational quadratic FIF in Fig. 2f which shows considerable difference in the last subinterval.

Our computational results suggest that the changes in spline parameters pertaining to a particular subinterval do not extremely influence the shape of the curve in other subintervals, and hence the globality of the FIF does not limit its applications in the field of shape preservation and shape modification. It is worth to recall that when scaling factors are taken as zero, our positivity preserving rational quadratic FIF reduces to the positivity preserving classical rational quadratic interpolant, which is completely local. From this it follows that our scheme is local or global depending on the values of the scaling factors. From our experiments with various data sets, the shape parameters and the scaling factors, we remark that the sensitivity of the rational quadratic FIF to the changes in the scaling factors versus changes in the shape parameters is data dependent. For instance, changes in r_3 with fixed values for all other parameters do not make considerable changes in the curve whereas perturbation in α_3 gets reflected in the shape of the curve (see Fig. 2a, b, e). On the other hand, changes in α_7 with all other parameters held constant do not modify any important geometric property of the curve in that subinterval, whereas a modification in r_7 changes the convexity nature of the curve (see Fig. 2a, d, f). Among the scaling and shape parameters, we can elect the one which offers more flexibility in the allowable range, for a shape modification. Thus, the rational quadratic FIF offers more flexibility compared to the classical quadratic interpolant, and a proper balancing between these scaling, shape parameters enable us to produce desired shape modification and shape preservation. The added flexibility offered by the fractal approach in conjunction with suitable optimization techniques can be effectively utilized in various engineering and industrial problems. As observed earlier, the classical rational quadratic interpolant $C \in \mathcal{C}(I)$ is differentiable except possibly at the knot points whereas the rational quadratic FIF S may be nondifferentiable in a dense subset of I. It is felt that the fractal dimension of S or S^1 may be used to study the complexity of the associated signal.

Note that the data set considered in these examples also possesses monotonicity property and hence it would be desirable to produce monotonic and positive FIFs. However, we reserve the monotonicity of the proposed quadratic FIF for a future contribution.

6 Conclusions

A new kind of rational quadratic FIF with one family of shape parameters is introduced as a tool for the positivity preserving interpolation. The proposed scheme expands the positivity preserving interpolation and approximation model, including functions for which functions themselves or their derivatives can even be non-differentiable in a dense set of points of the domain. It is shown that the problem of preserving positivity with the rational quadratic FIF depends only on a finite set of inequalities. These inequalities are shown to be solvable if the nonnegative scaling factors lie below and the shape parameters lie above some explicitly computable bounds. Our approach encompasses the classical shape preserving rational quadratic function analyzed earlier in the literature [15], and thus enlarges the field of shape preserving rational quadratic interpolants. Uniform convergence of the quadratic FIF to the data generating function $\Phi \in \mathcal{C}(I)$ is established. The developed FIF fails to be $\mathcal{C}^1$-smooth for arbitrary values of shape parameters. However, the parameters are constrained to increase the order of continuity from zero to one. In general, the conditions on the parameters that yield $\mathcal{C}^1$-smoothness to the rational quadratic FIF may not be consistent with the conditions for positivity. Hence, the user may have to compromise between the smoothness and the shape preserving requirements. This necessitates the development of higher degree rational spline FIFs, where the additional degree of freedom can be utilized to obtain required $\mathcal{C}^1$-smoothness with suitable values of the scaling factors and seemingly arbitrary values of the shape parameters. These parameters can then be constrained for preserving fundamental shape properties inherent in the data.

Acknowledgements The first author is thankful to the SERC DST Project No. SR/S4/MS: 694/10 for this work. The second author is partially supported by the Council of Scientific and Industrial Research India, Grant No. 09/084(0531)/2010-EMR-I.

References

1. Barnsley, M.F.: Fractal functions and interpolation. Constr. Approx. **2**, 303–329 (1986)
2. Barnsley, M.F.: Fractals Everywhere. Academic, Orlando (1988)
3. Barnsley, M.F., Harrington, A.N.: The calculus of fractal interpolation functions. J. Approx. Theory **57**, 14–34 (1989)
4. Chand, A.K.B., Kapoor, G.P.: Generalized cubic spline fractal interpolation functions. SIAM J. Numer. Anal. **44**, 655–676 (2006)
5. Chand, A.K.B., Navascués, M.A.: Generalized Hermite fractal interpolation. Rev. R. Acad. Cienc. Zaragoza **64**, 107–120 (2009)
6. Chand, A.K.B., Navascués, M.A.: Natural bicubic spline fractal interpolation. Nonlinear Anal. **69**(11), 3679–3691 (2008)
7. Chand, A.K.B., Viswanathan, P.: Cubic Hermite and cubic spline fractal interpolation functions. AIP Conf. Proc. **1479**, 1467–1470 (2012)
8. Chand, A.K.B., Viswanathan, P.: A constructive approach to cubic Hermite fractal interpolation function and its constrained aspects. BIT Numer. Math. **53**, 841–865 (2013)

9. Delbourgo, R., Gregory, J.A.: Determination of derivative parameters for a monotonic rational quadratic interpolant. IMA J. Numer. Anal. **5**(1), 397–406 (1985)
10. Fritsch, F.N., Butland, J.: A method for constructing local monotone piecewise cubic interpolants. SIAM J. Sci. Stat. Comput. **5**, 303–304 (1984)
11. Fritsch, F.N., Carlson, R.E.: Monotone piecewise cubic interpolation. SIAM J. Numer. Anal. **17**, 238–246 (1980)
12. Gregory, J.A., Delbourgo, R.: Shape preserving piecewise rational interpolation. SIAM J. Stat. Comput. **6**(4), 967–976 (1985)
13. Gregory, J.A., Delbourgo, R.: Piecewise rational quadratic interpolation to monotonic data. IMA J. Numer. Anal. **2**(1), 123–130 (1982)
14. Gregory, J.A.: Shape preserving spline interpolation. Comput. Aided Des. **18**(1), 53–57 (1986)
15. Hussain, M.Z., Ayub, N., Irshad, M.: Visualization of 2D data by rational quadratic functions. J. Inf. Comput. Sci. **2**(1), 17–26 (2007)
16. Mandelbrot, B.B.: The Fractal Geometry of Nature. Freeman, New York (1983)
17. Navascués, M.A.: Fractal polynomial interpolation. Z. Anal. Anwend. **25**, 401–418 (2005)
18. Navascués, M.A., Chand, A.K.B., Viswanathan, P., Sebastián, M.V.: Fractal interpolation functions: a short survey. Appl. Math. **5**, 1834–1841 (2014)
19. Passow, E., Roulier, J.A.: Monotone and convex spline interpolation. SIAM J. Numer. Anal. **14**, 904–909 (1977)
20. Sarfraz, M., Mulhem, M.A., Ashraf, F.: Preserving monotonic shape of the data using piecewise rational cubic function. Comput. Graph. **21**(1), 5–14 (1997)

Interval Wavelet Sets Determined by Points on the Circle

Divya Singh

Abstract Having observed that an interval wavelet set corresponds to the points in a circle, we obtain points in the circle which characterize two-interval wavelet sets and also those points which characterize three-interval wavelet sets for dilation $d \geq 2$. Further points in the circle characterizing one-interval and two-interval H^2-wavelet sets for dilation $d \geq 2$ are obtained. In addition, we discuss three-interval wavelet sets of $\mathbb{R}$ in respect of being associated with a multiresolution analysis (MRA).

Keywords Wavelet set • H^2-wavelet set • MSF wavelet • Multiresolution analysis

AMS Subject Classification 2000: 42C15, 42C40

1 Introduction

Dai and Larson [5] called a measurable set W of the real line $\mathbb{R}$ to be a *wavelet set* if the characteristic function χ_W on W is equal to the modulus of the Fourier transform $\hat{\psi}$ of some orthonormal wavelet ψ on $L^2(\mathbb{R})$. By an *orthonormal wavelet* ψ, we mean a function in $L^2(\mathbb{R})$, whose successive dilates by a scalar d other than 0, 1 and -1, followed by all integral translates, form an orthonormal basis for $L^2(\mathbb{R})$. These definitions were generalized to higher dimensions in [2, 4, 6, 7, 14]. Fang and

D. Singh (✉)
Department of Mathematics, National Institute of Technology, Rourkela, Rourkela 769008, Odisha, India
e-mail: singhd@nitrkl.ac.in

C. Bandt et al. (eds.), *Fractals, Wavelets, and their Applications*, Springer Proceedings in Mathematics & Statistics 92, DOI 10.1007/978-3-319-08105-2_19,

$$W = [\beta + 2m\pi\,,\, \alpha + 2\pi + 2m\pi] \cup [\alpha + 2l\pi\,,\, \beta + 2l\pi]\,.$$

As before, (3) gives

$$\alpha = \frac{2\left(l - d^j - d^j m\right)\pi}{d^j - 1} \quad \text{and} \quad \beta = \frac{2\left(l - d^{j+1} m\right)\pi}{d^{j+1} - 1}.$$

On applying the condition $0 < \alpha < \beta \le 2\pi$, we get $m = 0$ and $\left(d^j - 1\right) < l - 1 < \frac{d\left(d^j - 1\right)}{d-1}$.

Thus

$$W = \left[\frac{2\,(p+1)\,\pi}{d^{j+1} - 1}\,,\, \frac{2p\pi}{d^j - 1}\right] \cup \left[\frac{2d^j p\pi}{d^j - 1}\,,\, \frac{2d^{j+1}\,(p+1)\,\pi}{d^{j+1} - 1}\right],$$

where $p = l - 1$, $j \in \mathbb{N}$ and $\left(d^j - 1\right) < p < \frac{d\left(d^j - 1\right)}{d-1}$. □

A similar procedure gives the following result for two-interval H^2-wavelet sets, in case of dilation $d = 2$.

For $j \in \mathbb{N}$, define $E'_{j_1} = \left\{1, 2, \ldots, 2^j - 1\right\}$ and $E'_{j_2} = \left\{2^j, 2^j + 1, \ldots, 2^{j+1} - 3\right\}$.

Theorem 5.2. (1) *For $m \in E'_{j_1}$, the points*

$$\alpha = \frac{2\,(m+1)\,\pi}{2^{j+1} - 1} \quad \textit{and} \quad \beta = \frac{2m\pi}{2^j - 1}$$

in S^1 are such that $0 < \alpha < \beta \le 2\pi$, and $[\alpha\,,\,\beta] \cup [\beta + 2m\pi\,,\, \alpha + 2\pi + 2m\pi]$ forms an H^2-wavelet set.

(2) *For $m \in E'_{j_2}$, the points*

$$\alpha = \frac{2\left(m + 1 - 2^j\right)\pi}{2^j - 1} \quad \textit{and} \quad \beta = \frac{2\,(m+1)\,\pi}{2^{j+1} - 1}$$

in S^1 are such that $0 < \alpha < \beta < 2\pi$, and $[\beta\,,\, \alpha + 2\pi] \cup [\alpha + 2\,(m+1)\,\pi\,,\, \beta + 2\,(m+1)\,\pi]$ forms an H^2-wavelet set.

6 MRA Associated Three-Interval Wavelet Sets

In this Section, we consider three-interval wavelet sets and determine if these are associated with a MRA. In [10], it has been shown that if p is an odd natural number, then the wavelet set $W\,(j, p)$ is not associated with an MRA, for $j \ge 2$. Using a result stated below obtained in [9], we provide an alternative proof of this result. In addition, we prove that if $p = 2^j - 2$, then the wavelet set $W(j, p)$ is associated with an MRA, where $j \geq 2$. That each of the two-interval wavelet sets is associated with a MRA can be settled on the same lines.

Theorem 6.1 ([9]). *For an MSF wavelet ψ with $|\hat{\psi}| = \chi_W$, the following are equivalent:*

(i) ψ is associated with an MRA.
(ii) $\mu(W^S \cap (W^S + 2k\pi)) = 2\pi\delta_{k,0}$, $k \in \mathbb{Z}$.
(iii) $\{W^S + 2k\pi : k \in \mathbb{Z}\}$ forms a partition of $\mathbb{R}$,

where $W^S = \bigcup_{n=1}^{\infty} 2^{-n} W$.

Consider the three-interval wavelet set

$$W(j,p) = \left[-2\left(1 - \frac{2p+1}{2^{j+1}-1}\right)\pi, -\left(1 - \frac{2p+1}{2^{j+1}-1}\right)\pi\right] \cup \left[\frac{2(p+1)\pi}{2^{j+1}-1}, \frac{2(2p+1)\pi}{2^{j+1}-1}\right]$$

$$\cup \left[\frac{2^{j+1}(2p+1)\pi}{2^{j+1}-1}, \frac{2^{j+2}(p+1)\pi}{2^{j+1}-1}\right]$$

denoted by W, for simplicity. Then

$$W^S = \left[-\left(1 - \frac{2p+1}{2^{j+1}-1}\right)\pi, 0\right) \cup \left(0, \frac{2(p+1)\pi}{2^{j+1}-1}\right]$$

$$\cup \left(\bigcup_{k=1}^{j} 2^{j+1-k}\left[\frac{(2p+1)\pi}{2^{j+1}-1}, \frac{2(p+1)\pi}{2^{j+1}-1}\right]\right),$$

where $j \geq 2$ and $1 \leq p \leq 2^j - 2$.

Theorem 6.2. *The wavelet set $W(j, p)$, where p is odd and $j \geq 2$ is not associated with a MRA.*

Proof. Since

$$\frac{2^j(2p+1)\pi}{2^{j+1}-1} < (p+1)\pi < \frac{2^{j+1}(p+1)\pi}{2^{j+1}-1}$$

for $j \geq 2$ and $1 \leq p \leq 2^j - 2$, we have

$$(p+1)\pi \in 2^{j+1-k}\left[\frac{(2p+1)\pi}{2^{j+1}-1}, \frac{2(p+1)\pi}{2^{j+1}-1}\right],$$

when $k = 1$. Therefore, the 2π-translation map $\tau : W^S \longrightarrow [0, 2\pi)$ cannot be a bijection, if p is odd. It is because, when $p = 2m + 1$, where $m = 0, 1, \ldots, (2^{j-1} - 2)$,

$$\tau\left(\left[(p+1)\pi\,,\,\frac{2^{j+1}(p+1)\pi}{2^{j+1}-1}\right]\right)=\left[(p+1)\pi\,,\,\frac{2^{j+1}(p+1)\pi}{2^{j+1}-1}\right]-(p+1)\pi$$

intersects

$$\tau\left(\left(0\,,\,\frac{2(p+1)\pi}{2^{j+1}-1}\right]\right)=\left(0\,,\,\frac{2(p+1)\pi}{2^{j+1}-1}\right].$$

Thus by (iii) of Theorem 6.1, $W(j,p)$ is not associated with a MRA. □

Theorem 6.3. *The wavelet set $W(j,p)$, where $p=2^j-2$ and $j\geq 2$, is associated with a MRA.*

Proof. Writing W for $W(j,2^j-2)$, we have

$$\begin{aligned}W^S&=\left[-\frac{2\pi}{2^{j+1}-1}\,,\,0\right)\cup\left(0\,,\,\left(\frac{2^{j+1}-2}{2^{j+1}-1}\right)\pi\right]\\&\quad\cup\bigcup_{k=1}^{j}2^k\left[\left(\frac{2^{j+1}-3}{2^{j+1}-1}\right)\pi\,,\,\left(\frac{2^{j+1}-2}{2^{j+1}-1}\right)\pi\right].\end{aligned}$$

Since the map $\tau: W^S\longrightarrow[0\,,\,2\pi)$ defined by

$$\tau(x)=\begin{cases}x+2\pi, & \text{if } x\in\left[-\frac{2\pi}{2^{j+1}-1}\,,\,0\right)\\ x, & \text{if } x\in\left(0\,,\,\left(\frac{2^{j+1}-2}{2^{j+1}-1}\right)\pi\right]\cup 2\left[\left(\frac{2^{j+1}-3}{2^{j+1}-1}\right)\pi\,,\,\left(\frac{2^{j+1}-2}{2^{j+1}-1}\right)\pi\right]\\ x-(2^m-2)\pi, & \text{if } x\in 2^m\left[\left(\frac{2^{j+1}-3}{2^{j+1}-1}\right)\pi\,,\,\left(\frac{2^{j+1}-2}{2^{j+1}-1}\right)\pi\right]\end{cases}$$

where $m=2,\ 3,\ \ldots,\ j$ is a bijection, $\{W^S+2k\pi : k\in\mathbb{Z}\}$ forms a partition of $\mathbb{R}$. The proof follows from (iii) of Theorem 6.1. □

Remark 6.1. It can be worked out manually that when j is 3 or 5, there is no three-interval wavelet set which is associated with an MRA except when $p=2^j-2$. In case j is 7, $W\,(7,36)$ and $W\,(7,90)$ are the other two three-interval wavelet sets associated with a MRA in addition to that of $W\,(7,2^7-2)$.

Acknowledgement The author thanks to Professor K.K. Azad for his help and encouragement.

References

1. Arcozzi, N., Behera, B., Madan, S.: Large classes of minimally supported frequency wavelets of $L^2(\mathbb{R})$ and $H^2(\mathbb{R})$. J. Geom. Anal. **13**, 557–579 (2003)
2. Baggett, L.W., Medina, H.A., Merrill, K.D.: Generalized multi-resolution analyses and a construction procedure for all wavelet sets in $\mathbb{R}^n$. J. Fourier Anal. Appl. **5**, 563–573 (1999)
3. Bownik, M., Hoover, K.: Dimension functions of rationally dilated GMRAs and wavelets. J. Fourier Anal. Appl. **15**(5), 585–615 (2009)
4. Bownik, M., Rzeszotnik, Z., Speegle, D.: A characterization of dimension functions of wavelets. Appl. Comput. Harmon. Anal. **10**, 71–92 (2001)
5. Dai, X., Larson, D.: Wandering vectors for unitary systems and orthogonal wavelets. Mem. Amer. Math. Soc. **134**(640), MR 98m:47067. (1998)
6. Dai, X., Larson, D., Speegle, D.: Wavelet sets in $\mathbb{R}^n$. J. Fourier Anal. Appl. **3**, 451–456 (1997)
7. Dai, X., Larson, D., Speegle, D.: Wavelet sets in $\mathbb{R}^n$ II. Contemp. Math. **216**, 15–40 (1998)
8. Singh, D.: On wavelet induced isomorphisms. Int. J. Wavelets Multiresolut. Inf. Process. **8**(3), 359–371 (2010)
9. Fang, X., Wang, X.: Construction of minimally-supported-frequency wavelets. J. Fourier Anal. Appl. **2**, 315–327 (1996)
10. Ha, Y., Kang, H., Lee, J., Seo, J.: Unimodular wavelets for L^2 and the Hardy space H^2. Michigan Math. J. **41**, 345–361 (1994)
11. Hernández, E., Weiss, G.: A First Course on Wavelets. CRC Press, Boca Raton (1996)
12. Ionascu, E.: A new construction of wavelet sets. Real Anal. Exch. **28**, 593–610 (2002/2003)
13. Ionascu, E., Larson, D., Pearcy, C.: On wavelet sets. J. Fourier Anal. Appl. **4**, 711–721 (1998)
14. Merrill, K.D.: Simple wavelet sets for scalar dilations in $\mathbb{R}^2$. In: Representations, Wavelets and Frames: A Celebration of the Mathematical Work of Lawrence W. Baggett, pp. 177–192. Birkhäuser, Boston (2008)

Inverse Representation Theorem for Matrix Polynomials and Multiscaling Functions

M. Mubeen and V. Narayanan

Abstract Wavelet analysis provides suitable bases for the class of L^2 functions. The function to be represented is approximated at different resolutions. The desirable properties of a basis are orthogonality, compact supportedness and symmetricity. In the scalar case, the only wavelet with these properties is Haar wavelet. Theory of multiwavelets assumes significance since it offers symmetric, compactly supported, orthogonal bases for $L^2(R)$. The properties of a multiwavelet are determined by the corresponding Multiscaling Function. A multiscaling function is characterized by its symbol function which is a matrix polynomial in complex exponential. The inverse representation theorem of matrix polynomials provides a method to construct a matrix polynomial from its Jordan pair. Our objective is to find the properties that characterize a Jordan pair of a symbol function of a multiscaling function with desirable properties.

Keywords Multiwavelet • Multiscaling function • Jordan pair

1 Introduction

Wavelet analysis is widely used in different areas of science and engineering. The task of finding a suitable basis to represent the data effectively with lower risk arises in different types of theoretical as well as practical problems [3, 10, 18]. Wavelet analysis provides better algorithms for such problems. Bases for the class of square integrable functions $L^2(\mathbb{R})$ can be constructed based on the concept of MultiResolution Analysis, i.e. approximating the given data at different resolutions. An MRA can be defined as follows:

M. Mubeen (✉) • V. Narayanan
Department of Mathematics, National Institute of Technology Calicut, Kerala, India
e-mail: mubeen@nitc.ac.in; vna@nitc.ac.in

C. Bandt et al. (eds.), *Fractals, Wavelets, and their Applications*, Springer Proceedings in Mathematics & Statistics 92, DOI 10.1007/978-3-319-08105-2_20,

Definition 1.1 (Keinert [11, p. 10]). An MRA (Multiresolution Analysis) of L^2 is a doubly infinite nested sequence of subspaces of L^2

$$\cdots \subset V_{-1} \subset V_0 \subset V_1 \subset V_2 \cdots$$

with properties

1. $\bigcup_n V_n$ is dense in L^2
2. $\bigcap_n V_n = 0$
3. $f(x) \in V_n \Leftrightarrow f(mx) \in V_{n+1}$ for all $n \in Z$. Here $m \in \mathbb{R}$ is the dilation factor and usually taken as 2.
4. $f(x) \in V_n \Leftrightarrow f(x - m^{-n}k) \in V_n$ for all $n, k \in Z$
5. There exists a function $\phi \in L^2$,$\phi : \mathbb{R} \to \mathbb{C}$ so that

$$\{\phi(x-k) : k \in Z\}$$

   form a basis for V_0.

Then we say that ϕ generates the MRA.

Definition 1.2 (Keinert [11, p. 3]). A function $\phi : \mathbb{R} \to \mathbb{C}$ is called a scaling function or refinable function if it satisfies the scaling equation

$$\phi(x) = \sqrt{m} \sum_{k=0}^{l} h_k \phi(mx - k) \tag{1}$$

where $k \in \mathbb{Z}$, $h_k \in \mathbb{C}$ and $m \in \mathbb{R}$ is the dilation factor. Usually we take the value for the dilation factor m as 2.

Following are some of the desirable properties of a scaling function ϕ.

Definition 1.3 (Keinert [11, p. 3]). The refinable function ϕ is orthogonal if

$$< \phi(x), \phi(x-t) >= \int \phi(x)\overline{\phi(x-t)}dx = \delta_{0t} \tag{2}$$

Definition 1.4 (Keinert [11, p. 191]). The refinable function ϕ is symmetric about some point $a \in \mathbb{R}$ if

$$\phi(a + x) = \phi(a - x) \forall x \in \mathbb{R} \tag{3}$$

Definition 1.5 (Keinert [11, p. 8]). A function ϕ is said to be L^2 *stable* if for any function f such that $f(x) = \sum_{k \in Z} f_k \phi(x-k)$ there exist constants $0 < A \leq B < \infty$ such that

$$A \sum_k \| f_k \|_2^2 \leq \| f \|_2^2 \leq B \sum_k \| f_k \|_2^2 \tag{4}$$

Definition 1.6 (Keinert [11, p. 8]). Support of ϕ is the closure of the set

$$\{x \in \mathbb{R} : \phi(x) \neq 0\}$$

A function $\phi \in L^2$ generates an MRA if it satisfies certain properties, of them the key property is that it satisfies the scaling equation. A class of functions constructed using ϕ called wavelet functions acts as a basis for $L^2(\mathbb{R})$. Desirable properties of a good basis can be achieved by a suitable choice of ϕ. The desirable properties of ϕ include compact supportedness, symmetricity, orthogonality, higher approximation order, etc. A compactly supported orthogonal wavelet is symmetric only if it is the well-known Haar Wavelet function. It is not possible to construct a compactly supported, orthogonal and symmetric wavelet function other than Haar wavelet [3]. But if we increase the multiplicity, i.e. generate MRA with more than one scaling function, we can solve this issue [1, 2, 4, 14, 15, 19, 20, 23, 24]. Even though the increase in multiplicity increases computational as well as theoretical complexity, we have several advantages. The function vector $\Phi \in L^2, \Phi : \mathbb{R} \to \mathbb{C}^n$ so that

$$\{\Phi_i(x-k) : 1 \leq i \leq n, k \in \mathbb{Z}\}$$

form a stable basis for V_0 will be called a multiscaling function and it satisfies the matrix version of Eq. (1).

Definition 1.7 (Keinert [11, p. 124]). A function vector $\Phi : \mathbb{R} \to \mathbb{C}^n$ is called a multiscaling function or refinable function if it satisfies the equation

$$\Phi(x) = \sqrt{m} \sum_{k=0}^{l} H_k \Phi(mx - k) \tag{5}$$

where $k \in \mathbb{Z}$, $H_k \in \mathbb{C}^{n\times n}$ and $m \in \mathbb{R}$ is the dilation factor.

Following are some of the desirable properties of a multiscaling function Φ.

Definition 1.8 (Keinert [11, p. 124]). The refinable function vector Φ is orthogonal if

$$< \Phi(x), \Phi(x-t) >= \int \Phi(x)\Phi(x-t)^* dx = \delta_{0t} I_n \tag{6}$$

Definition 1.9 (Keinert [11, p. 192]). The refinable function vector Φ is symmetric if each component function ϕ_i, $1 \leq i \leq n$ is symmetric about some point a_i. That is

$$\phi_i(a_i + x) = \phi_i(a_i - x) \forall x, 1 \leq i \leq n \tag{7}$$

Definition 1.10 (Keinert [11, p. 130]). A function vector Φ is said to have linearly independent shifts if for all sequences of vectors $\{c_k\}$ in $\mathbb{C}^n$,

$$\sum_k c_k^* \Phi(x-k) = 0 \Rightarrow c_k = 0 \forall k \tag{8}$$

Definition 1.11 (Keinert [11, p. 130]). A function vector Φ is said to be L^2 *stable* if for any function f such that $f(x) = \sum_{k \in Z} f_k^* \Phi(x-k)$ there exist constants $0 < A \leq B < \infty$ such that

$$A \sum_k \| f_k \|_2^2 \leq \| f \|_2^2 \leq B \sum_k \| f_k \|_2^2 \tag{9}$$

Definition 1.12 (Keinert [11, p. 129]). The support of a function vector Φ is defined as the union of the supports of its component functions.

$$supp\Phi = \bigcup_k supp\phi_k \tag{10}$$

A wavelet basis of multiplicity n will be constructed from Φ which is called a Multiwavelet Basis. Theory of Multiwavelets is developed analogous to the theory of scalar wavelets and most of the results in scalar case can be directly extended to the vector case. To find a solution vector Φ for Eq. (5), we usually switch over to the frequency domain where the above equation becomes

$$\hat{\Phi}(\xi) = H(\xi/2)\hat{\Phi}(\xi/2) \tag{11}$$

where

$$H(\xi) = \frac{1}{\sqrt{m}} \sum_{k=0}^{l} H_k e^{-ik\xi} \tag{12}$$

which is called a Symbol Function or Mask Function.

A function vector Φ generates an MRA if it is an L^2 stable, compactly supported, refinable function. Φ is a compactly supported L^2 solution of Eq. (5) with linearly independent shifts and nonzero integral only if the corresponding symbol function satisfies certain conditions called Basic Feasibility Conditions as follows.

Theorem 1.1 (Keinert [11, p. 131]). *A refinable function vector $\Phi \in L^2(\mathbb{R})$ is compactly supported with linearly independent shifts and nonzero integral only if the corresponding symbol function satisfies the conditions*

1. *H(0) has an eigenvalue 1 and all other eigenvalues are less than 1 in absolute value*
2. *There exists a nonzero vector $y_o \in \mathbb{C}^n$ such that*

$$\sum_k y_o^* \Phi(x-k) = c \tag{13}$$

where c is a nonzero constant

3. The same vector y_o satisfies

$$y_o^* H(\frac{2\pi t}{m}) = \delta_{0t} y_o^* \qquad t = 0, 1 \ldots m-1 \tag{14}$$

4. The same vector y_o satisfies

$$y_o^* \sum_p H_{pm+t} = \frac{1}{\sqrt{m}} y_o^* \qquad t = 0, 1 \ldots m-1 \tag{15}$$

The properties of a multiscaling function is dependent on the corresponding symbol function. Thus our main focus changes to the symbol function $H(\xi)$ which is a Matrix Polynomial restricted to the unit circle in the Complex Plane. If $H(\xi)$ is the mask function, then its spectral data, eigenvalues and generalized eigenvectors can be found. Theory of the spectral data of matrix polynomials is discussed in Sect. 2. We can construct a matrix polynomial from the given spectral data using the inverse representation theorem which is stated in Sect. 3 For details on the underlying theory, see [6, 8, 9, 12, 13, 16, 21, 22]. In Sect. 3, we have explained the necessary as well as sufficient conditions a spectral data must possess so that the corresponding matrix polynomial is the symbol function of a multiscaling function. We have explained a method to construct the symbol function $H(\xi)$ by choosing a suitable spectral data with example.

2 Preliminaries: Spectral Data of Matrix Polynomials

Let

$$L(\lambda) = \sum_{k=0}^{l} A_k \lambda^k \qquad A_k \in \mathbb{C}^{n \times n} \quad \lambda \in \mathbb{C} \tag{16}$$

be a matrix polynomial of degree l. Then $\lambda_0 \in \mathbb{C}$ is said to be an eigenvalue of L(λ) if Det L(λ_0) $= 0$. Then there exists a nonzero $x_0 \in \mathbb{C}^n$ such that $L(\lambda_0)x_0 = 0$.

Definition 2.1 (Gohberg, Lancaster [5, p. 23]). The chain of vectors $x_0, x_1 \ldots x_k$ $\in \mathbb{C}^n$, $x_0 \neq 0$, is a Jordan Chain of length k+1 of the matrix polynomial $L(\lambda)$ if

$$\sum_{p=0}^{i} \frac{L^p(\lambda_0)}{p!} x_{i-p} = 0 \qquad i = 0, 1, 2 \ldots k \tag{17}$$

where $L^p(\lambda_0)$ is the pth derivative of $L(\lambda)$ at λ_0

This is a generalization of the usual notion of a Jordan chain of a square matrix. A matrix polynomial $L(\lambda)$ is said to be a regular matrix polynomial if Det$L(\lambda) \not\equiv 0$[Gohberg, Lancaster [5, p. 181]]. Throughout this paper, we will assume that the matrix polynomial $L(\lambda)$ is regular.

Proposition 2.1 (Gohberg, Lancaster [5, p. 27]). The vectors $x_0, x_1 \dots x_k$ form a Jordan Chain of the matrix polynomial $L(\lambda) = \sum_{t=0}^{l} A_t \lambda^t$ corresponding to the eigenvalue λ_0 iff $x_0 \neq 0$ and

$$A_0 X_0 + A_1 X_0 J_0 + A_2 X_0 J_0^2 + \dots + A_l X_0 J_0^l = 0 \qquad X_0 = [x_0 x_1 \dots x_k] \quad (18)$$

X_0 is an $n \times (k+1)$ matrix and J_0 is a Jordan Block of size $(k+1) \times (k+1)$ with λ_0 on the main diagonal.

Definition 2.2 (Gohberg, Lancaster [5, p. 35]). A set of Jordan chains

$$\Phi_{j0}^{(i)}, \Phi_{j1}^{(i)} \dots \Phi_{j\mu_j^i - 1}^{(i)} j = 1, 2, 3 \dots S_i$$

of a matrix polynomial $L(\lambda) = \sum_j A_j \lambda^j$ corresponding to the eigenvalue λ_i with geometric multiplicity S_i and algebraic multiplicity α_i, is said to be a canonical set if the eigenvectors $\Phi_{10}^{(i)}, \Phi_{20}^{(i)} \dots \Phi_{S_i 0}^{(i)}$ are linearly independent and $\sum_{j=1}^{S_i} \mu_j^{(i)} = \alpha_i$.

For a given matrix polynomial $L(\lambda)$, the canonical set corresponding to an eigenvalue is not unique. But the set of values of the lengths of the Jordan chains corresponding to a particular eigenvalue is unique (Gohberg, Lancaster [5, p. 32]). Now we will define the Jordan pair of a given matrix polynomial $L(\lambda)$. Det$L(\lambda)$ is a scalar polynomial of degree nl. Let $\lambda_1, \lambda_2 \dots \lambda_r$ be the distinct zeroes of $L(\lambda)$ with algebraic multiplicities $\alpha_1, \alpha_2 \dots \alpha_r$, respectively, and $\sum_i \alpha_i = nl$. Then $\{\lambda_1, \lambda_2 \dots \lambda_r\}$ form $\sigma(L)$, the spectrum of $L(\lambda)$. For every λ_i, choose a canonical set of Jordan Chains of $L(\lambda)$ corresponding to λ_i, namely $\Phi_{j0}^{(i)}, \Phi_{j1}^{(i)} \dots \Phi_{j\mu_j^i - 1}^{(i)}, j = 1, 2, 3 \dots S_i$, where S_i is the geometric multiplicity of λ_i. The number of Jordan chains of $L(\lambda)$ corresponding to λ_i is given by S_i and length of the jth chain is equal to $\mu_j^{(i)}$. Also the equality $\sum_{j=1}^{S_i} \mu_j^{(i)} = \alpha_i$ holds. Consider the pair (X_i, J_i), where

$$X_i = [\underbrace{\Phi_{10}^{(i)} \dots \Phi_{1\mu_1^i - 1}^{(i)}}_{1}, \underbrace{\Phi_{20}^{(i)} \dots \Phi_{2\mu_2^i - 1}^{(i)}}_{2} \dots\dots \underbrace{\Phi_{S_i 0}^{(i)} \dots \Phi_{S_i \mu_{S_i}^i - 1}^{(i)}}_{S_i}]$$

is a matrix of size $n \times \alpha_i$ and

$$J_i = \begin{pmatrix} J_{i1} & & & & \\ & J_{i2} & & & \\ & & J_{i3} & & \\ & & & \dots & \\ & & & & J_{iS_i} \end{pmatrix}$$

is a block diagonal matrix of size α_i. The pair of matrices (X_i, J_i) is called a Jordan pair of $L(\lambda)$ corresponding to λ_i. For a given $L(\lambda)$, a Jordan pair corresponding to a particular eigenvalue λ_i is not unique since the canonical set is not uniquely defined.

Now we consider a pair of matrices (X_F, J_F) such that $X_F = (X_1 X_2 \ldots X_r)$ is a matrix of order $n \times nl$ and $J_F = Diag[J_1, J_2 \ldots J_r]$ is a block diagonal matrix of order nl and each (X_i, J_i) is a Jordan pair of $L(\lambda)$ corresponding to any finite eigenvalue λ_i. Then (X_F, J_F) is called a finite Jordan pair of $L(\lambda)$. The finite Jordan pair (X_F, J_F) does not determine $L(\lambda)$ uniquely. If a matrix polynomial $L(\lambda)$ has a finite Jordan pair (X_F, J_F), then any matrix polynomial of the form $Q(\lambda)L(\lambda)$ has the same finite Jordan pair (X_F, J_F) if $Q(\lambda)$ is a matrix polynomial with $DetQ(\lambda) \equiv constant \neq 0$ (Gohberg, Lancaster [5, p. 184]). In order to determine the matrix polynomial uniquely, we consider an additional Jordan pair (X_∞, J_∞) corresponding to $\lambda = \infty$. (X_∞, J_∞) is a Jordan pair of the matrix polynomial $\tilde{L}(\lambda) = \lambda^l L(\lambda^{-1})$ corresponding to the eigenvalue 0. Thus $L(\lambda)$ has an eigenvalue at ∞ if the coefficient matrix of the highest power of λ is singular.

Definition 2.3 (Gohberg, Lancaster [5, p. 188]). A pair of matrices

$$X = [X_1 X_2] \; and \; T = \begin{pmatrix} T_1 & 0 \\ 0 & T_2 \end{pmatrix}$$

where $X_1 \in \mathbb{C}^{n\times m}$,$X_2 \in \mathbb{C}^{n\times(nl-m)}$ and $T_1 \in \mathbb{C}^{m\times m}$,$T_2 \in \mathbb{C}^{(nl-m)\times(nl-m)}$ with $0 \le m \le nl$ is called a decomposable pair of degree l if the matrix

$$S_{l-1} = Col[X_1 T_1^i X_2 T_2^{l-1-i}]_{i=0}^{l-1}$$

is nonsingular. A pair (X, T) satisfying this property is called a decomposable pair of the regular $n \times n$ matrix polynomial $L(\lambda) = \sum_{i=0}^{l} A_i \lambda^i$ if

$$\sum_{i=0}^{l} A_i X_1 T_1^i = 0, \; \sum_{i=0}^{l} A_i X_2 T_2^{l-i} = 0 \tag{19}$$

Let (X_1, T_1) and (X_2, T_2) denote the finite and the infinite Jordan Pair of $L(\lambda)$, respectively. Then $([X_1 X_2], T_1 \oplus T_2)$ is a decomposable pair of $L(\lambda)$ (Gohberg, Lancaster [5, p. 189, Theorem 7.3]). For each matrix polynomial $L(\lambda)$, we can find the Jordan pair $(X, T) = ([X_1 X_2], T_1 \oplus T_2)$. The inverse question, whether it is possible to determine $L(\lambda)$ for a given Jordan pair (X, T) has been answered by the *Representation and the inverse theorem* of matrix polynomials.

3 The Inverse Problem

We will state the inverse representation theorem and try to explore the necessary and sufficient properties a spectral data must possess so that the corresponding matrix polynomial represents a multiscaling function. The notation $[Col(X_1T_1^iX_2T_2^{l-1-i})_{i=0}^{l-1}]$ represents the matrix

$$\begin{pmatrix} X_1X_2 & T_2^{l-1} \\ X_1T_1 & X_2T_2^{l-2} \\ \dots & \dots \\ X_1T_1^{l-1} & X_2 \end{pmatrix}.$$

Theorem 3.1 (Gohberg, Lancaster [5, p. 197]). *Let* $(X,T) = ([X_1X_2], T_1 \oplus T_2)$ *be a decomposable pair of degree* l*, and let*

$$S_{l-2} = Col[X_1T_1^iX_2T_2^{l-2-i}]_{i=0}^{l-2}$$

Then, for every $n \times nl$ *matrix V such that the matrix* $\begin{pmatrix} S_{l-2} \\ V \end{pmatrix}$ *is nonsingular, the matrix polynomial*

$$L(\lambda) = V(I-P)((I \oplus T_2)\lambda - (T_1 \oplus I))(U_0 + U_1\lambda + U_2\lambda^2 + \dots + U_{l-1}\lambda^{l-1}) \quad (20)$$

where

$$P = (I \oplus T_2)[Col(X_1T_1^iX_2T_2^{l-1-i})_{i=0}^{l-1}]^{-1}\begin{pmatrix} I \\ 0 \end{pmatrix} S_{l-2} \quad (21)$$

and

$$[U_0U_1U_2 \dots U_{l-1}] = [Col(X_1T_1^iX_2T_2^{l-1-i})_{i=0}^{l-1}]^{-1} \quad (22)$$

has (X,T) as its decomposable pair

The matrix polynomial determined by $([X_1X_2], T_1 \oplus T_2)$ is unique up to the multiplication by a nonsingular constant matrix Q, i.e. if L(λ) and $\tilde{L}(\lambda)$ have the same decomposable pair, then

$$\tilde{L}(\lambda) = Q \times L(\lambda) \quad (23)$$

for some constant nonsingular matrix Q. The converse is also true, if $\tilde{L}(\lambda)$ is defined as in Eq. (23), then L(λ) and $\tilde{L}(\lambda)$ have the same decomposable pair. Now we will state the necessary properties a Jordan pair must possess so that the corresponding matrix polynomial satisfies the Basic Feasibility Conditions (Theorem 1.1).

Proposition 3.1. *Let $([X_1X_2], T_1 \oplus T_2)$ be a given Jordan pair, then the corresponding matrix polynomial is the mask function of a compactly supported multiscaling function vector in L^2 with nonzero integral and linearly independent shifts only if there exists an $n \times nl$ matrix V such that*

$$\begin{pmatrix} S_{l-2} \\ V \end{pmatrix}$$

is nonsingular, where

$$S_{l-2} = Col[X_1 T_1^i X_2 T_2^{l-2-i}]_{i=0}^{l-2}$$

and

1. *The matrix*

$$[V(I-P)((I \oplus T_2) - (T_1 \oplus I))(U_0 + U_1 + U_2 + \ldots + U_{l-1})]$$

has a simple eigenvalue 1 and all other eigenvalues are less than 1 in absolute value, where the matrices P and $[U_0U_1U_2\ldots U_{l-1}]$ are given by Eqs. (21) and (22), respectively. Hence there exists a nonzero column vector y_0 such that

$$y_0^*[V(I-P)((I \oplus T_2) - (T_1 \oplus I))(U_0 + U_1 + U_2 + \ldots + U_{l-1})] = y_0^* \quad (24)$$

2. *The same vector y_0 satisfies*

$$y_0^*[V(I-P)(-(I \oplus T_2)-(T_1 \oplus I)) \\ (U_0-U_1+U_2-\ldots+(-1)^n U_n+\ldots+(-1)^{l-1}U_{l-1})]=0 \quad (25)$$

In general, for $\lambda_t = e^{\frac{-2\pi i t}{m}}, t = 0, 1 \ldots m-1$

$$y_0^*[V(I-P)((I \oplus T_2)\lambda_t-(T_1 \oplus I))(U_0+U_1\lambda_t+U_2\lambda_t^2+\ldots+U_{l-1}\lambda_t^{l-1})] = \delta_{0t} y_0^* \quad (26)$$

3. *The vector y_0 satisfies*

$$y_0^* \sum_k L_{km+t} = \frac{1}{\sqrt{m}} y_0^* t = 0, 1 \ldots m-1 \text{ where}$$

$$L_k = V(I-P)[(I \oplus T_2)U_{k-1} - (T_1 \oplus I)U_k] \quad (27)$$

Proof. From the inverse representation Theorem 3.1, it is known that L(λ) given by

$$L(\lambda) = V(I-P)((I \oplus T_2)\lambda - (T_1 \oplus I))(U_0 + U_1\lambda + U_2\lambda^2 + \ldots + U_{l-1}\lambda^{l-1})$$

has the decomposable pair $([X_1 X_2], T_1 \oplus T_2)$. From Theorem 1.1, a matrix polynomial $H(\xi)$ is a mask function of a multiscaling function vector Φ with properties stated above only if it satisfies the Basic Feasibility Conditions.

1. H(0) must have a simple eigenvalue 1 and all other eigenvalues are less than 1 in absolute value. Consider the matrix polynomial L(λ) corresponding to the given decomposable pair where $\lambda = e^{i\xi}$. Now, $L(e^{i\xi})|_{\xi=0} = L(1)$. Thus the matrix L(1) should have a simple eigenvalue 1 and all other eigenvalues are less than 1 in absolute value. But L(1) is given by

$$L(1) = V(I-P)((I \oplus T_2) - (T_1 \oplus I))(U_0 + U_1 + U_2 + \ldots + U_{l-1})$$

   The matrix L(1) must have a simple eigenvalue 1 and all other eigenvalues are less than 1 in absolute value. Also there must exist a nonzero vector y_0 such that $y_0^* L(1) = y_0^*$, i.e.,

$$y_0^*[V(I-P)((I \oplus T_2) - (T_1 \oplus I))(U_0 + U_1 + U_2 + \ldots + U_{l-1})] = y_0^*$$

2. For the matrix polynomial $H(\xi)$, the nonzero vector y_0 satisfies

$$y_0^* H(\frac{2\pi t}{m}) = \delta_{0t} y_0^* \qquad t = 0, 1 \ldots m-1 \tag{28}$$

   Take $\lambda_t = e^{-\frac{2\pi i t}{m}}, t = 0, 1 \ldots m-1$. For the matrix polynomial L(λ) corresponding to the given decomposable pair, Eq. (28) can be written as

$$y_0^* L(\lambda_t) = \delta_{0t} y_o^* \; t = 0, 1 \ldots m-1 \tag{29}$$

   But

$$L(\lambda_t) = V(I-P)((I \oplus T_2)\lambda_t - (T_1 \oplus I))(U_0 + U_1\lambda_t + U_2\lambda_t^2 + \ldots + U_{t-1}\lambda_t^{l-1}) \tag{30}$$

   Combining Eqs. (29) and (30), we get

$$y_0^*[V(I-P)((I \oplus T_2)\lambda_t - (T_1 \oplus I))(U_0 + U_1\lambda_t + U_2\lambda_t^2 + \ldots + U_{t-1}\lambda_t^{l-1})] = \delta_{0t} y_0^*$$

3. The same vector y_0 satisfies

$$y_0^* \sum_k H_{km+t} = \frac{1}{\sqrt{m}} y_0^* \qquad t = 0, 1 \ldots m-1$$

   or

$$y_0^* \sum_k L_{km+t} = \frac{1}{\sqrt{m}} y_0^* \qquad t = 0, 1 \ldots m-1$$

But

$$L(\lambda) = \sum_k L_k \lambda^k = V(I - P)((I \oplus T_2)\lambda - (T_1 \oplus I))$$
$$(U_0 + U_1\lambda + U_2\lambda^2 + \ldots + U_{l-1}\lambda^{l-1})$$
$$\Rightarrow L_k = V(I - P)[(I \oplus T_2)U_{k-1} - (T_1 \oplus I)U_k]$$

Thus we get

$$y_0^* \sum_k L_{km+t} = \frac{1}{\sqrt{m}} y_0^*, t = 0, 1 \ldots m - 1 \; where$$
$$L_k = V(I - P)[(I \oplus T_2)U_{k-1} - (T_1 \oplus I)U_k]$$

□

We have to find the sufficient conditions on the spectral pair so that it represents a symbol function $H(\xi)$ for which there exists a solution vector Φ for the refinement equation (5). We will briefly discuss the conditions on $H(\xi)$ such that a solution vector exists. From Eq. (11), the fourier transform of multiscaling equation is given by

$$\hat{\Phi}(\xi) = H(\xi/2)\hat{\Phi}(\xi/2)$$

$$\Rightarrow \hat{\Phi}(\xi) = \Pi_{i=1}^{j} H(\xi/2^i)\hat{\Phi}(\xi/2^j) \tag{31}$$

Taking

$$P_j(\xi) = \Pi_{i=1}^{j} H(\xi/2^i) \tag{32}$$

We get

$$\Rightarrow \hat{\Phi}(\xi) = \lim_{j\to\infty} P_j(\xi)\hat{\Phi}(0) \tag{33}$$

Taking

$$P(\xi) = \lim_{j\to\infty} P_j(\xi) = \Pi_{i=1}^{\infty} H(\xi/2^i) \tag{34}$$

We get

$$\hat{\Phi}(\xi) = P(\xi)\hat{\Phi}(0) \tag{35}$$

Taking $\xi = 0$ in Eq. (35), we get

$$\Rightarrow \hat{\Phi}(0) = P(0)\hat{\Phi}(0) \tag{36}$$

It has been established in Heil and Colella [7] that, when the matrix $H(0)$ has a nondegenerate eigenvalue 1 and all other eigenvalues are less than 1 in absolute value, the infinite product of matrices $P(0)$ in Eq. (36) converges and the product $P_j(\xi)$ (32) converges uniformly on compact sets to $P(\xi)$ (34). Also, the infinite product $P(\xi)$ has polynomial growth and $\hat{\Phi}(\xi)$ is the Fourier transform of a distributional solution of Eq. (5). Let p be the multiplicity of 1 as an eigenvalue of $H(0)$, then there exists p independent solution vectors Φ corresponding to p linearly independent 1-eigenvectors of $H(0)$.

Definition 3.1 (Keinert [11, p. 131]). A matrix is said to satisfy condition $E(p)$ if it has a p-fold nondegenerate eigenvalue 1 and all other eigenvalues are less than 1 in absolute value.

Theorem 3.2 (Keinert [11, p. 220]). *The equation*

$$\hat{\Phi}(\xi) = H(\xi/2)\hat{\Phi}(\xi/2)$$

corresponding to a symbol function $H(\xi)$ has a solution vector Φ such that $\hat{\Phi}$ is continuous at 0 with $\hat{\Phi}(0) \neq 0$ if $H(0)$ satisfies condition $E(p)$.

Proposition 3.2. *Let $(X, T) = ([X_1 X_2], T_1 \oplus T_2)$ be a given Jordan pair, then there exists a symbol function $H(\xi)$ with Jordan pair (X,T) such that the corresponding multiscaling equation (5) has a solution vector Φ such that $\hat{\Phi}$ is continuous at 0 with $\hat{\Phi}(0) \neq 0$, if there exists an $n \times nl$ matrix V such that the $n \times n$ matrix*

$$[V(I - P)((I \oplus T_2) - (T_1 \oplus I))(U_0 + U_1 + U_2 + \ldots + U_{l-1})]$$

satisfies condition $E(p)$, with

$$\begin{pmatrix} S_{l-2} \\ V \end{pmatrix}$$

is nonsingular, where

$$S_{l-2} = Col[X_1 T_1^i X_2 T_2^{l-2-i}]_{i=0}^{l-2}$$

and the matrices P and $[U_0 U_1 U_2 \ldots U_{l-1}]$ are given by Eqs. (21) and (22), respectively.

Proof. Given that the $n \times n$ matrix

$$[V(I - P)((I \oplus T_2) - (T_1 \oplus I))(U_0 + U_1 + U_2 + \ldots + U_{l-1})]$$

satisfies condition $E(p)$. From Eq. (20), the matrix polynomial $H(\xi)$ determined by $([X_1 X_2], T_1 \oplus T_2)$ is given by

$$H(\xi) = V(I-P)((I \oplus T_2)e^{i\xi} - (T_1 \oplus I))(U_0 + U_1 e^{i\xi} + U_2 e^{2i\xi} + \ldots + U_{l-1} e^{(l-1)i\xi})$$

(we have taken $\lambda = e^{i\xi}$). Then

$$H(0) = [V(I-P)((I \oplus T_2) - (T_1 \oplus I))(U_0 + U_1 + U_2 + \ldots + U_{l-1})]$$

which satisfies condition $E(p)$. By Theorem 3.2, Eq. (5) corresponding to the symbol function $H(\xi)$ has a solution vector Φ such that $\hat{\Phi}$ is continuous at 0 with $\hat{\Phi}(0) \neq 0$.

We will construct a symbol function by selecting suitable spectral data using the inverse representation theorem. We need the following Lemmas to find the $n \times nl$ matrix V mentioned in the inverse representation theorem for the construction of symbol function.

Lemma 3.1 (Predrag Stanimirovic, Miomir Stankovic [17]). *Let A be an $m \times n$ rectangular matrix where $m > n$. Then A has a unique Moore Penrose generalized inverse A^g such that $A^g A = I_n$, if A is of full rank, i.e. A has rank n.*

Lemma 3.2. *Let A be an $m \times m$ square matrix and B be an $m \times n$ rectangular matrix, $m > n$. Then the $m \times n$ product matrix AB has full rank, i.e. rank(AB)=n, if A has rank m and B has rank n.*

Proof. Given that the $m \times m$ matrix A has rank m and the $m \times n$ rectangular matrix B has rank n, $m > n$. Consider the linear transformation $T_A : \mathbb{R}^m \rightarrow \mathbb{R}^m$ and $T_B : \mathbb{R}^n \rightarrow \mathbb{R}^m$. Since A and B are of full rank, $Kernel(T_A)$ and $Kernel(T_B)$ are $\{0\}$. Now let $T_A T_B(x) = 0$, since A is of full rank we have

$$T_A(T_B x) = 0 \Rightarrow T_B x = 0$$

But the $m \times n$ matrix B is also of full rank. Then

$$T_B x = 0 \Rightarrow x = 0$$

Thus

$$T_A T_B(x) = 0 \Rightarrow x = 0$$

or

$$T_{AB}(x) = 0 \Rightarrow x = 0$$

i.e. Kernel of T_{AB} is $\{0\} \Rightarrow$ The matrix AB has full rank.

Now we will construct a symbol function $H(\xi)$ from a Jordan pair $(\mathrm{X}, \mathrm{T}) = ([X_1 X_2], T_1 \oplus T_2)$ such that $H(0)$ satisfies condition $E(p)$. Select a Jordan pair $(\mathrm{X}, \mathrm{T}) = ([X_1 X_2], T_1 \oplus T_2)$ such that the $nl \times nl$ matrix $(I \oplus T_2) - (T_1 \oplus I)$ is of full rank. Since $([X_1 X_2], T_1 \oplus T_2)$ is a Jordan pair, the columns of $nl \times nl$ matrix

$$S_{l-1} = [Col(X_1 T_1^i X_2 T_2^{l-1-i})_{i=0}^{l-1}]$$

are independent. Then columns of the matrix S_{l-1}^{-1} are also independent. From Eq. (22), we have

$$[U_0 U_1 U_2 \ldots U_{l-1}] = S_{l-1}^{-1}$$

i.e. columns of the matrix $[U_0 U_1 U_2 \ldots U_{l-1}]$ are independent. Then the columns of

$$U = [U_0 + U_1 + U_2 + \ldots + U_{l-1}]$$

are also independent, i.e. the $nl \times n$ matrix U is of full rank. Since $(I \oplus T_2) - (T_1 \oplus I)$ is of full rank, from Lemma 3.2 it follows that the product matrix

$$F = ((I \oplus T_2) - (T_1 \oplus I))(U_0 + U_1 + U_2 + \ldots + U_{l-1}) \tag{37}$$

of size $nl \times n$ is of rank n. By Lemma 3.1, there exists a unique Moore Penrose generalized inverse matrix $F^g \in \mathbb{C}^{n \times nl}$ such that

$$F^g \times F = I_n$$

Now we have to show that

$$\begin{pmatrix} S_{l-2} \\ F^g \end{pmatrix}$$

is nonsingular. For that we will prove two Lemmas.

Lemma 3.3. *Let A be an $(n-m) \times n$ matrix with linearly independent row vectors $v_1, v_2 \ldots v_{n-m} \in \mathbb{R}^n$ and B be an $n \times m$ matrix with linearly independent column vectors $u_1, u_2 \ldots u_m \in \mathbb{R}^n$ where $m, n \in \mathbb{N}$. If $A \times B = 0$, then the $n \times n$ matrix*

$$\begin{pmatrix} A \\ B^T \end{pmatrix}$$

is nonsingular.

Proof. *Given that* $A \times B = 0$, *then* $v_i \perp u_j \ \forall\, i, j$ *where* $i \in \{1, 2 \ldots n - m\}$ *and* $j \in \{1, 2 \ldots m\}$. *Then for any* j, $u_j \in \mathbb{R}^n$ *cannot be written as a linear combination of the vectors in* $\{v_i \in R^n / i = 1, 2 \ldots n - m\}$. *Thus the* $n \times n$ *matrix*

$$\begin{pmatrix} A \\ B^T \end{pmatrix}$$

is of rank n, i.e. the matrix is nonsingular.

Now, we will prove that $S_{l-2}F = 0$

Lemma 3.4. *Let the matrices* S_{l-2} *and* F *are given by*

$$S_{l-2} = Col[X_1 T_1^i X_2 T_2^{l-2-i}]_{i=0}^{l-2}$$

and

$$F = ((I \oplus T_2) - (T_1 \oplus I))(U_0 + U_1 + U_2 + \ldots + U_{l-1}) \tag{38}$$

where

$$[U_0 U_1 U_2 \ldots U_{l-1}] = S_{l-1}^{-1} Eq. (22)$$

Then

$$S_{l-2}F = 0 \tag{39}$$

Proof. *Define a matrix K as*

$$K = S_{l-2}((I \oplus T_2) - (T_1 \oplus I))S_{l-1}^{-1} \tag{40}$$

i.e.

$$K = S_{l-2}((I \oplus T_2) - (T_1 \oplus I))[U_0 U_1 U_2 \ldots U_{l-1}] \tag{41}$$

Define the $nl \times n$ *matrices* $K_0, K_1, K_2 \ldots K_{l-1}$ *such that*

$$K = [K_0 K_1 K_2 \ldots K_{l-1}]$$

Then from Eqs. (38) *and* (41), *we get*

$$S_{l-2}F = K_0 + K_1 + K_2 + \ldots + K_{l-1} \tag{42}$$

Now

$$K = \begin{pmatrix} X_1X_2T_2^{l-2} \\ X_1T_1X_2T_2^{l-3} \\ \dots\dots \\ X_1T_1^{l-2}X_2 \end{pmatrix} ((I \oplus T_2) - (T_1 \oplus I)) \begin{pmatrix} X_1X_2T_2^{l-1} \\ X_1T_1X_2T_2^{l-2} \\ \dots\dots \\ X_1T_1^{l-1}X_2 \end{pmatrix}^{-1}$$

Take $X_1T_1^i = A_i$ *and* $X_2T_2^i = B_i$ *where* $i = 0, 1, 2 \dots l-1$. *Then the above product will be of the form*

$$\begin{pmatrix} A_0 - A_1B_{l-1} - B_{l-2} \\ A_1 - A_2B_{l-2} - B_{l-3} \\ \dots\dots \\ A_{l-2} - A_{l-1}B_1 - B_0 \end{pmatrix} \begin{pmatrix} A_0B_{l-1} \\ A_1B_{l-2} \\ \dots\dots \\ A_{l-1}B_0 \end{pmatrix}^{-1}$$

$$= \begin{pmatrix} A_0B_{l-1} \\ A_1B_{l-2} \\ \dots\dots \\ A_{l-2}B_1 \end{pmatrix} \begin{pmatrix} A_0B_{l-1} \\ A_1B_{l-2} \\ \dots\dots \\ A_{l-1}B_0 \end{pmatrix}^{-1} - \begin{pmatrix} A_1B_{l-2} \\ A_2B_{l-3} \\ \dots\dots \\ A_{l-1}B_0 \end{pmatrix} \begin{pmatrix} A_0B_{l-1} \\ A_1B_{l-2} \\ \dots\dots \\ A_{l-1}B_0 \end{pmatrix}^{-1}$$

$$= \begin{pmatrix} 1\,0\,0\,.\,.\,0 & -1 & 0 & 0 & 0\,.\,.\,.\,.\,. & 0 \\ 0\,1\,0\,.\,.\,0 & 0 & -1 & 0 & 0\,.\,.\,.\,.\,. & 0 \\ 0\,0\,1\,.\,.\,0 & 0 & 0 & -1 & 0\,.\,.\,.\,.\,. & 0 \\ .\,.\,.\,.\,.\,. & . & . & . & .\,.\,.\,.\,.\,. & . \\ .\,.\,.\,.\,.\,. & . & . & . & .\,.\,.\,.\,.\,. & . \\ 0\,0\,0\,.\,.\,0 & 0 & 0 & 1 & 0\,.\,.\,.\,.\,-1 & 0 \\ 0\,0\,0\,.\,.\,0 & 0 & 0 & 0 & 1\,0\,0\,.\,.\,0 & -1 \end{pmatrix}$$

Thus

$$K = \begin{pmatrix} 1\,0\,0\,.\,.\,0 & -1 & 0 & 0 & 0\,.\,.\,.\,.\,. & 0 \\ 0\,1\,0\,.\,.\,0 & 0 & -1 & 0 & 0\,.\,.\,.\,.\,. & 0 \\ 0\,0\,1\,.\,.\,0 & 0 & 0 & -1 & 0\,.\,.\,.\,.\,. & 0 \\ .\,.\,.\,.\,.\,. & . & . & . & .\,.\,.\,.\,.\,. & . \\ .\,.\,.\,.\,.\,. & . & . & . & .\,.\,.\,.\,.\,. & . \\ 0\,0\,0\,.\,.\,0 & 0 & 0 & 1 & 0\,.\,.\,.\,.\,-1 & 0 \\ 0\,0\,0\,.\,.\,0 & 0 & 0 & 0 & 1\,0\,0\,.\,.\,0 & -1 \end{pmatrix}$$

But from Eq. (42), *we have*

$$S_{l-2}F = K_0 + K_1 + K_2 + \dots + K_{l-1}$$

where

$$K_0 = \begin{pmatrix} 1\,0\,0\,.\,.\,0 \\ 0\,1\,0\,.\,.\,0 \\ 0\,0\,1\,.\,.\,0 \\ .\,.\,.\,.\,.\,. \\ 0\,0\,0\,.\,.\,0 \\ 0\,0\,0\,.\,.\,0 \end{pmatrix}$$

$$K_1 = \begin{pmatrix} -1 & 0 & 0 & .\,. & 0 \\ 0 & -1 & 0 & .\,. & 0 \\ 0 & 0 & -1 & .\,. & 0 \\ . & . & . & .\,. & . \\ 0 & 0 & 0 & .\,. & 0 \\ 0 & 0 & 0 & .\,. & 0 \end{pmatrix}$$

and so on.

We get $K_0 + K_1 + K_2 + \ldots + K_{l-1} = 0$, *i.e.*

$$S_{l-2}F = 0$$

Corollary 3.1. *If*

$$P = (I \oplus T_2)[Col(X_1 T_1^i X_2 T_2^{l-1-i})_{i=0}^{l-1}]^{-1} \begin{pmatrix} I \\ 0 \end{pmatrix} S_{l-2},$$

then $PF = 0$.

Proof. *From Lemma 3.4, we have*

$$S_{l-2}F = 0$$

But Kernel(S_{l-2}*) = Kernel(*P*) (Gohberg, Lancaster [5, p. 191, Proposition 7.4])*
Hence

$$PF = 0$$

From Lemmas 3.3 and 3.4 , we get

$$\begin{pmatrix} S_{l-2} \\ F^T \end{pmatrix}$$

is nonsingular. The Moore–Penrose generalized inverse of F is given by $F^g = (F^T F)^{-1} F^T$[Predrag Stanimirovic, Miomir Stankovic [17]]. Since $(F^T F)^{-1}$ is of rank n and

$$\begin{pmatrix} S_{l-2} \\ F^T \end{pmatrix}$$

is nonsingular, we get

$$\begin{pmatrix} S_{l-2} \\ F^g \end{pmatrix}$$

is nonsingular. Define $V = E_n F^g$ where E_n is an $n \times n$ full rank diagonal matrix where 1 occurs once and other entries are less than one absolute value. Since $\begin{pmatrix} S_{l-2} \\ F^g \end{pmatrix}$ is nonsingular, we get

$$\begin{pmatrix} S_{l-2} \\ V \end{pmatrix}$$

is nonsingular. Also

$$[V(I - P)((I \oplus T_2) - (T_1 \oplus I))(U_0 + U_1 + U_2 + \ldots + U_{l-1})] = V(I - P)F$$

$$= VF - VPF = VF(Corollary\ 3.1)$$

Then

$$VF = E_n F^g F = E_n$$

But E_n satisfies condition $E(p)$ for $p = 1$. Thus we have found V such that

$$[V(I - P)((I \oplus T_2) - (T_1 \oplus I))(U_0 + U_1 + U_2 + \ldots + U_{l-1})]$$

satisfies condition $E(p)$. By Proposition 3.2, there exists a symbol function $H(\xi)$ with Jordan pair (X,T) such that the corresponding multiscaling equation (5) has a solution vector Φ such that $\hat{\Phi}$ is continuous at 0 with $\hat{\Phi}(0) \neq 0$. We will state this as a theorem as follows.

Theorem 3.3. *Let* (X, T) $= ([X_1 X_2], T_1 \oplus T_2)$ *be a Jordan pair such that the* $nl \times nl$ *matrix* $(I \oplus T_2) - (T_1 \oplus I)$ *is of full rank. Then there exist a symbol function* $H(\xi)$ *with Jordan pair (X,T) such that the corresponding multiscaling equation* (5) *has a solution vector* Φ *such that* $\hat{\Phi}$ *is continuous at 0 with* $\hat{\Phi}(0) \neq 0$.

If we do not choose the matrix V appropriately, then any matrix polynomial $\tilde{L}(\lambda)$ having the same Jordan pair (X,T) may not necessarily generate a multiscaling function which is proved by the following theorem.

Theorem 3.4. *Given a Jordan pair $(X, T) = ([X_1 X_2], T_1 \oplus T_2)$. Let the matrix polynomial*

$$L(\lambda) = V(I - P)((I \oplus T_2)\lambda - (T_1 \oplus I))(U_0 + U_1\lambda + U_2\lambda^2 + \ldots + U_{l-1}\lambda^{l-1})$$

having (X,T) as its Jordan pair, generate a solution vector Φ of Eq. (5) *such that $\hat{\Phi}$ is continuous at 0 with $\hat{\Phi}(0) \neq 0$. Then any matrix polynomial $\tilde{L}(\lambda)$ having the same Jordan pair (X,T) may not necessarily generate such a multiscaling function.*

Proof. *Given that*

$$L(\lambda) = V(I - P)((I \oplus T_2)\lambda - (T_1 \oplus I))(U_0 + U_1\lambda + U_2\lambda^2 + \ldots + U_{l-1}\lambda^{l-1})$$

generate a solution vector Φ of Eq. (5) *such that $\hat{\Phi}$ is continuous at 0 with $\hat{\Phi}(0) \neq 0$. Then L(1) will have an eigenvalue 1 (Keinert [11, Theorem (11.1), p. 220]). Choose $a > 1$ such that any of the eigenvalue of L(1) not equal to $\frac{1}{a}$. Consider*

$$\tilde{L}(\lambda) = \bar{V}(I - P)((I \oplus T_2)\lambda - (T_1 \oplus I))(U_0 + U_1\lambda + U_2\lambda^2 + \ldots + U_{l-1}\lambda^{l-1})$$

with $\bar{V} = aV$. Then $\tilde{L}(\lambda)$ also have the same Jordan pair (X,T). But $\tilde{L}(1)$ does not have an eigenvalue 1 and will have an eigenvalue a which is greater than 1 in absolute value (by the choice of a), which violates the necessary condition that $\tilde{L}(1)$ must have an eigenvalue 1 to generate a multiscaling function Φ [Keinert [11, Theorem (11.1), p. 220]]. Thus $\tilde{L}(\lambda)$ does not generate a solution vector Φ of Eq. (5) *such that $\hat{\Phi}$ is continuous at 0 with $\hat{\Phi}(0) \neq 0$ even though it has the same Jordan pair (X,T).*

Definition 3.2 (Keinert [11, p. 143]). The recursion coefficients H_k of a matrix refinement equation with dilation factor m satisfy the sum rules of order p if there exist vectors $y_0, y_1 \ldots y_{p-1}$ with $y_0 \neq 0$ such that

$$\sum_{t=0}^{n} \binom{n}{t} m^t (-i)^{n-t} y_t D^{n-t} H(\frac{2\pi s}{m}) = \delta_{0s} y_n s = 0, 1, 2 \ldots m - 1 \qquad (43)$$

for n=0...p-1.

We can construct a matrix polynomial which satisfies sum rules of order 1 using the procedure of construction of a matrix polynomial for a given Jordan pair we explained above.

Theorem 3.5. *Let (X,T)=($[X_1 X_2], T_1 \oplus T_2$) be a Jordan pair such that $X_1 \in \mathbb{C}^{n\times m}$, $X_2 \in \mathbb{C}^{n\times(nl-m)}$ and $T_1 \in \mathbb{C}^{m\times m}$, $T_2 \in \mathbb{C}^{(nl-m)\times(nl-m)}$ and the $nl \times nl$ matrix*

$(I \oplus T_2) - (T_1 \oplus I)$ is of full rank and -1 is a diagonal entry in T_1. Then there exist a matrix polynomial $L(\lambda) = \sum_{k=0}^{l} A_k\lambda^k$, $A_k \in \mathbb{C}^{n\times n}$ with Jordan pair (X,T) such that $L(\lambda)^T = \sum_{k=0}^{l} A_k^T\lambda^k$ satisfies the sum rules of order 1.

Proof. Given that $(I \oplus T_2) - (T_1 \oplus I)$ is of full rank and -1 is a diagonal entry in T_1. Let y_0 be the eigenvector corresponding to the eigenvalue -1. Now construct a matrix polynomial $L(\lambda) = \sum_{k=0}^{l} A_k\lambda^k$ with Jordan pair (X,T) using the procedure above by selecting the matrix E_n in such a way that y_0^T is a left eigenvector of E_n^T corresponding to the eigenvalue 1 (which is always possible since we can choose E_n to be the identity matrix of order n). Since y_0 is an eigenvector of $L(\lambda)$ corresponding to the eigenvalue -1, we get

$$L(-1)y_0 = 0$$

Taking transpose, we get

$$y_0^T L(-1)^T = 0 \tag{44}$$

Also

$$y_0^T L(1)^T = y_0^T E_n^T = y_0^T \tag{45}$$

From Eqs. (44) and (45), we get that $L(\lambda)^T$ satisfies sum rules of order 1.

4 Conclusion

We have constructed a symbol function $H(\xi)$ by selecting a suitable Jordan pair $([X_1X_2], T_1 \oplus T_2)$ such that the corresponding multiscaling equation (5) has a solution vector Φ such that $\hat{\Phi}$ is continuous at 0 with $\hat{\Phi}(0) \neq 0$. It is worth noting that any matrix polynomial having the same Jordan pair $([X_1X_2], T_1 \oplus T_2)$ may not generate such a multiscaling function. But a suitable choice of the matrix V in the construction will guarantee a solution. We have also constructed a symbol function which satisfies the sum rules of order 1 using this approach.

References

1. Bownik, M.: On characterizations of multiwavelets in $L^2(R^n)$. Proc. Am. Math. Soc. **129**(11), 3265–3274 (2001)
2. Chui, C.K., Jiang, Q.: Balanced Multiwavelets in R. Math. Comput. **74**(251) 1323–1344 (2004)
3. Daubechies, I.: Ten Lectures on Wavelets, vol. 61. Society for Industrial and Applied Mathematics, Philadelphia (1992)

4. Geronimo, J.S., Hardin, D.P., Massopust, P.R.: Fractal functions and wavelet expansions based on several scaling functions. J. Approx. Theory **78**(3), 373–401 (1994)
5. Gohberg, I., Rodman, L., Lancaster, P.: Matrix Polynomials. Academic, London, (1982)
6. Gripenberg, G.: Computing the joint spectral radius. Linear Algebra Appl. **234**, 43–60 (1996)
7. Heil, C., Colella, D.: Matrix refinement equations: existence and uniqueness. J. Fourier Anal. Appl. **2**, 363–378 (1996)
8. Heil, C., Strang, G.: Continuity of the joint spectral radius: application to wavelets. In: Linear Algebra for Signal Processing, pp. 51–61. Springer, New York (1995)
9. Heil, C., Strang, G., Strela, V.: Approximation by translates of refinable functions. Numer. Math. **73**(1), 75–94 (1996)
10. Kaiser, G.: A Friendly Guide to Wavelets. Birkhauser, Basel (1994)
11. Keinert, F.: Wavelets and Multiwavelets. Chapman and Hall/CRC, London (2004)
12. Laub, A.J.: Matrix Analysis for Scientists and Engineers. SIAM, Philadelphia (2005)
13. Markus, A.S.: Introduction to the Spectral Theory of Polynomial Operator Pencils, vol. 71. AMS Bookstore, Providence (1988)
14. Plonka, G., Strela, V.: From Wavelets to Multiwavelets. Mathematical Methods for Curves and Surfaces II (eds- M. Dahlen, T. Lyche, L.L. Schumaker), Vanderbilt University Press 1998, Nashville, TN, pp. 1–25 (1997)
15. Plonka, G., Strela, V.: Construction of multiscaling functions with approximation and symmetry. SIAM J. Math. Anal. **29**(2), 481–510 (1998)
16. Sjoestrand, J.: Spectral properties of non-self-adjoint operators. arXiv preprint arXiv:1002.4844 (2010)
17. Stanimirovic, P., Stankovic, M.: Determinants of rectangular matrices and Moore-Penrose inverse. Novi Sad J. Math **27**(1), 53–69 (1997)
18. Strang, G.: Wavelets and dilation equations: a brief introduction. SIAM Rev. **31**(4), 614–627 (1989)
19. Strang, G., Strela, V.: Orthogonal multiwavelets with vanishing moments. Opt. Eng. **33**(7), 2104–2107 (1994)
20. Strela, V.: Multiwavelets: regularity, orthogonality and symmetry via two scale similarity transform. Stud. Appl. Math. **98**(4), 335–354 (1997)
21. Theys, J.: Joint spectral radius: theory and approximations. Diss. Ph.D. thesis, Universite catholique de Louvain (2005)
22. Thielemann, H.: Polynomial functions are refinable. arXiv preprint arXiv:1012.2453 (2010)
23. Turcajova, R.: An algorithm for the construction of symmetric orthogonal multiwavelets. SIAM J. Matrix Anal. Appl. **25**(2), 532–550 (2003)
24. Turcajova, R., Kautsky, J.: Block Toeplitz-like operators and multiwavelets. In: SPIE's 1995 Symposium on OE/Aerospace Sensing and Dual Use Photonics. International Society for Optics and Photonics (1995)

A Remark on Reconstruction of Splines from Their Local Weighted Average Samples

P. Devaraj and S. Yugesh

Abstract In this paper, we study the reconstruction of cardinal spline functions from their weighted local average samples $y_n = f \star h(n), n \in \mathbb{Z}$, where the weight function $h(t)$ has support in $[-\frac{1}{2}, \frac{1}{2}]$. We prove that there exists a unique solution for the following problem: for the given data y_n and given degree d, find a cardinal spline $f(t)$ of degree d satisfying $y_n = f \star h(n), n \in \mathbb{Z}$.

Keywords Regular average sampling • Cardinal spline interpolation

2010 MSC: 42C15, 94A20

1 Introduction and Preliminaries

The sampling theorem is one of the most powerful results in signal analysis. The classical sampling theorem of Shannon gives the following reconstruction formula

$$f(x) = \sum_{k \in \mathbb{Z}} f\left(\frac{k\pi}{\Omega}\right) \frac{\sin(\Omega x - k\pi)}{\Omega x - k\pi.},$$

provided $supp(\hat{f}) \subset [-\Omega, \Omega]$. The assumption that a signal is bandlimited necessitates the signal to be of infinite duration which is not always realistic. Hence it is natural to consider and investigate other classes of signals for which sampling theorem holds. We consider the simple model functions.

P. Devaraj (✉) • S. Yugesh
Department of Mathematics, College of Engineering Guindy,
Anna University, Chennai 600 025, India
e-mail: devaraj@annauniv.edu; devap2001@yahoo.com

C. Bandt et al. (eds.), *Fractals, Wavelets, and their Applications*, Springer Proceedings in Mathematics & Statistics 92, DOI 10.1007/978-3-319-08105-2_21,

Let us introduce some notations. The cardinal central B-spline of degree d is denoted by β_d, and is defined as

$$\beta_d = \chi_{[-\frac{1}{2},\frac{1}{2}]} \star \chi_{[-\frac{1}{2},\frac{1}{2}]} \star \cdots \star \chi_{[-\frac{1}{2},\frac{1}{2}]} (d+1 \text{ terms}).$$

Let S_d denote the set of all functions of the form

$$f(t) = \sum_{n\in\mathbb{Z}} a_n \beta_d(t-n)$$

with suitable coefficients a_n, let

$$S_{d,\gamma} = \{f(t) \in S_d : f(t) = O(|t|^\gamma) \; as \; t \to \pm\infty\}$$

and

$$D_\gamma = \{\{y_n\} : y_n = O(|n|^\gamma) \; as \; n \to \pm\infty\}.$$

Consider the following cardinal spline interpolation problem: Given a sequence of real numbers $\{y_n\}_{n\in\mathbb{Z}}$, find a spline $f \in S_d$ such that $f(n) = y_n, n \in \mathbb{Z}$. When $d = 1$ this problem has a unique solution but for $d > 1$, it has infinitely many solutions. Moreover, the solution space is a $d-1$ dimensional subspace of S_d when d is odd and a d dimensional subspace of S_d when d is even. By imposing growth conditions, Shoenberg [2] proved that for a given sequence of numbers $\{y_n\}_{n\in\mathbb{Z}} \in D_\gamma$ the following problem:

Find a spline $f \in S_{d,\gamma}$, satisfying $f(n) = y_n, n \in \mathbb{Z}$,

has a unique solution.

In practical applications, the measurements y_n of the signal are not always exact but local weighted average of the signal at n. In [1], the authors have considered one such average sampling problem:

Problem 1. Given a sequence of numbers $\{y_n\}_{n\in\mathbb{Z}}$, find $f \in S_{d,\gamma}$, such that $f \star h(n) = y_n, n \in \mathbb{Z}$, where h satisfies

$$supp(h) \subseteq [-\frac{1}{2}, \frac{1}{2}], \qquad h(t) \geq 0, t \in \mathbb{R} \tag{1}$$

$$0 < \int_{-\frac{1}{2}}^{0} h(t)dt < \infty \quad \text{and} \quad 0 < \int_{0}^{\frac{1}{2}} h(t)dt < \infty. \tag{2}$$

They have shown that there is a unique solution when $d = 1, 2, 3, 4$. The same problem for $d \geq 5$ has been posed as an open problem in [1]. The problem becomes more complex if the method of [1] is used. We show that the solution to this problem is unique for all d with some additional realistic conditions on h.

2 Existence and Uniqueness Theorem

Theorem 2.1 (Main Theorem). *Let $d \in \mathbb{N}$ and h be a nonnegative measurable function satisfying conditions (1), (2) and the polynomial $\sum_{n\in\mathbb{Z}} h \star \beta_d(n)z^{-n}$ has no root on the unit circle, then for a given sequence of real numbers $\{y_n\}_{n\in\mathbb{Z}} \in D_\gamma$, there exists a unique spline $f \in S_{d,\gamma}$ satisfying $f \star h(n) = y_n, n \in \mathbb{Z}$. This unique solution can be expressed as $f(t) = \sum_{n\in\mathbb{Z}} y_n L_{h,d}(t-n)$, where the reconstruction spline $L_{h,d}$ is given by $L_{h,d} = \sum_{n\in\mathbb{Z}} c_n \beta_d(t-n)$ and c_n are the coefficients of the expansion $\frac{1}{G_{h,d}(z)} = \sum_{n\in\mathbb{Z}} c_n z^{-n}$. There exists a constant $\mu_{h,d} \in (0,1)$ such that $L_{h,d}(t) = O(|t|^l \mu_{h,d}^{|t|})$.*

We define the function

$$G_{h,d}(z) := \sum_{n\in\mathbb{Z}} h \star \beta_d(n)z^{-n},$$

where h is a nonnegative measurable function satisfying conditions (1) and (2). Then $G_{h,d}(z)$ is a Laurent polynomial and can be rewritten in terms of exponential splines considered in [2] as

$$G_{h,d}(z) := \int_{-\frac{1}{2}}^{\frac{1}{2}} h(t)\Upsilon_{z,d}(t)dt,$$

where

$$\Upsilon_{z,d}(t) := \sum_{n\in\mathbb{Z}} z^{-n}\beta_d(n-t).$$

For the proof of the theorem, we need some properties of $\Upsilon_{z,d}(t)$ discussed in [1]:

Lemma 2.1 ([1]). *For $d \in \mathbb{N}, n \in \mathbb{Z}$ and $z \in \mathbb{C}\setminus\{0\}$, we have:*

(i) $\Upsilon_{z^{-1},d}(-t) = \Upsilon_{z,d}(t)$,
(ii) $\Upsilon_{z,d}(t+n) = z^{-n}\Upsilon_{z,d}(t)$.

Proof. (i) We have

$$\Upsilon_{z^{-1},d}(-t) = \sum_{n\in\mathbb{Z}}(z^n)\beta_d(n+t) = \sum_{n\in\mathbb{Z}} z^n\beta_d(-n-t) = \sum_{n\in\mathbb{Z}} z^{-n}\beta_d(n-t) = \Upsilon_{z,d}(t).$$

(ii)

$$\Upsilon_{z,d}(t+n) = \sum_{k\in\mathbb{Z}} z^{-k}\beta_d(k-t-n) = z^{-n}\sum_{k\in\mathbb{Z}} z^{-k}\beta_d(k-t) = z^{-n}\Upsilon_{z,d}(t).$$

□

We also need some additional properties of $\Upsilon_{z,d}$:

Lemma 2.2. *For $d, l \in \mathbb{N}, n \in \mathbb{Z}$ and $z \in \mathbb{C} \setminus \{0\}$, the following holds:*

$$\frac{d^l}{dz^l}(\Upsilon_{z,d}(t+n)) = z^{-n}\Bigg[(-1)^{2l}\frac{d^l}{dz^l}\Upsilon_{z,d}(t) + \sum_{k=1}^{l-1}\binom{l}{k}(-1)^{2l+k}$$
$$\frac{n(n+1)\cdots(n+k-1)}{z^k}\frac{d^{l-k}}{dz^{l-k}}\Upsilon_{z,d}(t) + (-1)^{3l}\frac{n(n+1)\cdots(n+l-1)}{z^l}\Upsilon_{z,d}(t)\Bigg]$$

Proof. We prove by induction on l. First we shall prove the statement for $l = 1$.

$$\begin{aligned}\frac{d}{dz}(\Upsilon_{z,d}(t+n)) &= -\sum_{k\in\mathbb{Z}}\frac{k}{z}z^{-k}\beta_d(k-t-n)\\ &= \sum_{k\in\mathbb{Z}}\frac{(-k-n)}{z}z^{-k-n}\beta_d(k+n-t-n)\\ &= z^{-n}\sum_{k\in\mathbb{Z}}(-k)z^{-k-1}\beta_d(k-t) - \frac{n}{z}z^{-n}\sum_{k\in\mathbb{Z}}z^{-k}\beta_d(k-t)\\ &= z^{-n}\left[\frac{d}{dz}\Upsilon_{z,d}(t) - \frac{n}{z}\Upsilon_{z,d}(t)\right].\end{aligned}$$

Assuming the statement for $l-1$ and using some simple manipulations of the same, we obtain

$$\frac{d^l}{dz^l}(\Upsilon_{z,d}(t+n)) = \frac{d}{dz}\left(\frac{d^{l-1}}{dz^{l-1}}(\Upsilon_{z,d}(t+n))\right)$$
$$= \frac{d}{dz}\Bigg(z^{-n}\Bigg[(-1)^{2(l-1)}\frac{d^{l-1}}{dz^{l-1}}\Upsilon_{z,d}(t) + \sum_{k=1}^{l-2}\binom{l-1}{k}(-1)^{2l-2+k}$$
$$\frac{n(n+1)\cdots(n+k-1)}{z^k}\frac{d^{l-1-k}}{dz^{l-1-k}}\Upsilon_{z,d}(t) + (-1)^{3l-3}\frac{n(n+1)\cdots(n+l-2)}{z^{(l-1)}}\Upsilon_{z,d}(t)\Bigg]\Bigg)$$
$$= \frac{d}{dz}\Bigg(z^{-n}(-1)^{2(l-1)}\frac{d^{l-1}}{dz^{l-1}}\Upsilon_{z,d}(t) + \sum_{k=1}^{l-2}\binom{l-1}{k}(-1)^{2l-2+k}$$
$$\frac{n(n+1)\cdots(n+k-1)}{z^{k+n}}\frac{d^{l-1-k}}{dz^{l-1-k}}\Upsilon_{z,d}(t) + (-1)^{2l-2}\frac{n(n+1)\cdots(n+l-2)}{z^{(n+l-1)}}\Upsilon_{z,d}(t)\Bigg)$$
$$= z^{-n}\Bigg[(-1)^{2l}\frac{d^l}{dz^l}\Upsilon_{z,d}(t) + \sum_{k=1}^{l-1}\binom{l}{k}(-1)^{2l+k}\frac{n(n+1)\cdots(n+k-1)}{z^k}\frac{d^{l-k}}{dz^{l-k}}\Upsilon_{z,d}(t)$$
$$+(-1)^{3l}\frac{n(n+1)\cdots(n+l-1)}{z^l}\Upsilon_{z,d}(t)\Bigg].$$

□

In [1], the authors have shown that if h satisfies conditions (1) and (2), then the set of solutions of $f \star h(n) = y_n$ in S_d is a d+1 dimensional subspace of S_d when d is odd and a d dimensional subspace when d is even.

Lemma 2.3. *Let $\Lambda = \{f \in S_d : f \star h(n) = 0, n \in \mathbb{Z}\}$ and $z_1, z_2, \ldots, z_r$ be the roots of $G_{h,d}$ with corresponding multiplicities $m_1, m_2, \ldots, m_r$, respectively. Then the set of functions $\frac{d^i}{dz^i}\Upsilon_{z_j^{-1},d}$, where $i = 0, 1, 2, \ldots, m_j - 1$ and $j = 1, 2, \ldots, r$ forms a basis of Λ.*

Proof. Proof of the lemma follows from Lemmas 2.1, 2.2 and some simple manipulations. □

Lemma 2.4. *If $G_{h,d}(z)$ has no root on the unit circle and if $\sum_{n\in\mathbb{Z}} c_n z^{-n}$ is the Laurent's series expansion of $\dfrac{1}{G_{h,d}(z)}$, then their exists $\mu_{h,d} \in (0,1)$ such that $c_n = o\left(|n|^l (\mu_{h,d})^{|n|}\right)$.*

Proof. $G_{h,d}(z)$ can be written as $G_{h,d}(z) = z^{-\frac{l}{2}} p(z)$, where $p(z)$ is a polynomial of degree l. Let $z_1, z_2, \ldots, z_r$ be the roots of $G_{h,d}(z)$ with multiplicities $m_1, m_2, \ldots, m_r$, respectively. Then

$$\frac{1}{G_{h,d}(z)} = z^{\frac{l}{2}} \sum_{i=1}^{r}\sum_{j=1}^{m_i} \frac{A_{ij}}{(z-z_i)^j}.$$

First we consider the case $|z_i| < 1$.

$$\begin{aligned} z^{\frac{l}{2}} \frac{A_{ij}}{(z-z_i)^j} &= z^{\frac{l}{2}} A_{ij} \sum_{n=j}^{\infty} n(n-1)\cdots(n-j+1)(z_i)^{n-j} \frac{1}{z^n} \\ &= A_{ij} \sum_{n=j}^{\infty} n(n-1)\cdots(n-j+1)(z_i)^{n-j} \frac{1}{z^{n-\frac{l}{2}}}. \end{aligned}$$

In the above series expansion,

$$\begin{aligned} \text{coefficient of } \frac{1}{z^{|n|}} &= A_{ij}(n+\frac{l}{2})(n+\frac{l}{2}-1)\cdots(n+\frac{l}{2}-j+1)(z_i)^{n+\frac{l}{2}-j} \\ &= o(|n|^l |z_i|^n (|z_i|)^{-l} \\ &= o(|n|^l z_i^{|n|}). \end{aligned}$$

Next we consider the case $|z_i| > 1$.

$$z^{\frac{L}{2}}\frac{A_{ij}}{(z-z_i)^j} = z^{\frac{L}{2}}A_{ij}\sum_{n=j}^{\infty} n(n-1)\cdots(n-j+1)(z)^{n-j}\frac{1}{z_i^n}$$
$$= A_{ij}\sum_{n=j}^{\infty} n(n-1)\cdots(n-j+1)(z)^{n+\frac{L}{2}-j}\frac{1}{z_i^n}.$$

$$\text{Coefficient of } z^{|n|} = A_{ij}(n-\frac{l}{2}+j)(n-\frac{l}{2}+j-1)\cdots(n-\frac{l}{2})\frac{1}{(z_i)^{n-\frac{L}{2}+j}}$$
$$\leq n^l\frac{1}{z_i^n}(z_i)^{\frac{L}{2}}$$
$$= o(|n|^l\frac{1}{z_i^{|n|}}).$$

The above expansions are valid inside the annulus $r_1 < |z| < r_2$, where $r_1 = Max\{|z_i| : |z_i| < 1\}$ and $r_2 = Min\{|z_i| : |z_i| > 1\}$. Combining the two cases, we obtain $c_n = o\left(|n|^l(\mu_{h,d})^{|n|}\right)$, for some $\mu_{h,d} \in (0,1)$.

□

Proof (Main Theorem). Define

$$L_{h,d} = \sum_{n\in\mathbb{Z}} c_n\beta_d(t-n).$$

By Lemma 2.4

$$c_n = O\left(|n|^l(\mu_{h,d})^{|n|}\right).$$

As β_d has compact support,

$$O(L_{h,d}) = O(|t|^l\mu_{h,d}^{|t|}).$$

Now for $|t| > 2$,

$$\frac{\sum_{n\in\mathbb{Z}}|n|^\gamma|t-n|^l\mu_{h,d}^{|t-n|}}{(|t|+1)^\gamma} \leq \frac{\sum_{n\in\mathbb{Z}}|n|^\gamma(|[t]-n|+1)^l\mu_{h,d}^{|[t]-n|-1}}{(|[t]|)^\gamma}$$
$$= \frac{\sum_{n\in\mathbb{Z}}(|[t]-n|)^\gamma(|n|+1)^l\mu_{h,d}^{|n|-1}}{(|[t]|)^\gamma}$$

$$\leq \sum_{n\in\mathbb{Z}} (1+|n|)^{\gamma} (|n|+1)^{l} \mu_{h,d}^{|n|-1}$$
$$< \infty.$$

Using this inequality, we obtain

$$f(t) = \sum_{n\in\mathbb{Z}} y_n L_{h,d}(t-n) = O(|t|^{\gamma}),$$

as $t \to \pm\infty$.

Since

$$y_n L_{h,d}(t-n) = O(|t-n|^{l} \mu_{h,d}^{|t-n|} |n|^{\gamma}),$$

the series

$$\sum_{n\in\mathbb{Z}} y_n L_{h,d}(t-n)$$

converges uniformly and absolutely on every compact subset of $\mathbb{R}$. Now

$$\begin{aligned} f(t) &= \sum_{n\in\mathbb{Z}} y_n L_{h,d}(t-n) \\ &= \sum_{n\in\mathbb{Z}} y_n \sum_{k\in\mathbb{Z}} c_k \beta_d(t-n-k) \\ &= \sum_{k\in\mathbb{Z}} \left(\sum_{n\in\mathbb{Z}} y_n c_{k-n} \right) \beta_d(t-k). \end{aligned}$$

Therefore $f \in S_{d,\gamma}$.

Define

$$g_n = h \star \beta_d(n).$$

Then

$$G_{h,d}(z) = \sum_{n\in\mathbb{Z}} g_n z^{-n}.$$

Using

$$\left(\sum_{n\in\mathbb{Z}} g_n z^{-n} \right) \left(\sum_{n\in\mathbb{Z}} c_n z^{-n} \right) = 1,$$

we obtain

$$(h \star L_{h,d})(n) = \sum_{k\in\mathbb{Z}} c_k h \star \beta_d(n-k) = \sum_{k\in\mathbb{Z}} c_k g_{n-k} = \delta_0(n).$$

Hence

$$f(t) = \sum_{n\in\mathbb{Z}} y_n L_{h,d}(t-n)$$

satisfies

$$(h \star f)(n) = y_n, n \in \mathbb{Z}.$$

Next we show the uniqueness.

Suppose that $f, g \in S_{d,\gamma}$ and are solutions of Problem 1. Then $f - g \in \Lambda$. Using Lemma 2.3, there exist constants c_{ji} such that

$$f(t) - g(t) = \sum_{j=1}^{r} \sum_{i=0}^{m_j-1} c_{ji} \frac{d^i}{dz^i}\left(\Upsilon_{z_j^{-1},d}\right).$$

As $|z_i| \neq 1$, we get $f(t) - g(t) = o(|t|^{\gamma})$.

Using Lemma 2.1 and the behavior of

$$\frac{d^i}{dz^i}\left(\Upsilon_{z_j^{-1},d}\right)(t)$$

at $\pm\infty$, we obtain $c_{ji} = 0$ and hence $f = g$.

□

Acknowledgements The second author was supported by the Anna Centenary fellowship, Anna University, Chennai-25.

References

1. Perez-Villalon, G., Portal, A.: Reconstruction of splines from local average samples. Appl. Math. Lett. **25**, 1315–1319 (2012)
2. Schoenberg, I.J.: Cardinal spline interpolation. In: SIAM Regional Conference Series in Applied Mathematics (1973)

$\mathcal{C}^1$-Rational Cubic Fractal Interpolation Surface Using Functional Values

A.K.B. Chand and N. Vijender

Abstract Fractal interpolation is a modern technique for fitting of smooth/non-smooth data. In the present article, we develop the $\mathcal{C}^1$-rational cubic fractal interpolation surface (FIS) as a fixed point of the Read-Bajraktarević (RB) operator defined on a suitable function space. Our $\mathcal{C}^1$-rational cubic FIS is effective tool to stich surface data arranged on a rectangular grid. Our construction needs only the functional values at the grids being interpolated, therefore implementation is an easy task. We first construct the x-direction rational cubic FIFs (x-direction fractal boundary curves) to approximate the data generating function along the grid lines parallel to x-axis. Then we form a rational cubic FIS as a blending of these fractal boundary curves. An upper bound of the uniform distance between the rational cubic FIS and an original function is estimated for the convergence results. A numerical illustration is provided to explain the visual quality of our rational cubic FIS. An extra feature of this fractal surface scheme is that it allows subsequent interactive alteration of the shape of the surface by changing the scaling factors and shape parameters.

Keywords Fractals • Iterated function systems • Fractal interpolation functions • Fractal interpolation surfaces • Blending function

AMS Classification: 28A80, 65D05, 14Q10, 34D45, 37C25.

A.K.B. Chand • N. Vijender (✉)
Department of Mathematics, Indian Institute of Technology Madras, Chennai 600036, India
e-mail: chand@iitm.ac.in; vijendernallapu@gmail.com

C. Bandt et al. (eds.), *Fractals, Wavelets, and their Applications*, Springer Proceedings in Mathematics & Statistics 92, DOI 10.1007/978-3-319-08105-2_22,

1 Introduction

Data visualization is a tool to convert data into visual display for gaining understanding and insight into the data. Data visualization has proved its dominant role in many areas including computer graphics, reverse engineering, aerospace industries, earth and atmospheric science, medical imaging, and architecture design. The data that is available is only a sample of a phenomena, and it may not explain the phenomena completely. To overcome this difficulty, data is visualized in the form of curves/surfaces. In computer graphics there is often the need to construct a curve/surface from an experimental data whose form can be interactively adjusted by means of suitable parameters. This requirement cannot ever be reached using a polynomial spline interpolation because the curve/surface representation is unique in this procedure. In order to overcome this disadvantage, many investigations have been directed towards the rational splines with shape parameters. To construct the polynomial splines, the derivative values are usually needed, along with the functional values as input data. Unfortunately, in many practical problems, such as the description of the rainfall in some rainy region and some geometric shapes, the derivative values are difficult to get. Based only on the values of the function being interpolated, Duan et al. [10] constructed a rational cubic interpolants with shape parameters to tackle this problem. Duan et al. [11] extended this univariate rational spline interpolant to the bivariate rational interpolant.

Fractal functions, namely the functions whose graphs are fractal sets [15], offer an adequate tool to approximate non-smooth and irregular data. Utilizing the iterated function system (IFS) theory [13], Barnsley [1] proposed the concept of a FIF such that it is the attractor of a specific IFS. In general, FIFs are fixed points of the Read-Bajraktarević operator, which are defined on suitable function spaces. By imposing suitable conditions on the scaling factors, Barnsley and Harrington [2] introduced the construction of k- times differentiable FIFs, if up to k^{th} order derivative values of the original function are known at the initial end point of the interval. However, it is difficult to get all types of boundary conditions for fractal splines in this iterative construction. Fractal splines with general boundary conditions have studied recently [4, 8, 16]. The problem of shape preserving interpolation has important role in various engineering problems, for example car modeling, construction of mask surfaces, aero-plane and ship design, etc. Including aforementioned FIFs, all existing polynomial FIFs are not ideal for shape preservation. Owing to this reason, our group has introduced the shape preserving cubic spline FIF and rational spline FIF in the literature [6, 7, 9].

Fractal surfaces are proved to be useful to approximate various type of surfaces in chemistry, physics, image processing, material science, geology, and ocean engineering. Fractal surfaces are a natural outgrowth of fractal sets and fractal functions. Fractal interpolation surface was first constructed by Massopust [14]. In this construction he assumed the surface as triangular simplex and interpolation points on the boundary to be co-planar. To overcome lack of flexibility in this construction, Geronimo and Hardin [12] and Zhao [18] have generalized the construction of FIS by allowing more general boundary data. Later, with arbitrary contraction factors and without any condition on boundary data, Xie and Sun [17] constructed bivariate FIS to stich the interpolation data arranged on the rectangular grid. Further research and developments in this direction using various thoughts are carried out by Chand and Navascués [5], and Chand [3]. However, a $\mathcal{C}^1$ FIS based only on the functional values being interpolated has not been studied. Thus we attempt to construct a new class of $\mathcal{C}^1$-rational cubic FISs based only on the functional values being interpolated.

The rest of this paper is organized as follows: In Sect. 2, we review the basics of IFS theory and its connection with fractal interpolation. In Sect. 3, first we construct the fractal boundary curves ($\mathcal{C}^1$-rational cubic spline FIFs) in x-direction, then we form a $\mathcal{C}^1$-rational cubic FIS using these fractal boundary curves and blending functions. The approximation properties of the rational cubic FIS are studied in Sect. 4. The developed rational cubic FIS is applied on a surface data to construct the desired rational cubic FISs in Sect. 5.

2 Fractal Interpolation Functions

The basics of IFS theory is discussed in Sect. 2.1, and the construction of a FIF from an IFS is given in Sect. 2.2.

2.1 IFS Theory

Let $(\mathbb{X},\sigma),(\mathbb{Y},\sigma^*)$ be two metric spaces. A function $\varrho:\mathbb{X}\to\mathbb{Y}$ is said to be a contraction map with contractive factor $0\le c<1$ if

$$\sigma^*(\varrho(x),\varrho(y))\le c\,\sigma(x,y) \text{ for all } x,y\in\mathbb{X}.$$

An IFS is considered as a complete metric space $(\mathbb{X},\sigma)$ together with a finite set of continuous transformations $w_i:\mathbb{X}\to\mathbb{X}, i=1,2,\ldots,M-1$. If w_i are contractions with contraction factors $|s_i|, i=1,2,\ldots,M-1$, then the IFS is called hyperbolic. Let $\mathbb{H}(\mathbb{X})$ be the set of all non-empty compact subsets of $\mathbb{X}$. Now $\mathbb{H}(\mathbb{X})$ is a complete metric space with respect to the Hausdorff metric "h", where $h(A,B)=\max\{d_\sigma(A,B),d_\sigma(B,A)\}, d_\sigma(A,B)=\max_{x\in A}\min_{y\in B}\sigma(x,y)$. For $E\in\mathbb{H}(\mathbb{X})$, it can be shown that the Hutchinson map on $\mathbb{H}(\mathbb{X})$ defined as $W(E)=\bigcup_{i=1}^{m-1}w_i(E)$ is a contraction map with the contractive factor $s=\max\{|s_i|:i=1,2,\ldots,M-1\}$ [1]. By the Banach Fixed Point Theorem, the sequence $\{W^k(E)\}_{k=1}^{\infty}$ has unique limit, say $\mathcal{A}$ i.e., $\lim_{k\to\infty}W^k(E)=\mathcal{A}$ forany $E\in\mathbb{H}(\mathbb{X})$, and this $\mathcal{A}$ is called the deterministic fractal or the attractor of the hyperbolic IFS. Such an attractor of a suitable IFS matches with the graph of a fractal interpolation function, and the details are given in the following.

2.2 *Fractal Interpolation Functions*

Let $x_1 < x_2 < \cdots < x_{M-1} < x_M$ be a partition of the real compact interval $I = [x_1, x_M]$. Let $\{(x_i, f_i) \in I \times \mathbb{R} : i = 1, 2, \ldots, M\}$ be a given data set. Set $I_i = [x_i, x_{i+1}]$ and let $\phi_i : I \to I_i$, $i = 1, 2, \ldots, M-1$, be the contractive homeomorphisms such that

$$\phi_i(x_1) = x_i, \ \phi_i(x_M) = x_{i+1}, \tag{1}$$

$$|\phi_i(c_1) - \phi_i(c_2)| \le l_i \ |c_1 - c_2| \quad \forall \ c_1, c_2 \in I,$$

for some $0 \le l_i < 1$. Denote $C = I \times D$, D is a compact subset of $\mathbb{R}$ such that $f_i \in D$ for $i = 1, 2, \ldots, M$. Define the continuous mappings $F_i : C \to D, \ i = 1, 2, \ldots, M-1$, such that

$$F_i(x_1, f_1) = f_i, \ \ F_i(x_M, f_M) = f_{i+1}, \tag{2}$$

$$|F_i(x, t^*) - F_i(x, t^{**})| \le |\lambda_i||t^* - t^{**}|, \quad x \in I, \quad t^*, t^{**} \in D, \quad -1 < \lambda_i < 1.$$

For the construction of a desired IFS, now define the functions $w_i : C \to I_i \times D, i = 1, 2, \ldots, M-1$, as $w_i(x, t) = (\phi_i(x), F_i(x, t))$. The construction of a FIF is based on the following result:

Proposition 2.1 ([1]). *The IFS* $\{C;\ w_i, i = 1, 2, \ldots, M-1\}$ *defined above admits a unique attractor* G. G *is the graph of a continuous function* $h^* : I \to \mathbb{R}$ *such that* $h^*(x_i) = f_i, i = 1, 2, \ldots, M$.

The above function h^* is called a FIF corresponding to the IFS $\{C;\ w_i, i = 1, 2, \ldots, M-1\}$, and the functional equation of h^* is based on the following discussion that is given in detail in [1].

Suppose $\mathcal{G} = \{g : I \to \mathbb{R} : g$ is continuous, $g(x_1) = f_1$ and $g(x_M) = f_M\}$. Then $\mathcal{G}$ is a complete metric space with respect to the metric d induced by the uniform norm on $\mathcal{C}[x_1, x_M]$. Define the Read-Bajraktarević operator T on $(\mathcal{G}, d)$ by

$$(Tg)(x) = F_i(\phi_i^{-1}(x), g(\phi_i^{-1}(x))), \ x \in I_i, i = 1, 2, \ldots, M-1. \tag{3}$$

Using (1) and (2), it is easy to show that Tg is continuous on the intervals $I_i, i = 1, 2, \ldots, M-1$, and at each of the points $x_2, \ldots, x_{M-1}$. Also, it easy to verify that T is a contraction map on the metric space $(\mathcal{G}, d)$, i.e.,

$$d(Tg, Tg^*) \le |\lambda|_\infty d(g, g^*), \tag{4}$$

where $|\lambda|_\infty = \max\{|\lambda_i| : i = 1, 2, \ldots, M-1\} < 1$. Therefore, by the Banach Fixed Point Theorem, T possesses a unique fixed point, say h^* on $\mathcal{G}$, i.e., $(Th^*)(x) = h^*(x)\ \forall\ x \in I$. According to (3), the FIF h^* satisfies the following functional equation:

$$h^*(x) = F_i(\phi_i^{-1}(x), h^* \circ \phi_i^{-1}(x)),\ \ x \in I_i,\ \ i = 1, 2, \ldots, M-1. \tag{5}$$

The most widely studied fractal interpolation functions so far are defined by the IFS

$$\left.\begin{aligned} \phi_i(x) &= a_i x + b_i, \\ F_i(x, f) &= \lambda_i f + r_i(x), \end{aligned}\right\} \tag{6}$$

where $-1 < \lambda_i < 1$, and $r_i : I \to \mathbb{R}$ are suitable continuous functions such that (2) are satisfied. λ_i is called a scaling factor of the transformation w_i, and $\lambda = (\lambda_1, \lambda_2, \ldots, \lambda_{M-1})$ is the scaling vector of the FIF. The existence of a spline FIF is given in [2], and that result can be extended for the existence of rational spline FIFs in the following:

Theorem 2.1. *Let $\{(x_i, f_i), i = 1, 2, \ldots, M\}$ be a given data set, where $x_1 < \cdots < x_M$. Suppose that $\phi_i(x) = a_i x + b_i, F_i(x, f) = \lambda_i f + r_i(x), r_i(x) = \frac{p_i(x)}{q_i(x)}$, $p_i(x), q_i(x)$ are suitably chosen polynomials in x of degree t_1, t_2, respectively, and $q_i(x) \neq 0$ for every $x \in [x_1, x_M]$. Suppose for some integer $p \geq 0, |\lambda_i| < a_i^p, i = 1, 2, \ldots, M-1$. Let $F_i^m(x, f) = \frac{\lambda_i f + r_i^{(m)}(x)}{a_i^m}, r_i^{(m)}(x)$ represents the mth derivative of $r_i(x)$,*

$$f_1^m = \frac{r_1^{(m)}(x_1)}{a_1^m - \lambda_1},\quad f_N^m = \frac{r_{M-1}^{(m)}(x_M)}{a_{M-1}^m - \lambda_{M-1}},\ \ m = 1, 2, \ldots, p.$$

If $F_i^m(x_M, f_M^m) = F_{i+1}^m(x_1, f_1^m), i = 1, 2, \ldots, M-2,\ m = 1, 2, \ldots, p,$

then $\{\mathbb{R}^2; w_i(x, f) = (\phi_i(x), F_i(x, f)),\ i = 1, 2, \ldots, M-1\}$ determines a rational FIF $\Phi \in \mathcal{C}^p[x_1, x_M]$, and $\Phi^{(m)}$ is the rational FIF determined by $\{\mathbb{R}^2; w_i(x, f) = (\phi_i(x), F_i^m(x, f)),\ i = 1, 2, \ldots, M-1\}$.

Since $q_i(x) \neq 0$ for all $x \in [x_1, x_M]$, the proof of the above theorem follows through the suitable modifications of the arguments in [2].

3 $\mathcal{C}^1$-Rational Cubic FIS

Let $\{(x_i, y_j, z_{i,j}), i = 1, 2, \ldots, M+1, j = 1, 2, \ldots, N+1\}$ be given surface data. Let

$$x_1 < x_2 < \cdots < x_{M-1} < x_M < x_{M+1},$$

$$y_1 < y_2 < \cdots < y_{N-1} < y_N < y_{N+1},$$

be the partition of $[x_1, x_{M+1}] \times [y_1, y_{N+1}]$. We construct a $\mathcal{C}^1$-rational cubic FIS $\Phi : [x_1, x_M] \times [y_1, y_N] \to \mathbb{R}$ such that

$$\Phi(x_i, y_j) = z_{i,j}, i = 1, 2, \ldots, M, j = 1, 2, \ldots, N.$$

First in Sect. 3.1, we construct x-direction fractal boundary curves $\Psi_j : [x_1, x_m] \to \mathbb{R}$, $j = 1, 2, \ldots, N+1$, such that for $j = 1, 2, \ldots, N+1$, Ψ_j interpolates the data set $\Sigma_j = \{(x_i, z_{i,j}), i = 1, 2, \ldots, M\}$. Then as a blending of Ψ_j, $j = 1, 2, \ldots, N+1$, we construct our $\mathcal{C}^1$-rational cubic FIS in Sect. 3.2. Our $\mathcal{C}^1$-rational cubic FIS is a generalization to the rational cubic FIS with two-families of shape parameters obtained by Duan et al [11].

3.1 Construction of x-Direction Fractal Boundary curves

Based on the condition on the scaling factors given in the Theorem 2.1, we now construct the $\mathcal{C}^1$-rational cubic FIFs in the following manner.

Theorem 3.1. *Let Σ_j, $j = 1, 2, \ldots, N+1$, be a given data sets. Consider the IFSs $\mathcal{I}_j^* = \{[x_1, x_M] \times K_{1,j}; w_{i,j}(x, f) = (\phi_i(x), F_{i,j}(x, \zeta)), i = 1, 2, \ldots, M-1\}$, $j = 1, 2, \ldots, N+1$, where $K_{1,j}$ is a suitable compact subset of $\mathbb{R}$, $\phi_i(x) = a_i x + b_i$ satisfies (1), $F_{i,j}(x, \zeta) = \lambda_{i,j}\zeta + r_{i,j}(x)$, $r_{i,j}(x) = \dfrac{p_{i,j}(x)}{q_{i,j}(x)}$,*

$$\begin{aligned} p_{i,j}(x) &= A_{i,j}(\lambda_{i,j})(1-\theta)^3 + B_{i,j}(\lambda_{i,j})(1-\theta)^2\theta + C_{i,j}(\lambda_{i,j})(1-\theta)\theta^2 \\ &\quad + D_{i,j}(\lambda_{i,j})\theta^3, \\ q_{i,j}(x) &= (1-\theta)\alpha_{i,j} + \theta, \theta = \frac{x - x_1}{x_M - x_1}, x \in [x_1, x_M], \end{aligned} \tag{7}$$

$|\lambda_{i,j}| < a_i$, and $|\bar{\lambda}_j|_\infty = \max\{\frac{|\lambda_{i,j}|}{a_i}; i = 1, 2, \ldots, M-1\}$. Let $F_{i,j}^1(x, \zeta) = \frac{\lambda_{i,j}\zeta + r'_{i,j}(x)}{a_i}$. For $i = 1, 2, \ldots, M-1$, $j = 1, 2\ldots, N+1$, if

$$\begin{aligned} &F_{i,j}(x_1, z_{1,j}) = z_{i,j}, F_{i,j}(x_M, z_{M,j}) = z_{i+1,j}, F_{i,j}^1(x_1, \Delta_{1,j}) = \Delta_{i,j}, \\ &F_{i,j}^1(x_M, \Delta_{M,j}) = \Delta_{i+1,j}, \Delta_{i,j} = \frac{z_{i+1,j} - z_{i,j}}{h_i}, h_i = x_{i+1} - x_i, \end{aligned} \tag{8}$$

then for $j = 1, 2, \ldots, N+1$, there exists a rational cubic FIF Ψ_j such that $\Psi_j(x_i) = z_{i,j}, i = 1, 2, \ldots, M$. Moreover, for $j = 1, 2, \ldots, N+1$, the attractor of the IFS $\mathcal{I}_j^*$ is the graph of Ψ_j over $[x_1, x_M]$.

Proof. For $j = 1, 2, \ldots, N+1$, suppose

$$\mathcal{F}_j = \{\tau \in \mathcal{C}^1[x_1, x_M] \mid \tau(x_1) = z_{1,j}, \tau(x_M) = z_{M,j}, \tau'(x_1) = \Delta_{1,j}, \text{and } \tau'(x_M) = \Delta_{M,j}\}.$$

Then for $j = 1, 2, \ldots, N+1$, $(\mathcal{F}_j, d_{\mathcal{C}^1})$ is a complete metric space, where $d_{\mathcal{C}^1}$ is the metric induced by the $\mathcal{C}^1$-norm on $\mathcal{C}^1[x_1, x_M]$, namely

$$d_{\mathcal{C}^1}(f, g) = ||f - g||_\infty + ||(f - g)'||_\infty.$$

Define the Read-Bajraktarević operator T_j^* on $\mathcal{F}_j$ as

$$(T_j\tau)(x) = \lambda_{i,j}\tau(\phi_i^{-1}(x)) + r_{i,j}(\phi_i^{-1}(x)),\ x \in I_i, i = 1, 2, \ldots, M-1. \quad (9)$$

Since $|\lambda_{i,j}| < a_i = \frac{h_i}{x_M - x_1} < 1$ and using (8), it is easy to see that

(*i*) $T_j\tau$ and $(T_j\tau)'$ are continuous on $(x_i, x_{i+1}), i = 1, 2, \ldots, M-1$,
(*ii*) $(T_j^*\tau)(x_i-) = (T_j\tau)(x_i+), (T_j\tau)'(x_i-) = (T_j\tau)'(x_i+), i = 1, 2, \ldots, M-1$,
(*iii*)

$$\begin{aligned} d_{\mathcal{C}^1}(T_j f, T_j g) &= ||T_j f - T_j g||_\infty + ||(T_j f - T_j g)'||_\infty \\ &\le |\lambda_{i,j}|||f - g||_\infty + \frac{|\lambda_{i,j}|}{a_i}||(f - g)'||_\infty \\ &\le |\bar{\lambda}_j|_\infty d_{\mathcal{C}^1}(f, g). \end{aligned} \quad (10)$$

Owing to the above reasons, for $j = 1, 2, \ldots, N+1$, T_j^* is contraction on $\mathcal{F}_j$ with contraction factor $|\lambda_j|_\infty$. Therefore by the Banach Fixed Point Theorem, there exists a fixed point (say) Ψ_j of T_j, i.e., $(T_j\Psi_j)(x) = \Psi_j(x)$ for all $x \in [x_1, x_M]$. Since Ψ_j is fixed point of T_j, all the properties of T_j automatically transfer to the Ψ_j. In other words, Ψ_j interpolates Σ_j and $\Psi_j \in \mathcal{C}^1[x_1, x_M]$. Also it is verify that Ψ_j satisfies the functional equation

$$\Psi_j(x) = \lambda_{i,j}\Psi_j(\phi_i^{-1}(x)) + r_{i,j}(\phi_i^{-1}(x)), x \in I_i,\ i = 1, 2 \ldots, M-1. \quad (11)$$

Since Ψ_j a $\mathcal{C}^1$-interpolant over $[x_1, x_M]$, following properties are valid :

$$\Psi_j(x_i) = z_{i,j}, \Psi_j(x_{i+1}) = z_{i+1,j}, \Psi_j'(x_i) = \Delta_{i,j}, \Psi_j'(x_{i+1}) = \Delta_{i+1,j}.$$

Now we determine arbitrary constants in $p_{i,j}(x)$, namely $A_{i,j}(\lambda_{i,j}), B_{i,j}(\lambda_{i,j})$, $C_{i,j}(\lambda_{i,j})$ and $D_{i,j}(\lambda_{i,j})$ using the above four properties.

Substituting $x = x_i$ in (11), we have

$$\begin{aligned} \Psi_j(x_i) &= \lambda_{i,j}\Psi_j(x_1) + \frac{p_{i,j}(0)}{q_{i,j}(0)} \\ \Rightarrow A_{i,j}(\lambda_{i,j}) &= \alpha_{i,j}(z_{i,j} - \lambda_{i,j}z_{1,j}). \end{aligned}$$

Substituting $x = x_{i+1}$ in (11), we have

$$\Psi_j(x_{i+1}) = \lambda_{i,j}\Psi_j(x_M) + \frac{p_{i,j}(1)}{q_{i,j}(1)}$$
$$\Rightarrow D_{i,j}(\lambda_{i,j}) = z_{i+1,j} - \lambda_{i,j} z_{M,j}.$$

The condition $\Psi_j'(x_i) = \Delta_{i,j}$ in (11) leads to

$$a_i\Delta_{i,j} = \lambda_{i,j}\Psi_j'(x_1) + \frac{q_{i,j}(0)p_{i,j}'(0) - q_{i,j}'(0)p_{i,j}(0)}{(q_{i,j}(0))^2}.$$

The algebraic manipulation of the above expression gives that

$$B_{i,j}(\lambda_{i,j}) = (\alpha_{i,j}+1)z_{i,j} + \alpha_{i,j}z_{i+1,j} - \lambda_{i,j}\left((2\alpha_{i,j}+1)z_{1,j} - \alpha_{i,j}(x_M - x_1)\frac{z_{2,j}-z_{1,j}}{h_1}\right).$$

The condition $\Psi'(x_{i+1}) = \Delta_{i+1,j}$ in (11) leads to

$$a_i\Delta_{i+1,j} = \lambda_{i,j}\Psi'(x_M) + \frac{q_{i,j}(1)p_{i,j}'(1) - q_{i,j}'(1)p_{i,j}(1)}{(q_{i,j}(1))^2}.$$

Simplifying this expression for $C_{i,j}$, we have

$$C_{i,j}(\lambda_{i,j}) = (\alpha_{i,j}+2)z_{i+1,j} - \lambda_{i,j}(\alpha_{i,j}+2)z_{M,j} - h_i\big(\Delta_{i+1,j} - \lambda_{i,j}\Delta_{M,j}\big).$$

By using similar arguments as in [1], it can be shown that for each $j = 1, 2, \ldots, N+1$, the IFS $\mathcal{I}_j^*$ has a unique attractor, and it is the graph of the rational cubic FIF $\Psi_j \in \mathcal{C}^1[x_1, x_M]$. □

Remark 3.1. The rational cubic FIFs $\Phi_j, j = 1, 2, \ldots, N+1$, are called x-direction fractal boundary curves.

Remark 3.2. If $\lambda_{i,j} = 0, i = 1, 2, \ldots, M-1, j = 1, 2, \ldots, N+1$, then for $j = 1, 2, \ldots, N+1$, the rational cubic FIF Φ_j reduces to the classical rational cubic interpolation function [10] $S_j(x)$ as

$$S_j(x) = \frac{A_{i,j}(0)(1-\rho)^3 + B_{i,j}(0)(1-\rho)^2\rho + C_{i,j}(0)(1-\rho)\rho^2 + D_{i,j}(0)\rho^3}{(1-\rho)\alpha_{i,j} + \rho},$$
$$\rho = \frac{x - x_i}{x_{i+1} - x_i}, x \in [x_i, x_{i+1}]. \tag{12}$$

3.2 Construction of $\mathcal{C}^1$-Rational Cubic FISs

Theorem 3.2. *Let* $\{(x_i, y_j, z_{i,j}), i = 1, 2, \ldots, M+1, j = 1, 2, \ldots, N+1\}$ *be a given surface data. Let*

$$x_1 < x_2 < \cdots < x_{M-1} < x_M < x_{M+1},$$

$$y_1 < y_2 < \cdots < y_{N-1} < y_N < y_{N+1},$$

be the partition of $[x_1, x_{M+1}] \times [y_1, y_{N+1}]$. *Consider the IFS*

$$\begin{aligned}\mathcal{I}^* \equiv \Big\{&[x_1, x_M] \times [y_1, y_N] \times K_c; W_{i,j}(x, y, z) = (x, \psi_j(y), G_{i,j}(x, y, z)),\\ &x \in [x_i, x_{i+1}], y \in [y_1, y_N], i = 1, 2, \ldots, M-1, j = 1, 2, \ldots, N-1\Big\},\end{aligned}$$

where K_c *is a suitable compact subset of* $\mathbb{R}$, $\psi_j(y) = c_j y + d_j : [y_1, y_N] \to [y_j, y_{j+1}]$ *is a contractive homeomorphism such that*

$$\psi_j(y_1) = y_j, \psi_j(y_N) = y_{j+1},$$

$$G_{i,j}(x, y, z) = \lambda_j^* z + r_{i,j}^*(x, y),\ r_{i,j}^*(x, y) = \frac{p_{i,j}^*(x, y)}{q_{i,j}^*(x, y)},$$

$$\begin{aligned}p_{i,j}^*(x, y) =& A_{i,j}^*(x, \lambda_{i,j}, \lambda_j^*)(1-\psi)^3 + B_{i,j}^*(x, \lambda_{i,j}, \lambda_j^*)(1-\psi)^2\psi\\ &+ C_{i,j}^*(x, \lambda_{i,j}, \lambda_j^*)(1-\psi)\psi^2 + D_{i,j}^*(x, \lambda_{i,j}, \lambda_j^*)\psi^3,\end{aligned}$$

$$\begin{aligned}&A_{i,j}^*(x, \lambda_{i,j}, \lambda_j^*) = \beta_j(\Psi_j(x) - \lambda_j^*\Psi_1(x)),\\ &B_{i,j}^*(x, \lambda_{i,j}, \lambda_j^*) = (\beta_j + 1)\Psi_j(x) + \beta_j\Psi_{j+1}(x) - \lambda_j^*\Big((2\beta_j + 1)\Psi_1(x)\\ &\qquad - \beta_j(y_N - y_1)\frac{\Psi_2(x) - \Psi_1(x)}{l_1}\Big),\\ &C_{i,j}^*(x, , \lambda_{i,j}, \lambda_j^*) = (\beta_j + 2)\Psi_{j+1}(x) - \lambda_j^*(2 + \beta_j)\Psi_M(x) - l_j\big(\Delta_{j+1}^*(x)\\ &\quad - \lambda_j^*\Delta_N^*(x)\big), D_{i,j}^*(x, \lambda_{i,j}, \lambda_j^*) = \Psi_{j+1}(x) - \lambda_j^*\Psi_N(x),\\ &q_{i,j}^*(x) = (1-\psi)\beta_j + \psi, \psi = \frac{y - y_1}{y_N - y_1}, y \in [y_1, y_N],\end{aligned}$$

$|\lambda_j^*| < c_j, \beta_j > 0$, $|\lambda^*|_\infty = \max\{\frac{|\lambda_j^*|}{c_j}; j = 1, 2, \ldots, N-1\} < 1$, $\Delta_j^*(x) = \frac{\Psi_{j+1}(x) - \Psi_j(x)}{l_j}, l_j = y_{j+1} - y_j$, *then there exists a* $\mathcal{C}^1$*-rational cubic FIS* Φ *such that* $\Phi(x_i, y_j) = z_{i,j}, i = 1, 2, \ldots, M,\ j = 1, 2, \ldots, N$. *Moreover the attractor of the IFS* $\mathcal{I}^*$ *is the graph of* Φ *over* $[x_1, x_M] \times [y_1, y_N]$.

Remark 3.3. The functions $A^*_{i,j}(x, \lambda_{i,j}, \lambda^*_j)$, $B^*_{i,j}(x, \lambda_{i,j}, \lambda^*_j)$, $C^*_{i,j}(x, \lambda_{i,j}, \lambda^*_j)$, and $D^*_{i,j}(x, \lambda_{i,j}, \lambda^*_j)$ are obtained in view of $A_{i,j}(\lambda_{i,j})$, $B_{i,j}(\lambda_{i,j})$, $C_{i,j}(\lambda_{i,j})$, and $D_{i,j}(\lambda_{i,j})$, respectively, in Theorem 3.2.

Remark 3.4. The scaling factors and shape parameters involved in the rational cubic IFS $\mathcal{I}^*$ are arranged in the matrix form as follows:
$\boldsymbol{\lambda} = [\boldsymbol{\lambda_1}, \boldsymbol{\lambda_2}, \boldsymbol{\lambda_3}, \ldots, \boldsymbol{\lambda_{N+1}}]$, $\boldsymbol{\lambda_j} = [\lambda_{1,j}, \lambda_{2,j}, \ldots, \lambda_{M-1,j}]^T$, $j = 1, 2, \ldots, N+1$,
$\boldsymbol{\lambda}^* = [\lambda^*_1, \lambda^*_2, \lambda^*_3, \ldots, \lambda^*_{N-1}]$,
$\boldsymbol{\alpha} = [\alpha_{i,j}]_{(M-1)\times(N+1)}$,
$\boldsymbol{\beta}^* = [\beta_1, \beta_2, \beta_3, \ldots, \beta_{N-1}]$

Proof. Suppose

$$\mathcal{F} = \Big\{h \in \mathcal{C}^1(I \times J) |\ h(x, y_1) = \Psi_1(x),\ h(x, y_N) = \Psi_N(x),\ \frac{\partial h(x,y)}{\partial y}\Big|_{(x,y_1)} = \Psi_1^{(1)}(x), \text{ and } \frac{\partial h(x,y)}{\partial y}\Big|_{(x,y_N)} = \Psi_N^{(1)}(x)\Big\}.$$

Then $(\mathcal{F}, d^\dagger_{\mathcal{C}^1})$ is a complete metric space, where $d^\dagger_{\mathcal{C}^1}$ is the metric induced by the $\mathcal{C}^1$-norm on $\mathcal{C}^1([x_1, x_M] \times [y_1, y_N])$, namely

$$d^\dagger_{\mathcal{C}^1}(h, g) = ||h - g||_\infty + ||(h - g)'||_\infty.$$

Define the Read-Bajraktarević operator $T_{\boldsymbol{\lambda},\boldsymbol{\lambda}^*}$ on $\mathcal{F}$ as

$$\begin{aligned}(T_{\boldsymbol{\lambda},\boldsymbol{\lambda}^*}h)(x, y) = \lambda^*_j h(x, \psi_j^{-1}(y)) + r^*_{i,j}(x, \psi_j^{-1}(y)),\ (x, y) \in \\ D_{i,j} = [x_i, x_{i+1}] \times [y_j, y_{j+1}], \\ i = 1, 2, \ldots, M-1,\ j = 1, 2, \ldots, N-1.\end{aligned} \tag{13}$$

Note that $T_{\boldsymbol{\lambda},\boldsymbol{\lambda}^*}$ is continuous inside every $D_{i,j}$. For (x_i, y_j) and $(x_{i+1}, y_{j+1}) \in D_{i,j} \cap D_{i,j+1}$, it is verified that

$$\left.\begin{aligned}\lim_{\substack{(x,y)\to(x_{i+1},y_{j+1})\\(x,y)\in D_{i,j}}} (T_{\boldsymbol{\lambda},\boldsymbol{\lambda}^*}h)(x, y) = \lim_{\substack{(x,y)\to(x_{i+1},y_{j+1})\\(x,y)\in D_{i,j+1}}} (T_{\boldsymbol{\lambda},\boldsymbol{\lambda}^*}h)(x, y) = z_{i+1,j+1}, \\ \lim_{\substack{(x,y)\to(x_i,y_{j+1})\\(x,y)\in D_{i,j}}} (T_{\boldsymbol{\lambda},\boldsymbol{\lambda}^*}h)(x, y) = \lim_{\substack{(x,y)\to(x_i,y_{j+1})\\(x,y)\in D_{i,j+1}}} (T_{\boldsymbol{\lambda},\boldsymbol{\lambda}^*}h)(x, y) = z_{i,j+1}.\end{aligned}\right\} \tag{14}$$

Since at $y = y_N$, $\psi = 1$, for $(x^*, y_{j+1}) \in (x_i, x_{i+1}) \times y_{j+1}$, from (13), it is easy to verify that

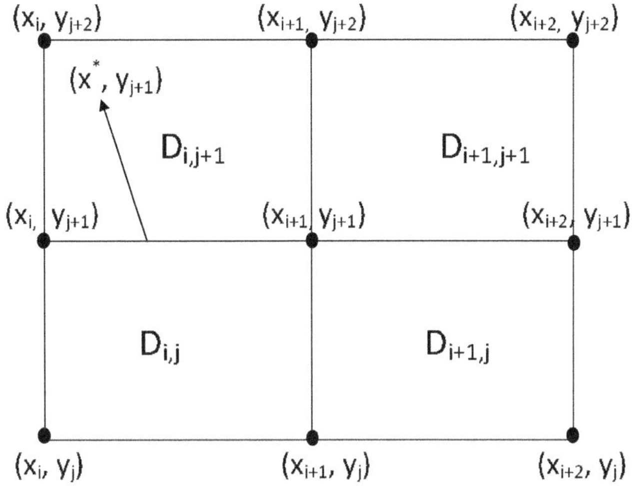

Fig. 1 Continuity domain

$$\lim_{\substack{(x,y)\to(x^*,y_{j+1})\\(x,y)\in D_{i,j}}} (T_{\lambda,\lambda^*}h)(x,y) = \lambda_j^* h(x^*, \psi_j^{-1}(y_{j+1})) + r_{i,j}^*(x^*, \psi_j^{-1}(y_{j+1}))$$

$$= \lambda_j^* h(x^*, y_N) + \Psi_{j+1}(x^*) - \lambda_j^* \Psi_N(x^*)$$

$$= \Psi_{j+1}(x^*) = z_{*,j+1} \text{ (say)}.$$

Similarly, since at $y = y_1$, $\psi = 0$, for $(x^*, y_{j+1}) \in (x_i, x_{i+1}) \times y_{j+1}$, from (13), it is easy to see that

$$\lim_{\substack{(x,y)\to(x^*,y_{j+1})\\(x,y)\in D_{i,j+1}}} (T_{\lambda,\lambda^*}h)(x,y)$$

$$= \lambda_j^* h(x^*, \psi_{j+1}^{-1}(y_{j+1})) + r_{i,j+1}^*(x^*, \psi_{j+1}^{-1}(y_{j+1}))$$

$$= \lambda_j^* h(x^*, y_1) + \Psi_{j+1}(x^*) - \lambda_j^* \Psi_1(x^*)$$

$$= \Psi_{j+1}(x^*) = z_{*,j+1} = \lim_{\substack{(x,y)\to(x^*,y_{j+1})\\(x,y)\in D_{i,j}}} (T_{\lambda,\lambda^*}h)(x,y).$$

Therefore we conclude that the Read-Bajraktarević operator T_{λ,λ^*} is continuous on $D_{i,j} \cap D_{i,j+1}, i = 1,2,\ldots,M-1, j = 1,2,\ldots,N-2$. In the similar way, we can verify the continuity of the Read-Bajraktarević operator T_{λ,λ^*} on $D_{i,j} \cap D_{i+1,j}, i = 1,2,\ldots,M-2, j = 1,2,\ldots,N-1$. Thus the Read-Bajraktarević operator T_{λ,λ^*} is continuous on the domain D. Similarly, we can verify that $\frac{\partial T_{\lambda,\lambda^*}h(x,y)}{\partial x}$ and $\frac{\partial T_{\lambda,\lambda^*}h(x,y)}{\partial y}$ are continuous over the domain D. Since both $\frac{\partial T_{\lambda,\lambda^*} h(x,y)}{\partial x}$ and $\frac{\partial T_{\lambda,\lambda^*} h(x,y)}{\partial y}$ are continuous over the domain D, $(T_{\lambda,\lambda^*} h)'(x,y)$ exist and it is continuous over the domain D. Also

$$\begin{aligned} d^{\dagger}_{C^1}(T_{\lambda,\lambda^*}h, T_{\lambda,\lambda^*}g) &= \|T_{\lambda,\lambda^*}h - T_{\lambda,\lambda^*}g\|_\infty + \|(T_{\lambda,\lambda^*}h - T_{\lambda,\lambda^*}g)'\|_\infty \\ &\le |\lambda_j^*|\|h-g\|_\infty + \frac{|\lambda_j^*|}{c_j}\|(h-g)'\|_\infty \\ &\le |\lambda^*|_\infty d^{\dagger}_{C^1}(h,g). \end{aligned} \tag{15}$$

Thus the Read-Bajraktarević operator T_{λ,λ^*} is a contraction on $\mathcal{F}$. Therefore by the Banach Fixed Point Theorem, there exists a fixed point (say) Φ of T_{λ,λ^*}, i.e., $(T_{\lambda,\lambda^*}\Phi)(x,y) = \Phi(x,y)$ for all $x \in [x_1, x_M] \times [y_1, y_N]$. Since Φ is fixed point of T_{λ,λ^*}, all the properties of T_{λ,λ^*} automatically transfer to the Φ. In other words, $\Phi(x_i, y_j) = z_{i,j}, i = 1,2,\ldots,M, j = 1,2,\ldots,N$ and C^1 over the domain D. Also it is easy to verify that Φ satisfies the functional equation

$$\begin{aligned} &\Phi(x,y) = \lambda_j^*\Phi(x,\psi_j^{-1}(y)) + r_{i,j}^*(x,\psi_j^{-1}(y)), (x,y) \in D_{i,j}, \\ &\quad i = 1,2,\ldots,M-1, j = 1,2,\ldots,N-1. \end{aligned} \tag{16}$$

□

4 Approximation Properties of Rational Cubic FIS

In this section, we study the convergence properties of rational FIS Φ, by calculating an upper bound for $\|\Phi - F\|_\infty$, where F is a data generating function. The following theorem provides an upper bound for $\|\Psi_j - S_j\|_\infty$, $j = 1,2,\ldots,N+1$ based on the assumptions $h = h_i, i = 1,2,\ldots,M-1, l = l_j, j = 1,2,\ldots,N-1$.

Lemma 4.1. *Let Ψ_j and S_j, respectively, be the rational cubic FIF and classical rational cubic interpolant with respect to the data Σ_j, $j=1,2,\ldots,N+1$. Suppose $V_{M-1,j} = \bigotimes_{i=1}^{M-1}[-\kappa_j a_i, \kappa_j a_i], 0 < \kappa_j < 1$, $|\boldsymbol{\lambda}_j|_\infty = \max\{|\lambda_{i,j}| : i = 1,2,\ldots,M-1\}$, $h = \max\{h_i : i = 1,2,\ldots,M-1\}$, $\alpha_{i,j} = \alpha_i, j = 1,2,\ldots,N-1$, and*

$$\left| \frac{\partial \left(\frac{p_{i,j}(\phi_i^{-1}(x), \eta_i)}{q_{i,j}(\phi_i^{-1}(x))} \right)}{\partial \lambda_{i,j}} \right| \le U_j. \tag{17}$$

Then

$$\|\Psi_j - S_j\|_\infty \le \Lambda_j(h) = \frac{|\boldsymbol{\lambda}_j|_\infty \left(\frac{9M^*}{4} + U_j\right)}{1 - |\boldsymbol{\lambda}_j|_\infty},$$

where $M^* = \max\{M^*_{i,j} : i = 1,2,\ldots,M-1, j = 1,2,\ldots,N-1\}$, $M^*_{i,j} = \max\{|z_{i+r,j+s}| : r = 0,1,2; s = 0,1,2\}$.

Proof. From (9), the Read-Bajraktarević operator $T^*_j : V_{M-1,j} \times \mathcal{F}^*_j \to \mathcal{F}^*_j$ is also viewed as

$$(T^*_{j,\boldsymbol{\lambda}_j}\tau)(x) = \lambda_{i,j}\tau(\phi_i^{-1}(x)) + \frac{p_{i,j}(\phi_i^{-1}(x),\lambda_{i,j})}{q_{i,j}(\phi_i^{-1}(x))}, \; x \in I_i, i = 1,2,\ldots,M-1. \tag{18}$$

Since Ψ_j and S_j, respectively, are the rational cubic FIF and classical cubic spline with respect to the data set Σ_j, it easy to see that Ψ_j and S_j are fixed points of Read-Bajraktarević operator $T^*_{j,\boldsymbol{\lambda}_j}$ for $\boldsymbol{\lambda}_j \neq \mathbf{0}$ and $\boldsymbol{\lambda}_j = \mathbf{0}$, respectively. For $\boldsymbol{\lambda}_j \neq \mathbf{0}$, from (18), it is easy to verify that

$$\|T^*_{j,\boldsymbol{\lambda}_j}\Psi_j - T^*_{j,\boldsymbol{\lambda}_j}S_j\|_\infty \leq |\boldsymbol{\lambda}_j|_\infty \, \|\Psi_j - S_j\|_\infty. \tag{19}$$

Also

$$|T^*_{j,\boldsymbol{\lambda}_j}S_j(x) - T^*_{j,\mathbf{0}}S_j(x)| = \left|\lambda_{i,j}\; S_j \circ \phi_i^{-1}(x) + \frac{p_{i,j}(\phi_i^{-1}(x),\lambda_{i,j})}{q_{i,j}(\phi_i^{-1}(x))} - \frac{p_{i,j}(\phi_i^{-1}(x),0)}{q_{i,j}(\phi_i^{-1}(x))}\right|$$

$$\leq |\lambda_{i,j}|\,\|S_j\|_\infty + \frac{|p_{i,j}(\phi_i^{-1}(x),\lambda_i) - p_{i,j}(\phi_i^{-1}(x),0)|}{q_{i,j}(\phi_i^{-1}(x))}. \tag{20}$$

Using Mean Value Theorem for functions of several variables, there exists $\eta_j = (\eta_{1,j},\eta_{2,j},\ldots,\eta_{M-1,j}) \in V_{M-1,j}$, such that

$$p_{i,j}(\phi_i^{-1}(x),\lambda_{i,j}) - p_{i,j}(\phi_i^{-1}(x),0) = \frac{\partial p_{i,j}(\phi_i^{-1}(x),\eta_{i,j})}{\partial \lambda_{i,j}}\,\lambda_{i,j}, |\eta_{i,j}| < |\lambda_{i,j}|. \tag{21}$$

From (20) and (21), we have

$$|T^*_{j,\boldsymbol{\lambda}_j}S_j(x) - T^*_{j,\mathbf{0}}S_j(x)| \leq |\boldsymbol{\lambda}_j|_\infty \left(\|S_j\|_\infty + \left|\frac{\partial\left(\frac{p_{i,j}(\phi_i^{-1}(x),\eta_i)}{q_{i,j}(\phi_i^{-1}(x))}\right)}{\partial\lambda_{i,j}}\right|\right). \tag{22}$$

From [11], it is known that

$$\|S_j\|_\infty \leq \frac{9M^*}{4}, j = 1,2,\ldots,N+1. \tag{23}$$

Substituting (17) and (23) in (22), we obtain

$$|T^*_{j,\boldsymbol{\lambda}_j}S_j(x) - T^*_{j,\mathbf{0}}S(x)| \leq |\boldsymbol{\lambda}_j|_\infty \,(\frac{9M^*}{4} + U_j).$$

This implies

$$\|T^*_{j,\boldsymbol{\lambda}_j} S_j - T^*_{j,\mathbf{0}} S_j\|_\infty \le |\boldsymbol{\lambda}_j|_\infty \left(\frac{9M^*}{4} + U_j\right). \tag{24}$$

Using (19) and (24) together with the inequality

$$\|\Psi_j - S_j\|_\infty = \|T^*_{j,\boldsymbol{\lambda}_j}\Psi_j - T^*_{j,\mathbf{0}} S_j\|_\infty \le \|T^*_{j,\boldsymbol{\lambda}_j}\Psi_j - T^*_{j,\boldsymbol{\lambda}_j} S_j\|_\infty + \|T^*_{j,\boldsymbol{\lambda}_j} S_j - T^*_{j,\mathbf{0}} S_j\|_\infty \tag{25}$$

we obtain the desired bound for $\|\Psi_j - S_j\|_\infty$. □

Theorem 4.1. *Let Φ be the rational cubic FIS with respect to the surface data $\{(x_i, y_j, z_{i,j}), i = 1,2,\ldots,M+1, j = 1,2,\ldots,N+1\}$ generated from an original function $F \in \mathcal{C}^1(D)$. Let $h = \max\{h_i : i = 1,2,\ldots,M-1\}$, $l = \max\{l_j : j = 1,2,\ldots,N-1\}$. Let $h = h_i$, $i = 1,2,\ldots,M-1$, $l = l_j$, $j = 1,2,\ldots,N-1$, and $\alpha_{i,j} = \alpha_i$ for $j = 1,2,\ldots,N-1$. Then*

$$||\Phi - F||_\infty \le \frac{\Lambda^{**}(h) + |\lambda^{**}|_\infty \frac{9}{4} M^*}{1 - |\lambda^{**}|_\infty} + \frac{9}{2}\left(h\left|\left|\frac{\partial F}{\partial x}\right|\right|_\infty + l\left|\left|\frac{\partial F}{\partial y}\right|\right|_\infty\right), \tag{26}$$

where $\Lambda^{**}(h) = \max\{\Lambda^*_j(h) : j = 1,2,\ldots,N-1\}$,

$$\begin{aligned}\Lambda^*_j(h) =& \Lambda_j(h)(K+1) + \Lambda_{j+1}(h)(2K+2) + |\lambda^{**}|_\infty\big(2\Lambda_1(h)K + \Lambda_2(h)K \\ &+ \Lambda_N(h)(K+2) + \Lambda_{N+1}(h)\big),\end{aligned}$$

$$\begin{aligned}K &= \max\{\beta_j, 2\beta_j + 1, 2 + \beta_j : j = 1,2,\ldots,N+1\},\\ |\lambda^{**}|_\infty &= \max\{|\lambda^*_j| : j = 1,2,\ldots,N-1\},\\ M^* &= \max\{M^*_{i,j} : i = 1,2,\ldots,M-1, j = 1,2,\ldots,N-1\},\\ M^*_{i,j} &= \max\{|z_{i+r,j+s}| : r = 0,1,2; s = 0,1,2\}.\end{aligned}$$

Proof. Since Ψ and S, respectively, are the fixed points of the Read-Bajraktarević operator T_{λ,λ^*} for $\boldsymbol{\lambda} \neq \mathbf{0}$, $\boldsymbol{\lambda^*} \neq \mathbf{0}$ and $\boldsymbol{\lambda} = \mathbf{0}$, $\boldsymbol{\lambda^*} = \mathbf{0}$, respectively, it is easy to see that

$$\begin{aligned}||\Phi - S||_\infty &= ||T_{\lambda,\lambda^*}\Phi - T_{0,0}S||_\infty \\ &\le ||T_{\lambda,\lambda^*}\Phi - T_{0,\lambda^*}\Phi||_\infty + ||T_{0,\lambda^*}\Phi - T_{0,\lambda^*}S||_\infty + ||T_{0,\lambda^*}S - T_{0,0}S||_\infty.\end{aligned} \tag{27}$$

Now we wish to calculate the bound of each term in the right-hand side of above inequality. For this purpose it is easy to see that

$$T_{\boldsymbol{\lambda},\boldsymbol{\lambda}^*}\Phi(x,y) - T_{\boldsymbol{0},\boldsymbol{\lambda}^*}\Phi(x,y) = \frac{\begin{aligned}\beta_j\big(A^*_{i,j}(x,\lambda_{i,j},\lambda^*_j) - A^*_{i,j}(x,0,\lambda^*_j)\big)(1-\psi)^3 \\ + \big(B^*_{i,j}(x,\lambda_{i,j},\lambda^*_j) - B^*_{i,j}(x,0,\lambda^*_j)\big)\psi(1-\psi)^2 \\ + \big(C^*_{i,j}(x,\lambda_{i,j},\lambda^*_j) - C^*_{i,j}(x,0,\lambda^*_j)\big)\psi^2(1-\psi) \\ + \big(D^*_{i,j}(x,\lambda_{i,j},\lambda^*_j) - D^*_{i,j}(x,0,\lambda^*_j)\big)\psi^3\end{aligned}}{(1-\psi)\beta_j + \psi}. \tag{28}$$

This implies

$$\begin{aligned}|T_{\boldsymbol{\lambda},\boldsymbol{\lambda}^*}\Phi(x,y) - T_{\boldsymbol{0},\boldsymbol{\lambda}^*}\Phi(x,y)| \le\ & |A^*_{i,j}(x,\lambda_{i,j},\lambda^*_j) - A^*_{i,j}(x,0,\lambda^*_j)| \\ & + |B^*_{i,j}(x,\lambda_{i,j},\lambda^*_j) - B^*_{i,j}(x,0,\lambda^*_j)| \\ & + |C^*_{i,j}(x,\lambda_{i,j},\lambda^*_j) - C^*_{i,j}(x,0,\lambda^*_j)| \\ & + |D^*_{i,j}(x,\lambda_{i,j},\lambda^*_j) - D^*_{i,j}(x,0,\lambda^*_j)|.\end{aligned} \tag{29}$$

Now it is easy to verify that

$$\begin{aligned}|A^*_{i,j}(x,\lambda_{i,j},\lambda^*_j) - A^*_{i,j}(x,0,\lambda^*_j)| &\le |\Psi_j(x) - S_j(x)| + |\lambda^*_j||\Psi_1(x) - S_1(x)| \\ &\le ||\Psi_j - S_j||_\infty + |\lambda^{**}|_\infty||\Psi_1 - S_1||_\infty \\ &\le \Lambda_j(h) + |\lambda^{**}|_\infty\Lambda_1(h) (\text{Using Lemma 4.1}).\end{aligned} \tag{30}$$

Similarly, it is easy to obtain that

$$\left.\begin{aligned}|B^*_{i,j}(x,\lambda_{i,j},\lambda^*_j) - B^*_{i,j}(x,0,\lambda^*_j)| &\le K\big(\Lambda_j(h) + \Lambda_{j+1}(h) + |\lambda^{**}|_\infty(2\Lambda_1(h) + \Lambda_2(h))\big), \\ |C^*_{i,j}(x,\lambda_{i,j},\lambda^*_j) - C^*_{i,j}(x,0,\lambda^*_j)| &\le (K+1)\Lambda_{j+1}(h) + |\lambda^{**}|_\infty(K\Lambda_N(h) + \Lambda_{N+1}(h)), \\ |D^*_{i,j}(x,\lambda_{i,j},\lambda^*_j) - D^*_{i,j}(x,0,\lambda^*_j)| &\le \Lambda_{j+1}(h) + |\lambda^{**}|_\infty\Lambda_N(h).\end{aligned}\right\} \tag{31}$$

Using (30) and (31) in (29), we get

$$|T_{\boldsymbol{\lambda},\boldsymbol{\lambda}^*}\Phi(x,y) - T_{\boldsymbol{0},\boldsymbol{\lambda}^*}\Phi(x,y)| \le \Lambda^*_j(h). \tag{32}$$

Since the above inequality is true for every $(x,y) \in D_{i,j}, i = 1,2,\ldots,M-1, j = 1,2,\ldots,N-1$, we get the following estimation:

$$||T_{\boldsymbol{\lambda},\boldsymbol{\lambda}^*}\Phi - T_{\boldsymbol{0},\boldsymbol{\lambda}^*}\Phi||_\infty \le \Lambda^{**}(h). \tag{33}$$

Also it is easy to see that

$$\left.\begin{aligned} ||T_{0,\lambda^*}\Phi - T_{0,\lambda^*}S||_\infty &\le |\lambda^{**}|_\infty ||\Phi - S||_\infty, \\ ||T_{0,\lambda^*}S - T_{0,0}S||_\infty &\le |\lambda^{**}|_\infty ||S||_\infty. \end{aligned}\right\} \tag{34}$$

Substituting (33) and (34) in (27), we obtain

$$||\Phi - S||_\infty \le \frac{\Lambda^{**}(h) + |\lambda^{**}|_\infty ||S||_\infty}{1 - |\lambda^{**}|_\infty}. \tag{35}$$

Also from [11], it is known that

$$||S||_\infty \le \frac{9}{4} M^* \tag{36}$$

Using (36) in (35), we have the following estimation:

$$||\Phi - S||_\infty \le \frac{\Lambda^{**}(h) + |\lambda^{**}|_\infty \frac{9}{4} M^*}{1 - |\lambda^{**}|_\infty}. \tag{37}$$

Again from [11], it is known that

$$||F - S||_\infty \le \frac{9}{2}\left(h||\frac{\partial F}{\partial x}||_\infty + l||\frac{\partial F}{\partial y}||_\infty\right). \tag{38}$$

Using (37) and (38) together with inequality

$$||\Phi - F||_\infty \le ||\Phi - S||_\infty + ||F - S||_\infty,$$

we obtain the desired bound for $||\Phi - F||_\infty$. □

Convergence results: Since $\Lambda^{**} = O(h)$ and $|\lambda^{**}|_\infty = O(l)$, Theorem 4.1 gives that the rational cubic FIS Φ converges uniformly to the original function F as $h \to 0^+$ and $l \to 0^+$.

5 Examples

Consider the surface data with random 25 points given in Table 1. According to the developed theory, first we construct the x-direction fractal boundary curves Ψ_j, $j = 1, 2, 3, 4, 5$, for the univariate data sets $\Sigma_1 = \{(1.2, 1265.4), (3.6, 2324), (6, 567.7), (7.2, 775.01)\}$, $\Sigma_2 = \{(1.2, 1), (3.6, 3784), (6, 1342), (7.2, 1543)\}$, $\Sigma_3 = \{(1.2, -4), (3.6, 11.3), (6, 11.1), (7.2, 15.088)\}$, $\Sigma_4 = \{(1.2, 7985.8), (3.6, -24), (6, -17), (7.2, 1118)\}$, and $\Sigma_5 = \{(1.2, -99), (3.6, -89), (6, 37), (7.2, 17)\}$,

Table 1 Interpolation data for C^1-rational cubic FISs

$\downarrow x/y \rightarrow$	4.3	7.1	8.5	11.3	18
1.2	1265.4	1	−4	7985.8	−99
3.6	2324	3784	11.3	−24	−89
6	567.7	1342	11.1	−17	37
7.2	775.01	1543	15.088	1118	17
8	0.01	−3	15	1.1	−180

respectively. Then using these fractal boundary curves in (16), we generate C^1-rational cubic FIS Φ such that $\Phi(x_i, y_j) = z_{i,j}, i = 1, 2, 3, 4; j = 1, 2, 3, 4$. With the choice of scaling and shape parameters $\boldsymbol{\lambda} = [0.19]_{(3\times5)}$ and $\boldsymbol{\alpha} = [6]_{(3\times5)}$, the x-direction fractal boundary curves are generated in Fig. 2a. By modifying only scaling matrix $\boldsymbol{\lambda}$ as $[0.01]_{(3\times5)}$ with respect to the matrices of scaling and shape parameters of Fig. 2a, we obtain one more set of fractal boundary curves in Fig. 2b. By comparing fractal boundary curves in Fig. 2b with fractal boundary curves in Fig. 2a, we observe the sensitivity of fractal boundary curves with respect to scaling factors.

By utilizing fractal boundary curves in Fig. 2a, and taking $\boldsymbol{\lambda}^* = [0.15]_{(1\times3)}$, $\boldsymbol{\beta} = [12]_{(3\times3)}$ in (16) we have generated the C^1-rational cubic FIS in Fig. 2c. Next by availing fractal boundary curves in Fig. 2b, and taking $\boldsymbol{\lambda}^* = [0.15]_{(1\times3)}$, $\boldsymbol{\beta} = [12]_{(3\times3)}$ in (16), we have generated a C^1-rational cubic FIS in Fig. 2d. We found a visually pleasing changes in the C^1-rational cubic FIS in Fig. 2d in comparison with the rational cubic FIS in Fig. 2c even if $\boldsymbol{\lambda}^*$ and $\boldsymbol{\beta}$ are same in their construction. From this we observe the sensitivity of rational cubic FIS with respect to x-direction fractal boundary curves. Finally we constructed rational cubic FISs in Fig. 2e, f, respectively, by taking $\boldsymbol{\lambda}^* = [0.01]_{(1\times3)}$ and $\boldsymbol{\beta} = [1000]_{(3\times3)}$ with respect to the Fig. 2c. An examination of rational cubic FISs in Fig. 2e, f with respect to the rational cubic FIS in Fig. 2c explains the sensitivity of rational cubic FIS with respect to the scaling matrix $\boldsymbol{\lambda}^*$ and matrix of shape parameters $\boldsymbol{\beta}$. By analyzing rational FISs in Fig. 2c–f, we reveal that these four rational cubic FISs are different to each other in nature. By proceeding in the same way we can construct a wide variety of rational cubic FISs for the given surface data using our method.

6 Conclusion and Remarks

In this work, we have constructed C^1-rational cubic FIS to interpolate a surface data arranged on a rectangular grid. Our surface interpolant needs only the functional values being interpolated. More important is that the value of the rational cubic FIS at any point in the interpolation region can be modified under the condition that the interpolation data is not changed by selecting suitable scaling factors and shape parameters. A uniform error bound has been calculated between the rational cubic FIS and an original function. It is observed that the developed rational cubic FIS

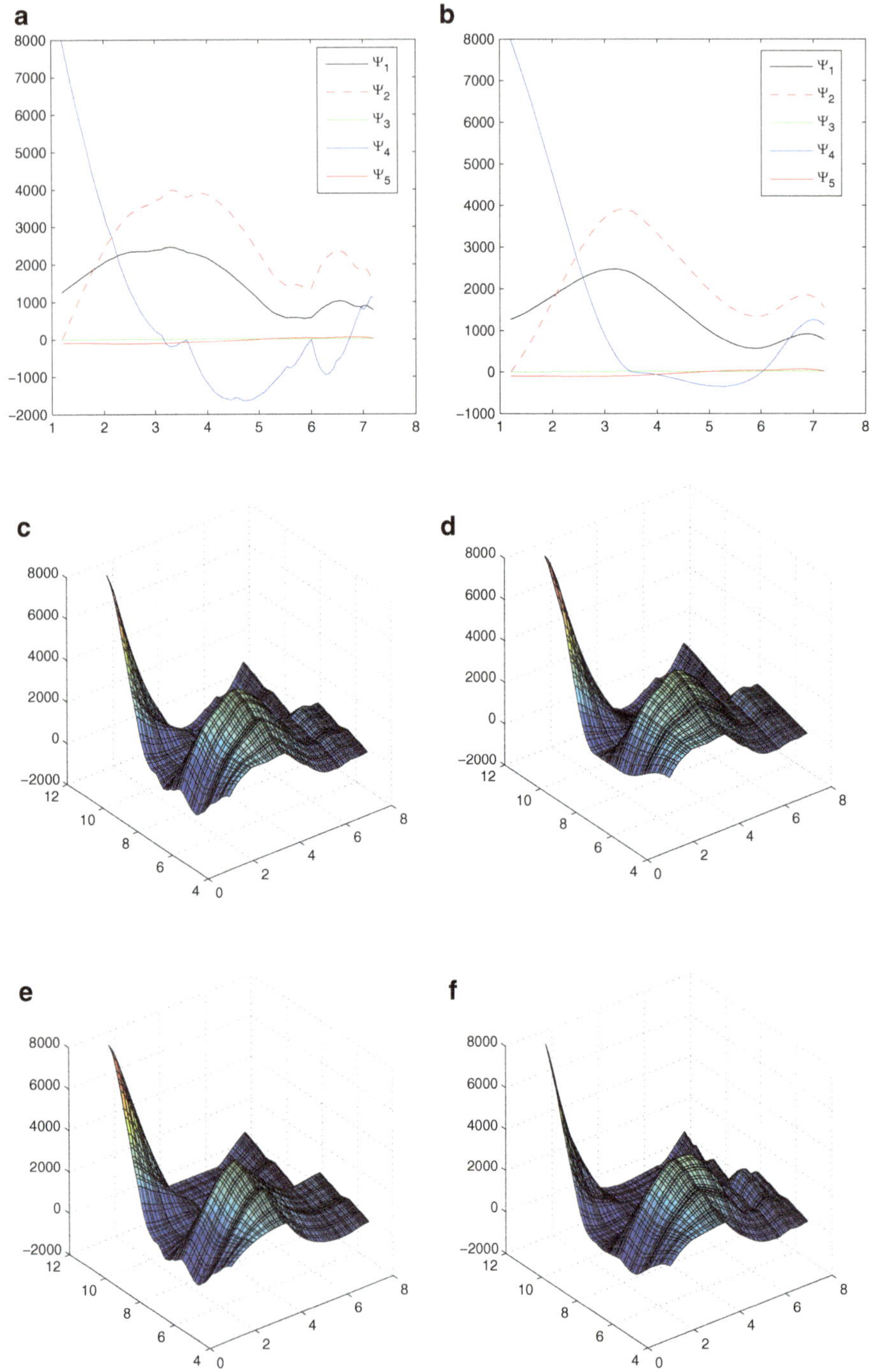

Fig. 2 Fractal boundary curves and $\mathcal{C}^1$-rational cubic FISs. (**a**) Fractal boundary curves in x-direction. (**b**). Fractal boundary curves in x-direction. (Effects of change in $\boldsymbol{\lambda}$ in Fig. 2a). (**c**) $\mathcal{C}^1$-rational cubic FIS. (**d**) $\mathcal{C}^1$-rational cubic FIS (Effects of change in fractal boundary curves in Fig. 2c). (**e**) $\mathcal{C}^1$-rational cubic FIS (Effects of change in $\boldsymbol{\lambda^*}$ in Fig. 2c). (**f**) $\mathcal{C}^1$-rational cubic FIS (Effects of change in $\boldsymbol{\beta}$ in Fig. 2c)

has linear convergence rate towards the original function. In this paper, the rational cubic FIS is developed with x-direction fractal boundary curves. Similarly, one can construct rational cubic FIS with y-direction fractal boundary curves. As a future research work, the shape preserving aspects of rational cubic FISs are under active consideration.

References

1. Barnsley, M.F.: Fractal functions and interpolation. Constr. Approx. **2**, 303–329 1986
2. Barnsley, M.F., Harrington, A.N.: The calculus of fractal interpolation functions. J. Approx. Theory **57**(1), 14–34 (1989)
3. Chand, A.K.B.: Natural cubic spline coalescence hidden variable fractal interpolation surfaces. Fractals **20**(2), 117–131 (2012)
4. Chand, A.K.B., Kapoor, G.P.: Generalized cubic spline fractal interpolation functions. SIAM J. Numer. Anal. **44**(2), 655–676 (2006)
5. Chand, A.K.B., Navascués, M.A.: Natural bicubic spline fractal interpolation. J. Nonlinear Anal. **69**, 3679–3691 (2008)
6. Chand, A.K.B., Vijender, N., Navascués, M.A.: Shape preservation of scientific data through rational fractal splines. Calcolo (2014). doi:10.1007/s10092-013-0088-2
7. Chand, A.K.B., Vijender, N.: Monotonicity preserving rational quadratic fractal interpolation functions. Adv. Numer. Anal. In Press
8. Chand, A.K.B., Viswanathan, P.: Cubic Hermite and cubic spline fractal interpolation functions. In: AIP Conf. Proc., vol. 1479, pp. 1467–1470 (2012)
9. Chand, A.K.B., Viswanathan, P., A constructive approach to cubic Hermite fractal interpolation function and its constrained aspects. BIT Numer. **53**, 841–865 (2013)
10. Duan, Q., Djidjeli, K., Price, W.G., Twizell, E.H.: Rational cubic spline based on function values. Comput. Graph. **22**(4), 479–486 (1998)
11. Duan, Q., Zhang, H., Zhang, Y.: Bounded property and point control of a bivariate rational interpolating surface. Comput. Math. Appli. **52**, 975–984 (2006)
12. Geronimo, J.S., Hardin, D.P.: Fractal interpolation functions from $\mathbb{R}^n \rightarrow \mathbb{R}^m$ and their projections. Z. Anal. Anwend. **12**, 535–548 (1993)
13. Hutchinson, J.E.: Fractals and self similarity. Indiana Univ. Math. J. **30**(5), 713–747 (1981)
14. Massopust, P.R.: Fractal surfaces. J. Math. Anal. Appl. **151**, 275–290 (1990)
15. Massopust, P.R.: Fractal Functions, Fractal Surfaces and Wavelets. Academic Press, Orlando (1994)
16. Navascués, M.A., Sebastián, M.V.: Smooth fractal interpolation. J. Inequal. Appl. **2006**, 1–20 (2006)
17. Xie, H., Sun, H.: The study of bivariate fractal interpolation functions and creation of fractal interpolated surfaces. Fractals **5**(4), 625–634 (1997)
18. Zhao, N., Construction and application of fractal interpolation surfaces. Vis. Comput. **12**, 132–146 (1996)

On Fractal Rational Functions

P. Viswanathan and A.K.B. Chand

Abstract This article introduces fractal perturbation of classical rational functions via α-fractal operator and investigates some aspects of this new function class, namely, the class of fractal rational functions. Its specific aims are: (1) to define the fractal rational functions along the lines of the fractal polynomials (2) to extend the Weierstrass theorem of uniform approximation to fractal rational functions (3) to deduce a fractal version of the classical Müntz theorem on rational functions (4) to prove the existence of a Schauder basis for $\mathcal{C}(I)$ consisting of fractal rational functions.

Keywords Fractal interpolation function • Fractal rational function • Fractal Müntz rational function • Positive approximant • Schauder basis

AMS Subject Classification: 28A80, 41A20, 41A29, 41A30

1 Introduction

Classical Approximation Theory concerns the description of functions using approximation sets consisting of smooth functions, sometimes infinitely differentiable. However, real-world sampled signals such as financial series, seismic data, speech signals, and bioelectric recordings show certain degree of irregularity. Fractal functions, an area championed by Barnsley, and taken up in earnest by a host of followers, provide an alternative technique for interpolation and approximation. Fractal functions aim mainly at data which present details at different scales or some

P. Viswanathan (✉) • A.K.B. Chand
Department of Mathematics, Indian Institute of Technology Madras, Chennai 600036, India
e-mail: amritaviswa@gmail.com; chand@iitm.ac.in

C. Bandt et al. (eds.), *Fractals, Wavelets, and their Applications*, Springer Proceedings in Mathematics & Statistics 92, DOI 10.1007/978-3-319-08105-2_23,

degree of self-similarity. These characteristics imply irregular, non-smooth structure that is inconvenient to describe by the traditional interpolation/approximation techniques.

Barnsley first introduced fractal functions as continuous functions interpolating a given set of data points [2]. Further, he has proposed the generalization of a continuous function, say f, defined on a real compact interval I by means of Fractal Interpolation Functions (FIFs) to obtain α-fractal function f^α. This function f^α retains some properties such as continuity and integrability of f, but in general does not possess differentiability. However, if the problem is of differential type, the parameters can be chosen in a specific way and f^α can be made so as to share the regularity of f. The parameter α present in f^α can be adjusted to modify or preserve the properties of the germ function f. The act of perturbing a continuous function f to obtain the α-fractal function f^α determines an operator, termed α-fractal operator: $\mathcal{F}^\alpha : \mathcal{C}(I) \to \mathcal{C}(I);\ \ f \mapsto f^\alpha$.

Having obtained an operator, it is natural to query on its properties like linearity, boundedness, invertibility, etc. Through a series of papers [5–10] Navascués and group have studied several properties of the α-fractal operator $\mathcal{F}^\alpha$, and extended it to more general spaces like L_p, $0 < p < \infty$. In particular, by considering the $\mathcal{F}^\alpha$-image of the most fundamental function class in $\mathcal{C}(I)$, namely, the space of all polynomials, Navascués has defined the α-fractal polynomials. Subsequently, some basic results including the approximation properties of the traditional polynomial space are extended to the space of α-fractal polynomials. In spite of the difference in smoothness property, a remarkable analogy between the space of traditional polynomials and the space of α-fractal polynomials is observed.

The problem of approximation of given functions not by polynomials but by more general rational functions is studied in the classical approximation theory. The study of approximation by rational functions is carried out not merely as a generalization of the problem of approximation by polynomials, but as a problem involving larger resources than the other. Wide applicability of rational functions may be attributed to their ability to accommodate a wider range of shapes than the polynomial family, excellent asymptotic properties, capability to model complicated structures, better interpolation properties, and excellent extrapolating powers. The flexibility and versatility gained by extending the traditional polynomials to the fractal polynomials motivates one to study the non-smooth fractal version of the class of traditional rational functions. The current article intends to be a contribution in this vein.

Following the procedure for defining the α-fractal polynomials adapted in reference [5], we define the α-fractal rational functions. Thus, the present article may be viewed as a sequel to [5]. While the two papers share a natural kinship, the reader will also discern a considerable degree of disparity between them. To be precise, as in [5], the basic idea is to consider the $\mathcal{F}^\alpha$-image of a classical function class (here the class of rational functions) to obtain a new class of functions with fractal characteristics (here the class of fractal rational functions). The Weierstrass theorem for the fractal rational functions is proved exactly as in the case of the fractal polynomials treated in [5]. Note that the fractal polynomial approximant for $f \in \mathcal{C}(I)$ developed in [5] may not preserve "shape" properties of f. In this article, we prove that a positive (nonnegative) continuous function can be uniformly well approximated by a positive (nonnegative) fractal rational function. Up to our knowledge, the aforementioned theorem is the first result of its kind that invites fractal methodology to the field of shape-preserving approximation. Further, in contrast to the fractal polynomials, the fractal rational functions allow the uniform approximation of continuous functions defined on an infinite interval as well.

The rest of the material of this article is organized as follows. The following section aims to set out the necessary preliminaries, while in Sect. 3 we introduce the class of fractal rational functions. In Sect. 4, the Weierstrass theorem for the fractal rational functions is developed. Further, the existence of a rational fractal approximant that is copositive with a continuous function and the uniform rational fractal approximation of a continuous function defined on an infinite interval are enunciated. The theme of Sect. 5 is to establish the existence of a Schauder basis consisting of fractal rational functions for $\mathcal{C}(I)$. In Sect. 6, a fractal version of the Müntz theorem for the rational functions is proven.

2 Background and Preliminaries

In this section we shall reintroduce the notion of FIF, and define the corresponding α-fractal operator which lay the requisite groundwork. Our sources for this material are [2, 4, 5, 7].

2.1 *Rudiments of FIF Theory*

Let $x_1 < x_2 < \cdots < x_N$, $N > 2$ be real numbers and $I = [x_1, x_N]$ be a closed interval that contains them. Let a set of interpolation points $\{(x_n, y_n) : n = 1, 2, \ldots, N\}$ be given. Set $J = \{1, 2, \ldots, N-1\}$ and $I_n = [x_n, x_{n+1}]$, $n \in J$. For $n \in J$, let $L_n : I \to I_n$ be contraction homeomorphisms such that:

$$L_n(x_1) = x_n, \;\; L_n(x_N) = x_{n+1}, \;\; n \in J. \tag{1}$$

Let D be a large enough compactum in $\mathbb{R}$ and set $K = I \times D$. For $n \in J$, let $F_n : K \to D$ be mappings satisfying:

$$\left.\begin{aligned} F_n(x_1, y_1) = y_n, \;\; F_n(x_N, y_N) = y_{n+1} \\ |F_n(x, y) - F_n(x, y^*)| \le |\alpha_n||y - y^*| \end{aligned}\right\}, \; n \in J, \tag{2}$$

where $(x, y), (x, y^*) \in K$, and $0 \le |\alpha_n| < 1$ for all $n \in J$.

Define $W_n(x, y) = \big(L_n(x), F_n(x, y)\big)$ for all $n \in J$. It is known [2] that there exists a metric on $\mathbb{R}^2$, equivalent to the Euclidean metric, with respect to which W_n, $n \in J$, are contractions. The collection $\{K; W_n, n \in J\}$ is termed an Iterated Function System (IFS). On $\mathcal{H}(K)$, the set of all non-empty compact subsets of K, endowed with the Hausdorff metric, define the set valued Hutchinson map $W(A) = \bigcup_{n \in J} W_n(A)$. Then, W is a contraction map on the complete metric space $\mathcal{H}(K)$. Thanks to the Banach Fixed Point Theorem, there exists a unique set $G \in \mathcal{H}(K)$ such that $W(G) = G$. This set G is termed the attractor or deterministic fractal corresponding to the IFS $\{K; W_n, n \in J\}$. For any choices of L_n and F_n satisfying the conditions prescribed in (1) and (2), the following theorem holds.

Theorem 2.1 (Barnsley [2]). *The IFS $\{K; W_n, n \in J\}$ defined above admits a unique attractor G. Further, G is the graph of a continuous function $g : I \to \mathbb{R}$ which obeys $g(x_n) = y_n$ for $n = 1, 2, \ldots, N$.*

The function g in Theorem 2.1 is called a FIF corresponding to the IFS $\{K; W_n, n \in J\}$. The characterization of the graph of g by means of an IFS leads to a recursive construction of g via the following functional equation:

$$g(x) = F_n\big(L_n^{-1}(x), g \circ L_n^{-1}(x)\big), \quad x \in I_n, \quad n \in J. \tag{3}$$

The following special class of IFS is well-studied in the literature:

$$\left.\begin{aligned} L_n(x) &= a_n x + b_n \\ F_n(x, y) &= \alpha_n y + q_n(x) \end{aligned}\right\}, n \in J, \tag{4}$$

where $q_n : I \to \mathbb{R}$, $n \in J$, are suitable continuous functions satisfying (2). The multiplier α_n is called a scaling factor of the transformation W_n and $\alpha = (\alpha_1, \alpha_2, \ldots, \alpha_{N-1})$ is the scale vector of the IFS. If q_n, $n \in J$, are polynomials, then the corresponding FIF is termed *polynomial FIF* [5]. In an analogous fashion, in case q_n, $n \in J$, are rational functions, we christen the corresponding FIF as *rational FIF*.

2.2 α- Fractal Operator

Let $f \in \mathcal{C}(I)$. In the IFS (4) we consider, in particular,

$$q_n(x) = f \circ L_n(x) - \alpha_n b(x), \tag{5}$$

where $b : I \to \mathbb{R}$ is a continuous map that fulfills the conditions $b(x_1) = y_1 = f(x_1)$, $b(x_N) = y_N = f(x_N)$ and $b \neq f$. Here the set of data points is $\big\{\big(x_n, f(x_n)\big) : n = 1, 2, \ldots, N\big\}$. In view of Theorem 2.1, the corresponding IFS provides an attractor, which is the graph of a continuous function $f^\alpha : I \to \mathbb{R}$ such that $f^{\alpha}(x_n) = f(x_n), n = 1, 2, \ldots, N$. This case is proposed by Barnsley [2] as a "generalization" of a continuous function f. Often f^{α} is nondifferentiable and its graph has a noninteger Hausdorff–Besicovitch dimension. Whence, we can treat f^{α} as a "fractal generalization" of f.

In particular, we consider the case

$$b = Lf, \tag{6}$$

where $L : \mathcal{C}(I) \to \mathcal{C}(I)$ is an operator which is linear, and bounded with respect to the uniform norm $\|f\|_\infty := \max\{|f(x)| : x \in I\}$ on $\mathcal{C}(I)$ satisfying $Lf(x_1) = f(x_1)$, $Lf(x_N) = f(x_N)$, $L \neq$ Identity. In what follows, two examples for such an operator are provided. Let us recall here that the operator norm $\|L\| := \sup\{\|Lf\|_\infty : \|f\|_\infty \leq 1\}$.

(1) $Lf(x) = f(x)v(x)$, where v is continuous on I, $v(x_1) = v(x_N) = 1$, and v is not identically equal to the constant function $\mathbf{1}$ defined by $\mathbf{1}(x) = 1$ for all $x \in I$. In this case, $\|L\| \leq \|v\|_\infty$.
(2) $Lf(x) = f \circ c\ (x)$, where c is a fixed continuous map satisfying $c(x_1) = x_1$, $c(x_N) = x_N$ and $c \neq$ Identity. Here $\|L\| = 1$.

Definition 2.1. Let $\Delta : x_1 < x_2 < \cdots < x_N$ be a partition of $I = [x_1, x_N]$. A scale vector associated with Δ is a vector $\alpha \in (-1, 1)^{N-1}$. Let f^{α} be the continuous function associated with the IFS (4)–(6). The map f^{α} is called a α-fractal function associated with f with respect to L and the partition Δ. For a fixed partition Δ, a scale vector α, and an operator L, the map f^{α} is unique. Further, in view of (3), f^{α} satisfies the functional equation:

$$f^{\alpha}(x) = f(x) + \alpha_n (f^{\alpha} - Lf) \circ L_n^{-1}(x) \ \ \forall\, x \in I_n,\ n \in J. \tag{7}$$

Definition 2.2. The transformation which assigns f^{α} to the function f is called a α-fractal operator denoted by $\mathcal{F}^{\alpha}_{\Delta,L}$. That is,

$$\mathcal{F}^{\alpha}_{\Delta,L} : \mathcal{C}(I) \to \mathcal{C}(I); \quad \mathcal{F}^{\alpha}_{\Delta,L}(f) = f^{\alpha}. \tag{8}$$

For brevity, we may suppress the dependence of $\mathcal{F}$ on Δ and L and denote $\mathcal{F}^{\alpha}_{\Delta,L}$ by $\mathcal{F}^{\alpha}$.

The following properties of the fractal function f^{α} and the α-fractal operator $\mathcal{F}^{\alpha}$ can be read in the references [7, 8].

Theorem 2.2. *Let $|\alpha|_\infty := \max\{|\alpha_n| : n \in J\}$, and let I_d be the identity operator on $\mathcal{C}(I)$.*

(a) For any $f \in \mathcal{C}(I)$, the perturbation error $\|f^{\alpha} - f\|_\infty \leq \frac{|\alpha|_\infty \|I_d - L\|}{1 - |\alpha|_\infty} \|f\|_\infty$.
(b) The operator $\mathcal{F}^{\alpha}$ is linear and bounded with respect to the uniform norm on $\mathcal{C}(I)$.
(c) For $|\alpha|_\infty < \|L\|^{-1}$, $\mathcal{F}^{\alpha}$ is bounded below.

(d) If $|\alpha|_\infty < \big(1 + \|I_d - L\|\big)^{-1}$, $\mathcal{F}^\alpha$ *has a bounded inverse. Further,* $\|(\mathcal{F}^\alpha)^{-1}\| \le \frac{1+|\alpha|_\infty}{1-|\alpha|_\infty \|L\|}$.

3 α-Fractal Rational Functions

For $k \in \mathbb{N} \cup \{0\}$, let $\mathcal{P}_k(I)$ be the set of polynomials of degree less than or equal to k defined on I, and $\mathcal{P}(I) = \underset{k}{\cup} \mathcal{P}_k(I)$. Let $\mathcal{R}_{mn}(I) = \big\{r = \frac{p}{q} : p \in \mathcal{P}_m(I), q \in \mathcal{P}_n(I); q > 0 \text{ on } I\big\}$ be the set of all real-valued rational functions of type (m,n), and $\mathcal{R}(I) = \underset{m,n}{\cup} \mathcal{R}_{mn}(I)$.

Definition 3.1. Let Δ, L be as given in Sect. 2.2. An α-fractal rational function is an element $r^\alpha \in \mathcal{C}(I)$ such that there is a rational function $r \in \mathcal{R}(I)$ with $\mathcal{F}^\alpha(r) = r^\alpha$. Denote $\mathcal{R}^\alpha_{mn}(I) = \mathcal{F}^\alpha\big(\mathcal{R}_{mn}(I)\big)$ and $\mathcal{R}^\alpha(I) = \mathcal{F}^\alpha\big(\mathcal{R}(I)\big)$. Then, $\mathcal{R}^\alpha(I)$ represents the class of all α-fractal functions associated with $\mathcal{R}(I)$, the family of rational functions of arbitrary type. Since $\mathcal{P}(I) \subset \mathcal{R}(I)$, it follows that the class of fractal rational functions contains all fractal polynomials.

4 Weierstrass-Type Theorem for α-Fractal Rational Functions

The celebrated Weierstrass approximation theorem states that continuous functions on compact intervals can be uniformly approximated by algebraic polynomials. More generally, we have the Stone–Weierstrass theorem which states: *suppose X is a compact Hausdorff space and A is a subalgebra of $\mathcal{C}(X)$ which contains a non-zero constant function. Then, A is dense in $\mathcal{C}(X)$ if and only if it separates points* $\big($*i.e., for any two different points x and y in X there exists a function p in A with $p(x) \neq p(y)$*$\big)$. The denseness of the class of rational functions in $\mathcal{C}(I)$ can be deduced either by using the fact that polynomials, which are subsets of rational functions, are dense in $\mathcal{C}(I)$ or by directly appealing to the Stone-Weierstrass theorem. However, since the class of fractal rational functions defined on I is not, in general, a subalgebra of $\mathcal{C}(I)$, we cannot adapt the Stone–Weierstrass theorem directly to obtain the denseness of fractal rational functions in $\mathcal{C}(I)$. Since the fractal polynomials constitute a subset of the fractal rational functions, the Weierstrass approximation theorem for the fractal rational functions in fact follows from the Weierstrass theorem for the fractal polynomials. However, for the sake of completeness and record we give an expanded rendition of the argument in the following theorem. Our proof relies heavily on the machinery and methods developed in [5] for establishing fractal polynomial analogue of the Weierstrass theorem.

Theorem 4.1. *Suppose f is a continuous real valued function defined on a real interval I. For every $\epsilon > 0$, any partition Δ of the interval with N points, and for a bounded linear operator $I_d \neq L : \mathcal{C}(I) \to \mathcal{C}(I)$ satisfying $Lf(x_1) = f(x_1)$, $Lf(x_N) = f(x_N)$, there exists an α- fractal rational function r^α with $\mathbf{0} \neq \alpha \in \mathbb{R}^{N-1}$ generated via the IFS (4)–(6) such that*

$$\left|f(x) - r^\alpha(x)\right| < \epsilon \ \text{ for all } \ x \in I.$$

Equivalently, $\|f - r^\alpha\|_\infty < \epsilon$.

Proof. Let $\epsilon > 0$ is given. By Stone–Weierstrass theorem it follows that there is a rational function $r \in \mathcal{C}(I)$ with

$$|f(x) - r(x)| < \frac{\epsilon}{2} \quad \text{for all } \ x \in I. \tag{9}$$

For a partition $\Delta : x_1 < x_2 < \cdots < x_N$ of I, select $\mathbf{0} \neq \alpha \in \mathbb{R}^{N-1}$ such that $|\alpha|_\infty < 1$ and

$$\frac{|\alpha|_\infty}{1 - |\alpha|_\infty}\|I_d - L\| \ \|r\|_\infty < \frac{\epsilon}{2}. \tag{10}$$

Now

$$\begin{aligned} |f(x) - r^\alpha(x)| &\leq |f(x) - r(x)| + |r(x) - r^\alpha(x)|, \\ &\leq |f(x) - r(x)| + \|r - r^\alpha\|_\infty, \\ &\leq |f(x) - r(x)| + \frac{|\alpha|_\infty}{1 - |\alpha|_\infty}\|I_d - L\| \ \|r\|_\infty, \\ &< \frac{\epsilon}{2} + \frac{\epsilon}{2} = \epsilon. \end{aligned}$$

The first step in the preceding analysis used the triangle inequality, the second step used the definition of uniform norm, and the third step borrowed Theorem 2.2(a). The last two steps were consequent upon inequalities (9) and (10). □

Note 4.1. If in $Lf = f \circ c$, $c \neq I_d$ is a rational function satisfying $c(x_1) = x_1$, $c(x_N) = x_N$, then the FIF r^α is a rational FIF.

The following statement is a simple consequence of Theorem 4.1.

Theorem 4.2. *The set of rational FIFs with non-null scale vector is dense in $\mathcal{C}(I)$.*

In the following theorem we establish the denseness of a class of fractal functions which is a proper subset of the one given in Theorem 4.2. We prove that one single scale vector is sufficient to obtain a fractal rational function approximant to a continuous function. Our proof is patterned after [8].

Theorem 4.3. *If $|\alpha|_\infty < \big(1 + \|I_d - L\|\big)^{-1}$, $\mathcal{R}^\alpha(I)$ is dense in $\mathcal{C}(I)$.*

Proof. Let $g \in \mathcal{C}(I)$. Since $|\alpha|_\infty < \big(1 + \|I_d - L\|\big)^{-1}$, Theorem 2.2 (d) ensures that there exists $f \in \mathcal{C}(I)$ with $\mathcal{F}^\alpha(f) = g$. For f in $\mathcal{C}(I)$, the Stone–Weierstrass theorem implies the existence of a sequence of rational functions r_n converging uniformly to f. Boundedness of the operator $\mathcal{F}^\alpha$ now yields $g = \mathcal{F}^\alpha(f) = \mathcal{F}^\alpha(\lim r_n) = \lim(r_n^\alpha)$, and with it the proof. □

Note 4.2. If $|\alpha|_\infty < \big(1 + \|I_d - L\|\big)^{-1}$, then $\mathcal{P}^\alpha(I) \subseteq \mathcal{R}^\alpha(I)$ is dense in $\mathcal{C}(I)$, and consequently $\mathcal{R}^\alpha(I)$ is dense in $\mathcal{C}(I)$.

The uniform fractal polynomial approximant for $f \in \mathcal{C}(I)$ established in [5] and the uniform fractal rational approximant for $f \in \mathcal{C}(I)$ obtained in Theorem 4.1 may not respect the shape property of f. For some reasons, perhaps the physical situation which r is intended to model, finding an approximant r from a prescribed subset of $\mathcal{C}(I)$ to $f \in \mathcal{C}(I)$ so as to inherit certain properties of f is always of interest. We now describe a theorem in this direction which is the main offering of the current section and which is expected to pave way to "shape preserving" fractal approximants to a continuous function.

Theorem 4.4. *Let $f \in \mathcal{C}(I)$ be such that $f(x) \geq 0$ for all $x \in I$. For any $\epsilon > 0$, there exists a nonnegative α- fractal rational function r^α with $\mathbf{0} \neq \alpha \in \mathbb{R}^{N-1}$ generated through a suitable IFS such that $\|f - r^\alpha\|_\infty < \epsilon$.*

Proof. Let $\epsilon > 0$ and $f \in \mathcal{C}(I)$ be such that $f(x) \geq 0$ for all $x \in I$. Consider the IFS defined by (4)–(6). We assume further that the operator L involved in the definition of the IFS fixes the constant function $\mathbf{1}$ defined by $\mathbf{1}(x) = 1$ for all $x \in I$. That is, $L\mathbf{1} = \mathbf{1}$.

First we prove that the α-fractal operator $\mathcal{F}^\alpha$ corresponding to L fixes the function $\mathbf{1}$. From the functional equation of f^α $\big($cf. (7)$\big)$ we obtain:

$$\|f^\alpha - f\|_\infty \leq |\alpha|_\infty \|f^\alpha - Lf\|_\infty.$$

Substituting $f = \mathbf{1}$, the fixed point of L, the above inequality manifests:

$$\|f^\alpha - \mathbf{1}\|_\infty \leq |\alpha|_\infty \|f^\alpha - \mathbf{1}\|_\infty.$$

Since $0 < |\alpha|_\infty < 1$, we see from the foregoing inequality that $\|f^\alpha - \mathbf{1}\|_\infty = 0$. Therefore the α-fractal function corresponding to $f = \mathbf{1}$ is $f^\alpha = \mathbf{1}$. In other words, $\mathbf{1}$ is a fixed point of $\mathcal{F}^\alpha$.

For $\epsilon > 0$ and $f \in \mathcal{C}(I)$, Theorem 4.1 ensures the existence of a rational function s and a scale vector $\mathbf{0} \neq \alpha \in \mathbb{R}^{N-1}$ for which $s^\alpha = \mathcal{F}^\alpha(s)$ satisfies $\|f - s^\alpha\|_\infty < \frac{\epsilon}{2}$.

Define:

$$r^\alpha(x) = s^\alpha(x) + \frac{\epsilon}{2} \quad \text{for all } x \in I.$$

We note that

$$r^\alpha(x) = s^\alpha(x) + \frac{\epsilon}{2}\mathbf{1}(x) = s^\alpha(x) + \frac{\epsilon}{2}\mathbf{1}^\alpha(x).$$

The linearity of the map $\mathcal{F}^\alpha$ stipulates:

$$r^\alpha = s^\alpha + \frac{\epsilon}{2}\mathbf{1}^\alpha = \mathcal{F}^\alpha(s) + \frac{\epsilon}{2}\mathcal{F}^\alpha(\mathbf{1}) = \mathcal{F}^\alpha\big(s + \frac{\epsilon}{2}\mathbf{1}\big).$$

Consequently, r^α is a fractal rational function, and satisfies

$$r^\alpha(x) = s^\alpha(x) + \frac{\epsilon}{2} = f(x) + \frac{\epsilon}{2} + s^\alpha(x) - f(x) \geq f(x) + \frac{\epsilon}{2} - \|s^\alpha - f\|_\infty \geq 0.$$

The result follows from the triangle inequality $\|f - r^\alpha\|_\infty \leq \|f - s^\alpha\|_\infty + \|s^\alpha - r^\alpha\|_\infty$. □

Note 4.3. The operator $L : \mathcal{C}(I) \to \mathcal{C}(I)$ defined by $Lf = f \circ c$, where c is a continuous map satisfying $c(x_1) = x_1$, $c(x_N) = x_N$ and $c \neq I_d$ fixes all constant functions, in particular, $\mathbf{1}$.

In the classical Weierstrass theorem on the uniform approximation of a continuous function by algebraic polynomials, compactness of the interval plays a crucial role. In fact, since no sequence of polynomials (with the exception of a stationary sequence) can be uniformly convergent in an infinite interval, not a single function other than a polynomial can be the limit of a uniformly convergent sequence of such polynomials in an infinite interval. The position is quite different, if instead of the polynomials as approximants on the infinite interval, use is made of the traditional/fractal rational functions. It may be supposed that the infinite interval is the whole real axis $-\infty < x < +\infty$. We have the following proposition.

Proposition 4.1 ([13] p. 12). *Let f be continuous on the whole real axis and has the finite limit $\lim_{x\to\pm\infty} f(x) = k$. For every $\epsilon > 0$ there exists a continuous rational function r such that $\|f - r\|_\infty < \epsilon$.*

Using the above proposition we shall establish the uniform approximation of continuous functions on an infinite interval by fractal rational functions.

Theorem 4.5. *Let f be a function continuous on the whole real axis such that $\lim_{x\to\pm\infty} f(x) = k$. For every $\epsilon > 0$ there exists a piecewise defined fractal rational function r^α such that $\|f - r^\alpha\|_\infty < \epsilon$.*

Proof. Let $\epsilon > 0$. By Proposition 4.1, there exists a continuous rational function r such that:

$$\|f - r\|_\infty < \frac{\epsilon}{2}. \tag{11}$$

For this continuous rational function $r : \mathbb{R} \to \mathbb{R}$, let $r_m = r\big|_{[m,m+1]}$, $m \in \mathbb{Z}$. Choose a partition Δ_m with N points on the interval $[m, m+1]$ and a scale vector

$\mathbf{0} \neq \alpha^m \in \mathbb{R}^{N-1}$. Now we can define α-fractal function of r_m as $r_m^{\alpha^m} = \mathcal{F}^{\alpha^m}(r_m)$ in the interval $[m, m+1]$. Let us denote the bounded maps used in the definition of $\mathcal{F}^{\alpha^m}$ by $L_m : \mathcal{C}[m, m+1] \to \mathcal{C}[m, m+1]$ and identity operator on $\mathcal{C}[m, m+1]$ by I_{dm}. Further, we assume that $\sup\big\{\|L_m\| : m \in \mathbb{Z}\big\} = L^*$ is finite. The fractal function of r is defined in a piecewise manner by $r^\alpha\big|_{[m,m+1]} = r_m^{\alpha^m}$, where α is an infinite matrix whose rows are the scale vectors α^m. Since $r_m^{\alpha^m}(m) = r_m(m) = r(m)$ for all $m \in \mathbb{Z}$, it follows that the fractal function r^α is continuous on $\mathbb{R}$.

Choose the scaling matrix α such that $|\alpha|_\infty = \sup\limits_{m\in\mathbb{Z}} |\alpha^m|_\infty$ is finite, and $\frac{|\alpha|_\infty}{1-|\alpha|_\infty}(1 + L^*)\|r\|_\infty < \frac{\epsilon}{2}$. With the aid of a series of self-explanatory steps we have

$$\begin{aligned}
\|r^\alpha - r\|_\infty &= \sup\big\{|r^\alpha(x) - r(x)| : x \in \mathbb{R}\big\},\\
&= \sup_{m\in\mathbb{Z}}\big\{|r_m^{\alpha^m}(x) - r(x)| : x \in [m, m+1]\big\},\\
&\le \sup_{m\in\mathbb{Z}} \frac{|\alpha^m|_\infty}{1 - |\alpha^m|_\infty}\|I_{dm} - L_m\|\;\|r_m\|_\infty, \qquad (12)\\
&\le \frac{|\alpha|_\infty}{1 - |\alpha|_\infty}(1 + L^*)\|r\|_\infty,\\
&< \frac{\epsilon}{2}.
\end{aligned}$$

Now the triangle inequality implies that

$$\|f - r^\alpha\| \le \|f - r\|_\infty + \|r^\alpha - r\|_\infty, \qquad (13)$$

and a combination of (11)–(13) completes the proof. □

5 Schauder Basis for $\mathcal{C}(I)$ Consisting of Fractal Rational Functions

As a consequence of the Weierstrass theorem for the fractal rational functions established in the previous section, we shall prove the existence of a Schauder basis of fractal rational functions for $\mathcal{C}(I)$. Our idea for constructing a basis of fractal rational functions stems from the well-known idea of a small perturbation argument. As a prelude to the desired result, let us recall the following definitions and result.

Definition 5.1 ([3] p. 1). A sequence $\{x_i\}_{i=1}^\infty$ in a Banach space $\big(X, \|.\|\big)$ is called a Schauder basis of X if for every $x \in X$ there is a unique sequence of scalars $\{a_i\}_{i=1}^\infty$ so that $x = \sum\limits_{i=1}^{\infty} a_i x_i$. A basis $\{x_i\}_{i=1}^\infty$ is called normalized if $\|x_i\| = 1$ for all i.

Definition 5.2 ([3] p. 2). Let $(X, \|.\|)$ be a Banach space with a basis $\{x_i\}_{i=1}^{\infty}$. The projections $P_i : X \to X$ defined by $P_i\Big(\sum_{k=1}^{\infty} a_k x_k\Big) = \sum_{k=1}^{i} a_k x_k$ are linear operators and $\sup_i \|P_i\| < \infty$. The projections $\{P_i\}_{i=1}^{\infty}$ are called the natural projections associated with $\{x_i\}_{i=1}^{\infty}$; the number $\sup_i \|P_i\|$ is called the basis constant of $\{x_i\}_{i=1}^{\infty}$.

Definition 5.3 ([3] p. 5). Two bases $\{x_i\}_{i=1}^{\infty}$ and $\{y_i\}_{i=1}^{\infty}$ are called equivalent provided a series $\sum_{i=1}^{\infty} a_i x_i$ converges if and only if $\sum_{i=1}^{\infty} a_i y_i$ converges.

Proposition 5.1 ([3] p. 5). *Let $\{x_i\}_{i=1}^{\infty}$ be a normalized basis of a Banach space X with basis constant K. Let $\{y_i\}_{i=1}^{\infty}$ be a sequence of vectors in X with $\sum_{i=1}^{\infty} \|x_i - y_i\| < \frac{1}{2K}$. Then, $\{y_i\}_{i=1}^{\infty}$ is a basis of X which is equivalent to $\{x_i\}_{i=1}^{\infty}$.*

We are now well equipped to prove the desired theorem.

Theorem 5.1. *There exists a Schauder basis of fractal rational functions in $\mathcal{C}(I)$.*

Proof. Let $\{x_i\}_{i=1}^{\infty}$ be a Schauder basis for $\mathcal{C}(I)$ consisting of the traditional rational functions. The existence of such a basis is already established in the literature (see, for instance, [12]). Assume further that the basis $\{x_i\}_{i=1}^{\infty}$ is normalized and has the basis constant K. By Theorem 4.1, for each x_i there exists a fractal rational function $r_i^{\alpha^i}$ such that $\sum_{i=1}^{\infty} \|x_i - r_i^{\alpha^i}\|_\infty < \sum_{i=1}^{\infty} \frac{1}{2^{i+1}K} = \frac{1}{2K}$. Appealing to Proposition 5.1, it follows that $\{r_i^{\alpha^i}\}_{i=1}^{\infty}$ is a Schauder basis of fractal rational functions for $\mathcal{C}(I)$. Further, $\{r_i^{\alpha^i}\}_{i=1}^{\infty}$ is equivalent to $\{x_i\}_{i=1}^{\infty}$. □

Recall that a subset A of a metric space (X, d) is said to be nowhere dense if its closure has empty interior. Further, A is said to be of first category or meagre in X if it is a union of countably many nowhere dense subsets and of second category or nonmeagre in X if it is not of first category in X. It would be of interest to examine the set $\mathcal{R}_{mn}^{\alpha}(I)$ with respect to its Baire property, that is, whether it is first or second category in $\mathcal{C}(I)$. The next remark addresses this for suitable values of the scaling vector.

Remark 1. We know that $\mathcal{R}_{mn}(I)$ is of first category in $\mathcal{C}(I)$ endowed with the uniform norm. In fact, the set of all continuous functions defined on I that are differentiable at least at one point in I is of first category. Since the property of being first category is a topological invariant, for $|\alpha|_\infty < (1 + \|I_d - L\|)^{-1}$, the set $\mathcal{R}_{mn}^{\alpha}(I)$ is of first category in $\mathcal{C}(I)$.

6 Fractal Müntz Rational Functions

Recall that a set S is fundamental in a normed linear space X if the set of linear combinations of elements of S is dense in X. The classical Müntz theorem says that for a sequence $\Lambda := \{\lambda_i\}_{i=0}^{\infty}$ with $0 = \lambda_0 < \lambda_1 < \lambda_2 < \ldots$, the Müntz space $M(\Lambda) = \text{Span}\{x^{\lambda_0}, x^{\lambda_1}, \ldots\}$ is dense in $\mathcal{C}[0,1]$ if and only if $\sum\limits_{i=1}^{\infty} \frac{1}{\lambda_i} = \infty$. By taking $\lambda_i = i$ we can recover the classical Weierstrass theorem on $[0,1]$, which states simply that $\{1, x, x^2, \ldots\}$ is fundamental in $\mathcal{C}[0,1]$. Furthermore, Müntz theorem is all the more interesting because it traces a logical connection between two apparently unrelated facts: the fundamentality of $\{1, x, x^2, \ldots\}$ and the divergence of the reciprocal exponents $\sum\limits_{i=1}^{\infty} \frac{1}{i}$. In fact, if we wish to delete functions from this set whilst maintaining its fundamentality, this divergence is precisely the property that must be preserved. We would like to remark that the point 0 is special in the study of Müntz spaces. Even replacing $[0,1]$ by an interval $[a,b] \subset [0,\infty)$ in Müntz theorem is a nontrivial issue.

Concerning the rational case, it has been proved that (see, for instance, [1]) $\mathcal{R}(\Lambda) := \{\frac{p}{q} : p, q \in M(\Lambda) q \neq 0\}$ is always dense in $\mathcal{C}[0,1]$. This surprising result says that while the set $M(\Lambda)$ of Müntz polynomials may be far from dense, the set $\mathcal{R}(\Lambda)$ of Müntz rationals is always dense in $\mathcal{C}[0,1]$, no matter what the underlying sequence Λ. In words of Newmann [11]: "*apparently, rational functions always want to be dense. There is something magical about performing that one division.*"

Next theorem addresses the fractal version of the Müntz theorem concerning the rational case.

Theorem 6.1. *Let $f \in \mathcal{C}[0,1]$ be given. For all $\epsilon > 0$, there exists a fractal Müntz rational function $r^{\alpha} = \mathcal{F}^{\alpha}(r)$, where $\mathbf{0} \neq \alpha \in \mathbb{R}^{N-1}$ and $r(x) = \frac{\sum_{i=0}^{M} \beta_i x^{\lambda_i}}{\sum_{i=0}^{M} \gamma_i x^{\lambda_i}}$, such that $\|f - r^{\alpha}\|_{\infty} < \epsilon$.*

Proof. Let $\epsilon > 0$. For $f \in \mathcal{C}[0,1]$, it is known [1] that there exists a Müntz rational function $r(x) = \frac{\sum_{i=0}^{M} \beta_i x^{\lambda_i}}{\sum_{i=0}^{M} \gamma_i x^{\lambda_i}}$ such that:

$$|f(x) - r(x)| < \frac{\epsilon}{2} \quad \text{for all } x \in [0,1]. \tag{14}$$

For a partition $\Delta : 0 = x_1 < x_2 < \cdots < x_N = 1$, we choose $\alpha \in \mathbb{R}^{N-1}$, $\alpha \neq \mathbf{0}$ such that $|\alpha|_{\infty} < 1$ and

$$\frac{|\alpha|_{\infty}}{1 - |\alpha|_{\infty}} \|I_d - L\| \, \|r\|_{\infty} < \frac{\epsilon}{2}. \tag{15}$$

With the aforementioned choice of partition and scale vector, let $r^\alpha = \mathcal{F}^\alpha(r)$.

The norm estimate $\|f - r^\alpha\|_\infty$ can be now derived exactly as we did in Theorem 4.1; the details follow strictly anticipated lines and are therefore omitted.

□

A conclusion is that the theory of approximation by rational functions can be extended to the fractal rational functions through the α-fractal operator. Further, this new class of functions shares many of the properties such as denseness in $\mathcal{C}(I)$ and constitution of bases for $\mathcal{C}(I)$ with its classical counterpart, while differs in some others, for instance, the smoothness of the elements in the class. In general, the article intends to be a small step forward in the knowledge of fractal functions from the perspective of approximation theory by taking class of rational functions as a medium.

Let us round off the paper with a few remarks. The fact that $\mathcal{P}_n^\alpha(I)$ is a finite dimensional linear space ensures the existence of a best approximant from $\mathcal{P}_n^\alpha(I)$ to $f \in \mathcal{C}(I)$. That is, there exists a $p^{\alpha *} \in \mathcal{P}_n^\alpha(I)$ such that $\|f - p^{\alpha *}\|_\infty = \inf\{\|f - p^\alpha\|_\infty : p^\alpha \in \mathcal{P}_n^\alpha(I)\}$, in other words $\mathcal{P}_n^\alpha(I)$ is an existence set. On the contrary, the set $\mathcal{R}_{mn}^\alpha(I)$ of all α-fractal rational functions of type (m,n) is not linear space. As a consequence, the following questions arise: is $\mathcal{R}_{mn}^\alpha(I)$ an existence set? In case $\mathcal{R}_{mn}^\alpha(I)$ is an existence set, whether the corresponding best approximation operator (possibly a multifunction) is continuous? To keep the size of this article within reasonable limits we have not been able to include the answers for these questions that naturally arise. We shall return to these questions elsewhere.

Acknowledgements The first author is partially supported by the Council of Scientific and Industrial Research India (Grant No. 09/084 (0531)/2010-EMR-I) for this work. The second author is thankful to the Science and Engineering Research Council, Department of Science and Technology India (Project No. SR/S4/MS: 694/10).

References

1. Bak, J., Newmann, D.J.: Rational combinations of x^{λ_k}, $\lambda_k \geq 0$, are always dense in $C[0, 1]$. J. Approx. Theory **23**, 155–157 (1978)
2. Barnsley, M.F.: Fractal functions and interpolations. Constr. Approx. **2**, 303–329 (1986)
3. Lindenstrauss, J., Tzafriri, L.: Classical Banach Spaces I. Springer, Berlin (1977)
4. Massopust, P.R.: Fractal functions and their applications. Chaos Solitons & Fractals **8**(2), 171–190 (1997)
5. Navascués, M.A.: Fractal polynomial interpolation. Z. Anal. Anwend **25**(2), 401–418 (2005)
6. Navascués, M.A., Sebastián, M.V.: Construction of affine fractal functions close to classical interpolants. J. Comput. Anal. Appl. **9**(3), 271–285 (2007)
7. Navascués, M.A., Chand, A.K.B.: Fundamental sets of fractal functions. Acta Appl. Math. **100**(3), 247–261 (2008)
8. Navascués, M.A.: Fractal approximation. Complex Anal. Oper. Theory **4**(4), 953–974 (2010)
9. Navascués, M.A.: Reconstruction of sampled signals with fractal functions. Acta Appl. Math. **110**(3), 1199–1210 (2010)
10. Navascués, M.A.: Fractal bases of L_p spaces. Fractals **20**(2), 141–148 (2012)

11. Newmann, D.J.: Approximation with rational functions. In: CBMS Regional Conf. Ser. in Math., vol. 41. Amer. Math. Soc., Providence (1979)
12. Petrushev, P.: Bases consisting of rational functions of uniformly bounded degrees or more general functions. J. Funct. Anal. **174**, 18–75 (2000)
13. Timan, A.F.: Theory of Approximation of Functions of a Real Variable. Pergamon Press, Oxford (1963)

Part III
Applications of Fractals and Wavelets

Innovation on the Tortuous Path: Fractal Electronics

Nathan Cohen

Abstract We describe the innovation environment that has limited the onset of innovations in technology, choosing as an example the use of fractals in electronics. There is a considerable time delay between the production of these innovations and their widespread implementation with end-users; this delay is typical in innovation; underlying and predictable pitfalls in the adoption process are inherent to the progress of innovation through a "tortuous path." The paradox of rapid adoption in pure science and math versus delay in practical applications should be considered a normal aspect of innovation in fractal technology.

Keywords Innovation • Fractal antennas • Tortuous path • Hibervation

1 Introduction: Understanding Innovation

Innovation has always been a major contributor to the world economy, based upon the needs of efficiency, low cost, and increasing functionality of processes. It is this long-term acculturation to innovation that has allowed U.S. President Obama to assert that the USA is an "innovation economy" [1], and this has in turn set an example for a global interest and emphasis on innovation, presumably for the progress of nations.

However, little is understood about the innovation process as it progresses to the end-user, particularly from the innovator's point of view. A working contemporary hypothesis is the notion that innovation succeeds through "disruption," whereby standard processes are usurped by usually unanticipated innovations, that first manifest as lower performance drivers in smaller markets outside those of the

N. Cohen (✉)
Fractal Antenna Systems, Inc., Bedford, MA 01730, USA
e-mail: ncohen@fractenna.com

C. Bandt et al. (eds.), *Fractals, Wavelets, and their Applications*, Springer Proceedings in Mathematics & Statistics 92, DOI 10.1007/978-3-319-08105-2_24,

larger and economically dominant ones [2]. In fact, most key innovations have not been accepted through disruption, including such obvious ones as: airplanes; automobiles; wireless and radio; printing; and so on. Such key innovations were dramatic departures from anything remotely similar and they entered key markets with no "bootstrap" through smaller ones. Disruption clearly has importance as a solution to getting the innovation to a dominant set of end-users. However it is not an all-inclusive explanation that defines innovation's path to success, nor does it explain why there exist impediments to reaching the end-user. Rather, disruption appears as a mechanism for innovation to gather niche end-users until the innovation passes to a mere improvement of the innovation, when it then can tackle existing market leaders at their own game.

A further failure of disruption is it does not account for how innovations themselves can fail, only to *come back later*, thus leading to the staggering and common euphemism of "a technology before its time." Does the lack of immediate adoption of an innovation indicate its inherent folly? Finally, the theory of disruption has been primarily used as a means of understanding how large companies get unexpectedly toppled by outside innovation, as opposed to the perspective presented here, where the issue *of how innovators get their innovations to the end-user, and what impedes that process*.

In the context of fractals, despite wide acceptance in science and math, technology in hardware has been lacking. Why has this occurred and what is the outcome of innovation using fractals? Here, these issues will be addressed by defining the problems inherent to innovation adoption, and presentation of examples in fractals in electronics.

2 A Summary of Innovation and Aspects of Innovators

It is important to define innovation and the characteristics of innovators in order to explore this path to innovation acceptance, particularly with this example of fractals in electronics. A more detailed discussion of this path will be found elsewhere [3]. And although there are always exceptions to generalized attributes, such attributes will be enumerated here for a commonality of definition and description.

Starting with innovation, the following definitions and attributes will be used:

- *IMPROVEMENT* is a process change that follows as *slight variation* of the existing art based on feedback from the end-users of the process. It is a reductionist, incremental change built upon a conventional existing process structure;
- *INNOVATION* is a process change that takes the state of the art well beyond an incremental improvement;
- *INNOVATION* is not reductionist, it is a *jump* that does not follow obviously from the existing art.

The dynamic between improvement and innovation is the driving force controlling if and when actual innovative processes get accepted and adopted by end-users. Improvement, with some exception, is the domain of extant technologies and their champions (companies; end-users; non-profits' governments; and so on) while innovation occurs from outside the arenas that are improvement-driven.

U.S. domination of innovation, at least historically, is squarely manifest by the enabling of personal traits of the innovators which historically may not have been tolerated, let alone encouraged, in other cultural or political contexts. As such, a profile of innovators [3], while encompassing diversity beyond these generalities, includes a commonality of some key characteristics:

- *INNOVATORS MAKE CONNECTIONS*—they see the process as being flawed and incomplete and see unrelated processes from outside the art, and bring them together;
- *INNOVATORS ARE THE "WRONG" PEOPLE*—they are bit players inside the art or come from outside of it. Oftentimes they view the art themselves without being inculcated into it by its peers. Innovators are (at first) not "leaders in the field";
- *INNOVATORS ARE END-USERS*—They use the process and understand it well and want to use it differently;
- *INNOVATORS ARE SOLO, OR IN SMALL GROUPS*—the steps leading the innovation are not generally a group assemblage;
- *INNOVATORS HAVE A NEED*—They are interested in how others use the process, but they themselves feel the process "just has to be a lot better" for their own use;
- *INNOVATORS SELDOM AT/STAY WITH LARGE COMPANIES*—their innovation gets no development internally at large companies because it is "off focus" and any IP dies or gets sold;
- *INNOVATORS ARE "KNOW IT ALL's"*—they have an encyclopedic knowledge of many subjects that extend far being the knowledge related to the process. They are more "generalists" than "experts";
- *INNOVATORS ARE NAÏVE*—they do not know about the obstacles to improvement in the state of the art and assume the effort will be minor, and the value lies in the innovation; they assume the innovation will be obvious in benefit to others, and that other end-users will want it as much as they do.

Innovators are thus driven to solve problems that they have with the extant process, finding it inadequate and produce far more advanced processes as innovation. Their emphasis is on the solution rather than the effort and method of reaching the other end-users. Innovators thus misjudge the effort needed, and the dangers inherent, to the path to the end-users. This becomes the key factor in whether innovations become adopted by the processes they enhance or replace.

3 The Tortuous Path

While some innovations require a great deal of money and time to bring the innovation to reality, most innovators are able to bring the innovation to fruition with modest to moderate resources. Many would-be innovators assume that the need for such resources is the entry barrier to acceptance to innovation. The barrier to acceptance, however, is not the availability of resources, but distinct classes of deleterious conditions, or *pitfalls* that transpire externally, that metaphorically comprise a "tortuous path." This tortuous path is the *process of getting the innovation to, adopted by, a large end-user base*. It is this tortuous path that kills most innovations during the period of interest or existence of the innovator.

The tortuous path is defined and constricted by the fact that very few innovations come about without some extant process or technology which dominates in achieving end-user needs. To wit: There is already a squatter in the territory. For example, the innovation of electric lighting eventually replaced arc and oil lamp lighting; cars replaced horse driven transportation; radio replaced wired telegraphy [4]. Because these processes already possessed vast end-users, work forces dependent upon their existence; and companies and governments that depend on continued use of the process, albeit with some opportunity of improvement, innovations are not met with great interest or desire for implantation of change. In fact, quite the opposite: the convention is to thwart the innovation.

Surprisingly, these efforts to thwart innovation—pitfalls on the tortuous path—are neither arbitrary nor so nebulous as to avoid categorization. They all share the ability to stop innovation, at least for the innovators themselves.

Some key pitfall categories include:

- "**Cinderella Syndrome**" (**FAIL-FAST**)—Investors are a key factor in thwarting innovation by demanding its success with end-users in very short time periods, often as little as 2 years. Analogous to Cinderella returning to her humble roots by midnight, this is called the "Cinderella syndrome," also known as "Fail-Fast." By example, as over 90 % of venture capital investments fail from the "Cinderella syndrome" VCs appear poor at choosing innovations with rapid time to market acceptance (RTMA) over the short time scale they defined for return on investment (ROI).
- **Fear Factor Catastrophies**—This tortuous path pitfall has become so ingrained in American culture that few remember times when it was rare. Here, a competitor(s) or competitive champion takes an innovation and demonstrates an apparent high risk or danger that stresses the disadvantage to the innovation. The fact that ALL innovations have some small risk is not discussed: instead the notion of "fear of use" is driven by an accident (often staged) that causes damage using the innovation.
- **Patent Napping**—The objective of patents is to secure time-limited rights on exclusivity to make, use, offer, and sell an inventive innovation. Ironically, patents can and are used (in many cases) to "sleep" innovation so that the competitors cannot step in, and the owner can continue with existing offerings

without being usurped by the innovation. Patent napping is also unintentionally done by patent offices who can take extraordinarily long execution times in deciding novelty, creating uncertainty on the part of the innovator to push the innovation forward.

- "**Statis Buy-out**" by large company—here, a large company buys out the innovation or the company who owns it, and then summarily stops its competitive progress by freezing its progress or not taking it to market.
- **Paradigm Killers**—Leaders in field give innovation a "kiss off." An "Endorsement of the status Quo."
- "**Statis Buy-in**"—is similar to the buy-out, but the large company makes a partnership with the innovator, who buys-in, but no execution to market occurs for the innovation.
- **Sirens of Titan**—Disinformation and defamation are spread in media for "reasonable cause for doubt." Intent is to erode credibility of the innovation and the innovator.

The effect of these tortuous path pitfalls is to slow down, stop, or kill the innovation for the innovators, long before the ultimate reach and decision by end-users is even possible.

Oddly, most innovations may die for the innovators, but as contracts and patents expire, as memory fades, and as corporate priorities change, the innovation goes out of hibernation—that is "*hibervation*"—and are resurrected by totally different groups than the innovators.

The time scales [3] may be quite long; hibervation is typically from 10 to 30 years. A large percent of modern innovations went through a hibervation period. This is an important point: most innovations do not die, but only appear to be dead for a period of time after the creation of the innovation.

4 Quick Review of Fractal Electronics

It seems puzzling that applications of fractals should be limited in acceptance despite their ubiquity in nature [5]. In the context of the tortuous path, it is enlightening to view an important applied segment of applied fractals in electronics.

In Table 1, several examples of the deliberate and intentional use of *fractal shaping* in electronics are given. They include some (otherwise) standard electronic components, such as transistors, and antennas. Resonators are a generic term for a tuned circuit and include filters. Notice the year of the innovation's start and a brief benefit.

The list is short but select: each of these innovations has been rendered to utility; poses clear benefits over the existing art; and may ultimately be dominant in the fields and markets so intended. Figures 1 and 2 show several examples of the above.

Table 1 Fractal electronic innovations

Device	Year	References	Benefits description
Antenna	1988	[6]	Smaller size, broader bands
Resonator	1988	[6]	Smaller size, broader bands
Capacitor	1998	[7]	Greater capacitance
Transistor	1998	[8]	Greater power handling
Diplexer	2009	[9]	Greater bandwidth, efficiency
Invisibility cloak	2008	[10]	Harry Potter effect

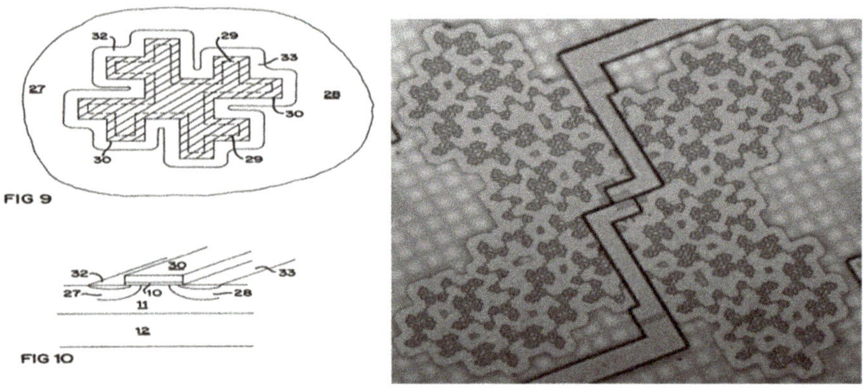

Fig. 1 Fractal transistor, and fractal capacitor (see [8] and [7]—(courtesy Michael Frame Yale University))

Fig. 2 Fractal antenna, cloak, and resonators (courtesy author)

The inspiration for all these devices appears to start with the application of fractals to antennas [6]. All share the benefit of fractals making smaller sized components based on greater perimeters in a given area, or areas in a given volume. None of them are difficult to fabricate; more expensive to make; and so on. Yet only one on the list could be considered, now, a successful innovation: fractal antennas.

Table 2 Emergence from Hibervation

Device	Year	Hibervation period
Antenna	2008	20 years
Resonator	–	25+ years
Capacitor	–	16+ years
Transistor	–	16+ years
Diplexer	–	4+ years
Invisibility cloak	–	6+ years

You cannot buy an electronic gadget loaded with fractal transistors; you cannot buy a power storage device using a fractal capacitor. These useful devices are in-waiting. They are in hibervation. Indeed, we can track the hibervation time scales. This is listed in Table 2, with the year of general use, and the hibervation period to date.

In other words, there is a wealth of valuable fractal based technology in electronics which has yet to be used on a large scale, based upon the hibervation periods shown from Table 2.

5 Discussion

A small data sample may define a trend, but does not prove it. Still, the devices of Tables 1 and 2 tell some important stories.

For example, fractal antennas and fractal resonators were created as innovations at the same time, patent applied for at the same time, and published at the same time, by the author as innovator. For fractal antennas, the author responded to many of the pitfalls described, but through persistence and fact, finally saw innovation acceptance, with substantial commercial application, by 2008.

In contrast, fractal resonators were clearly placed into a patent-nap, via dilatory prosecution at the patent office: 12 years transpired for issuance. This created a dilemma whereby the innovator was incapable of pushing for product level applications, as there was no assurance of the holder of the technology. By 2009, the first step in acceptance of fractal resonators—an entire book [11] on the subject—occurred, more than 20 years from creation. An increasing number of scientific papers on fractal resonators may signal the inchoate emergence from hibervation. The innovator (author) is now identifying niches of use for market products, that is, a disruptive strategy, to overcome the prolonged hibervation.

In a curious way, the comparison of the story of fractal antennas and fractal resonators reveals that a key distinction in the navigation of the tortuous path must be the innovator fighting off the pitfalls. No innovator navigating, and battling the pitfalls, means no traveling to the destination. Thus the fractal resonators acted as a "control" in an innovation experiment on the tortuous path, demonstrating that the innovator must push through the tortuous path for acceptance of the innovation.

For the fractal transistor, it is unusual that the technology has not entered the market place. This may be an additional case of patent-napping by the patent holder; Fairchild Semiconductor, the patent holder, does not have any products with fractal transistors, despite the patent claims of distinct and unique advantages.

For fractal capacitors, the "sirens" may be preventing cautious acceptance of the premise, or the paradigm killers may be maintaining the status quo. Clearly there is some aspect of the Cinderella syndrome occurring. Also, efforts by the author to discover the hibervation issue suggested a stasis buy-in by some research groups with corporate sponsors, but such efforts are not recent. Indeed, the patent on fractal capacitors has lapsed due to non-payment of maintenance fees (see US patent 6084285) so Cinderella has passed the striking of midnight for the innovators. The basic innovation of fractal capacitors is thus available to any and all parties. There is clearly a demonstrable need for capacitors and supercapacitors with higher power storage and higher voltage breakdown. This may be a case where disruption will be invoked to get the stasis overcome and to establish an initial niche for use.

Diplexers are still too new to have moved out of a possible hibervation stage. Invisibility cloaks, despite working as advertised, are well on their way to fighting tortuous path pitfalls and are ongoing.

6 Conclusions

The value of fractals may be self-evident, but accepted fractal innovation, as filtered by the tortuous path, should not be expected to occur for roughly a generation. Although this bodes well for future fractal technology, great caution should be practiced in assumptions about rejection at the present time. To wit: "if fractals are so good, then why aren't people using them"? The answer is clear: Some are and some will.

Acknowledgments I am grateful to (the late) Thomas Kuhn at MIT for feedback on the early days of my understanding the "tortuous path"; Clay Christensen at HBS for some opposing and enlightening discussions on innovation and disruption; Patrick McHugh at Brown for feedback on "hibervation," Barry Unger at BU for helpful discussions on adoption and markets; and Alexander Shelman-Cohen at Vanderbilt/Bain for a critical review of the manuscript.

References

1. Executive Office of the President. A strategy for American innovation: driving towards sustainable growth and quality. http://www.whitehouse.gov/assetsdocuments/SEPT_20__Innovation_Whitepaper_FINAL.pdf (2009)
2. Christiansen, C.: The Innovator's Dilemma. Harvard Business School Press, Boston (1997)
3. Cohen, N.: The Tortuous Path: Innovation's Perils in Progress. Belmont Hill Publishing, Belmont (in press, 2015)

4. Marvin, C.: When Old Technologies Were New. Oxford, Oxford (1990)
5. Mandelbrot, B.: Fractal Geometry of Nature. Freeman, San Francisco (1982)
6. Cohen, N.: Fractal antennas. Commun. Q. 3–25 (1995) (summer)
7. Samavati, H., Hajimiri, A., Shahani, A., Nasserbakht, G., Lee, T.: Fractal capacitors. IEEE J. Solid State Circ. **33**, 2035–2043 (1998)
8. Daniel, S.: Very high aspect ratio semiconductor devices using fractal based topologies. US Patent 6,229,163 (2001)
9. Xu, H., Wang, G., An, H.: Hilbert fractal curves form compact diplexer. Microwaves RF Magazine **8**, 38 (2010)
10. Cohen, N.: Electromagnetic cloaking system. US Patent 8,253,639 (2012)
11. Jarry, P., Beneaut, J.: Design and Realizations of Miniaturized Fractal Microwave and RF Filters. Wiley, New York (2009)

Permutation Entropy Analysis of EEG of Mild Cognitive Impairment Patients During Memory Activation Task

Leena T. Timothy, Bindu M. Krishna, Murali Krishna Menon, and Usha Nair

Abstract Permutation Entropy (PE) statistic is a measure of self-similarity of the time series estimated from its ordinal patterns. This measure is used to detect the dynamical differences between patients with mild cognitive impairment (MCI) and normal controls. The comparison of PE values of Electroencephalograph (EEG) of the two groups in the resting eyes closed (EC) state and the short-term memory task (STM) state reveals altered efficiency of the different lobes of MCI brain in the compensational dynamical mechanism for task management. In resting EC state, PE values of MCI group is significantly $(p\angle 0.05)$ lower than that of controls in the frontal, temporal, and anterior parietal regions. In the STM task state, entropy levels of MCI group are significantly $(p\angle 0.05)$ lower than that of controls in the frontal region and the left parietal region. These findings suggest that nonlinear analysis of EEG using PE can provide important information about EEG characteristic of cognitively impaired condition that can lead to Alzheimer's Disease(AD).

Keywords Permutation entropy • Mild cognitive impairment • Nonlinear time series analysis of EEG • Alzheimer's Disease

---

L.T. Timothy • U. Nair
School of Engineering, Cochin University of Science and Technology, Cochin 682022, India

B.M. Krishna (✉)
Sophisticated Test and Instrumentation Centre, Cochin University of Science and Technology, Cochin 682022, India
e-mail: bindum@cusat.ac.in

M.K. Menon
Department of neurology, Lakeshore Hospital and Research Centre, Cochin 682304, India

C. Bandt et al. (eds.), *Fractals, Wavelets, and their Applications*, Springer Proceedings in Mathematics & Statistics 92, DOI 10.1007/978-3-319-08105-2_25,

1 Introduction

Alzheimer's Disease (AD), the most common form of dementia, is a neurodegenerative disorder that affects mainly the aged population [14]. The disease course is characterized by initial mild memory disturbance followed by gradual global loss of cognition. MCI is a condition with memory problems greater than normal elderly, but do not fulfill the criteria for clinically probable AD. MCI subjects are at increased risk of developing AD ranging from 1–25 % per year [10]. Preclinical discrimination between MCI and normal subjects demands much attention, for once a subject is diagnosed for MCI and does not present with any features of AD, the onset of AD could be prevented successfully. A combination of extensive psychological and physiological tests as well as expensive imaging techniques are generally employed for the diagnosis of AD. EEG being a noninvasive and relatively inexpensive technique, several research initiatives have been taken up for studying the effectiveness of EEG analysis in AD diagnosis.

AD is expected to often produce characteristic changes in EEG as it is a form of cortical dementia. Linear techniques of spectral analysis have identified characteristic features of AD as decreased mean frequency and coherence [13, 18] which are found to be associated with the severity of disease. Investigations on the EEG band powers of MCI subjects have revealed significant increase in delta and theta power and decrease in alpha power of MCI EEG compared to controls [2, 3]. EEG analysis of MCI and controls during memory activation state has revealed less decrease in the lower alpha band power of MCI than controls [21]. The coherence between spectral bands of EEG of MCI subjects is found to be greater than that of controls during working memory task [24] indicating increased coordination between the cortical regions. Similar results are observed from the analysis of MCI EEG using synchronization likelihood [16]. However, no significant difference in synchronization likelihood was observed between these groups in resting condition [20].

Dynamics of brain is highly complex, involving large number of interrelated variables and their nonlinear interactions, which makes EEG highly nonlinear and nonstationary in nature. Therefore suitable advanced signal processing techniques are required for its analysis and with this aim nonlinear dynamical analysis has been widely used to study the complex nature of EEG. The basis of nonlinear time series analysis is delay reconstruction which involves the conversion of a scalar time series into vectors of the state space with appropriate embedding dimension [11]. The evolution of these vectors will represent the state space trajectory of the dynamical system which may be attracted to a subspace called the attractor. Most nonlinear techniques depend on analysis and characterization of the attractor properties to study the underlying dynamics of the system. Several algorithms have been proposed for the computation of characteristic invariant measures of the attractor. Conventional nonlinear measures like correlation dimension and Lyapunov exponent have revealed reduced complexity of AD EEG compared to that of controls [9, 23]. These conventional measures have the drawback of being sensitive to nonstationarity, noise contamination, and short data which reduces its applicability for analysis of real-world signals.

For better practical applicability, new techniques have been developed in the recent years which are suitable for the analysis and detection of dynamical changes from short noisy time series [1, 6]. Entropy is one such measure which estimates the complexity or predictability of a system from real-world time series [1,4,6,17]. Results of the existing studies on EEG signals of AD patients using such measures have identified lowered entropy values indicating lowered complexity or flexibility for information processing and transmission [1, 7, 8, 19]. Multi-scale entropy analysis has revealed scale discrepancies in MCI EEG as characteristic feature [15]. However, multi-scale entropy analysis provides a set of values for each time series under study which necessitates further parameterization for statistical analysis of participant groups. New improved measures like PE are found to be advantageous in terms of simplicity, fast calculation, efficiency, and robustness against nonstationarity and noise contamination of the signals. PE estimates the self-similarity of the time series based on the occurrence of order patterns in it [4, 6]. However, the effectiveness of such measures in characterizing cognitive deficit conditions is not widely studied.

In this study, we focus on nonlinear analysis of the dynamical aspects of MCI EEG during short-term memory (STM) task using PE. In the higher order cortical function, memory acts as a slave in the hierarchy. When memory gets affected, cognitive skills will also get affected, thus disabling the smooth performance of executive functions. When these functional disabilities are severe enough to interfere with everyday functioning, the patient is said to suffer from a dementing illness. Mild cognitive impairment (MCI) causes a slight but noticeable and measurable decline in cognitive abilities, including memory and thinking skills. STM being the first to get affected in early stages of dementia, it is important to decipher the dynamical aspects of MCI EEG during STM task and study the characteristic differences between control and MCI EEG. Most of the current studies on MCI EEG are conducted in resting eyes closed (EC) condition [2, 3]. Studies conducted on EEG of MCI subjects during working memory task have suggested the presence of compensatory processes in MCI which enhances the EEG coherence [16,21,24]. Here we aim to study the characteristic difference in the dynamics of EEG signals between control and MCI during STM task from the difference in PE of these groups.

2 Methods

2.1 Subjects

Fifteen controls and 12 MCI patients participated in this study. All subjects underwent thorough evaluation including clinical history, physical and neurological examination, and folstien mini mental state examination (MMSE) test. MMSE is a simple and quick test to evaluate cognitive dysfunction. MCI group consists of seven male and five female right-handed subject with mean age of 67.1 ± 7.8 years. The mean MMSE score of the MCI group is 26.3 ± 1.4. The control group consisted of 15 age matched right-handed controls (nine male and six female) with mean age 65.7 ± 6.0 years. None of the controls have any neurological disorders. All control subjects have MMSE score of 30 and are volunteers who gave informed written consent to participate in the study. Care-givers of patients also gave informed written consent of participation of their wards in the study.

2.2 *Experimental Protocol*

EEG background activity of the control and MCI subjects are recorded under two different cognitive states viz. (1) eyes closed (EC) (2) short-term memory task (STM). During both task states participants are seated in an armchair in a semi reclined fixed position to avoid muscle movements. They are also instructed to relax as much as possible and avoid eye ball movements. Initially, after giving a brief demo of memory activation task, 5 min of EEG recording is acquired when the subjects are in resting EC state. Giving an intertask interval of 60 s, the audio track of a small story narration is played back through head phones while the subjects are in EC state. After this, ten different questions based on the story are asked in a monotonous tone with 10 s gap while the subjects are still in the EC condition. The subjects are instructed to give the answer in this gap. This is repeated for two more stories giving 60 s gap between each story allowing them to relax.

2.3 *EEG Recording*

EEG are recorded from 19 electrode sites according to the international 10–20 system, with electrodes referenced to linked ear lobes. EEG are recorded using Neurocare Digital Wingraph EEG system with sampling frequency of 128 Hz and 16 bit A to D conversion. The recorded EEG is digitally filtered with a band pass filter of cut-off frequencies at 0.4 and 60 Hz.

2.4 *Data Analysis*

All EEG are visually inspected by a specialist physician for eye movement and muscle artifacts. Artifact free epochs of 10 s duration are chosen from all the electrodes simultaneously and stored in a PC for further off-line analysis using PE. Average value of PE of all the epochs for each of these channels is calculated for the analysis. As the fronto-polar channels FP1and FP2 are the most affected ones by even eye-ball movements, these two channels are not included in the study and only artifact free epochs from the channels F3, F4, F7, F8, T3,T4, T5, T6, C3, C4, P3, P4, O1 and O2 are used for analysis.

Computation of PE is based on comparison of neighbouring values in the time series of any dynamical variable of a system. According to the embedding theorem, any arbitrary time series $X = \{x_1, x_2, \ldots, x_T\}$ can be mapped on to an "n" dimensional space with vectors $X_i = \{x_i, x_{i+\tau}, x_{i+2\tau}, \ldots, x_{i+(n-1)\tau}\}$ where "n" is the embedding dimension and "τ" is the delay time for embedding calculated using appropriate methods like false nearest neighbour calculation and first minimum of autocorrelation function [16]. For any arbitrary vector X_i, the components are n number of real values of the time series $\{x_t, x_{t+\tau}, x_{t+2\tau}, \ldots, x_{t+(n-1)\tau}\}$ from time instant "t" to "$t + (n - 1)\tau$". Assuming $\tau = 1$ [4], each point in the "n" dimensional space represented by its corresponding vector will therefore be equivalent to a short sequence of the time series consisting of "n" number of real values as $\{x_t, x_{t+1}, x_{t+2}, \ldots, x_{t+(n-1)}\}$. If the components of each vector are arranged in ascending order, it will represent a pattern of evolution. Thus each of the vectors can be considered as a symbolic sequence which will be one of the n! possible permutations of "n" distinct symbols. The probability distribution of each pattern π can be represented as

$$p(\pi) = \frac{\#\{t | t \leq T - n, (x_{t+1}, \ldots . x_{t+n} hastype \pi)\}}{T - n + 1} \tag{1}$$

PE of order $n \geq 2$ is defined as the Shannon entropy of the n! patterns or symbolic sequences and can be written as

$$H(n) = -\sum p(\pi) log p(\pi) \tag{2}$$

where the sum runs over all n! permutations or sequences. $H(n)$ lies between 0 and $log(n!)$. For increasing or decreasing sequence of values, $H(n) = 0$, whereas for random series where all n! possible permutations appear with same probability, $H(n) = log(n!)$. For a time series representing some dynamics, $H(n) < log(n!)$. Therefore, normalised PE per symbol of order "n" is given by $H(n)/log(n!)$. Thus PE characterizes the system dynamics, with low values indicating regular behaviour. Any change in PE value will thus represent a change in irregularity in the dynamics of the system.

2.5 Statistical Analysis

SPSS for windows is used for statistical analysis. Differences in permutation entropy between MCI patients and normal controls are analysed on each cognitive state using One-Way ANOVA. Separate ANOVAs are conducted for each of the different electrodes.

3 Results

EEG data of the control and MCI groups are analysed using PE of order 5 and delay 5 for the artifact-free epochs of the channels F3, F4, F7, F8, T3,T4, T5, T6, C3, C4, P3, P4, O1 and O2. PE values are averaged for the artifact-free epochs of these channels.

PE analysis is carried out concentrating on the four lobes (1) Frontal represented by F3, F4, F7, F8 (2) Temporal by T3,T4, T5, T6 (3) Parietal by C3, C4, P3, P4 and (4) Occipital by O1 and O2. Analysis is carried out in two different cognitive states (1) resting eyes closed (EC) (2) short-term memory task (STM).

3.1 Resting EC Condition

Figure 1 shows the PE values of the frontal channels F3, F4, F7 and F8 during resting EC state. Similarly Fig. 2 shows the PE values of temporal channels T3,T4, T5, T6 of the above groups for the EC state and Fig. 3 shows that of the parietal channels C3, C4, P3, P4 and occipital channels O1 and O2. In the resting EC condition, the PE values of MCI subjects are found to be lower than that of controls in all the lobes. From these figures it can be observed that the PE of MCI group is globally lower than that of controls. Statistical analysis performed on the PE values of the different channels shows significant difference ($P \angle 0.05$) between MCI and control groups in all channels except P3, P4, O1 and O2. This result suggests that EEG activity of MCI patients have lowered complexity than that of controls in all the lobes except posterior parietal and occipital regions indicating lowered flexibility for information processing and transmission in the MCI stage also.

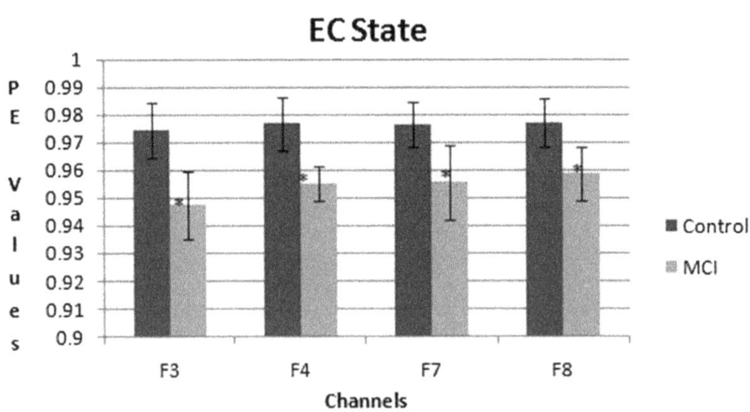

Fig. 1 PE values for control and MCI for frontal channels under EC state

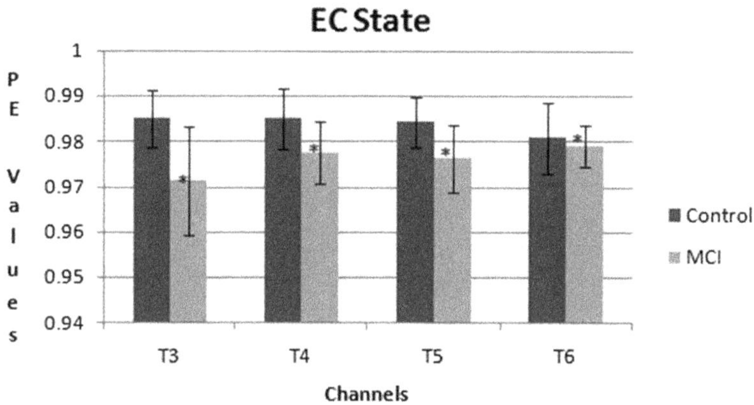

Fig. 2 PE values for control and MCI for temporal channels under EC state

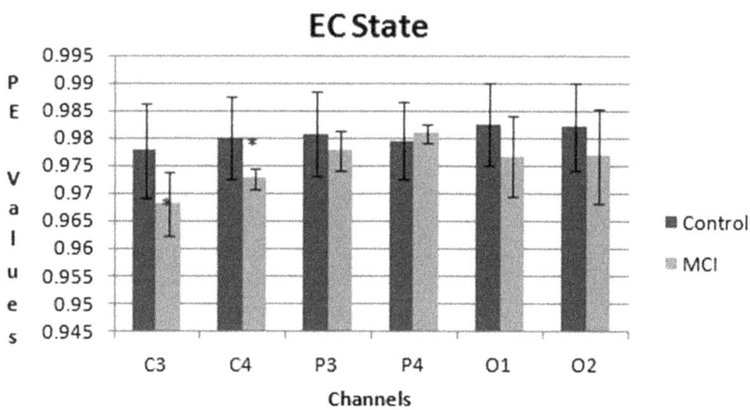

Fig. 3 PE values for control and MCI for parietal and occipital channels under EC state

3.2 *Short-Term Memory Task*

To study the dynamical aspects of MCI EEG and for its effective differentiation from that of controls, PE analysis is carried out during short-term memory (STM) task of story recollection in both groups. Figure 4 shows the PE values of the frontal channels F3, F4, F7, F8 of the control and MCI groups during STM state. Similarly Fig. 5 shows the PE values of the temporal channels T3, T4, T5, T6 and Fig. 6 that of parietal and occipital channels C3, C4, P3, P4, O1 and O2 of the above groups for the STM state. In this state of short-term memory recollection also, the entropy level of MCI group is found to be globally lower than that of controls. Statistical analysis is carried out on the PE values of each of the channels. Significant difference ($p \angle 0.05$) between the PE values of control and MCI groups is observed in all the frontal channels. In the case of parietal channels significant differences ($p \angle 0.05$) is

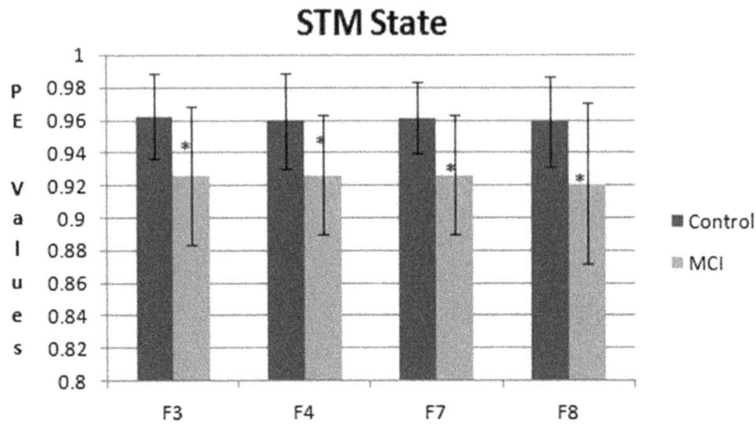

Fig. 4 PE values for control and MCI for frontal channels under STM state

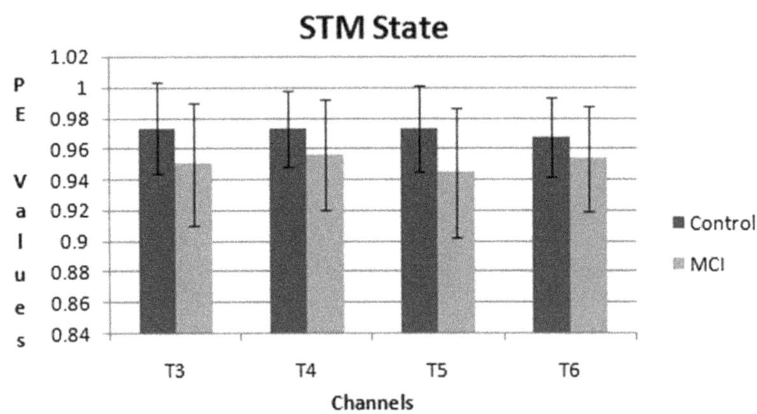

Fig. 5 PE values for control and MCI for temporal channels under STM state

observed only in the left channels of C3 and P. The right parietal channels do not show significant differences ($p\angle 0.05$) between MCI and control groups. Also, no significant difference ($p\angle 0.05$) is observed in any of the temporal channels between the MCI and control groups.

4 Discussion and Conclusion

EEG data of 15 controls and 12 MCI are analysed using PE of order 5 and delay 5 in two different cognitive states of resting EC and STM task conditions. PE is a complexity measure suitable for application to practical EEG monitoring systems as it is computationally simple and efficient and robust to artifacts [4].

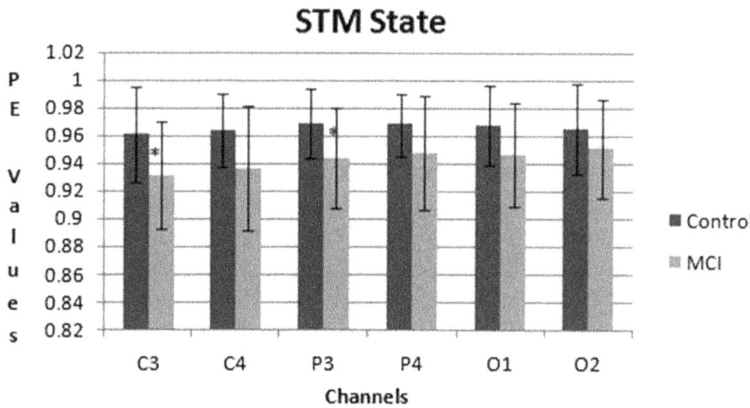

Fig. 6 PE values for control and MCI for parietal and occipital channels under STM state

The results of our analysis show that PE can effectively bring out the dynamical characteristics of MCI EEG. From the PE analysis carried out on resting EEG of control and MCI subjects, it is found that the complexity of MCI EEG is significantly lower than that of controls in all the channels of the frontal, temporal, and the anterior parietal regions. This confirms the findings of lowered complexity or increased regularity of MCI EEG [15]. No significant lowering of PE is observed in the occipital channels and the posterior parietal channels. Lack of any clear distinction between control and MCI group in the occipital region can be due to the fact that this lobe is involved in visual processing and in the resting EC state this region is not encountered with any task demand. However, the observed significant difference ($p\angle 0.05$) between the MCI and control EEG in the frontal, temporal, and the anterior parietal channels indicate that even in this task free resting EC condition, the different cortical regions are working in different activation levels. Therefore it can be concluded that more effective differentiation between the control and MCI EEG can be obtained during specific cognitive task states.

The results of PE analysis under STM task state shows significant difference ($p\angle 0.05$) between MCI and control groups only in the frontal and left parietal channels. In contrast to the resting EC condition, no significant difference ($p\angle 0.05$) is observed in any of the temporal channels in the STM task state. This indicates that when the MCI brain is given a task to recollect the recently acquired memory, the temporal lobes are able to get activated to cope with the task demand. However the frontal channels of MCI group show significant difference ($p\angle 0.05$) from that of controls indicating lack of enough activation during such a task demand. This lack of activation of the frontal lobe even during a task demand indicates the lack of attention effected by the inefficient compensational mechanism in the frontal lobe of MCI brain. Thus, in addition to the presence of compensational mechanism, another important feature of altered efficiency of compensational mechanism in different lobes of MCI EEG is also revealed from the PE analysis. In the case of parietal lobe also, no significant difference ($p \angle 0.05$) is observed between the MCI and control groups in the right channels C4 and P4. This indicates that the right hemisphere of the parietal lobe is also able to perform fairly well. The left parietal channels show significant differences ($p \angle 0.05$) between controls and MCI group during the STM task demand. This also points towards the altered efficiency in the compensational mechanism exhibited by different lobes and their hemispheres of the MCI brain.

The damage or reduced efficiency of information processing resulting from the connectional disturbances and/or localized damage is more evidenced in the dominant hemisphere of the brain. In our study, all the participants are right-handed persons and this suggests that both the control and MCI groups are comprised of left hemisphere dominant subjects. As no preferential dominance of a particular hemisphere is observed between the lobes during the resting EC state, this result of lack of activation of left parietal region signifies the importance of dominant hemisphere and its role in task management. Further studies on other groups with right hemisphere dominant subjects under the same task condition is necessary to clearly distinguish whether this is due to preferential damage to the left parietal lobe or due to the effect of dominant hemisphere in revealing the cortical atrophy. Parietal lobe acts as the site of multimodal afferent integration centre for the different inputs like sensory, auditory, etc. In recent studies it is found that inferior parietal lobe lesions are associated with STM loss and non-dominant parietal lobe lesions produce impaired attention [5] and that the parietal lobes are highly sensitive to cognitive tasks [12, 22]. Thus use of PE analysis on EEG of STM task state can effectively bring out the deficient information processing in the parietal lobe of MCI brain.

The results of our PE analysis confirm the findings of lowered complexity as well as compensational mechanisms exhibited by MCI brain during resting and memory task states, respectively. In addition our results of PE analysis have revealed that although temporal and parietal lobes play important role in STM state, in MCI condition the parietal lobe may be involved earlier than the temporal lobe. The lack of activation of dominant hemisphere of parietal lobes to normal levels and the compensational mechanisms of the temporal lobe evidenced here indicate the importance of electrophysiological testing and its PE analysis for both the dynamical investigation of the MCI brain and the characterization of MCI EEG.

Acknowledgements The Authors would like to thank Prof. (Late) R. Pratap for suggesting the problem and for the stimulating discussions during the initial stage of this work. One of the authors, (BMK) would like to acknowledge financial support from Science and Engineering Research Board, Department of Science and Technology (India), through Fast Track Scheme No.SR/FTP/PS-006/2010 and thank Prof. Jacob Philip, Director, STIC for all the support and encouragement.

References

1. Abasalo, D., Hornero, R., Espino, P., Alvarez, D., Poza, J.: Entropy analysis of the EEG background activity in Alzheimer'sdisease patients. Physiol. Meas. **27**, 241–253 (2006)
2. Alder, G., Bramesfeld, A., Jajcevic, A.: Mild cognitive impairment in old-age depression is associated with increased EEG slow-wave power. Neurophyschobiology **40**, 218–222 (1999)
3. Babiloni, C., Binetti, G., Cassetta, E., Dal Forno, G., Del Percio, C., Ferreri, F., Ferri, R., Frisoni, G., Hirata, K., Lanuzza, B., Miniussi, C., Moretti, D.V., Nobili, F., Rodriguez, G., Romani, G.L., Salinari, S., Rossini, P.M.: Sources of cortical rhythms change as a function of cognitive impairment in pathological aging: a multicenter study. Clin. Neurophysiol. **117**(2), 252–68 (2006). (Epub Dec 27, 2005)
4. Bandt, C., Pompe, B.: Permutation entropy: a natural complexity measure for time series. Phys. Rev. Lett. **88**, 174102 (2002)
5. Bradley, W.G.: In: Daroff, R.B., Fenichel, G., Jankovic, J (eds.) Bradley's Neurology in Clinical Practice, 5th edn. Elsevier, Philadelphia (2008)
6. Cao, Y., Tung, W.W., Gao, J.B., Protopopescu, V.A., Hively, L.M.: Detecting dynamical changes in time series using the permutation entropy. Phys. Rev. E **70**, 046217 (2004)
7. Gao, J., Hu, J., Tung, W.W.: Complexity measures of brain wave dynamics. **5**(2), 171–182 (2011)
8. Hogan, M.J., Kilmartin, L., Keane, M., Collins, P., Staff, R.T., Kaiser, J., Lai, R., Upton, N.: Electrophysiological entropy in younger adults, older controls and older cognitively declined adults. Brain Res. **1445**(22), 1–10 (2012)
9. Ikawa, M., Nakanishi, M., Furukawa, T., Nakaaki, S., Hori, S., Yoshida, S.: Relationship between EEG dimensional complexity and neuropsychological findings in Alzheimer's disease. Psychiartry Clin. Neurosci. **54**, 537–54 (2000)
10. Jha, S., Patel, R.: Some observations on the spectrum of Dementia. Neurol. Ind. **52**, 213 (2004)
11. Kantz, H., Schreiber, T.: Nonlinear time series analysis. Cambridge university Press, Cambridge (1997)
12. Kazui, H., Kitagaki, H., Mori, E.: Cortical activation during retrieval of arithmetic facts and actual calculation: a functional magnetic imaging study. Psychiat. Clin. Neuros. **54**, 479–485 (2000)
13. Locatelli, T., Cursi, M., Liberati, D., Franceschi, M., Comi, G.: EEG coherence in Alzheimer's disease. Electroencephhalogr. Clin Neurophysiol. **106**, 229–237 (1998)
14. Nestor, P.J., Scheltens, P., Hodges, J.R.: Advances in the early detection of Alzheimer's disease. Nat. Med. **10**, S34–41 (2004)
15. Park, J.H., Kim, S., Kim, C.H., Cichocki, A., Kim, K.: Multiscale entropy analysis of EEG from patients under different pathological conditions. Fractals **15**, 399–404 (2007)
16. Pijnenburg, Y.A.L., vd Made, Y., van Cappellen van Walsum, A.M., Knol, D.L., Scheltens, Ph., Stam, C.J.: EEG synchronization likelihood in mild cognitive impairmentand Alzheimer's disease during a working memory task. Clin. Neurophysiol. **115**, 1332–1339 (2004)
17. Pincus, S.M.: Approximate entropy as a measure of system complexity. Proc. Natl. Acad. Sci. **88**, 2297–2301 (1991)
18. Prichep, L.S., John, E.R., Ferris, S.H., Reisberg, B., Almas, M., Alper, K., Cancro, R.: Quantitative EEG correlates of cognitive deterioration in the elderly. Neurobiol. Aging **15**, 85–90 (1994)
19. Sneddon, R., Shankle, W., Hara, J., Fallon, J., Saha U.: The Tsllis entropy in EEGs of normal and demented individuals. In: 11th Joint Symposium on Neural Computation, University of Southern California, May 15 2004
20. Stam, C.J., Van Der Made, Y., Pijnenburg, Y.A.L., Scheltens, Ph.: EEG synchronization in mild cognitive impairment and Alzheimer's disease. Acta Neurol. Scand. doi:10.1034/j.1600-0404.2003.02067.x

21. van der Hiele, K., Vein, A.A., Kramer, C.G.S., Reijntjes, R.H.A.M., van Buchem, M.A., Westendorp, R.G.J., Bollen, E.L.E.M., van Dijk, J.G., Middelkoop, H.A.M.: Memory activation enhances EEG abnormality in mild cognitive impairment. Neurobiol. Aging **28**(1), 85–90 (2007)
22. Xie, S., Xiao, J. Jiang, X.: The fMRI study of the calculation tasks in normal aged volunteers. J Peking Univ. (Health Sci) **35**, 311–313 (2003)
23. Yagyu, T., Wackermann, J., Shigeta, M., Jelic, V., Kinoshita, T., Kochi, K., Julin, P., Almkvist, O., Wahlund, L.O., Kondakor, I., LehmannDonald, D., Echman, P.: Global dimensional complexity of multichannel EEG in mild Alzheimer's disease and age- matched cohorts. Dement. Geriatr. Cogn. Disord. **8**, 343–347 (1997)
24. Zheng-yan, J., Lei-lei, Z.: Inter and intra hemispheric EEG coherence in patients with mild cognitive impairment at rest and during working memory task. J. Zhejiang Univ. **7**, 357–364 (2006)

A Multifractal-Based Image Analysis for Cervical Dysplasia Classification

P. Singh, J. Jagtap, C. Pantola, A. Agarwal, and A. Pradhan

Abstract This paper presents a study on microscopic images to classify cervical precancers by a multifractal analysis. Since internal structure of tissue is non-deterministic, multifractal spectrum is required to characterize such structure. The periodic structure of collagen present in the stromal region of cervical tissue gets disordered with progress in grade of dysplasia. This disorder is investigated through the multifractal study, enabling us to discriminate between normal and abnormal human cervical tissue sections. Holder exponent classifies normal from abnormal dysplasia by capturing local irregularities present in the image. While mean of Hausdorff–Besicovich dimension which describes global regularity are used to classify various grades of dysplasia. The box-counting method is used to estimate the fractal dimension. The results show, remarkably, the classification feature of multifractal analysis.

Keywords Multifractal spectrum • Microscopic image analysis • Cervical dysplasia

1 Introduction

Cervical dysplasia term describes the early stage of cervical cancer, i.e. abnormal growth of precancerous cells on the surface of cervix. Depending upon growth of abnormal cells in tissue different grades of dysplasia are classified. Higher

P. Singh (✉) • J. Jagtap • A. Pradhan
Department of Physics, IIT of Kanpur, Kanpur 208016, India
e-mail: singhp@iitk.ac.in

A. Pradhan
Center for Laser Technology, IIT Kanpur, Kanpur 208016, India

C. Pantola • A. Agarwal
Department of Pathology, GSVM Medical College, Kanpur, India

C. Bandt et al. (eds.), *Fractals, Wavelets, and their Applications*, Springer Proceedings in Mathematics & Statistics 92, DOI 10.1007/978-3-319-08105-2_26,

grade of cervical dysplasia leads to cervical cancer. Statistically, 30–50 % cervical dysplasia converts into invasive cancer, so early detection of cancer, i.e. detection at dysplastic stage plays key role to reduce mortality rate. Gold standard technique, to classify different grades of dysplasia, is histopathology, but it is time consuming and also requires efficient pathologist, to examine tissue samples on microscope, so a quantitative computational technique is required. Recent developments in medical imaging [1, 2, 4, 10] show promise in discriminating different grades of human cancers due to rapid and continuing progress in image visualization and advanced computerized analysis. Changes in shape, size, refractive index and irregularity in fibrous structure in human cervical tissue sections can be probed through multi-fractal (MF) analysis. Specifically, the breaking of collagen bonds in stromal region results in disorders in fibrous structure with increasing grades of precancers. This may give rise to fluctuations in scattering intensity which can be probed through a multi-fractal study [3, 5, 7]. In general, multiple tasks will be needed to refine the data of interest from an image so as to de-mask vital information and classify among anomalies as well as from the normal one. Fluctuation in intensity has been observed in biopsy images caused by randomness and heterogeneous behavior in cell which was reported in our earlier study [8]. Multifractal analysis [6, 9, 11, 12] can extract a quantitative parameter from a 2-dimensional image. We report that MF spectra extracted from simple microscopic images show discriminating features among various grades of dysplasia as well as from normal human cervical tissue. This study may provide a new approach for early detection of precancerous stages.

2 Method

GSVM medical college, Kanpur provided H&E stained human cervical tissue sections which were subsequently imaged with a camera integrated simple microscope. From microscope we have taken an image of stroma region of tissue section. The original was 2048 × 2048 image, so the microscopic image of stromal region was cropped to obtain 3 or 4 sub-images (512 × 750). A digital image consists of intensity values distributed in matrix form and pixel corresponds to an index value of the matrix. Multifractal analysis of an irregular natural structure (image) needs to compute a spectrum of fractal dimension, i.e. multifractal spectrum, which is performed in two steps. At first step, microscopy images are converted to grayscale images and analysed to find local singularity coefficient, also known as Holder exponent or α-value, at each pixel. An intensity based measures[5] (μ), mutifractal measure, measured within the square box of size ϵ centered at p^{th} pixel, varies with scale ϵ as follows:

$$\mu_p(\epsilon) = A\epsilon^{(\alpha_p)}$$

$$\log(\mu_p(\epsilon)) = \alpha_p \log(\epsilon) + \log(A)$$

Thus for a set of measurement, i.e. ϵ =1, 2, 3, 4, 5, holder exponent is calculated by slope of linear fitting. In the holder exponent image, a range of α-values, with α_{min} and α_{max}, are distributed. In the next step, the whole range of α-values is discretized into 100 sub-ranges between α_{min} and α_{max}. Now we estimate Hausdorff–Besicovich fractal dimension for each sub-ranged α-value. There are many algorithms to compute fractal dimension, here we are using box-counting algorithm due to highly efficient and ease of implementation. This is done by covering the whole image with non-overlapping boxes of size ϵ and counting the number of boxes $N_\epsilon(\alpha)$ containing holder exponent value. From a set of sizes of boxes, we calculate fractal dimension by slope of linear fitting.

$$f_\epsilon(\alpha_i) = -\frac{\ln(N_\epsilon(\alpha_i))}{\ln(\epsilon)}, \text{ for } \epsilon = 1, 2, 3, 4, \ldots..$$

In this way for a set of Holder exponent values corresponding set of Hausdorff dimension values are calculated for each sub-image and the mean is taken. Thus the original stromal image will have a set of 3 or 4 values of mean fractal dimension corresponding to 3–4 sub-images. Finally this set of mean fractal dimension values can be expressed by an average value and variance of Hausdorff values for the original image which shows discriminating features among various grades of dysplasia and from normal dysplasia.

3 Results, Discussions and Conclusions

There are four types of multifractal measure: maximum measure, minimum measure, summation measure and iso measure. We have analysed with all measure and chose maximum measure because of its good classification characteristic. Maximum box size was fixed at side length 5 to determine the holder exponent at each pixel. Maximum box size has been selected on the basis of an iterative study of α-image for different sizes of the box.

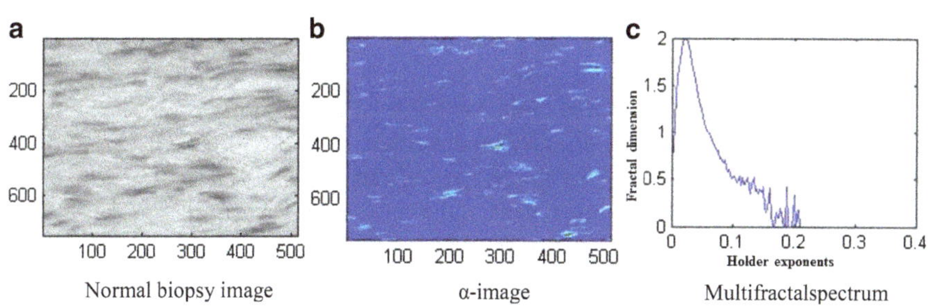

Fig. 1 (**a**) Microscopic image of normal tissue section (**b**) Holder exponent image (**c**) Multifractal spectrum

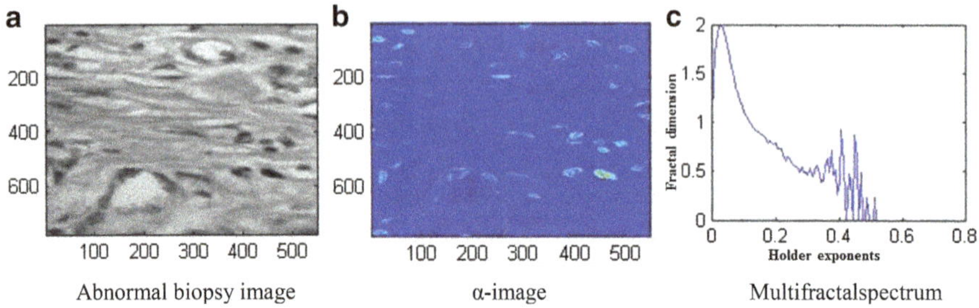

Fig. 2 (**a**) Microscopic image of abnormal tissue section (**b**) Holder exponent image (**c**) Multifractal spectrum

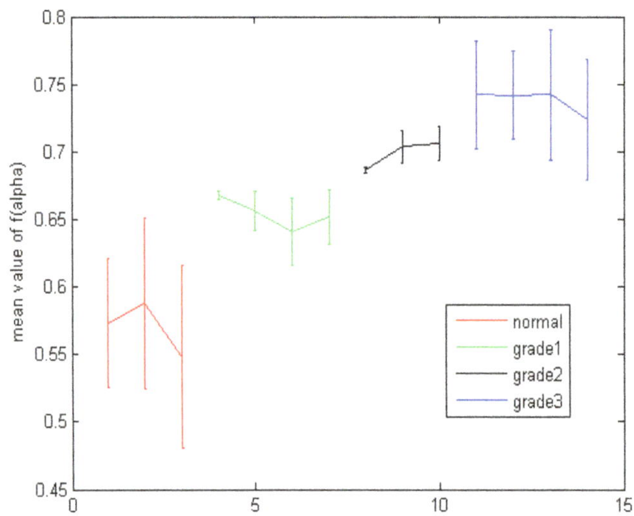

Fig. 3 Classification

Clearly multifractal spectrum shows the difference between normal and abnormal biopsy image. Abnormal tissue image has more irregular singularities than normal one. We have studied the different grades of dysplasia among abnormal tissues (Figs. 1 and 2).

Figure 3 displays the results of the mean value of fractal dimension for the three classes of dysplasia and one normal. Since we have cropped three sub-image from an original image, so mean of fractal dimension can be supposed to vary. The error bars correspond to variance of mean fractal dimension for the original image of each class. A clear difference in mean values of fractal dimension is seen for normal and the three grades of dysplasia. Analysis has been performed over three normal, four grade 1, three grade2, four grade3 patients.

In this study multifractal nature of stroma has been investigated by Hausdorff–Besicovich dimension. Our main aim is to discriminate the different grades of dysplasia by probing the internal structure of various grades and normal through its multifractal nature.

From the microscopic images, where collagen strands are not clearly visible, one infers that pattern of collagen fibers changes for different grades of dysplasia by using this technique (Multifractal analysis). So for higher the grades of dysplasia, the fractal nature would be higher.

For more accurate prediction of mean value of fractal dimension values for different grades of dysplasia and normal tissue sections, a large number of samples of each class are required. Further analysis is under process.

References

1. Atupelage, C., Nagahashi, H., Sakamoto, M., et al.: Computational grading of hepatocellular carcinoma using multifractal feature description. Comput. Med. Imaging Graph. **37**(1), 61–71 (2013)
2. Baish, J.W., Jain, R.K.: Fractals and cancer. Cancer Res. **60**, 3683–3688 (2000)
3. Das, N., Jagtap, J., Pradhan, A., Panigrahi, P.K., Ghosh, N., et al.: Probing multifractality in tissue refractive index: prospects for precancer detection. Opt. Lett. **38**(2), 211–213 (2013)
4. George, L.E., Mohammed, E.Z.: Cancer tissues recognition system using box counting method and artificial neural network. In: IEEE International Conference of Soft Computing and Pattern Recognition (SoCPaR), 978-1-4577-1195-4 (2011)
5. Ghosh, S., Jagtap, J., Pradhan, A., Ghosh, N., Panigrahi, P.K., et al.: Differing self-similarity in light scattering spectra: a potential tool for pre-cancer detection. Opt. Exp. **19**, 19717–30 (2011)
6. Hemsley, A., et al.: Multifractal measures for tissue image classification and retrieval. In: 11th IEEE International Symposium on Multimedia, pp. 618–623 (2009)
7. Jagtap, J., Pradhan, A., et al.: Wavelet-based multifractal analysis of laser biopsy imagery. In: Proc. SPIE, vol. 8222 (2012)
8. Jagtap, J., Singh, P., Pradhan, A., et al.: Study and discrimination of human cervical tissue images through multifractal analysis. In: SPIE BiOS, 85770W-85770W-9 (2013)
9. Nilsson, E.: Multifractal-based Image Analysis: with applications in Medical Imaging. Master's Thesis in Computing Science and Mathematics (2007)
10. Reljin, I., Reljin, B., Pavlovic, I., Rakoevic, I.: Multifractal analysis of gray-scale images. In: IEEE 10th Mediterranean Electrotechnical Conference, MEleCon (2000)
11. Stojic, T., Reljin, I., Reljin, B.: Adaptation of multifractal analysis to segmentation of microcalcifications in digital mammograms. Physica A **367**, 494–508 (2006)
12. Vasiljevic, J., Reljin, B., Reljin, I., et al.: Application of multifractal analysis on microscopic images in the classification of metastatic bone disease. Biomed Microdevices **14**, 541–548 (2012)

Self-Similar Network Traffic Modelling Using Fractal Point Process-Markovian Approach

Rajaiah Dasari, Ramesh Renikunta, and Malla Reddy Perati

Abstract Several recent Internet traffic measurement studies reported that traffic in modern high-speed networks is a self-similar process. If the stochastic self-similar network traffic models do not accurately represent the real traffic, then the network performance may be over estimated or underestimated, and it causes degradation of Internet router performance. Therefore, it is decisive for an appropriate design of a router. In this paper, we investigate mean waiting time and tail probability of network router with pseudo self-similar traffic input. We use Fractal Point Process (FPP) as input process since it emulates self-similar traffic. However, this process is asymptotic in nature and has less effective in queueing based performance analysis. Therefore, for queueing analysis Markov modulated Poisson process (MMPP) is fitted for FPP. FPP involves another parameter Fractal Onset Time (FOT) besides Hurst parameter. Effect of FOT on tail probability and mean waiting time is examined.

Keywords Network router • Self-similar • Fractal point process • Markov modulated Poisson process • Mean waiting time • Tail probability • Traffic intensity

---

R. Dasari • M.R. Perati (✉)
Department of Mathematics, Kakatiya University, Warangal, India
e-mail: dsreddy.hari@gmail.com; mperati@yahoo.com

R. Renikunta
Department of Mathematics, Kakatiya Institute of Technology and Science, Warangal, India
e-mail: rameshrenikuntla@gmail.com

C. Bandt et al. (eds.), *Fractals, Wavelets, and their Applications*, Springer Proceedings in Mathematics & Statistics 92, DOI 10.1007/978-3-319-08105-2_27,

1 Introduction

Rapid growth of Internet traffic and investigation of measurement studies have shown that the real time networks (e.g. Ethernet LAN, WAN, ISDN, and VBR Video) are clearly different from the synthesized traffics using traditional Poisson process or related models. Seminal studies of said traffic revealed the presence of self-similarity or fractal nature, and its impact on the network traffic [2, 9, 11]. This type of traffic exhibits statistical similarity over different time scales and is highly correlated, hence drawn significant research interest in modeling the Internet traffic. Characterizing the statistical behavior of traffic is crucial to proper buffer design of router in the network traffic to provide the quality of service (QoS). Various stochastic models have been proposed that emulates the statistical nature of self-similar network traffic. Traffic models such as Fractional Brownian Motion (FBM), Fractional Auto Regressive Integrated Moving Average (FARIMA), Chaotic maps are proposed to characterize the self-similarity. These models describe the self-similar behavior in a relatively simple manner. Although, these processes are parsimonious, but are less effective in the case of queueing based performance evaluation. In order to enhance the resilience to burstiness of traffic, a number of approaches have been proposed. If the self-similar network traffic models do not accurately represent the real traffic, then the network performance may be over estimated or underestimated [10]. On the other hand, the numbers of Markovian models are proposed to emulate the self-similar network traffic. In [1, 5, 18, 20], Markovian arrival process (MAP) is employed to model the self-similar behavior over the different time scales. These fitting models equate the second order statistics of self-similar traffic and that of superposition of several two-state Markov modulated Poisson processes (MMPP) over desired time-scales. However, in the paper [1], covariance function of resultant MMPP is approximated by suppressing the higher order terms in its Taylor's expansion. In the paper [20], MMPP emulating the self-similar traffic is fitted by matching the variance over the desired time-scales. Resultant MMPP here is superposition of several Interrupted Poisson process (IPPs) wherein two modulating parameters of each IPP are equal. The fitting method [5,20] is generalized in the paper [18] by taking distinct modulating parameters in each IPP. Paulo Salvador et.al [17] proposed a model to fit discrete time MMPP that matches both autocovariance and marginal distribution of the counting process in such a way that model can capture self-similar behavior up to the time-scales of interest. In the said papers, the Markov-modulated Poisson process (MMPP) emulating the self-similar traffic over the different time scale is fitted, however, the time scale where self-similar nature actually begins is not considered.

Fractal point process (FPP) is shown to be self-similar process and also provides novel tools for understanding, modeling, and analyzing the self-similar network traffic behavior [6, 15]. Self-similar processes based on fractal point processes (FPP) lead to natural models of network traffic that possess attractive properties. FPP presents a variety of economical computing-effective and very suitable self-similar processes of the second order. Fractal point process involves another parameter fractal onset time (FOT) besides Hurst parameter H. Fractal on set time (FOT) defines the time scale at which self-similar behavior begins and is denoted by T_0 [6]. The important fractal features such as long-range dependence and the slowly decaying variance can be characterized by T_0. According to the measurement studies, FOTs of the network traffic are at the scales in the order of a few hundreds of milliseconds. However, FPP is not useful in the context of queueing based performance evaluation.

On the other hand, in general, networks could be divided into two categories based on their operation modes: slotted synchronous and un-slotted asynchronous [19]. Initial research on networks focused on slotted synchronous for switching fixed-size packets [4]. Recently, the design of nodes that are capable of switching variable-size packets has been attracting the researchers. When the routers are operated under slotted synchronous mode, packets need to be aligned into a fixed size and are placed in time slots before entering the switching matrix. At the downside, synchronous networks require packet alignment and synchronization stages. Hence, implementation of synchronous nodes with high data rates is more costly. In an unslotted network, the packets may or may not have the same size. Packets arrive and enter into the router without being aligned in time. Therefore, the packet-by-packet switch operation could take place at any point of time. On the other hand, unslotted networks are more flexible compared with slotted networks, since they are better at accommodating packets with variable sizes. Since IP packets are, in general, variable in length. Therefore, asynchronous routers are more suitable to carry variable-size IP packets. For the said reasons, performance analysis of router by means of queueing system wherein service time is deterministic may not be appropriate. In the papers [12, 14], routers with the variable length packet traffic are modeled as $MMPP/M/1/K$ system wherein service time is exponential. That is, packet length is assumed to follow exponential distribution to make the performance analysis of router handling self-similar traffic with variable length packets. In the present paper, first, router is modeled as $MMPP/M/1$ queueing system; thereby mean waiting time and tail probability are computed. Here, the input process MMPP is fitted for FPP which is a pseudo self-similar process.

The rest of the paper is organized as follows. In Sect. 2, we first overview the fundamentals of Fractal point process. In Sect. 3, we outline the fitting procedure of MMPP. Analytical results of $MMPP/M/1/K$ queueing system are given in the Sect. 4. In Sect. 5, we demonstrate the accuracy of proposed model by means of numerical results. Finally, some conclusions are made in Sect. 6.

2 Fractal Point Process (FPP)

Self-similar processes based on fractal point processes (FPP) lead to natural models of network traffic that possess attractive properties [15, 16]. FPPs are well suited for modeling the self-similar traffic in packet data networks. The second order statistics IDC and ACF of FPP are relatively straightforward to fit the parameters of a model emulating self-similar traffic [1]. The parameter T_0 defines the lower limit for indication of scale behavior in IDC and ACF. For this reason, the parameter T_0 is often referred to as the fractal onset time. In order to determine the second-order statistical characteristics of the fractal point process it is sufficient to have only three parameters: the average intensity λ, the Hurst parameter H and the fractal onset time T_0, at the given time scale ranges. As a result, these statistics are exploited by several authors. In this section, we overview the fundamental concepts of the fractal point processes in terms of second order statistics.

The process $N(t)$ describes the number of packet arrivals between the time interval $(0, t]$. Define X_n as the number of packets arrive during the nth time interval of size T sec, i.e., $X_n \equiv N(nT) - N(n-1)T$ then $c(n, T) = cov(X_n, X_{n+k})$, is defined as the covariance between the number of arrivals in two counting windows of counting time T and separation kT. Then the index of dispersion for counts (IDC) is given by [16],

$$\begin{aligned} IDC(T) &= \frac{Var(N(T))}{E(N(T))} \\ &= 1 + (\tfrac{T}{T_0})^{\alpha} \end{aligned}$$

where $\alpha = 2H - 1$, and $0 < \alpha < 1$, $\frac{1}{2} < H < 1$ in the case of self-similar process. The autocorrelation function is given by [16]

$$Cov : C(k, T) = \lambda T \begin{cases} 1 + (\frac{T}{T_0})^{\alpha} & k = 0 \\ (\frac{T}{T_0})^{\alpha} \frac{1}{2} \nabla^2 (k^{\alpha+1}) & k > 0 \end{cases} \tag{1}$$

where $\nabla^2(.)$ is the second central difference operator.
The auto covariance function (ACF) denoted by $r(k; T)$ is given by

$$r(k : T) = \frac{C(k; T)}{C(0; T} \tag{2}$$

$$= \frac{T^{\alpha}}{T^{\alpha} + T_0^{\alpha}} \frac{1}{2} \nabla^2 (k^{\alpha+1}) \qquad (k > 0) \tag{3}$$

Recall that X_n represents the number of packets during the nth time interval of size T and $X_n^{(m)} = \frac{1}{m} \sum_{i=1}^{m} X_{(n-1)m+i}$, $n = 1, 2, 3 \ldots$, is the average of the original sequence in m non-overlapping blocks, then the covariance $C^m(k; T)$, ACF $r^m(k; T)$, and variance $Var(X^{(m)})$ of aggregated processes are, respectively, given by [16]

$$C^m(k : T) = m^{-2} C(k; mt), \tag{4}$$

$$r^{(m)}(k; T) = \frac{(mT)^{\alpha}}{(mT)^{\alpha} + T_0^{\alpha}} \frac{1}{2} \nabla^2 (k^{\alpha+1})(k > 0), \tag{5}$$

$$Var(X^{(m)} = \lambda T[m^{-1} + \frac{(}{T} T_0)^{\alpha} m^{-(1-\alpha)}]. \tag{6}$$

3 Fitting Procedure

In this section, we summarize the fitting model for FPP which generates the self-similar traffic using Markovian approach [13]. This procedure is based on second order statistics variance while taking FOT into consideration. This model is similar to that of the paper [18] involving superposition of two-state interrupted Poisson process (IPP) and Poisson process. IPP is a particular case of MMPP. We can describe i^{th} IPP as follows:

$$Q_i = \begin{bmatrix} -c_{1i} & c_{1i} \\ c_{2i} & -c_{2i} \end{bmatrix} \quad \Lambda_i = \begin{bmatrix} \lambda_i & 0 \\ 0 & 0 \end{bmatrix} \quad 1 \leq i \leq d.$$

Superposition of above d IPPs and a Poisson process is again an MMPP, and is given by

$$Q = Q_1 \oplus Q_2 \oplus \ldots \oplus Q_d, \wedge = \wedge_1 \oplus \wedge_2 \oplus \ldots \oplus \wedge_d, \tag{7}$$

where $\bigoplus$ denotes the Kronecker sum, and λ_p is the arrival rate of the Poisson process to be superposed. Then the whole arrival rate λ is given by

$$\lambda = \lambda_p + \sum_{i=1}^{d} \frac{c_{2i}}{c_{1i} + c_{2i}} \lambda_i. \tag{8}$$

Let $N_{t,i}$ be the number of arrivals from the ith IPP during the tth timeslot and $N_{t,p}$ be the corresponding number of arrivals from the Poisson process, and let $N_{t,i}^{(m)}$ and $N_{t,p}^{(m)}$ be the number of arrivals from the averaged process of ith IPP and Poisson process, respectively. The variance of this averaged process is given by [5, 18],

$$Var[N_{t,i}^{(m)}] = \frac{c_{2i}\lambda_i}{m(c_{1i} + c_{2i})} + \frac{c_{1i}c_{2i}}{m(c_{1i} + c_{2i})^3}[1 - \frac{1 - e^{-m(c_{1i}+c_{2i})}}{m(c_{1i} + c_{2i})}]\lambda_i^2 \quad 1 \leq i \leq d \tag{9}$$

and

$$Var[N_{t,p}^{(m)}] = \frac{\lambda_p}{m}. \tag{10}$$

Hence, the variance of whole process is

$$Var[X_t^{(m)}] = \frac{\lambda}{m} + \sum_{i=1}^{d} \eta_i \lambda_i^2 \tag{11}$$

where

$$\eta_i = \frac{2c_{1i}c_{2i}}{m(c_{1i}+c_{2i})^3}[1 - \frac{1-e^{-m(c_{1i}+c_{2i})}}{m(c_{1i}+c_{2i})}] \tag{12}$$

To fit MMPP, we match the expressions in (6) & (12) over a time interval [13].

4 The MMPP/G/1 Queueing System

Consider the infinite queueing system MMPP/G/1, where input process is MMPP parameterized by Q and $\wedge$ and service time follows general distribution. The queue length at departures may be studied from the embedded Markov chain at departures with transition probability matrix $\tilde{Q}$ [7]. The stationary probability vector $x = (x_0, x_1, \ldots)$ of transition probability matrix $\tilde{Q}$ [3], where $x_i = (x_{i,1}, x_{i,2}, \ldots, x_{i,m})$, and x_{ij}=Pr a departure leaves the system with i customers and the MMPP in state j. The probabilities $x_i^{'s}$ are obtained by solving the system of equations

$$x_i = x_0 B_i + \sum_{v=1}^{i} +1 x_v A_{i+1-v}, i \geq 0.$$

The quantities $x_i, i \geq 1$ could be determined once the initial vector x_0, matrices of counting function B_i, and A_{i+1-v} are known which can be computed using the following algorithm [3]:

Step 1. Compute $A_v = \sum_{n=v}^{\infty} \gamma_n K_v^{(n)}, \quad v \geq 0,$

where

$$K_0^{(0)} = I$$
$$K_v^{(0)} = 0,$$
$$K_0^{(n)} = K_0^{(n-1)}[\Theta^{-1}(Q - \wedge) + I],$$
$$K_v^{(n)} = K_v^{(n-1)}[\Theta^{-1}(Q - \wedge) + I] + K_{v-1}^{(n-1)}\Theta^{-1}\wedge,$$
$$K_v^{(n)} = 0.$$

Step 2. Compute $B_i = (\wedge - Q)^{-1} \wedge A_i$.

Step 3. Compute the stochastic matrix $G = (G_{ij})$, where G_{ij} is the probability that a busy period starting in state i and ends in state j.

Step 4. Compute the steady state vector g which satisfies

$$gG = g, \qquad ge = 1. \tag{13}$$

Step 5. Compute

$$x_0 = \frac{1-\rho}{\lambda_{tot}} g(\wedge - Q).$$

The queue length distribution at an arbitrary time y_i can be computed as follows:

$$\begin{aligned} y_0 &= (1-\rho)g, \\ y_i &= (y_{i-1} \wedge - \lambda_{tot}(x_{i-1} - x_i))(\wedge - Q)^{-1}. \end{aligned} \tag{14}$$

Mean waiting time could be computed by the formula [3]

$$MWT = \frac{1}{2(1-\rho)}[2\rho + h^{(2)}\lambda_{tot} - 2h((1-\rho)g + h\Pi\Lambda)(Q + e\Pi)^{-1}\lambda], \tag{15}$$

where h, and $h^{(2)}$ first and second moments of general distribution G.

5 Numerical Results

In this section, we investigate queueing based performance measures, namely mean waiting time (MWT) and tail probability (queue length distribution) against the traffic intensity in terms of fractal onset time T_0, Hurst parameter H, number of IPPs d, and the time scale. First, transition rate matrix Q and arrival rate matrix Λ of MMPP are fitted according to the method described in the last section for the self-similar internet traffic pertaining to the values $H = 0.6$, $\lambda = 1$, $\sigma^2 = 0.6$, (Sample 1), $H = 0.7$, $\lambda = 1$, $\sigma^2 = 0.6$, (Sample 2), and $H = 0.8, \lambda = 1$, $\sigma^2 = 0.6$, (Sample 3), $T = 1$, and for arbitrary values of FOT over the different time scales $[10^2, 10^6]$, $[10^2, 10^7]$, and $[10^2, 10^8]$. These samples are generated in the paper [8] by random midpoint displacement algorithm. We use matrix analytic methods [4, 18] to compute steady state probability distribution of the transition probability matrix $\tilde{Q}$ then we compute the tail probability and mean waiting time using the Eqs. (14), and (15), respectively. Numerical calculations are performed using MATLAB and results are shown in the Figs. 1, 2, 3, 4, 5, 6. Figure 1 illustrates the results for the case of Hurst parameter $H = 0.7$ for different values of T_0 over the time scale $[10^2, 10^8]$. From this figure, we conclude that MWT is increasing as fractal onset time T_0 is decreasing. Parameter setting for the Fig. 2 are $H = 0.7$ over the different time scales $[10^2, 10^6]$, $[10^2, 10^7]$, and $[10^2, 10^8]$, for the fixed value of T_0. Figure 3 depicts the results for the case of fractal onset time T_0 for the time scales $[10^2, 10^8]$ over the different Hurst parameters $H = 0.6$, $H = 0.7$, and $H = 0.8$. Figures 4, 5, 6 illustrate the result for the tail probability, PrL.10, for Sample 2. Figure 4 depicts the results for the case of Hurst parameter $H = 0.7$ for different values of T_0 over the time scale $[10^2, 10^7]$. Parameter setting for

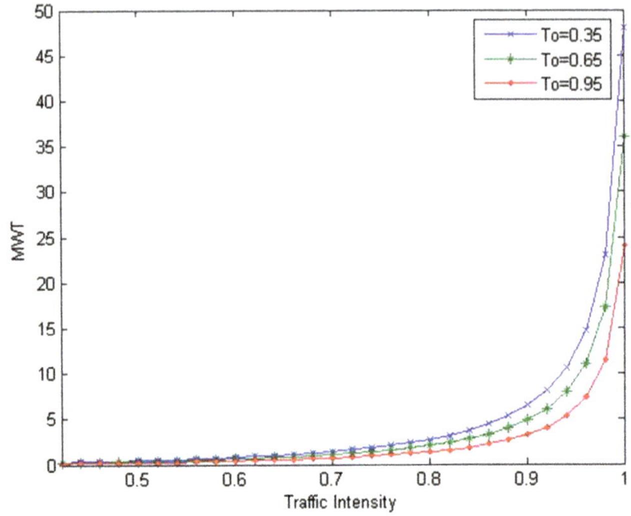

Fig. 1 Mean waiting time of the resultant $MMPP/M/1$ queues with $d = 4,\ H = 0.7,\ \lambda = 1,$ and $T = 1$ over the time scale $[10^2, 10^8]$

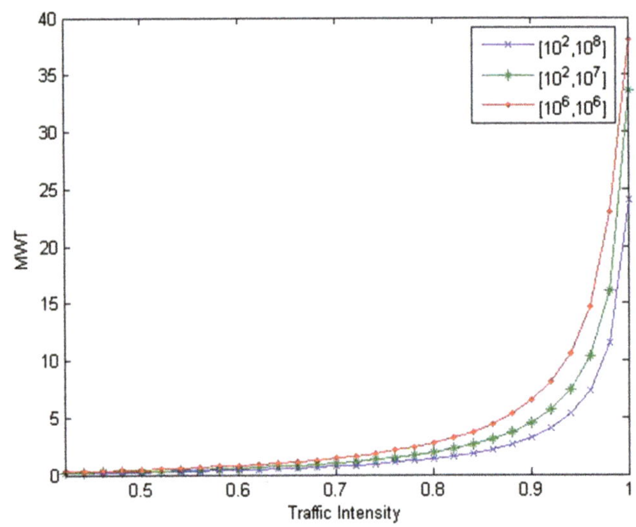

Fig. 2 Mean waiting time of the resultant $MMPP/M/1$ queues with $d = 4,\ H = 0.7, \lambda = 1,$ $T = 1$, and $T_0 = 0.95$

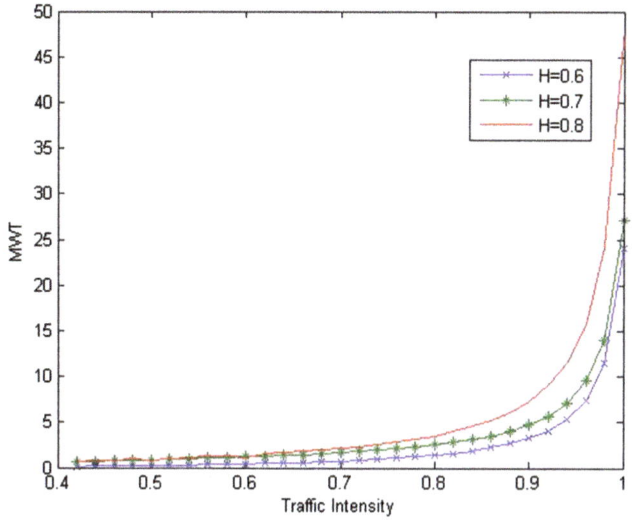

Fig. 3 Mean waiting time of the resultant $MMPP/M/1$ queues with $d = 4$, $\lambda = 1$, $T = 1$, and $T_0 = 0.95$ over the time scale $[10^2, 10^8]$

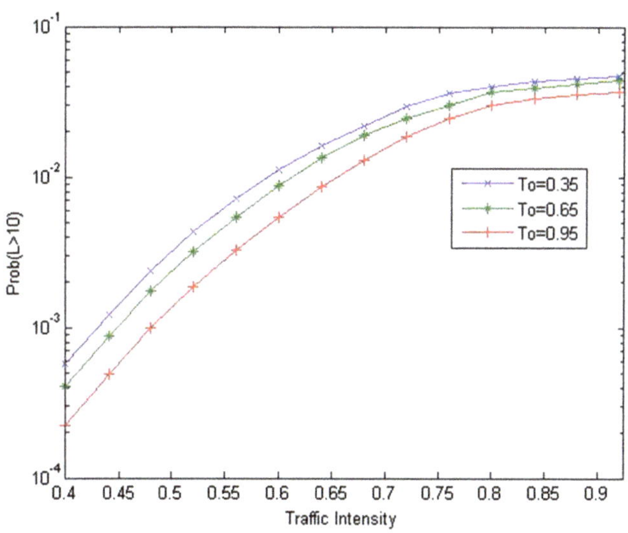

Fig. 4 $\Pr(L > 10)$ of $MMPP/M/1$ queues with $d = 4$, $H = 0.7$, and $T = 1$ over the time scale $[10^2, 10^7]$

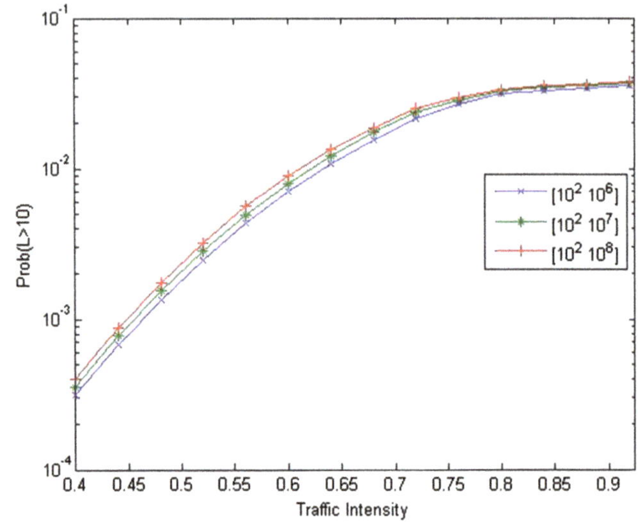

Fig. 5 Pr($L > 10$) of $MMPP/M/1$ queues with $d = 4, H = 0.7, T = 1$, and $T_0 = 0.95$

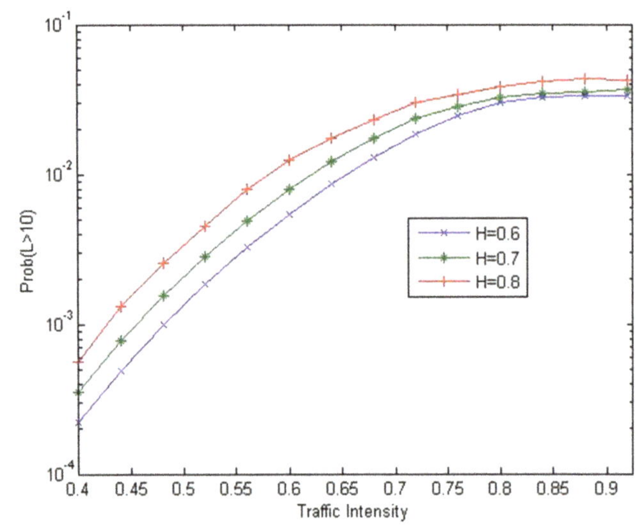

Fig. 6 Pr($L > 10$) of $MMPP/M/1$ queues with $d = 4$, and $T = 1$ over the time scale $[10^2, 10^7]$

the Fig. 5 are $H = 0.7$ over the different time scales $[10^2, 10^6], [10^2, 10^7]$, and $[10^2, 10^8]$, for the fixed value of T_0. Figure 6 depicts the results for the case of fractal onset time T_0 for the time scales $[10^2, 10^7]$ over the different Hurst parameters $H = 0.6$, $H = 0.7$, and $H = 0.8$. From this figure, we conclude that tail probability increases as T_0 and Hurst parameter is increase and decreases as the time scale increases. Moreover, real time data measured at AT&T Bell Labs are used for

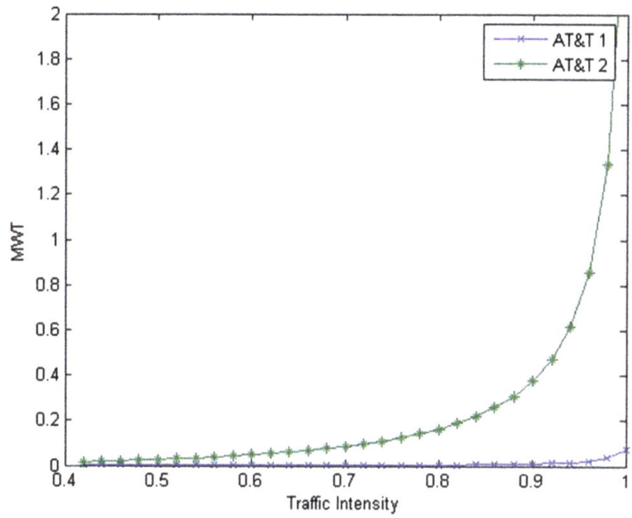

Fig. 7 Mean waiting time of the resultant $MMPP/M/1$ queues for Bellcore data over the time scale $[10^2, 10^8]$

the numerical process and results are given in Fig. 7. The Parameter setting for the Fig. 7 are $H = 0.87, \lambda = 318.2, \sigma^2 = 838$, and $T_0 = 0.006$ (say AT&T 1), and $H = 0.92, \lambda = 8.1, \sigma^2 = 133.5$, and $T_0 = 0.033$ (say AT&T 2). These parameter values are pertaining to the traffic measured at AT&T Bell Labs [16]. From this figure, we conclude that MWT increases as H increases.

6 Conclusion

In this paper, the asynchronous router with the self-similar variable length packet traffic is modeled as MMPP/M/1 queueing system where pseudo self-similar input process FPP is modeled as MMPP. In this model, service time distribution (packet length) is taken to be exponential rather than deterministic, since TCP/IP packets are variable in length and cost effective. We investigated the mean waiting time and tail probability of router over a time-scale. Our numerical results reveal the impact of FOT on mean waiting time and tail probability. Mean waiting time, tail probability increase as the traffic parameter H, T_0, and traffic intensity increase and decrease as time-scale increases. Based on the analysis presented here one could select the appropriate time-scale in the variance based fitting method [13] to provide QoS guaranteed. This kind of analysis is useful in dimensioning the router under self-similar input traffic.

Acknowledgements One of the three authors (DR) wishes to acknowledge Council of Scientific and Industrial Research (CSIR), Government of India for its funding under Senior Research Fellowship (CSIR-UGC-SRF) scheme.

References

1. Andersen, A., Nielsen, B.: A Markovian approach for modeling packet traffic with long-range dependence. IEEE J. Sel. Areas Commun. **16**(5), 719–73 (1998)
2. Crovella, M., Bestavros, A.: Self-similarity in world wide web traffic: evidence and possible causes. IEEE/ACM Trans. Netw. **5**, 835–846 (1997)
3. Fisher, W., Meier-Hellstern, K.S.: The Markov-modulated Poisson process (MMPP) cookbook. Perform. Eval. **18**, 149–171 (1992)
4. Gambini, P., Renaud, M., Guillemot, C., et al.: Transparent optical packet switching: network architecture and demonstrators in the KEOPS project. IEEE J. Sel. Areas Commun. **16**(7), 1245–1259 (1998)
5. Kasahara, S.: Internet traffic modelling: Markovian approach to self-similar traffic and prediction of loss probability for finite queues (Special Issue on Internet Technology). IEICE Trans. Commun. **E84-B**(8), 2134–2141 (2001)
6. Kim, G.Y., Shiravi, A., Min, P.S.: Congestion prediction of self-similar network through parameter estimation. In: Network Operations and Management Symposium, NOMS 2006. 10th IEEE/IFIP (2006)
7. Kleinrock, L.: Queueing Systems Vol. 1: Theory. Wiley, New York (1975)
8. Lau, W.C., Erramilli, A., Wang, J. L., Willinger, W.: Self-similar traffic generation: the random midpoint displacement algorithm and its properties. In: Proc. ICC'95, Seattle, pp. 466–472 (1995)
9. Leland, W.E., Taqqu, M.S., Willinger, W., Wilson, W.V.: On the self-similar nature of ethernet traffic (extended version). IEEE/ACM Trans. Netw. **2**, 1–15 (1994)
10. Michiel, H., Laevens, K.: Telegraphic engineering in a broad -band era. Proc. IEEE **85**(12), 2007–2033 (1997)
11. Paxson, V., Floyd, S.: Wide area traffic: the failure of Poisson modeling. IEEE/ACM Trans. Netw. **3**, 226–244 (1995)
12. Perati, M.R., Kumar, L.P.R., Sampath Kumar, K., Shao, S.K.: Analytical model for the switch handling self-similar traffic with variable packet length. In: Proceedings IEEE ICON-2008, IIT Roorkee, India, 978-1-4244-3805-1/08. IEEE (2008)
13. Rajaiah, D., Malla Reddy, P.: Internet traffic modelling-variance based Markovian fitting of fractal point process from self-similarity perspective. J. Inf. Comput. (Accepted)
14. Raj Kumar, L.P. Sampath Kumar, K., Mallikarjuna Reddy, D., Perati, M.R.: Analytical model for performance study of the switch under self-similar variable length packet traffic. In: Proceedings of The World Congress on Engineering and Computer Science-2010, WCECS 2010, San Francisco, pp. 243–247, 20–22 October 2010
15. Ryu, B.K., Lowen, S.B.: Point process approaches to the modeling and analysis of self similar traffic: Part I - Model construction. In: Proc. IEEE INFOCOM, San Francisco, pp. 1468–1475 (1996)
16. Ryu, B.K., Lowen, S.B.: Point process models for self-similar network traffic, with applications. Stoch. Models **14**(3), 735–761 (1998)
17. Salvador, P., Pacheco, A., Valadas, R.: Multiscale fitting procedure using Markov modulated Poisson processes. Telecommun. Syst. **23**(1–2), 123–148 (2003)

18. Shao, S.K., Malla Reddy, P., Tsai, M.G., Tsao, H.W., Wu, J.: Generalized variance-based Markovian fitting for self-similar traffic modeling. IEICE Trans. Commun. **E88-B**(12), 4659–4663 (2005)
19. Yoo, M., Qiao, C., Dixit, S.: IEEE Commun. Mag. **39**(2), 98–104 (2001)
20. Yoshihara, T., Kasahara, S., Takahashi, Y.: Practical time-scale fitting of self-similar traffic with Markov modulated Poisson process. Telecommun. Syst. **17**, 185–211 (2001)

Validation of Variance Based Fitting for Self-similar Network Traffic

Ramesh Renikunta, Rajaiah Dasari, Ranadheer Donthi, and Malla Reddy Perati

Abstract Most of the classical self-similar traffic models are asymptotic in nature. Hence, they are not suitable for queuing based performance evaluation. In this paper, we have validated further fitting method of CMMPP emulating self-similar traffic by means of IDC. We conclude from the numerical examples that self-similar traffic can be well represented by the proposed model.

Keywords Self-similarity • CMMPP • Variance • IDC • Fitting • Time scale • Markovian approach

1 Introduction

In recent years seminal studies have been shown the presence of self-similarity or long range dependence (LRD) in LAN, WAN, the variable bit rate (VBR) video traffic and its influence on network traffic. These types of traffic exhibit statistical similarity over different time scales and are highly correlated. The self-similarity in the network traffic has considerable impact on queueing performance. TCP/IP packet traffic over Ethernet and WAN network providing WWW service is much

R. Renikunta
Department of Mathematics, Kakatiya Institute of Technology and Science, Warangal, AP, India
e-mail: rameshrenikuntla@gmail.com

R. Dasari • M.R. Perati (✉)
Department of Mathematics, Kakatiya University, Warangal, AP, India
e-mail: dsreddy.hari@gmail.com; mperati@yahoo.com

R. Donthi
Department of Mathematics, Medak College of Engineering and Technology, Medak, AP, India
e-mail: ranadheer.phdku@gmail.com

C. Bandt et al. (eds.), *Fractals, Wavelets, and their Applications*, Springer Proceedings in Mathematics & Statistics 92, DOI 10.1007/978-3-319-08105-2_28,

better modeled using Markovian arrival process (MAP) [3,6,7]. If the traffic models do not accurately represent the real traffic, then the network performance may be overestimated or underestimated. Characterizing the statistical behavior of traffic is crucial to proper buffer design of router or switch in the network traffic to provide the quality of service (QoS). Several authors proposed the various stochastic models, techniques, and statistical nature of self-similar network traffic. Traffic models such as Fractional Brownian Motion (FBM), Fractional Auto Regressive Integrated Moving Average (FARIMA), and Chaotic maps are proposed to characterize the self-similarity. These models describe the self-similar behavior in a relatively simple manner. Although, these processes are parsimonious, but are less effective in the context of queuing based performance evaluation. In [1,5,9,10], MAP is employed to model the self-similar behavior over the desired time scales. These fitting models equate the second-order statistics of self-similar traffic and that of superposition of several 2-state Markov modulated Poisson Processes (MMPP). Covariance function of resultant MMPP is approximated by suppressing the higher order terms in its Taylor's expansion [1]. MMPP emulating the self-similar traffic is fitted by matching the variance over the desired time scales [10]. Resultant MMPP in the said paper is superposition of several Interrupted Poisson Processes (IPPs) wherein two modulating parameters of each IPP are equal. The fitting method [5, 10] is generalized in the paper [9] by taking distinct modulating parameters in each IPP. The said models hold good for voice and data traffic as IPP consists of two states: talkspurt and silence.

On the other hand, in the case of two state Circulant Markov Modulated Poisson Process (2-CMMPP) which is the restricted version of Switched Poisson process (2-state MMPP), two states are active unlike IPP. The Circulant Markov modulated Poisson process (CMMPP) has several advantages over MMPP in terms of computational complexity [4]. In the paper [8], CMMPP emulating self-similar is fitted by equating the variance of packet counts. In the present paper, the said fitting method is validated further by means of Index of dispersion for counts (IDC). Moreover, queueing behavior in terms of packet loss probability of resultant queueing system is examined.

The rest of the paper is organized as follows. In Sect. 2, we first overview the fundamentals of Self-similar process and CMMPP. In Sect. 3, IDC of the resultant CMMPP and that of self-similar traffic is derived. Numerical results of fitting model are given in Sect. 4. Finally, some conclusions are made in Sect. 5.

2 Self-similar Process and CMMPP

The second-order statistics, namely variance and IDC are relatively straightforward to fit the parameters of a model emulating self-similar traffic and give much information. As a result, these statistics are exploited by several authors. In this section, first we overview the definition of the self-similar process and the CMMPP in terms of the second-order statistics.

The definition of exact second-order self-similar process is given as follows. If we consider X as a second-order process with variance σ^2, and divide time axis into disjoint intervals of unit length, we could define $X = (X_t, t = 1, 2, \ldots.)$ to be the number of points (packet arrivals) in the tth interval. A new sequence $X^{(m)} = (X_t^{(m)})$ where $X_t^{(m)} = \frac{1}{m}\sum_{i=1}^{m} X_{(t-1)m+i}$, $t = 1, 2, 3 \ldots\ldots$ is the average of the original sequence in "m" nonoverlapping blocks. Then the process X is said to be exact second-order self-similar process with Hurst parameter $H = 1 - \frac{\beta}{2}$, if

$$Var(X^{(m)}) = \sigma^2 m^{-\beta}, \forall m \geq 1. \tag{1}$$

On the other hand, CMMPP is a doubly stochastic process in which arrival rate is given by $\lambda[J_t]$ where $J_t, t \geq 0$ is an m state Markov process. The arrival rate can therefore take on only m values, namely $\lambda_1, \lambda_2, \lambda_3 \ldots\ldots \lambda_m$. It is equal to λ_j whenever the Markov process is in the state j, $1 \leq j \leq m$. The CMMPP is fully parameterized by the infinitesimal generator Q (Circulant Markovian) of the Markov process and the vector $\lambda = (\lambda_1, \lambda_2, \lambda_3 \ldots\ldots \lambda_m)$ of the arrival rates. Let Λ be the diagonal matrix with $\Lambda_{jj} = \lambda_j$, $1 \leq j \leq m$. In the case of two state CMMPP, Q and Λ are given as follows:

$$Q = \begin{bmatrix} -c_1 & c_1 \\ c_1 & -c_1 \end{bmatrix}, \Lambda = \begin{bmatrix} \lambda_1 & 0 \\ 0 & \lambda_2 \end{bmatrix}. \tag{2}$$

The mean arrival rate λ of CMMPP is given by $\lambda = \vec{\pi}\Lambda e$, where $\vec{\pi}$ is the stationary probability vector of Q, i.e. $\vec{\pi} Q = 0$, $\vec{\pi} e = 1$ and e is an all 1 column vector with designated dimension. If we let N_t, $t \geq 0$ be the number of arrivals in (0,t], the mean of N_t is

$$E(N_t) = \frac{(\lambda_1 + \lambda_2)}{2} t. \tag{3}$$

The variance of N_t is given as follows:

$$var(N_t) = \frac{(\lambda_1 + \lambda_2)}{2} t + \frac{(\lambda_1 - \lambda_2)^2}{4c_1} t - \frac{(\lambda_1 - \lambda_2)^2}{8c_1^2}[1 - e^{-2c_1 t}]. \tag{4}$$

the IDC is defined as

$$IDC(t) = \frac{Var(N_t)}{E(N_t)}.$$

From (3) and (4), we can obtain

$$IDC(t) = 1 + \frac{(\lambda_1 - \lambda_2)^2}{(2c_1)(\lambda_1 + \lambda_2)} - \frac{(\lambda_1 - \lambda_2)^2(1 - e^{-2c_1 t})}{(4c_1^2)(\lambda_1 + \lambda_2)t}. \tag{5}$$

We then could obtain the following remarks:

(1) $IDC(t) \to 1$ as $t \to 0$, that is, CMMPP tends to a Poisson process.
(2) $IDC(t) \to 1 + \frac{(\lambda_1 - \lambda_2)^2}{(2c_1)(\lambda_1 + \lambda_2)}$, a constant, as t→ ∞
(3) IDC(t) is monotonic increasing over a finite time interval and is bounded.
(4) Steady state distribution of states of 2-state CMMPP is (0.5,0.5) which is constant and independent of transition rates.

The IDC of and exact second-order self-similar process is monotonic increasing and bounded when $t < \infty$ and $H \geq 0.5$. Hence, it is likely that we could use appropriate CMMPPs to emulate exact second-order self-similar processes over a specified time interval according to their second-order statistics.

3 Analytical Results of IDC

Generalized variance-based fitting method is a procedure to find out the traffic model parameters, in order to match the variance of self-similar traffic. Following [1, 5, 10], the fitted model emulating self-similar traffic consists of a superposition of $'d'$ 2-CMMPPs and one Poisson process. We describe the ith 2-CMMPP as follows:

$$Q_i = \begin{bmatrix} -c_i & c_i \\ c_i & -c_i \end{bmatrix} \quad , \Lambda_i = \begin{bmatrix} \lambda_{1i} & 0 \\ 0 & \lambda_{2i} \end{bmatrix}. \tag{6}$$

Superposition of above "d" CMMPSs and a Poisson process is a Transition rate matrix, and is given by

$$Q = Q_1 \oplus Q_2 \oplus \ldots \oplus Q_d, \;\; \wedge = \wedge_1 \oplus \wedge_2 \oplus \ldots \oplus \wedge_d \oplus \lambda_p, \tag{7}$$

In (7), $\oplus$ means the Kronecker sum and λ_p is the arrival rate of the Poisson process. The whole arrival rate λ is then given by

$$\lambda = \lambda_p + \sum_{i=1}^{d} \frac{\lambda_{1i} + \lambda_{2i}}{2}. \tag{8}$$

Let $N_{t,i}$, $N_{t,p}$ be the number of arrivals packets from the ith $CMMPP$ and Poisson process, respectively, during the tth time slot, and let $N_{t,i}^{(m)}$ and $N_{t,p}^{(m)}$ be the number of arrivals from the averaged processes of ith CMMPP and Poisson process, respectively.

$$put \;\; X_t^{(m)} = \sum_{i=1}^{d} (N_{t,i}^{(m)} + N_{t,p}^{(m)}). \tag{9}$$

Using (4), we obtain the variance of the ith $CMMPP$ as

$$Var(N_{t,i}^{(m)}) = \frac{\lambda_{1i} + \lambda_{2i}}{2m} + [\frac{1}{4mc_i} - \frac{1}{8m^2c_i^2}(1 - e^{-2mc_i})](\lambda_{1i} - \lambda_{2i})^2. \quad (10)$$

$$Also \quad Var(N_{t,p}^{(m)}) = \frac{\lambda_p}{m}. \quad (11)$$

From (9) to (11) and using the fact that superposition of independent sub-processes preserves the variance, we obtain

$$Var(X_t^m) = \frac{\lambda_p}{m} + \sum_{i=1}^{d} \frac{\lambda_{1i} + \lambda_{2i}}{2m} + \sum_{i=1}^{d} \eta_i(\lambda_{1i} - \lambda_{2i})^2, \quad (12)$$

$$Where \quad \eta_i = [\frac{1}{4mc_i} - \frac{1}{8m^2c_i^2}(1 - e^{-2mc_i})]. \quad (13)$$

Using (1) and (12), we can fit the CMMPP emulating self-similar traffic as in the paper [8]. And, we obtain the mean of the aggregated process is

$$Var(X_t^m) = \frac{\lambda}{m} \quad (14)$$

From (13), (15), the IDC of an aggregated process is

$$IDC(X_t^m) = \frac{Var(X_t^m)}{E(X_t^m)} \quad (15)$$

$$IDC(X_t^m) = \frac{\frac{\lambda_p}{m} + \sum_{i=1}^{d} \frac{\lambda_{1i}+\lambda_{2i}}{2m} + \sum_{i=1}^{d} \eta_i(\lambda_{1i} - \lambda_{2i})^2}{\frac{\lambda}{m}} \quad (16)$$

From (1) and (15), the Index of dispersion for counts of an exact self-similar process is

$$IDC(X_t^m) = \frac{Var(X_t^m)}{E(X_t^m)} = \frac{\sigma^2 m^{-\beta}}{\frac{\lambda}{m}} \quad (17)$$

Using (16) and (17), we would match the IDC at "d" different points "m_i", and is given by

$$m_i = m_{min}a^{i-1}, i = 1, 2, 3 \ldots..d,$$

where

$$a = (\frac{m_{max}}{m_{min}})^{\frac{1}{d-1}}, d > 1.$$

4 Numerical Results

In this section, we investigate the accuracy of the fitting model in terms of the variance and packet loss probability of resultant queueing system. We have fitted the CMMPP pertaining to the traffic parameters $H = 0.6, \lambda = 1, \sigma^2 = 0.6$ (Sample1), $H = 0.7, \lambda = 1, \sigma^2 = 0.6$ (Sample 2), and, $H = 0.8, \lambda = 1, \sigma^2 = 0.6$ (Sample 3). In all the above cases, the number of superposed two state SPPs, d, is taken to be 4 and the specified time scale ranges are $[10^2, 10^6]$, $[10^2, 10^7]$, $[10^2, 10^8]$, $[5 - 7]$. The IDC-time scale curves of the resultant CMMPP and self-similar traffic are depicted in Figs. 1, 2, and 3. From these figures, we conclude that proposed fitting is in agreement with that of exact self-similar traffic. Next, we investigate queueing behavior in terms of a performance measure, namely packet loss probability of the resultant MMPP/D/1/K queue using the method [5]. Following [9], and [2], we use matrix analytic methods to compute steady state probability distribution of the transition probability matrix of buffer occupancy that in turn gives the packet loss probability.

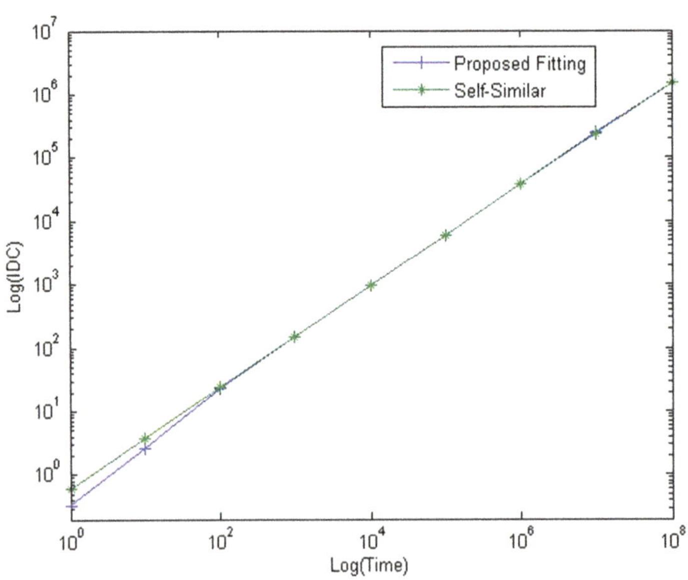

Fig. 1 Log(IDC)–log(time) curves of the resultant CMMPPs and self-similar over the time scale over the time scale range $[10^2, 10^8]$, and H = 0.6, $\sigma^2 = 0.6$

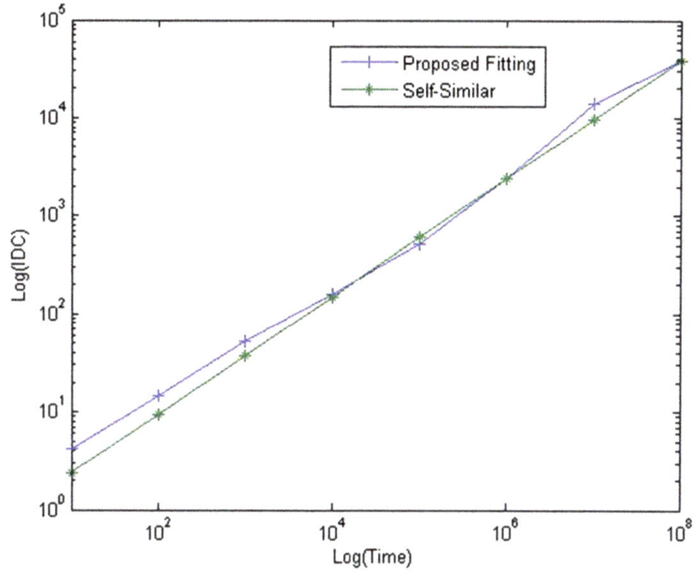

Fig. 2 Log(IDC)–log(time) curves of the resultant CMMPPs and self-similar over the time scale over the time scale range $[10^2, 10^8]$, and $\mathrm{H} = 0.7$, $\sigma^2 = 0.6$

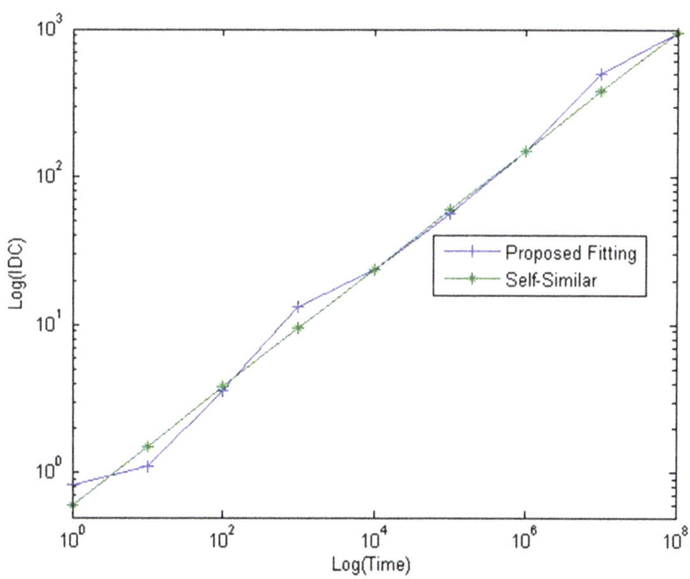

Fig. 3 Log(IDC)–log(time) curves of the resultant CMMPPs and self-similar over the time scale over the time scale range $[10^2, 10^8]$, and $\mathrm{H} = 0.8$, $\sigma^2 = 0.6$

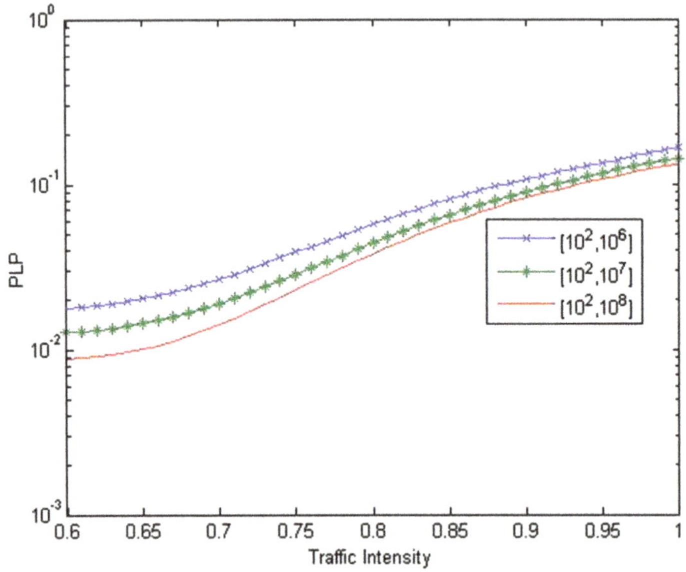

Fig. 4 Probability of the resultant MMPP/D/1/K queues with d = 4, $\lambda = 1$, H = 0.8, and K = 10 over $[10^2, 10^8]$

The buffer depth K is taken to be 10. Numerical calculations are performed using MATLAB and results are shown in Fig. 4. Figure 4 illustrates the results for the case of Hurst parameter $H = 0.8$ over the time scales $[10^2, 10^6], [10^2, 10^7], [10^2, 10^8]$. From this figure we conclude that packet loss probability decreases as the time scale increases.

Variance-time scale curves of the resultant CMMPP and self-similar traffic are depicted in Figs. 1, 2, and 3. From these figures, we conclude that proposed fitting is in agreement with that of exact self-similar traffic.

5 Conclusions

Earlier models for self-similar traffic are asymptotic and are less effective in queuing based performance evaluation. Therefore, Markovian models emulating self-similar traffic are proposed, as they hold good for queueing theory. These models are based on second-order statistics. In this paper, the fitted CMMPP emulating self-similar traffic is validated by means of IDC. Moreover, queueing behavior is examined by means of packet loss probability over different time scales over which self-similar traffic fitted. It is found from the numerical examples that self-similar can be well represented by the proposed model.

References

1. Anderson, A.T., Nielsen, B.F.: A Markovian approach for modeling packet traffic with long range dependence. IEEE J. Sel. Areas Commun. **16**(5), 719–732 (1998)
2. Blondia, C.: The N/G/l finite capacity queue. Commun. Stat. Stoch. Models **5**, 273–294 (1989)
3. Crovella, M., Bestavros, A.: Self-similarity in World Wide Web traffic: evidence and possible causes. IEEE/ACM Trans. Netw. **5**(6), 835–846 (1997)
4. De Cock, K., Van Gestel, T., De Moo, B.: Identification of circulant modulated poisson process a time domain approach. Proceedings of MTNS'98, (1998)
5. Kasahara, S.: "Internet traffic modelling: Markovian approach to self-similar traffic and prediction of loss probability for finite queues", Special Issue on Internet Technology. IEICE Trans. Commun. **E84-B**(8), 2134–2141 (2001)
6. Leland, W.E., Taqqu, M.S., Willinger, W., Wilson, D.V.: On the self-similar nature of ethernet traffic (extended version). IEEE/ACM Trans. Netw. **2**, 1–15 (1994)
7. Paxson, V., Floyd, S.: Wide area traffic: the failure of Poisson modeling. IEEE/ACM Trans. Netw. **3**, 226–244 (1995)
8. Ranadheer, D., Ramesh, R., Rajaiah, D., Malla Reddy, P.: Self-similar network traffic modeling using circulant Markov modulated poisson process (CMMPP)(Manuscript). Communicated to International conference on Fractals and Wavelets
9. Shao, S.K., Perati, M.R., Tsai, M.G., Tsao, H.W., Wu, J.: Generalized variance-based Markovian fitting for self-similar traffic modeling. IEICE Trans. Commun. **E88-B**(12), 4659–4663 (2005)
10. Yoshihara, T., Kasahara, S., Takahashi, Y.: Practical time-scale fitting of self-similar traffic with Markov-modulated Poisson process. Telecommun. Syst. **17**, 185–211 (2001)

Self-Similar Network Traffic Modeling Using Circulant Markov Modulated Poisson Process

Ranadheer Donthi, Ramesh Renikunta, Rajaiah Dasari, and Malla Reddy Perati

Abstract Most of the classical self-similar traffic models are asymptotic in nature. Hence, they are not suitable for queuing based performance evaluation. In this paper, we propose a model for self-similar traffic using Circulant Markov modulated Poisson process (CMMPP). This model is to match the variance of self-similar traffic and that of CMMPP over a time-scale. The resultant CMMPP consists of several two-state CMMPPs. We conclude from the numerical examples that self-similar traffic can be well represented by the proposed model.

Keywords Self-similarity • CMMPP • Variance • Fitting • Time-scale • Markovian approach

1 Introduction

Because of the growing diversity of services and applications with the network traffic, there is a strong requirement to measurements of packets and to describe the traffic through appropriate models. The search for accurate mathematical models of data streams in modern LAN, WAN, and WWW Internet traffic has attracted a considerable amount of interest in the last few years. Several studies have shown that said traffic exhibits properties of self-similarity or long-range dependence

R. Donthi (✉)
Department of Mathematics, Medak College of Engineering and Technology, Medak, AP, India
e-mail: ranadheer.phdku@gmail.com

R. Renikunta
Department of Mathematics, Kakatiya Institute of Technology and Science, Warangal, AP, India
e-mail: rameshrenikuntla@gmail.com

R. Dasari • M.R. Perati
Department of Mathematics, Kakatiya University, Warangal, AP, India
e-mail: dsreddy.hari@gmail.com; mperati@yahoo.com

C. Bandt et al. (eds.), *Fractals, Wavelets, and their Applications*, Springer Proceedings in Mathematics & Statistics 92, DOI 10.1007/978-3-319-08105-2_29,

(LRD) [2,5,6], which has significant impact on network performance. Characterizing the statistical behavior of traffic is crucial to proper buffer design of router or switch in the network traffic to provide the quality of service (QoS). Various stochastic models have been proposed that emulate the statistical nature of self-similar network traffic. Traffic models such as Fractional Brownian Motion (FBM), Fractional Auto Regressive Integrated Moving Average (FARIMA), and Chaotic maps are proposed to characterize the self-similarity. These models describe the self-similar behavior in a relatively simple manner. Although, these processes are parsimonious, but are less effective in the context of queuing based performance evaluation. In [1, 4, 7, 8], Markovian arrival process (MAP) is employed to model the self-similar behavior over the desired time scales. These fitting models equate the second order statistics of self-similar traffic and that of superposition of several 2-state Markov modulated Poisson Processes (MMPP). Covariance function of resultant MMPP is approximated by suppressing the higher order terms in its Taylor's expansion [1]. MMPP emulating the self-similar traffic is fitted by matching the variance over the desired time-scales [8]. Resultant MMPP in the said paper is superposition of several Interrupted Poisson Processes (IPPs) wherein two modulating parameters of each IPP are equal. The fitting method [4, 8] is generalized in the paper [7] by taking distinct modulating parameters in each IPP. The said models hold good for voice and data traffic as IPP consists of two-states talkspurt and silence.

On the other hand, in the case of two-state Circulant Markov Modulated Poisson Process (2-CMMPP) which is restricted version of Switched Poisson process (2-state MMPP), two-states are active unlike IPP. The CMMPP has several advantages over MMPP in terms of computational complexity [3]. In the present paper, CMMPP emulating Self-Similar is fitted by equating the variance of packet counts.

The rest of the paper is organized as follows. In Sect. 2, we first overview the fundamentals of Self-similar process and Circulant Markov modulated Poisson process. In Sect. 3, the generalized fitting procedure is given. Analytical results of fitting model are given in the Sect. 4. Finally, some conclusions are made in Sect. 5.

2 Self-Similar Process and Circulant Markov Modulated Poisson Process

The definition of exact second-order self-similar process is given as follows. If we consider X as a second-order process with variance σ^2, and divide time axis into disjoint intervals of unit length, we could define $X = (X_t, t = 1, 2, \ldots.)$ to be the number of points (packet arrivals) in the tth interval. A new sequence $X^{(m)} = (X_t^{(m)})$ where $X_t^{(m)} = \frac{1}{m} \sum_{i=1}^{m} X_{(t-1)m+i}, t = 1, 2, 3 \ldots..$ is the average of the original sequence in $'m'$ non-overlapping blocks. Then the process X is said to be exact second-order self-similar process with hurst parameter $2H = 2 - \beta$ and variance σ^2 if

$$Var(X^{(m)}) = \sigma^2 m^{-\beta}, \forall m \geq 1. \tag{1}$$

On the other hand, CMMPP is a doubly stochastic process in which arrival rate is given by $\lambda[J_t]$ where $J_t, t \geq 0$ is an m state Markov process. The arrival rate can therefore take on only m values, namely $\lambda_1, \lambda_2, \lambda_3 \ldots ..\lambda_m$. It is equal to λ_j whenever the Markov process is in the state j ,$1 \leq j \leq m$. The CMMPP is fully parameterized by the infinitesimal generator Q (Circulant Markovian) of the Markov process and the vector $\lambda = (\lambda_1, \lambda_2, \lambda_3 \ldots ..\lambda_m)$ of the arrival rates. Let Λ be the diagonal matrix with $\Lambda_{jj} = \lambda_j$, $1 \leq j \leq m$. In the case of two-state CMMPP, Q and Λ are given as follows:

$$Q = \begin{bmatrix} -c_1 & c_1 \\ c_1 & -c_1 \end{bmatrix}, \Lambda = \begin{bmatrix} \lambda_1 & 0 \\ 0 & \lambda_2 \end{bmatrix}. \tag{2}$$

The mean arrival rate λ of CMMPP is given by $\lambda = \vec{\pi}\Lambda e$, where $\vec{\pi}$ is the stationary probability vector of Q, i.e. $\vec{\pi}Q = 0$, $\vec{\pi}e = 1$ and e is an all 1 column vector with designated dimension. If we let N_t, $t \geq 0$ be the number of arrivals in (0,t], the mean of N_t is

$$E(N_t) = \frac{(\lambda_1 + \lambda_2)}{2} t. \tag{3}$$

The variance of N_t is given as follows:

$$var(N_t) = \frac{(\lambda_1 + \lambda_2)}{2} t + \frac{(\lambda_1 - \lambda_2)^2}{4c_1} t - \frac{(\lambda_1 - \lambda_2)^2}{8c_1^2}[1 - e^{-2c_1 t}]. \tag{4}$$

Since the index of dispersion for counts (IDC) is defined as

$$IDC(t) = \frac{Var(N_t)}{E(N_t)}.$$

From (3) and (4), we can obtain

$$IDC(t) = 1 + \frac{(\lambda_1 - \lambda_2)^2}{(2c_1)(\lambda_1 + \lambda_2)} - \frac{(\lambda_1 - \lambda_2)^2(1 - e^{-2c_1 t})}{(4c_1^2)(\lambda_1 + \lambda_2)t}. \tag{5}$$

We then could obtain the following remarks:

1. $IDC(t) \to 1$ as $t \to 0$, that is, CMMPP tends to a Poisson process.
2. $IDC(t) \to 1 + \frac{(\lambda_1 - \lambda_2)^2}{(2c_1)(\lambda_1 + \lambda_2)}$, a constant, as t$\to \infty$.
3. IDC(t) is monotonic increasing over a finite time interval and is bounded.
4. Steady state distribution of states of 2-state CMMPP is (0.5,0.5) which is constant and independent of transition rates.

The IDC of and exact second-order self-similar process is monotonic increasing and bounded when $t < \infty$ and $H \geq 0.5$. Hence, it is likely that we could use appropriate CMMPPs to emulate exact second-order self-similar processes over a specified time interval according to their second-order statistics.

3 Generalized Variance Based Fitting Procedure

Generalized variance based fitting method is a procedure to find out the traffic model parameters, in order to match the variance of self-similar traffic. Following [1, 4, 8], the fitted model emulating self-similar traffic consists of a superposition of $'d'$ 2-CMMPPs and one Poisson process. We describe the ith 2-CMMPP as follows.

$$Q_i = \begin{bmatrix} -c_i & c_i \\ c_i & -c_i \end{bmatrix}, \Lambda_i = \begin{bmatrix} \lambda_{1i} & 0 \\ 0 & \lambda_{2i} \end{bmatrix}. \tag{6}$$

Superposition of above $'d'$ CMMPSs and a Poisson process is a Transition rate matrix, and is given by

$$Q = Q_1 \oplus Q_2 \oplus \ldots \oplus Q_d, \quad \wedge = \wedge_1 \oplus \wedge_2 \oplus \ldots \oplus \wedge_d \oplus \lambda_p. \tag{7}$$

In (7), $\oplus$ means the Kronecker sum, and λ_p is the arrival rate of the Poisson process. The whole arrival rate λ is then given by

$$\lambda = \lambda_p + \sum_{i=1}^{d} \frac{\lambda_{1i} + \lambda_{2i}}{2}. \tag{8}$$

Let $N_{t,i}$, $N_{t,p}$ be the number of arrivals packets from the ith $CMMPP$ and Poisson process, respectively, during the tth time slot, and let $N_{t,i}^{(m)}$ and $N_{t,p}^{(m)}$ be the number of arrivals from the averaged processes of ith CMMPP and Poisson process, respectively.

$$put \quad X_t^{(m)} = \sum_{i=1}^{d} (N_{t,i}^{(m)} + N_{t,p}^{(m)}). \tag{9}$$

Using (4), we obtain the variance of the ith $CMMPP$ as

$$Var(N_{t,i}^{(m)}) = \frac{\lambda_{1i} + \lambda_{2i}}{2m} + [\frac{1}{4mc_i} - \frac{1}{8m^2c_i^2}(1 - e^{-2mc_i})](\lambda_{1i} - \lambda_{2i})^2. \tag{10}$$

$$Also \quad Var(N_{t,p}^{(m)}) = \frac{\lambda_p}{m}. \tag{11}$$

From (9), (10), (11) and using the fact that superposition of independent sub-processes preserves the variance, we obtain

$$Var(X_t^m) = \frac{\lambda_p}{m} + \sum_{i=1}^{d} \frac{\lambda_{1i} + \lambda_{2i}}{2m} + \sum_{i=1}^{d} \eta_i (\lambda_{1i} - \lambda_{2i})^2, \tag{12}$$

$$Where \quad \eta_i = [\frac{1}{4mc_i} - \frac{1}{8m^2c_i^2}(1 - e^{-2mc_i})]. \tag{13}$$

Using (1) and (12), we can match the variance at $'d'$ different points $m_i = 1, 2, 3, \ldots d$. Let $[m_{min}, m_{max}]$ $(m_{min} \leq m \leq m_{max})$ be the time interval over which we want the process to express self-similarity of the original process, then m_i is given by

$$m_i = m_{min}a^{i-1}, i = 1, 2, 3 \ldots ..d, \tag{14}$$

$$where \quad a = (\frac{m_{max}}{m_{min}})^{\frac{1}{d-1}}, d > 1. \tag{15}$$

Now, we assume the following relations between c_i and m_i

$$m_i c_i = const \quad (1 \leq i \leq d).$$

That is, c_i can be determined using

$$c_i = \frac{m_1}{m_i}c_1 \quad i = 1, 2, 3 \ldots ..d. \tag{16}$$

This assumption comes from the intuitive understanding that a self-similar process looks the same in any time-scale. By this assumption, we can reduce the number of parameters to be determined. That is, if we determine c_i, we can obtain the values of c_i $(2 \leq i \leq d)$ by using (16). Furthermore, we can obtain λ_p from (8) if we determine λ_{1i},λ_{2i}. Now the parameters we need to find are only c_i,λ_{1i} and λ_{2i}.

We describe the following algorithm for determining the said parameters.

3.1 Algorithm for Parameter Fitting

Step 1: Find the range of c_1 heuristically and fix c_1.
Step 2: Determine λ_{1i} and λ_{2i} as a function of c_i from (1) and (14), we have

$$\sigma^2 \begin{pmatrix} m_1^{-\beta} \\ m_2^{-\beta} \\ . \\ . \\ m_d^{-\beta} \end{pmatrix} = \lambda \begin{pmatrix} m_1^{-1} \\ m_2^{-1} \\ . \\ . \\ m_d^{-1} \end{pmatrix} + B \begin{pmatrix} (\lambda_{11} - \lambda_{21})^2 \\ (\lambda_{12} - \lambda_{22})^2 \\ . \\ . \\ (\lambda_{1d} - \lambda_{2d})^2 \end{pmatrix} \tag{17}$$

where B is dXd the matrix whose (i, j) elements is

$$B_{ij} = \frac{1}{4m_i c_j} - \frac{1}{8m_i^2 c_j^2}(1 - e^{-2m_i c_j}) \tag{18}$$

Equation (17) is a non-homogeneous system of equations in $(\lambda_{1i} - \lambda_{2i})^2$ for $(1 \leq i \leq d)$. Solving this system by any Matrix method, Cramer's rule (say), we could express λ_{1i} and λ_{2i} values in terms of c_1 from the range found in Step 1.

Step 3: Determine the values of c_i

Using (16) and the expressions for $(\lambda_{1i} - \lambda_{2i})^2$ obtained in Step 2, consider the Integral

$$\int_{m_{min}}^{m_{max}} [\sigma^2 m^{-\beta} - RHSof(12)]dm.$$

This integral is function of c_1. Determine the values of c_i such that value of the integral is minimum.

Step 4: Compute the values of $(\lambda_{1i} - \lambda_{2i})^2$ for $(1 \leq i \leq d)$ for from the equations obtained in Step 2.

4 Numerical Results

In this section, we investigate the accuracy of the fitting model in terms of the variance. We have fitted the CMMPP pertaining to the traffic parameters $H = 0.6$, $\lambda = 1, \sigma^2 = 0.6$ (Sample 1), $H = 0.7, \lambda = 1, \sigma^2 = 0.6$ (Sample 2), and, $H = 0.8$, $\lambda = 1, \sigma^2 = 0.6$ (Sample 3). In all the above cases, the number of superposed two-state SPPs, d, is taken to be 4 and the specified time scale range is $[10^2, 10^8]$ [4,7,8]. The variance-time scale curves of the resultant CMMPP and self-similar traffic are depicted in Figs. 1, 2, 3. From these figures, we conclude that proposed fitting is in agreement with that of exact self-similar traffic.

5 Conclusions

Earlier models for self-similar traffic are asymptotic and are less effective in queuing based performance evaluation. Therefore, Markovian models emulating self-similar traffic are proposed, as they hold good for queueing theory. These models are based on second-order statistics. In this paper, we have fitted CMMPP emulating self-similar traffic by equating the variance. It is found from the numerical examples that self-similar can be well represented by the proposed model.

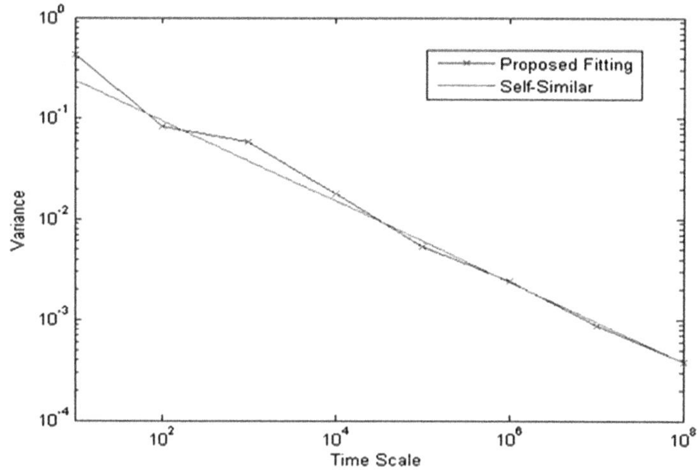

Fig. 1 Variance-time curves of the resultants of the self-similar traffic and CMMPPs with d = 4 over the time scale range $[10^2, 10^8]$, and H = 0.7

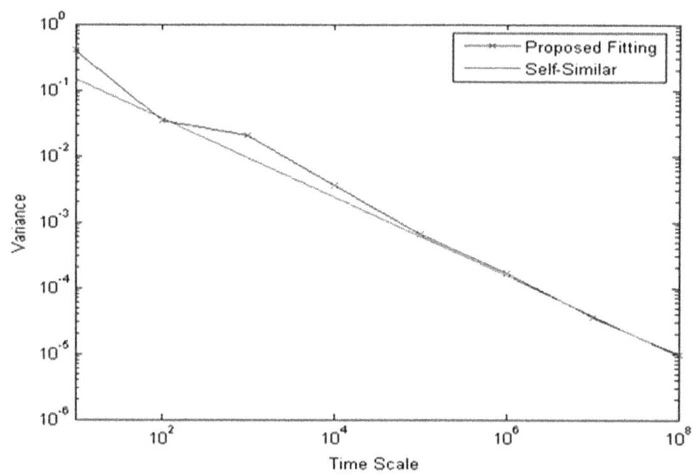

Fig. 2 Variance-time curves of the resultants of the self-similar traffic and CMMPPs with d = 4 over the time scale range $[10^2, 10^8]$, and H = 0.8

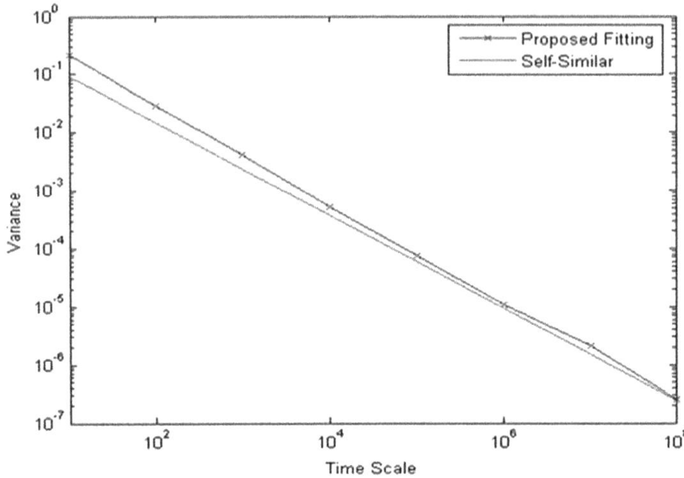

Fig. 3 Variance-time curves of the resultants of the self-similar traffic and CMMPPs with d = 4 over the time scale range $[10^2, 10^8]$, and H = 0.9

References

1. Anderson, A.T., Nielsen, B.F.: A Markovian approach for modeling packet traffic with long range dependence. IEEE J. Sel. Area Comm. **16**(5), 719–732 (1998)
2. Crovella, M., Bestavros, A.: Self-similarity in World Wide Web traffic: evidence and possible causes. IEEE/ACM Trans. Netw. 835–846 (1997)
3. De Cock, K., Van Gestel, T., De Moo, B.: Identification of circulant modulated poisson process a time domain approach, Proceeding of MTNS, 739–742 (1998)
4. Kasahara, S.: Internet traffic modelling: Markovian approach to self-similar traffic and prediction of loss probability for finite queues, special issue on internet technology. IEICE Trans. Comm. **E84-B**(8), 2134–2141 (2001)
5. Leland, W.E., Taqqu, M.S., Willinger, W., Wilson, D.V.: On the self-similar nature of ethernet traffic (extended version). IEEE/ACM Trans. Netw. **2**, 1–15 (1994)
6. Paxson, V., Floyd, S.: Wide area traffic: the failure of poisson modeling. IEEE/ACM Trans. Netw. **3**, 226–244 (1995)
7. Shao, S.K., Perati, M.R., Tsai, M.G., Tsao, H.W., Wu, J.: Generalized variance-based Markovian fitting for self-similar traffic modeling. IEICE Trans. Comm. **E88-B**(12), 4659–4663 (2005)
8. Yoshihara, T., Kasahara, S., Takahashi, Y.: Practical time-scale fitting of self-similar traffic with Markov-modulated Poisson process. Telecommun. Syst. **17**, 185–211 (2001)

Investigation of Priority Based Optical Packet Switch Under Self-Similar Variable Length Input Traffic Using Matrix Queueing Theory

Ravi Kumar Gudimalla and Malla Reddy Perati

Abstract In this paper, queueing behavior of the optical packet switch (*OPS*) employing priority based partial buffer sharing (*PBS*) mechanism under asynchronous self-similar variable length packet input traffic is investigated. Markov modulated Poisson process (*MMPP*) emulating self-similar traffic is used as input process. In view of wavelength division multiplexing (*WDM*)*OPS* output port of switch is modeled as multi-server ($MMPP/M/c/K$) queueing system. Service times (packet lengths) are assumed to be exponential distributed as traffic under consideration is unslotted asynchronous. Performance measures, namely, high priority packet loss probability and low priority packet loss probability against the system parameters and traffic parameters are computed by means of matrix-geometric solutions and approximate Markovian model. This kind of analysis is useful in dimensioning the switch employing *PBS* mechanism under self-similar variable length packet input traffic and to provide differentiated services (*DiffServ*) and quality of service (*QoS*) guarantee.

Keywords Optical packet switch • Self-similar • Partial buffer sharing • Multi-server queue • High priority packets • Low priority packets • Traffic intensity

1 Introduction

Internet Protocol (*IP*) traffic of both Ethernet and wide area network (*WAN*) traffic are shown to be self-similar and long-range dependent (*LRD*) [6, 10, 13]. Markov modulated Poisson process (*MMPP*) could be used to model the self-similar traffic

R.K. Gudimalla (✉) • M.R. Perati
Department of Mathematics, Kakatiya University, Warangal, AP 506009, India
e-mail: grk_maths@yahoo.co.in; mperati@yahoo.com

C. Bandt et al. (eds.), *Fractals, Wavelets, and their Applications*, Springer Proceedings in Mathematics & Statistics 92, DOI 10.1007/978-3-319-08105-2_30,

over the desired time-scales [2, 8, 19] to investigate the queueing behavior of network switches. These models hold good for queueing based performance analysis and are computationally tractable. Optical packet switch (*OPS*) with wavelength division multiplexing (*WDM*) technology is promising a good quality of service (*QoS*). In *OPS*, there are N input fiber lines, N output fiber lines, and each fiber line has K wavelength channels and a wavelength converter pool of size c $(0 \leq c \leq K)$, dedicated to each output fiber line. In general, *OPS* networks are classified into two categories: synchronous (*slotted*) and asynchronous (*unslotted*) [15]. In the case of first one, all the packets have same size [4, 14]. In asynchronous *OPS* networks all the packets have variable lengths and are not aligned before they enter the *OPS* [5, 16, 18]. The specific output port of asynchronous *OPS* with self-similar variable length packet input traffic is modeled as $MMPP/M/c/K$ queueing system. Therefore, performance analysis of *OPS* is equivalent to solving a problem of $MMPP/M/c/K$ queueing system.

The another issue is to provide differentiated service (*DiffServ*) as Internet traffic is moving towards *DiffServ* rather than integrated services. Broadband integrated services digital network (B-ISDN) has to support different kinds of communication services such as video phone, video conferencing, data traffic, and voice sources in more efficient manner. High demand of network traffic results in congestion problems. Congestion problem in network traffic can be dealt with some priority mechanisms. One of such mechanisms is buffer access control (*BAC*), also called space priority mechanism. There are several strategies by which one can implement this *BAC* mechanism; one of such strategies is partial buffer sharing (*PBS*) mechanism. In this scheme, a threshold is imposed on both high priority and low priority packets. The part of buffer on or below the threshold is shared by all arriving packets. When the buffer occupancy is above the threshold, the arriving low priority packets will be rejected. High priority packets will be lost only when buffer is full. If the threshold is relatively high, then the high priority packets will be lost more than expectedly. Whereas, if the threshold is relatively low, then low priority packets will be lost excessively [14, 16, 17, 21]. This way, there is a trade-off between threshold setting and packet loss. Hence, optimal threshold setting is very important in buffer dimensioning. Priority queue models are briefly discussed below. In the paper [20], the queueing analysis of infinite buffer priority system with *MMPP* as the input process is investigated with an assumption that the delay sensitive cells and non-delay sensitive cells arrive at two separate queues. This kind of scheme is not realistic as the buffers consist of limited number of fiber delay lines (*FDLs*) with fixed granularity. The loss behavior of finite buffer space priority queues with discrete batch Markovian arrival process ($D - BMAP$) has been analyzed in the paper [21] which is not the case, since the router under consideration is handling self-similar traffic with continuous time Markov process. In the papers [14, 16], switch is modeled as single server queueing system which is not appropriate for the reason mentioned in the first paragraph. To the best of our knowledge, *WDMOPS* with self-similar variable length packet input traffic employing *PBS* mechanism is not yet investigated. In this paper, we model *WDMOPS* with self-similar variable length packet input traffic as *MMPP*/*M*/*c*/*K* queueing system with *PBS* mechanism.

The rest of the paper is organized as follows: In Sect. 2, queueing model of the switch employing *PBS* mechanism is briefly introduced. In Sect. 3, numerical results are presented graphically. Finally, conclusion is given in Sect. 4.

2 Queueing Model of the Switch Employing Partial Buffer Sharing Mechanism

Consider the *WDM* asynchronous $N \times N$ *OPS* with each output fiber line consisting of K wavelength channels and a wavelength converter pool of size c. Buffer depth then is $K - c$. Such a switch with self-similar variable length packet input traffic can be modeled as *MMPP*/*M*/*c*/*K* queueing system. The operation and queueing model of the *OPS* employing *PBS* mechanism is shown in Fig. 1. As shown in the figure, the threshold is set at the level $K - 2c - d$, where d is a positive integer. The low priority packets can only access first $K - 2c - d - 1$ buffer spaces, whereas the high priority packets can utilize the whole buffer space [14, 20]. For the sake of simplicity, two priorities are considered. Each priority traffic is characterized by *MMPP*. Assume that high priority (*class1*) packets and low priority (*class2*) packets arrive at the system according to *MMPPs* with number of states m_1 and m_2, respectively. These *MMPPs* are characterized by the matrices $\{Q(1), \Lambda(1)\}$ and $\{Q(2), \Lambda(2)\}$, respectively. The service time is generally and identically distributed with distribution function $H(t)$. Let $B_m^{(k)}(t)$, $\{m \geq 0, k = 1, 2\}$ denote the matrices whose (i, j)th element is the probability that given departure of class k at time 0, there is at least one packet left in the system and the process is in state i, the next departure of class k occurs no later than time t with the arrival process in state k, and during that service time there are m arrivals. Then $B_m^k(t)$ satisfies the following equation

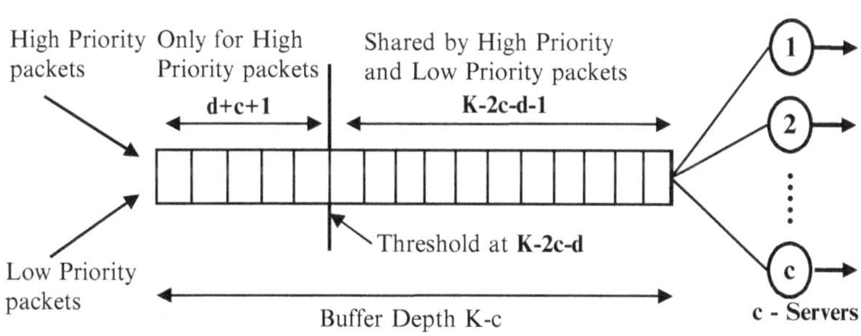

Fig. 1 The operation and the equivalent queueing model of a specific output port of OPS employing PBS mechanism with two different priority input traffic

$$\sum_{m=0}^{\infty} B_m^{(k)}(t)\, z^m = \int_0^t e^{[\,Q(k)-\Lambda(k)+\,\Lambda(k)\,z]\,\tau}\, dH(\tau), \quad k = 1, 2. \tag{1}$$

where *H(t)* is the service time distribution. When the service time is exponential with mean service rate μ, we have

$$\sum_{m=0}^{\infty} B_m^{(k)}\, z^m = \mu^2 \sum_{m=0}^{\infty} [\,\frac{Q(k) - \Lambda(k) + \Lambda(k)\, z}{\mu}]^m, \quad k = 1, 2. \tag{2}$$

It is obvious that the matrices $B_m^{(k)}$ can be evaluated by comparing the coefficient of z^m on both sides of the Eq. (2) and the procedure is outlined in the paper [16]. We consider the embedded Markov chain $\{L_n, J_n/n \geq 0\}$ at the departure epochs of the queueing system on the state space $U = \{(b, i, j)/0 \leq b \leq K - c, 1 \leq i \leq m_1, 1 \leq j \leq m_2\}$, where L_n denotes the buffer occupancy and J_n denotes the phase of superposed *MMPP*. For convenience, a queueing system is said to be at level b, if its buffer occupancy is equal to b (excluding the ones in service). The embedded Markov chain has an irreducible transition probability matrix P (with the dimension $(K - c + 1)m_1m_2 \times (K - c + 1)m_1m_2$) given as follows:

$$P = [P_1 : P_2] \tag{3}$$

where,

$$P_1 = \begin{bmatrix} B_0 & B_1 & \cdots & B_{K-3c-d-1} & B_{K-3c-d} & \cdots & B_{K-2c-d-1} & C_{K-2c-d} \\ \vdots & \vdots & \ddots & \vdots & \vdots & \ddots & \vdots & \vdots \\ B_0 & B_1 & \cdots & B_{K-3c-d-1} & B_{K-3c-d} & \cdots & B_{K-2c-d-1} & C_{K-2c-d} \\ 0 & B_0 & \cdots & B_{K-3c-d-2} & B_{K-3c-d-1} & \cdots & B_{K-2c-d-2} & C_{K-2c-d-1} \\ \vdots & \vdots & \ddots & \vdots & \vdots & \ddots & \vdots & \vdots \\ 0 & 0 & \cdots & B_0 & B_1 & \cdots & B_c & C_{c+1} \\ 0 & 0 & \cdots & 0 & B_0 & \cdots & B_{c-1} & C_c \\ \vdots & \vdots & \ddots & \vdots & \vdots & \ddots & \vdots & \vdots \\ 0 & 0 & \cdots & 0 & 0 & \cdots & B_0 & C_1 \\ 0 & 0 & \cdots & 0 & 0 & \cdots & 0 & B_0^{(1)} \otimes B_{-2} \\ \vdots & \vdots & \ddots & \vdots & \vdots & \ddots & \vdots & \vdots \\ 0 & 0 & \cdots & 0 & 0 & \cdots & 0 & 0 \end{bmatrix} \tag{4}$$

$$P_2 = \begin{bmatrix} B^{(1)}_{K-2c-d+1} \bigotimes B_{-2} & \cdots & B^{(1)}_{K-2c} \bigotimes B_{-2} & \cdots & B^{(1)}_{K-c-1} \bigotimes B_{-2} & E(K-c) \\ \vdots & \ddots & \vdots & \ddots & \vdots & \vdots \\ B^{(1)}_{K-2c-d+1} \bigotimes B_{-2} & \cdots & B^{(1)}_{K-2c} \bigotimes B_{-2} & \cdots & B^{(1)}_{K-c-1} \bigotimes B_{-2} & E(K-c) \\ B^{(1)}_{K-2c-d} \bigotimes B_{-2} & \cdots & B^{(1)}_{K-2c-1} \bigotimes B_{-2} & \cdots & B^{(1)}_{K-c-2} \bigotimes B_{-2} & E(K-c-1) \\ \vdots & \ddots & \vdots & \ddots & \vdots & \vdots \\ B^{(1)}_{c+2} \bigotimes B_{-2} & \cdots & B^{(1)}_{c+d+1} \bigotimes B_{-2} & \cdots & B^{(1)}_{2c+d} \bigotimes B_{-2} & E(2c+d+1) \\ B^{(1)}_{c+1} \bigotimes B_{-2} & \cdots & B^{(1)}_{c+d} \bigotimes B_{-2} & \cdots & B^{(1)}_{2c+d-1} \bigotimes B_{-2} & E(2c+d) \\ \vdots & \ddots & \vdots & \ddots & \vdots & \vdots \\ B^{(1)}_{2} \bigotimes B_{-2} & \cdots & B^{(1)}_{d+1} \bigotimes B_{-2} & \cdots & B^{(1)}_{c+d} \bigotimes B_{-2} & E(2c+d+1) \\ B^{(1)}_{1} \bigotimes B_{-2} & \cdots & B^{(1)}_{d} \bigotimes B_{-2} & \cdots & B^{(1)}_{c+d-1} \bigotimes B_{-2} & E(2c+d) \\ \vdots & \ddots & \vdots & \ddots & \vdots & \vdots \\ 0 & \cdots & B^{(1)}_{0} \bigotimes B_{-2} & \cdots & B^{(1)}_{c-1} \bigotimes B_{-2} & E(c) \end{bmatrix} \quad (5)$$

In (4) and (5),

$$\begin{aligned} B_i &= \sum_{i'=0}^{i} (B^{(1)}_{i'} \bigotimes B_{i-i'}{}^{(2)}) \\ B_{-2} &= \sum_{i=0}^{\infty} B^{(2)}_i \\ C_l &= \sum_{i=0}^{l} (B^{(1)}_i \bigotimes B^{(l-1)}_{-2}) \\ E(p) &= B^{(p)}_{-1} \bigotimes B_{-2}, \quad p = c, c+1, \ldots, K-c. \\ B^{(l)}_k &= \sum_{i=l}^{\infty} B^{(k)}_i, \quad k = 1, 2. \end{aligned}$$

In the above, the elements of the first $(c + 1)$ rows are identical. The fundamental arrival rate of class k packets is $\lambda^{(k)} = \pi(k)\Lambda(k)e,\ k = 1, 2$, where $\pi(k)$ is the steady state probability vector of $Q(k)$. The traffic intensity $\rho = (\lambda^{(1)} + \lambda^{(2)})\ E[H(t)]$. From the *PBS* mechanism, it is clear that high priority packet loss occurs if buffer is full, whereas low priority packet loss is due to the threshold mechanism. Let $x = (x_0, x_1, \ldots, x_{K-c})$, where $x_q = (x_{q,11}, x_{q,12}, \ldots, x_{q,m_1,m_2})$, $q = 0, 1, \ldots, K-c$, and $x_{q,l,m}$ is the conditional probability that there are q packets in the system given that embedded Markov chain is in the (l, m) state. Therefore, we have $xP = x,\ xe = 1$ where e is the column vector consisting of all 1. Then, in the steady state, high priority packet loss probability, P^{hp}, and low priority packet loss probability, P^{lp}, respectively, are [21],

$$P^{hp} = \frac{1}{\lambda^{(1)}}\{ \sum_{r=0}^{c} \sum_{i=1}^{\infty} i x_r\ (B^{(1)}_{K-c+i} \bigotimes B_{-2})e$$

$$+\sum_{r=1}^{K-2c}\sum_{i=1}^{\infty} i x_{c+r}\,(B^{(1)}_{K-c+i-r}\bigotimes B_{-2})e\} \tag{6}$$

$$P^{lp}=\frac{1}{\lambda^{(2)}}\{\sum_{r=0}^{c}(\sum_{i=1}^{\infty} i x_r[\sum_{j=0}^{K-2c-d}(B^{(1)}_j\bigotimes B^{(2)}_{K-2c-d+i-j})$$

$$+(B^{(K-2c-d+1)}_{-1}\bigotimes B_{-2})]e])$$

$$+\sum_{r=1}^{K-2c-d-1}(\sum_{i=1}^{\infty} i\, x_{c+r}[\sum_{j=0}^{K-2c-d-r}(B^{(1)}_j\bigotimes B^{(2)}_{K-2c-d+i-j})$$

$$+(B^{(K-2c-d+1+r)}_{-1}\bigotimes B_i{}^{(2)})]\,e)$$

$$+\sum_{r=K-2c-d}^{K-2c}(\sum_{i=1}^{\infty} i\, x_{c+r}\,(B_{-1}\bigotimes B^{(2)}_i)\,e)\} \tag{7}$$

Further, in view of threshold setting we decompose the state space U into two subsets:

$$U_{nc}=\{(b,i,j)/\,0\le b\le K-2c-d-1,\ 1\le i\le m_1,\ 1\le j\le m_2\}$$

$$U_c=\{(b,i,j)/\,K-2c-d\le b\le K-c,\ 1\le i\le m_1,\ 1\le j\le m_2\}$$

This partition of U makes the transition probability matrix Q decomposed as follows:

$$Q=\begin{bmatrix} Q_{nc} & Q_{nc,c}\\ Q_{c,nc} & Q_c\end{bmatrix} \tag{8}$$

The sub-matrices $Q_{nc}, Q_{nc,c}, Q_{c,nc}$, and Q_c are the left upper part, right upper part, left lower part, and right lower part of the matrix Q with dimensions of $(K-2c-d)m_1m_2\times(K-2c-d)m_1m_2$, $(K-2c-d)m_1m_2\times(d+c+1)m_1m_2$, $(d+c+1)m_1m_2\times(K-2c-d)m_1m_2$, and $(d+c+1)m_1m_2\times(d+c+1)m_1m_2$, respectively. These sub-matrices govern the transitions from U_{nc} into itself, from U_{nc} into U_c, from U_c into U_{nc}, and from U_c into itself, respectively. Following the algorithm in the paper [21], we compute the high priority packet loss probability and low priority packet loss probability.

3 Numerical Results

Following matrix-geometric solutions, [3, 9, 11, 12], we compute the steady state packet loss probabilities. We follow the generalized variance based Markovian fitting method proposed in [19] to emulate the second order self-similar variable length packet input traffic for both high priority and low priority packet traffic. The mean arrival rate and variance of the self-similar traffic is set to be 1 and 0.6, respectively [19], the interested time-scale range to emulate self-similarity is over $[10^2, 10^7]$, [2, 19]. It is shown from [2, 19] that in order to emulate self-similar traffic well, the minimum number of states of the resultant *MMPPs* must be greater than or equal to 16. That is, both m_1 and m_2 must be ≥ 16. Therefore, each class is characterized by 16×16 matrices. With such a high dimensional *MMPP* for both high priority and low priority traffic, it is a great challenge for the numerical process. In order to reduce the computation complexity, we use approximate model [14], which is based on the papers [1, 7]. The resultant 16-state *MMPP* of low priority packets is approximated by a 2-state *MAP*. By applying this approximated model, the computational complexity is reduced by $8^3 = 512$ times. We employ two different traffic corresponding to the Hurst parameter values $H = 0.7$ and $H = 0.8$, and the buffer depth, K, of the router and the number of servers, c, are set to be 24 and 4, respectively, and the results are presented in Figs. 2, 3, 4, 5. From Fig. 2, we observe that the high priority packet loss probability decreases and the low priority packet loss probability increases as the threshold increases. In order to find out the optimal level of the threshold, we illustrate a plot of high priority packet loss probabilities against the low priority ones at various d in Fig. 3. We could find out that the optimal level of the threshold is the one located nearest to the left lower corner of the plot, which is around $d = 6$. Figures 4, 5 depict the variation of packet loss probability against traffic intensity and buffer capacity, respectively. From Fig. 4, it is clear that packet loss probability increases as traffic intensity increases. From Fig. 5, we observe that of high priority and low priority packet loss probabilities both decrease as the buffer capacity increases.

4 Conclusion

We investigate the loss behavior of asynchronous router employing PBS mechanism to provide differentiated services under Markovian modeled self-similar variable length packet input traffic. The performance measures, namely, the steady state high priority and low priority packet loss probabilities are computed and presented graphically. To reduce the computation complexity, the original high dimensional of the low priority packets is approximated by 2-state. With this analysis, we could locate the optimal threshold position of buffer to obtain the greatest performance.

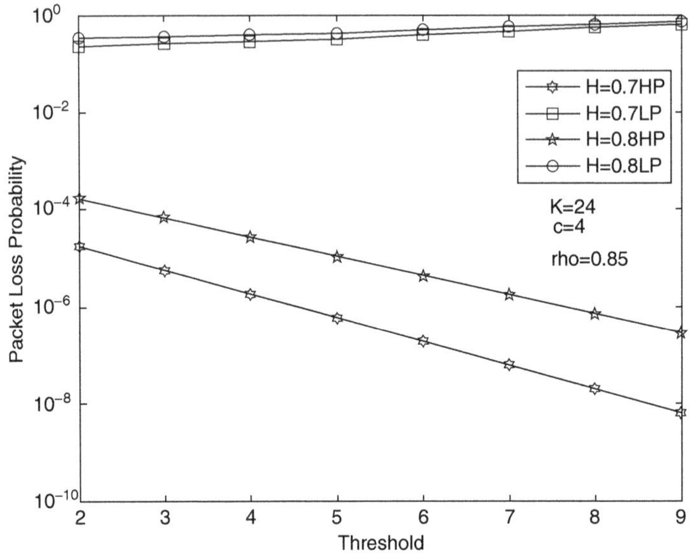

Fig. 2 Variation of packet loss probability with threshold

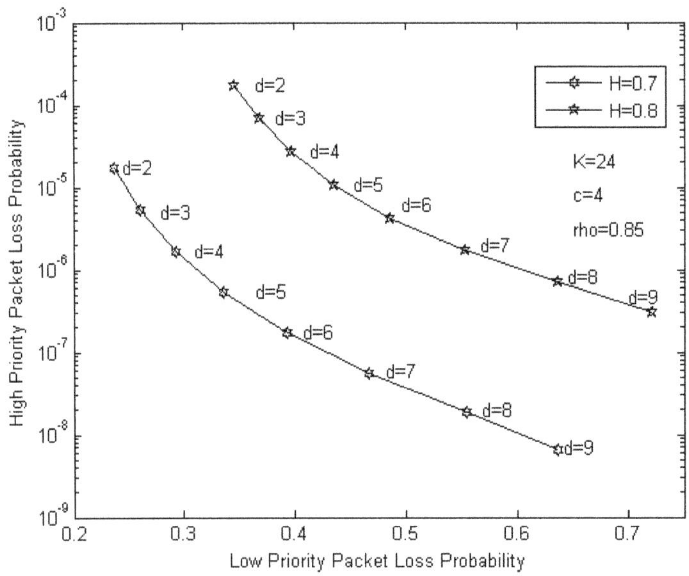

Fig. 3 Comparison of high priority PLP with low priority packet loss probability

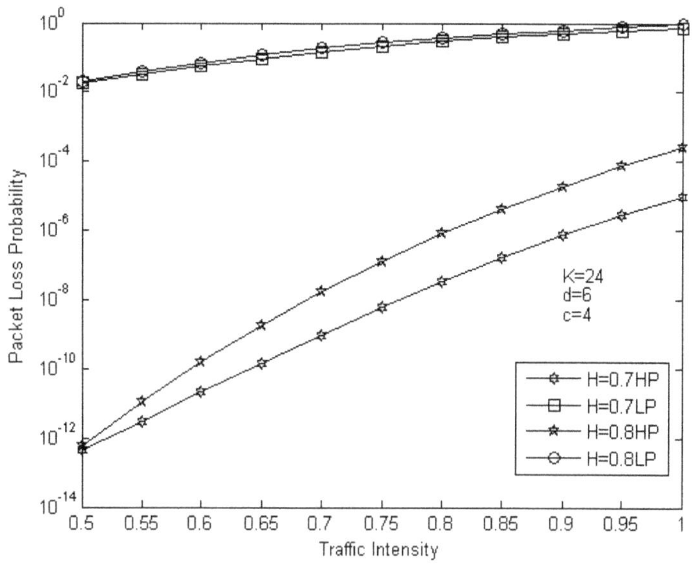

Fig. 4 Variation of packet loss probability with traffic intensity

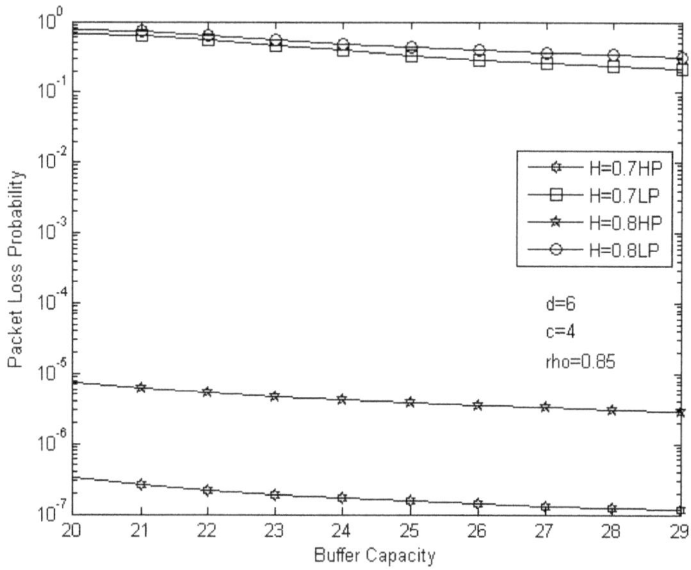

Fig. 5 Variation of packet loss probability with buffer capacity

Acknowledgements The authors wish to acknowledge the Department of Science and Technology (*DST*), Government of India for its funding under the major research project (*MRP*) scheme with the grant number: *SR/S4/MS : 530/08*.

References

1. Altiok, T.: On the phase-type approximations of general distributions. IIE Trans. **17**, 110–116 (1985)
2. Anderson, A.T., Niesen, B.F.: A Markovian approach for modeling packet traffic with long-range dependence. IEEE J. Sel. Area Comm. **16**, 719–732 (1998)
3. Blondia, C.: The N/G/1 finite capacity queue. Comm. Stat. Stoch. Model **5**, 273–294 (1989)
4. Chen, C.-Y., Chang, C.H., Perati, M.R., Shao, S.K., Wu, J.: Performance analysis of WDM OPS employing wavelength conversion under Markovian modeled self-similar traffic input. IEEE HPSR-2007, 1-ISBN 1-4244-1206 (2007)
5. Collegati, F.: Approximation modeling of optical buffers for variable length packets. Photonic Network Comm. **3**(4), 383–390 (2001)
6. Crovella, M., Bestavros, A.: Self-similarity in World Wide Web traffic: evidence and possible causes. IEEE/ACM Trans. Netw. **5**(6), 835–846 (1997)
7. Diamond, J.E., Alfa, A.S.: On approximating higher order MAPs with MAPs of order two. Queue. Syst. **34**, 269–288 (2000)
8. Kasahara, S.: Internet traffic modeling: Markovian approach to self-similar traffic and prediction of loss probability for finite queues. IEICE Trans. Comm. **E84-B**(8), 2134–2141 (2001)
9. Latouche, G., Ramaswami, V.: Introduction to Matrix Analytic Methods in Stochastic Modelling. SIAM Press, Philadelphia (1999)
10. Leland, W.E., Taqqu, M.S., Willinger, W., Wilson, D.V.: On the self-similar nature of ethernet traffic (extended version). IEEE/ACM Trans. Netw. **2**, 1–15 (1994)
11. Lucantoni, D.M., Meier-Hellstern, K.S., Neuts, M.F.: A single-server queue with server vacations and a class of nonrenewal arrival processes. Adv. Appl. Prob. **22**, 676–705 (1990)
12. Neuts, M.F.: Matrix-Geometric Solutions in Stochastic Models: An Algorithmic Approach. Dover Publications, New York (1995)
13. Paxson, V., Floyd, S.: Wide area traffic: The failure of Poisson modeling. IEEE/ACM Trans. Netw. **3**(3), 226–244 (1995)
14. Perati, M.R., Shao, S.-K., Chang, C.-H., Wu, J.: An efficient approximate Markovian model for optical packet switches employing partial buffer sharing mechanism under self-similar traffic input. IEEE-HPSR-2007, ISBN 1-4244-1206-4/07 (2007)
15. Qiao, C., Yooo, M., Yu, Z.X.: Optical burst switching (OBS); A new paradigm for an optical Internet. J. High Speed Networks (JHSN) **8**(1), 69–84 (1999)
16. Raj Kumar, L.P., Sampath Kumar, K., Mallikarjuna Reddy, D., Perati, M.R.: Performance analysis of Internet router employing partial buffer sharing mechanism under Markovian modeled self-similar variable length packet input traffic. Int. J. Pure Appl. Math. (IJPAM) **67**(4), 407–421 (2011)
17. Riberio, M.R.N., O'Mahony, M.J.: Improvements on performance of photonic packet switching nodes by priority assignment and buffer sharing. In: Proc. ICC'2000, vol. 3, pp.1738–172. New Orleans, USA (2000)
18. Sampath Kumar, K., Perati, M.R., Adilakshmi, T.: Performance study of WDM OPS employing tunable converter sharing under self-similar variable length packet traffic. IEEE-2012, ISBN 978-1-4673-4523-1/12 (2012)
19. Shao, S.K., Perati, M.R., Tsai, M.G., Tsao, H.W., Wu, J.: Generalized variance-based Markovian fitting for self-similar traffic modeling. IEICE Trans. Comm. E88-B **12**, 4659–4663 (2005)

20. Venkataramani, B., Bose, S.K., Srivathsan, K.R.: Queueing analysis of a non-pre-emptive MMPP/D/1 priority system. Comput. Comm. **20**, 999–1018 (1997)
21. Wang, Y.C., Lin, C.W., Lu, C.C.: Loss behaviour in space priority queue with batch Markovian arrival process-discrete time case. Perform. Eval. **41**(4), 269–293 (2000)

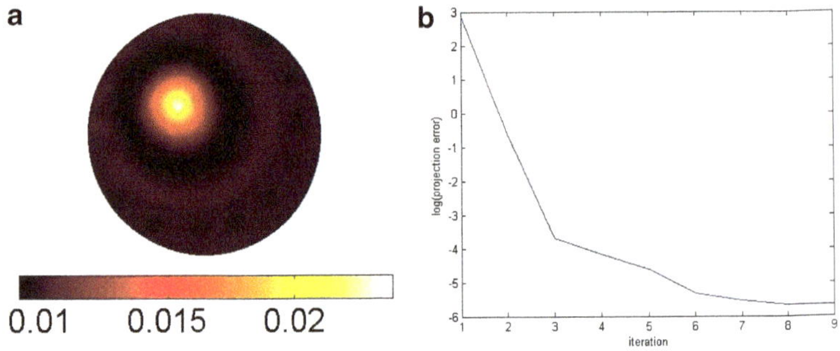

Fig. 8 (**a**) Shows reconstructed μ with CGD (**b**) Projection error for each iteration

Table 2 Comparison of multiresolution and single resolution solver

Parameter	Wavelet solver (multiresolution)	CGD solver (single resolution)
Number of iteration	9	26
Minimum error function	0.0034	0.0888
Computation time (s)	20.3114	47.5796
NRMS	0.9508×10^{-3}	1.0371×10^{-3}

Table 3 Optical properties used in the experiment

Experiment	Back ground μ_a (cm^{-1})	Constant scattering (cm^{-1})	Absorber value (cm^{-1})	Fluorescent dye absorption (cm^{-1})
1	0.01	2	0.02	0.02
2	0.01	2	0.02	0.015
3	0.01	2	0.02	0.03

Normalized Root Mean Square error is given by (22)

$$NRMS = \left(\frac{\sum_{i=1}^{N} (\widetilde{\mu_i} - \mu_i)^2}{\sum_{i=1}^{N} (\widetilde{\mu_i} - \overline{\mu}_i)^2} \right)^{1/2} \tag{22}$$

where μ_i and $\widetilde{\mu_i}$ are original absorption value and reconstructed absorption value at ith node point. $\overline{\mu}_i$ is the mean of original absorption value at all node points. Both Error function and NRMS is less in case of Multiresolution solver than single resolution solver (Table 2).

Reconstruction of fluorescent dye location and its absorption coefficient for different dye concentration is carried out. In the forward solver fluorescent quantum yield η is taken as 0.114 and fluorescent life time as 1 (Table 3).

Figures 9, 10, 11 show the independent reconstruction of $\mu_a + \epsilon c(r)$ and μ_a obtained through inversion of equation (4) and (5). The optical properties so obtained are subtracted and the background optical properties are added to obtain

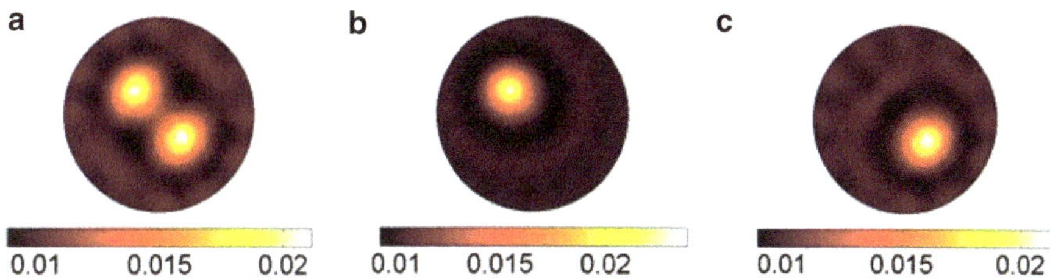

Fig. 9 Experiment 1. (**a**) and (**b**) are, respectively, independent reconstructed images of $\mu_a + \epsilon c(r)$ and μ_a. (**a**) minus (**b**) added with background absorption coefficient give fluorescent dye property given by (**c**). Here $\epsilon c(r)$ is 0.02

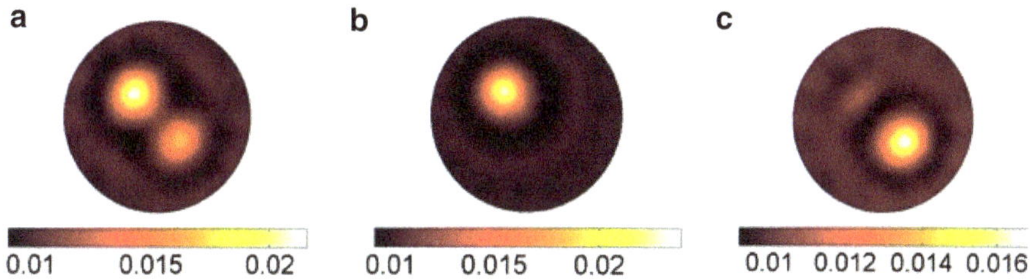

Fig. 10 Experiment 1. (**a**) and (**b**) are, respectively, independent reconstructed images of $\mu_a + \epsilon c(r)$ and μ_a. (**a**) minus (**b**) added with background absorption coefficient give fluorescent dye property given by (**c**). Here $\epsilon c(r)$ is 0.015

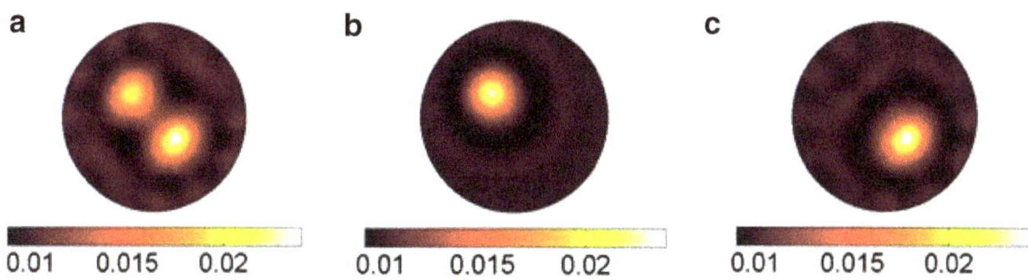

Fig. 11 Experiment 1. (**a**) and (**b**) are, respectively, independent reconstructed images of $\mu_a + \epsilon c(r)$ and μ_a. (**a**) minus (**b**) added with background absorption coefficient give fluorescent dye property given by (**c**). Here $\epsilon c(r)$ is 0.03

the exact absorption of fluorescent and hence the concentration of fluorescent dye. This independent inversion method to retrieve back fluorescent concentration is valid when the strokes' shift is small and the optical properties for excitation and emission do not change much.

When the fluorescent absorption is closer to the background the reconstructed value is more accurate and as it increases the reconstructed value will deviate from the desired linear path as shown in Fig. 12.

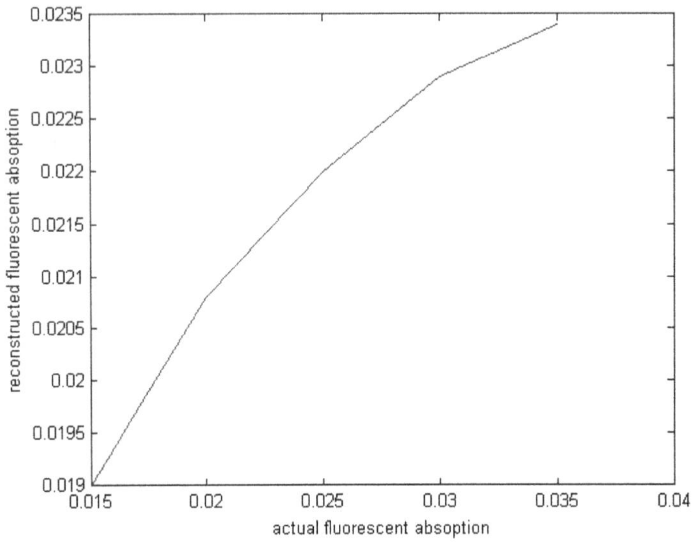

Fig. 12 Actual fluorescent value vs reconstructed value

5 Conclusions

In this work we present our pre-clinical study results of ongoing efforts to use the novel NIR fluorescent dye for early detection of breast cancer. Dye modeling and reconstruction of location and concentration is carried out. Validity of independent DOT inversion method to locate fluorescent dye is tested. The result proves that wavelet based multiresolution solver to solve large system of sparse matrix is very efficient and less time consuming than the single resolution solvers. In our future work we intend to develop a complete imaging system to collect experimental data instead of synthesis data for experiments.

References

1. Arridge, S.R., Chweiger, M.S., Hiruoka, M., Delpy, D.T.: A finite element approach for modeling photon transport in tissue. Inst. Phys. Publish. Med. Phys. **20**, 299–309 (1993)
2. Arridge, S.R., Schweige, M.: Photon-measurement density functions. Part 2: finite-element-method calculations. Appl. Optic **34**, 8016–8037 (1995)
3. Bhattacharjee, S., Jose, I.: Early detection of breast cancer: A moleculer Optical imaging approach using novel estrogen conjugate fluorescent dye. Proc. SPIE **7896**, F1–F15 (2011)
4. Dehgheni, H., et al.: Near infrared optical tomography using NIRFAST algorithm for numerical model and image reconstruction. Comm. Numer. Meth. Eng. **26**(6), 711–732 (2009)
5. Eamer, M.E., Pogue, B.W., Yalavarthy, P.K., Dehghani, H.: A efficient Jacobian reduction method for diffuse optical image reconstruction. Opt. Express **15**, 15908–15919 (2007)

6. Egger, H., Freiberger, M., Schlottbom, M.: On forward and inverse model in fluorescent diffuse optical tomography. Inverse Probl. Imag. **4**(3), 1–17 (2010)
7. Gibson, A.P., Hebden, J.C., Arridge, S.R.: Recent advances in diffuse optical imaging. Phy. Med. Bio. J. **50**, R1–R43 (2004)
8. Gupta, A., Karpis, G., Kumar, V.: Highly scalable parallel algorithm for sparse matrix factorization. IEEE Trans. Parallel Distr. Syst. **8**(5), 502–520 (1997)
9. Hawrysz, D.J., Sevick-Muraca, E.M.: Development towards diagnostic breast cancer imaging using near - infrared optical measurements and fluorescent contrast agents. Nat. Neoplasia **2**(5), 388–417 (2000)
10. Iven, J., Gargi, V., Kodand, D., Uday, D.: Non-inverse imaging of breast cancer: synthesis and study of novel near infrared fluorescent estrogen conjugate. Proc. SPIE **5693**, 521–527 (2005)
11. Jose, I., Dinakar, K., Desal, U.B., Bhattacharjee, S.: Early detection of breast cancer: synthesis and characterization of Novel target specific NIR-fluorescent easterogen conjugate for molecular optical imaging. J. Fluoresc. 795–781 (2011)
12. Ramanujan, N.: Fluorescent spectroscopy of neoplastic and non-neoplastic tissues. Nat. Neoplasia **2**(1–2), 89–117 (2000)
13. Schulz, R.B., Peter, J., Semmler, W.: Independent model of fluorescent excitation and emission with the finite element method. In: Proc. OSA, Biomedical Topical Meeting (2004)
14. Song, X., Yi, J., Bai, J.: A parallel reconstruction scheme in flourescence tomography based on contrast of independent inversed absorption properties. Int. J. Biomed. Imag. Article ID 70839, 1–7 (2006)
15. Zhu, W., Wang, Y., Deng, Y., Yao, Y., Barbour, R.L.: A wavelet-based multiresolution regularized least squares reconstruction approach for optical tomography. IEEE Med. Imag. **16**(2), 210–217 (1997)
16. Zou, W., Wang, J., Wu, K., Dagan Feng, D.: A wavelet-based mutiresolution reconstruction method for fluorescent molecular tomography. Int. J. Biomed. Imag. Article ID 294545, 1–11 (2009)

Implementation of Wavelet Based and Discrete Cosine Based Algorithm on Panchromatic Image

Jyoti Sarup, Jyoti Bharti, and Arpita Baronia

Abstract In the past few years, there has been a tremendous increase in the need for the amount of information stored in the form of images especially from Remote Sensing Satellites. Recently there has been a exponentially conversion of conventional analog images to digital images. The volume of digitized image being very high will considerably slow down the transmission and storage of such images. Therefore there is strong need of compression of the images by extracting the visible elements which are encoded and transmitted. This paper compares different image compression techniques such as JPEG (Joint Picture Expert Group), JPEG2000 (Joint Picture Expert Group-2000), and SPIHT (Set Partitioning in Hierarchical Tree) using a set of objective picture quality measures like Peak Signal to Noise Ratio (PSNR) and Mean Square Error (MSE) have been used to measure the picture quality and comparison has been done based upon the results of these quality measures. Standard test images were assessed with different compression ratios. It is found that the JPEG2000 based compression has achieved better results as compared to SPIHT and JPEG for all compressions and images were produced showing better image quality.

Keywords Image compression • JPEG • JPEG2000 • Lossless compression • Lossy compression • Remote sensing • SPIHT

J. Sarup (✉) • J. Bharti • A. Baronia
Civil, Computer Science, Remote Sensing and GIS, Maulana Azad National Institute of Technology, Bhopal, MP 462051, India
e-mail: jyoti.sarup@gmail.com; jyoti2202@gmail.com; arpita.baronia@gmail.com

C. Bandt et al. (eds.), *Fractals, Wavelets, and their Applications*, Springer Proceedings in Mathematics & Statistics 92, DOI 10.1007/978-3-319-08105-2_32,

1 Introduction

Image compression can be of two types (1) Lossless and (2) Lossy compressions. With lossless compression, every single bit of data that was originally in the image remains after the image is decompressed. On the other hand, lossy compression reduces an image by permanently eliminating certain information, especially redundant information [5]. In Still images there are many efficient compression techniques. Mainly they are JPEG [11] which is based on discrete cosine transform, SPIHT [8] and JPEG 2000 [10] which is based on discrete wavelet transform.

JPEG is a popular and continuous tone still image compression mechanism established by first Joint Photographic expert Group in 1992. JPEG is based on Discrete Cosine Transformation of encoder and decoder both. It is a block based technique where the original image is divided into small n×n (usually 8×8) blocks and then DCT transformation is applied. The data compression is achieved via quantization followed by Huffman coding. The disadvantage of JPEG is the blocking artifacts in reconstructed image [7].

Unlike the case of DCT is composed on cosine functions here as DWT can be composed on function (wavelet) which satisfies the multi resolutions. The choice of Wavelet depends on contents and resolution of image in recent time, much of the research activities in image coding have been focused on the Discrete Wavelet Transform (DWT). DWT offers adaptive spatial-frequency resolution (better spatial resolution at high frequencies and better frequency resolution at low frequencies) that is well suited to the properties of Human Visual System (HVS). It can provide better image quality than DCT, especially at higher compression ratio [4].

The SPIHT algorithm was introduced by [8]. It is a powerful, efficient and computationally simple image compression algorithm. By using this algorithm, the highest PSNR values for given compression ratios for a variety of images can be obtained. SPIHT was designed for optimal progressive transmission, as well as for compression. One of the important features of SPIHT is that at any point during the decoding of an image, the quality of the displayed image is the best that can be achieved for the number of bits input by the decoder up to that moment. The wavelet coefficients can be referred as ci,j. The main aim in progressive transmission is to transmit the most important image information at first priority [6].

JPEG 2000 is based on the idea that the coefficients of a transform that decorrelates the pixels of an image can be coded more efficiently than the original pixel themselves. If the transform basis function is wavelet then JPEG 2000 pack most of the visual information into a small number of coefficients, the remaining coefficients can be quantized coarsely or truncated to zero with little image distortion [2].

The following steps are followed for comparisons of compression techniques:

1. Select the images (like PANCHROMATIC image).
2. Apply the compression techniques (SPHIT, JPEG and JPEG-2000) on these images.

3. Evaluate the quality and impact of different technique on image interpretability, finally, a quantitative evaluation of compressed images in order to estimate the MSE and the PSNR comparisons with the original images.

2 Indentation and Equation

The quality measure for an image is evaluated by MSE and PSNR shown in Eqs. (1) and (2).

Mean Square Error (MSE): The mean square error measures the error with respect to the center of the image values, i.e. the mean of the pixel values of the image, and by averaging the sum of squares of the error between the two images.

$$MSE = \frac{1}{mn} \sum_{y=1}^{m} \sum_{x=1}^{n} [I(x, y - I(x, y)]^2 \tag{1}$$

where I(x,y) is the original image, I(x,y) is the approximated version (which is actually the decompressed image), and M,N are the dimensions of the images. A lower value of MSE signifies lesser error in the reconstructed image [1]. Peak signal-to-noise Ratio (PSNR): The peak signal-to-noise ratio (PSNR) measures the estimates of the quality of reconstructed image compared with the original image and is a standard way to measure image fidelity. Here signal corresponds to the original image and noise corresponds to the error in reconstructed image due to compression and decompression. The PSNR is a single number that reflects the quality of the reconstructed image and is measured in decibels (db) [9].

$$PSNR = 20 * log10(\frac{S}{sqrt(MSE)}) \tag{2}$$

where $PSNR$ is the Peak Signal to Noise Ratio,

S is the maximum pixel value, and RMSE is the root mean square error of the image. The actual value of the PSNR is not meaningful but the comparison between two values between different reconstructed images gives one measure of quality. As seen from inverse relation between the MSE and PSNR, a low value of MSE/RMSE translates to a higher value of PSNR, thereby signifying that a higher value of PSNR indicates higher reconstruction fidelity [3].

Table 1 Mean Square Error

Compression ratio	SPIHT	JPEG2000	JPEG
2:1	0.70723	0.34938	6.25778
10:1	1.5086	1.1506	16.0592
12:1	0.89324	0.50849	6.56521
14:1	0.66869	0.29009	6.36022
20:1	0.70716	0.37423	6.36022

Table 2 Peak Signal to Noise Ratio

Compression Ratio	SPIHT	JPEG2000	JPEG
2:1	12.354	13.5964	11.1657
10:1	11.2309	12.3603	10.1506
12:1	12.354	13.5964	11.1654
14:1	12.354	13.5964	11.1654
20:1	12.354	13.5964	11.1654

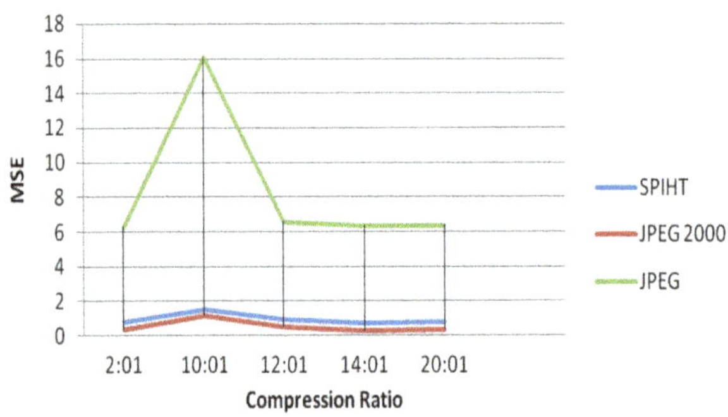

Fig. 1 MSE between original and decompressed images

3 Experimental Result and Analysis

Experiments are conducted on the test image (PANCHROMATIC) using MATLAB platform, coded with JPEG, JPEG 2000, SPIHT image compression coder for each test image with 5 different compression ratios (CR) as 2:1, 10:1, 12:1,14:1,20:1.

Table 1 shows the MSE of three algorithms between original and decompressed image, Table 2 Shows PSNR value which is always be greater than the MSE for good result. Figures 1 and 2 show the graphical representation of tables. Figure 3 shows the original image in Tiff Format. Figures 4, 5, and 6 show different decompressed images at compression ratio 12:1.

The result consists of comparison between three compression technique methods on the basis of calculation of MSE and PSNR of original image and decompressed image.

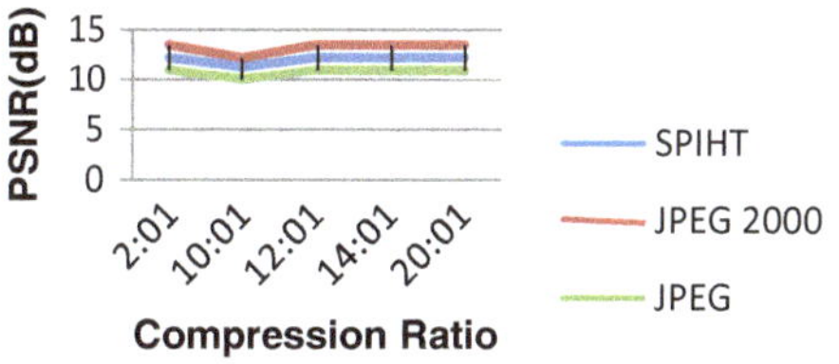

Fig. 2 PSNR between original and decompressed image

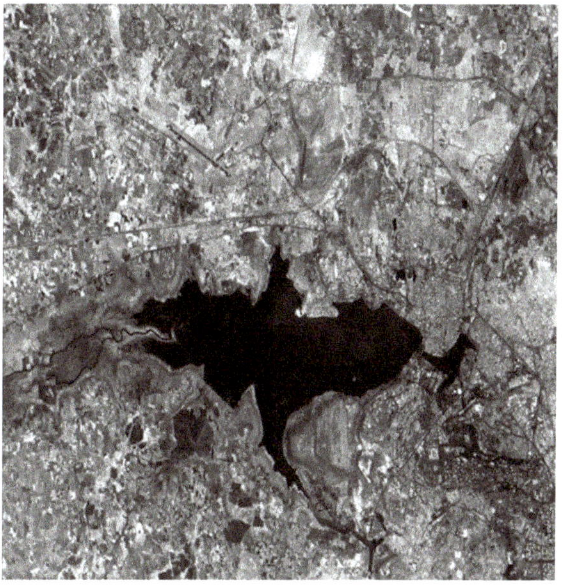

Fig. 3 Original image

4 Conclusion

Based on the limited testing results obtained in this study due to time constraints, it is to be concluded that there could be a decrease in image quality with compression ratio increase. JPEG2000 has better performance than SPIHT and JPEG. JPEG has poor performance than all the compression methods because all other methods are Wavelet based. Wavelet-based compression provides substantial improvement in picture quality because of overlapping basis functions and better energy compaction property of wavelet transforms that makes images smoother and preserves object edges, while DCT-based JPEG creates blocking artifacts. JPEG2000 is better than SPIHT compression technique. These techniques are scene dependent and for this study area JPEG2000 performs better than the SPIHT.

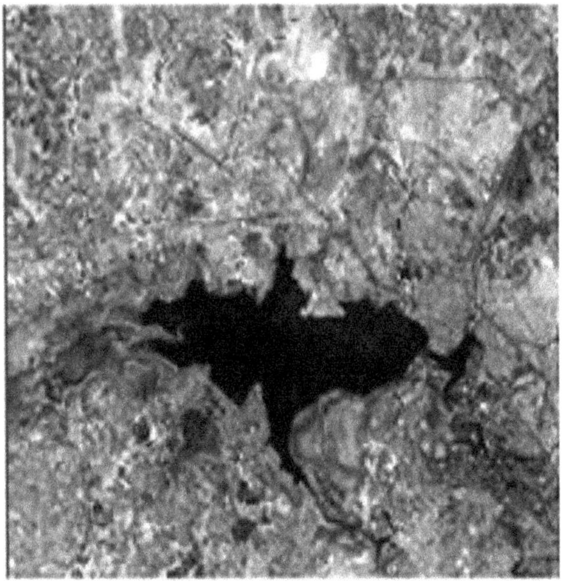

Fig. 4 JPEG2000 (CR12:1)

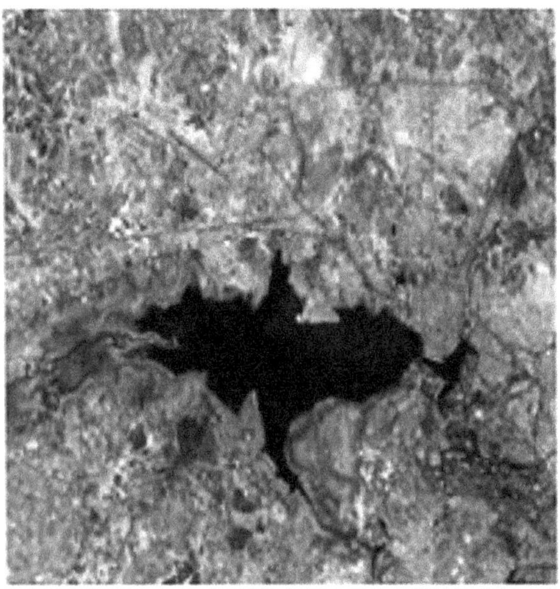

Fig. 5 SPIHT (CR 12:1)

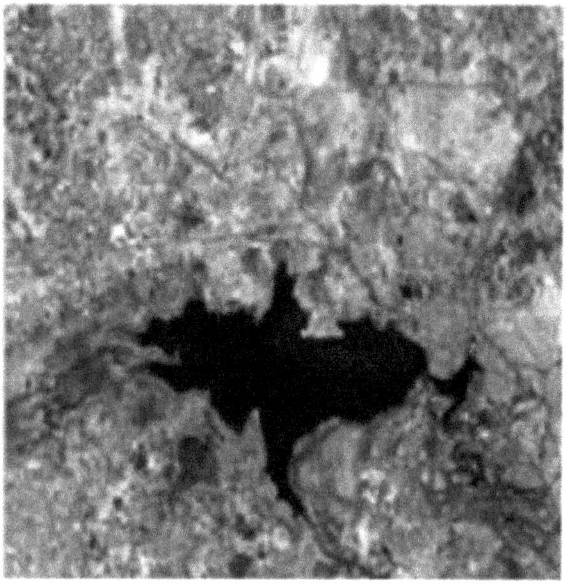

Fig. 6 JPEG (CR 12:1)

References

1. Avcbas, I., Sankur, B., Sayood, K.: Statistical evaluation of image quality measures. J. Electron. Imag. **11**(2), 206–223 (2002)
2. Christopoulos, C., Skodras, A., Ebrahimi, T.: The JPEG2000 still image coding system: an overview. IEEE Trans. Consum. Electron. 46(4), 1103–1127 (2000)
3. Gonzalez, R.C., Woods, R.E., Eddins, S.L.: Digital Image Processing Using Matlab, 2nd edn, p. 374. Tata Mc Graw Hill, New Delhi (2011)
4. Jose, G., Murugananth, N.: Image identification using compression technique. In: International Conference on VLSI. Communication and Instrumentation (ICVCI), pp. 29–31 (2011)
5. Paul, P.J., Koteswari, S., Rani, B.K.: A novel VLSI architecture of SOC for image compression model for multimedia applications. Int. J. Comput. Sci. Tech. **2**(3), 130–134 (2011)
6. Raja, S.P., Narayanan, P.N., Arif Abdul Rahuman, S., Kurshid Jinna, S., Princess, S.P.: Wavelet based image compression: a comparative study. In: International Conference on Advances in Computing, Control, and Telecommunication Technologies, pp. 545–549 (2009)
7. Rani, B., Bansal, R.K., Bansal, S.: Comparison of JPEG and SPIHT image compression algorithms using objective quality measures. In: International Conference on Muitmedia,Signal Processing and Communication Technology, IMPACT'09, pp. 90–93 (2009)
8. Said, A., Pearlman, W.A.: A new fast and efficient image codec based on set partitioning in hierarchical trees. IEEE Trans. Circ. Syst. Video Tech. 243–250 (1996)
9. Santa-Cruz, D., Ebrahimi, T., Askelof, J., Larsson, M., Christopoulos, C.: JPEG 2000 still image coding versus other standards. In: Proceedings of SPIE, SPIEs 45th Annual Meeting, Applications of Digital Image Processing XXIII. San Diego, California, vol. 4115, pp. 446–454 (2000)
10. Skodras, A.N., Christopoulos, C.A., Ebrahimi, T.: JPEG2000: The upcoming still image compression method. Pattern Rcognit. Lett. **22**, 1337–1345 (2001)
11. Wallace, G.K.: The JPEG still picture compression standard. IEEE Trans. Consum. Electron. **38**, xviii–xxxiv (1992)

Trend, Time Series, and Wavelet Analysis of River Water Dynamics

Kulwinder Singh Parmar and Rashmi Bhardwaj

Abstract Time series, trend, wavelet and statistical analysis of water quality parameters Chemical Oxygen Demand (COD), Biochemical Oxygen Demand (BOD), Dissolved Oxygen (DO) monitored for river Yamuna in India have been studied. It is observed that COD is highly correlated with BOD. For all auto regressive integrated moving average model (p,d,q) value of "d," i.e. middle value is zero thus process is stationary. It is also observed that RMSE values are comparatively very low, thus dependent series is closed with the model predicted level. MAPE, MaxAPE, MAE, MaxAE, Normalized BIC are calculated and have low value for all parameters. Trend is calculated by using auto correlation function, partial auto correlation function, and lag. Thus the predictive model is useful at 95 % confidence limits. 1-D discrete and continuous Daubechies Wavelet analysis explains that the parameters COD, BOD, DO have the maximum value 120, 50, 8 and amplitude (a5) varies between 52 to 78, 10 to 30, 0.2 to 1.4, respectively. The scale values of Db5, i.e. d5, d4, d3, d2, and d1 range between −20 and +20 for all parameters. All parameters cross the prescribed limits of WHO/EPA, thus water is not fit for drinking, agriculture, and industrial use.

Keywords Time series • Daubechies wavelet • Trend analysis • ARIMA • ACF • PACF

K.S. Parmar • R. Bhardwaj (✉)
Non-Linear Dynamics Lab, Department of Mathematics, University School of Basic and Applied Sciences, Guru Gobind Singh Indraprastha University, New Delhi 110078, India
e-mail: rashmib22@gmail.com

C. Bandt et al. (eds.), *Fractals, Wavelets, and their Applications*, Springer Proceedings in Mathematics & Statistics 92, DOI 10.1007/978-3-319-08105-2_33,

1 Introduction

Water is a base of life on earth and rivers are the most important resources of water. Adequate supply of safe water is essential for maintaining health and sanitary conditions. Water as a natural resource has influenced almost every aspect of development. Natural water always contains dissolved and suspended substances of organic and mineral origin. The river water draws attention of Government, Public, NGO's, and Environmentalists in India and world over. Domestic, industrial, and agricultural wastes pollute the river water.

Natural water always contains dissolved and suspended substances of organic and mineral origin [24]. Yamuna is the largest tributary river of the Ganga in northern India. It originates from Yamunotri glacier at a height of 6,387 m on southwestern slopes of Banderpooch peaks (38° 59′ N 78° 27′ E) in the lower Himalayas in Uttarakhand. It travels a total length of 1,376 km by crossing several states, Uttarakhand, Haryana, Himachal Pradesh, Delhi, Uttar Pradesh and has a mixing of drainage system of 366,233 km^2 before merging with Ganga at Allahabad, i.e. a total of 40.2 % of entire Ganga basin. Yamuna river accounts for more than 70 % of Delhi's water supplies and about 57 million people depend on river water for their daily usage (CPCB [7]). Nizamuddin is approximately 14 km downstream from Wazirabad barrage at Delhi and 410 km from Yamunotri. Pollution in river water is continuously increasing due to urbanization, industrialization, population growth etc. Many rivers are dying due to pollution which is an alarming signal. The water quality at Nizamuddin (Delhi) has the impact of industrial, sewerage, domestic discharge from Haryana and Delhi [3, 11, 13, 14, 17, 23].

An industrial, domestic and sewerage discharge mixes with river water and affects the quality of water. Treatment of domestic wastewater using laboratory scale Hybrid Upflow Anaerobic Sludge Blanket (HUASB) reactor reduced treatment cost significantly [1]. The oxidation treatment system constructed under a riverbed of Nan-men Stream located at Shin Chu City of Taiwan modeled such that it has significant efficiency [15]. Trihalomethane compounds were determined in the drinking water samples that were collected from the selected consumption sites and treatment plants of both Okinawa and Samoa Islands and observed that the Chloroform, Bromodichloromethane compound exceeded the level of Japan water quality and WHO standards [12, 30]. Water quality of watersheds is studied using hydrochemical data that mingle multiple linear regression and structural equation modeling [6]. Regression equations can be used to estimate constituent concentrations. Constituent concentrations can be used by water-quality managers for comparison of current water-quality conditions to water-quality standards. Examination of stream flow and physical properties of water that act as surrogates for constituents of interest also helps for collection of water-quality samples [18, 25, 29, 31].

Wavelets are considered as a special characteristic to study the abnormality, singularity, chaos, turbulence, fractal in the study of financial data, meteorological data, bioinformatics data such as DNA sequence, protein structure, micro array and medical data such as ECG data, mammogram data, etc. In recent centuries, to maintain quality of water is the main area of research both theoretical and experimental. Wavelet and wavelet based multifractal formalism have become very popular for analyzing meteorological data and related problems in different parts of the world for analysis of monthly and daily temperatures, wind speed simulations, monsoon variability, surface air temperature, atmospheric structure, and weather phenomena [5, 10]. The climatic dynamic has been studied using fractal dimensional, wavelet analysis and analyzed time series data of three major dynamic components of the climate, i.e. temperature, pressure, and precipitation. Fractal dimension and predictability analysis is used to predict the behavior of water quality parameters [2]. It has been observed that regional climatic models would not be able to predict local climate as it deals, with averaged quantities and that precipitation during the south-west monsoon is affected by temperature and pressure variability during the preceding winter [26]. Time series can be modeled by a stochastic process possessing long range correlation [16, 22, 27, 28].

In this paper, time series analysis, trend analysis, Daubechies wavelet analysis of water parameters have been estimated at Nizamuddin bridge-mid stream (Delhi) of Yamuna River in India. The river map is shown in Fig. 1.

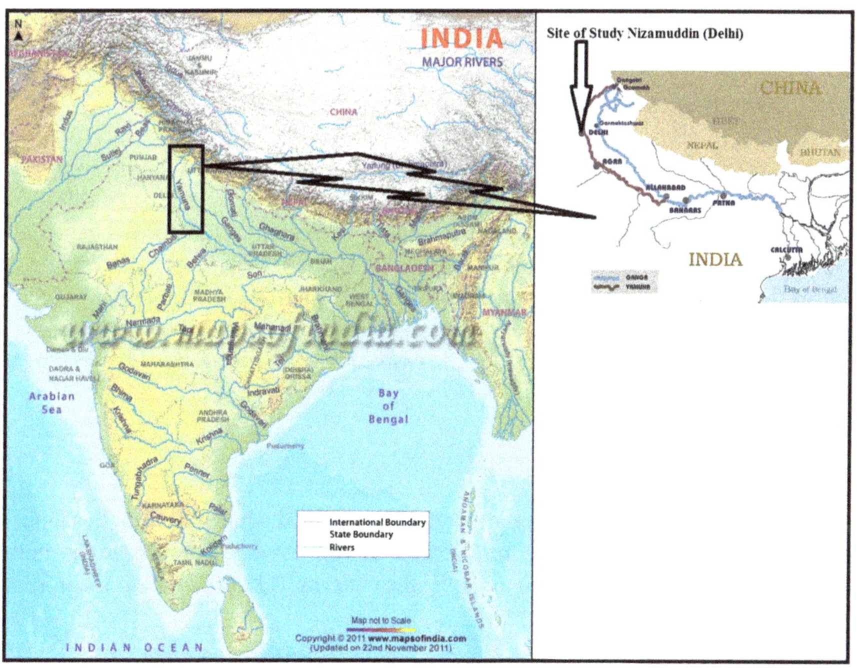

Fig. 1 Map of sample site

2 Methodology

The monthly average value of last 10 years of water quality parameters Chemical Oxygen Demand (COD), Biochemical Oxygen Demand (BOD), Dissolved Oxygen (DO) monitored at Nizamuddin bridge-mid Stream of Yamuna River in Delhi (India) has been considered for the present study.

2.1 Time Series

Time series is a sequence of data points, measured typically at successive times spaced at uniform time intervals. Time series analysis comprises methods for analyzing time series data in order to extract meaningful statistics and other characteristics of data and to forecast future events based on known past events to predict data points before these are measured. Time series model reflects that observations close together in time one closely related than observations further apart. In addition, time series models will often make use of natural one-way ordering of time so that values for a given period will be expressed as deriving in some way from past values, rather than from future values (Fig. 2).

2.1.1 Auto Regressive Integrated Moving Average (ARIMA)

ARIMA model of a time series is defined by three terms (p, d, q). Identification of a time series is process of finding integer, usually very small (e.g., 0, 1, or 2), values of p, d, and q model patterns in data. When value is 0, element is not needed in model. The middle element, d, is investigated before p and q. The goal is to determine if process is stationary and, if not, to make it stationary before determining the values of p and q. A stationary process has a constant mean and variance over time period of study. The representation of an autoregressive model in time series [4,9,21], well known as AR(p), is defined as

$$Y_t = \alpha_0 + \alpha_1 Y_{t-1} + \alpha_2 Y_{t-2} + \dots\dots\dots + \alpha_p Y_{p-1} + \varepsilon_t \tag{1}$$

where the last term is source of randomness and is called white noise, are constants. A series may have both auto-regressive and moving average components so both types of correlations are required to model the patterns. If both elements are present only at lag 1, the equation is

$$Y_t = \phi_1 Y_{t-1} - \theta_1 a_{t-1} + a_t \tag{2}$$

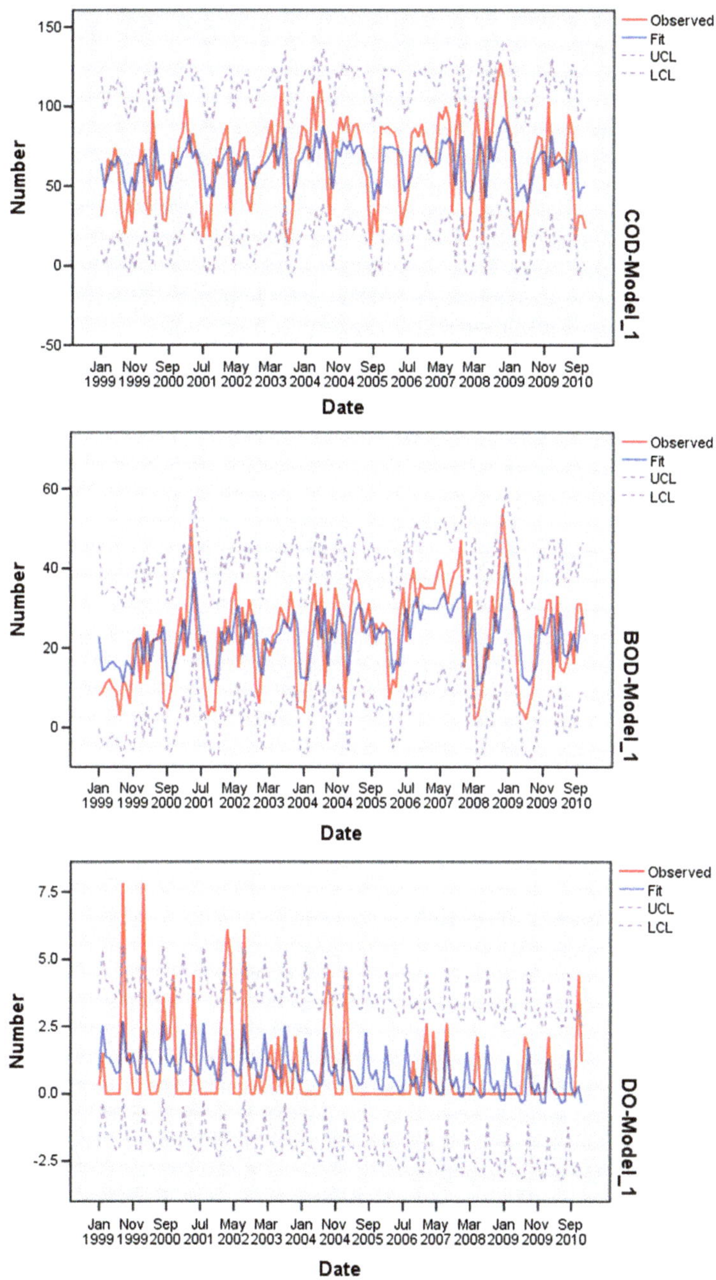

Fig. 2 Trend and time series analysis of water quality parameters

2.2 Auto Correlation Functions (ACF) and Partial Auto Correlation Functions (PACF)

Models are identified through patterns in their ACFs and PACFs. Both autocorrelations and partial autocorrelations are computed for sequential lags in a series. The first lag has an autocorrelation between Y_{t-1} and Y_t, the second lag has both an autocorrelation and partial autocorrelation between Y_{t-2} and Y_{t-1}, and so on. ACFs and PACFs are the functions across all lags. If plotted residuals are greater than standard errors and away from a zero mean, then it indicates statistically significant autocorrelation. Autocorrelation is defined as

$$r_k = \frac{\frac{1}{n-k}\sum_{t=1}^{n-k}\left(Y_t - \overline{Y}\right)\left(Y_{t-k} - \overline{Y}\right)}{\frac{1}{n-1}\sum_{t=1}^{n}\left(Y_t - \overline{Y}\right)^2} \tag{3}$$

where N is number of observations in a whole series, k is lag. $\overline{Y}$ is mean of whole series. At lag 1, there are no previous autocorrelations, so r_0^2 is set to be 0. Standard error of an autocorrelation depends on squared autocorrelations of previous lags and is defined as

$$SE_{r_k} = \sqrt{\frac{1 + 2\sum_{l=0}^{k-1} r_l^2}{N}} \tag{4}$$

Standard error for a partial autocorrelation is same at all lags.

$$SE_{pr} = \frac{1}{\sqrt{N}} \tag{5}$$

The relation between ACF and PACF for first three lags is given as [21]:

$$PACF\,(1) = ACF\,(1) \tag{6}$$

$$PACF\,(2) = \frac{ACF\,(2) - (ACF\,(1))^2}{1 - (ACF\,(1))^2} \tag{7}$$

If an autocorrelation at some lag is significantly different from zero, the correlation is included in ARIMA model. Similarly, if a partial autocorrelation at some lag is significantly different from zero, it is also included in ARIMA model. The significance of full and partial autocorrelations is assessed using their standard errors. The boundary lines around functions are 95 % confidence bounds. For large pattern, negative autocorrelation at lag 1 and a decaying PACF, an ARIMA (0, 0, 1) model is useful.

In order to detect seasonality, plot the autocorrelation function (ACF) by calculating and graphing the residuals. The graph of the residual against a specified time interval is called a lagged autocorrelation function. In time series analysis, lag k is defined as difference of an event occurring at time t + k (k $\succ$ 0) to an event occurring at time t. The partial autocorrelation function (PACF) is also used to detect trends and seasonality. PACF is the amount of correlation between a variable and its lag that is not explained by correlations at all lower-order lags [4,9,21].

2.3 Wavelet Analysis

Wavelet Transform (T_ψ) decomposes a signal into several groups (vectors) of coefficients. Different coefficients vectors contain information about characteristics of the sequence at different scales. The mother wavelet ψ satisfying the conditions (Fig. 3):

$$\psi_{a,b}(t) = a^{-\frac{1}{2}}\psi\left(\frac{t-b}{a}\right) \; where \quad a \succ 0 \tag{8}$$

where $T_\psi f\,(a,b)$ is called the wavelet transform of $f\,(t)$.

$$T_\psi f\,(a,b) = a^{-\frac{1}{2}}\int_{-\infty}^{+\infty} f\,(t)\,\psi\left(\frac{t-b}{a}\right)dt = \langle f, \psi_{a,b}\rangle \tag{9}$$

It may be observed that wavelet transform is a prism which exhibits properties of signal such as points of abrupt changes, seasonality, or periodicity. The wavelet transform is a function of "a" and "b" where a is the scale of frequency, b is spatial position (translation) or time. Discrete data is obtained in the form of signal. Let "S" represent the raw signal and an represent the corresponding amplitude at different levels for n $= 1, 2, 3, \ldots\ldots.., d_1, d_2, d_3, d_4, \ldots$ represent the detail of signal at different levels.

Let $f = (f_1, f_2, f_3, \ldots., f_N)$ N is an even integer. $a_m = \frac{f_{2m-1}+f_{2m}}{\sqrt{2}}$, m = 1, 2, 3, …, N/2. The first trend sub-signal (approximation) $d_m = \frac{f_{2m-1}-f_{2m}}{\sqrt{2}}$, m = 1, 2, 3, …, N/2. These wavelets are defined as $W_1^1 = \left[\frac{1}{\sqrt{2}}, -\frac{1}{\sqrt{2}}, 0, 0 \ldots\ldots., 0\right]$, $W_2^1 = \left[0, 0, \frac{1}{\sqrt{2}}, -\frac{1}{\sqrt{2}}, 0, \ldots., 0\right]$, and so on. $W_{\frac{N}{2}}^1 = \left[0, 0, 0, \ldots., 0, \frac{1}{\sqrt{2}}, -\frac{1}{\sqrt{2}}\right]$. Also $d_1 = f.W_1^1$, $d_2 = f.W_2^1$, $d_3 = f.W_3^1 \ldots$ so on $d_m = f.W_m^1$ for m = 1, 2, 3, …, N/2.

The plane defined by variables (a, b) is called scale-space or time frequency plane. Wavelet transform $T_\psi f\,(a,b)$ measures variation of f in neighborhood of b. For a compactly supported wavelet (for a wavelet vanishing outside a closed and bounded interval), value of $T_\psi f\,(a,b)$ depends upon value of f in a neighborhood of b of size proportional to scale a. At small scales, $T_\psi f\,(a,b)$ provides localized information such as localized regularity (smoothness) of f. The local regularity of a function (or signal) is often measured with Lipschitz exponent. The global and

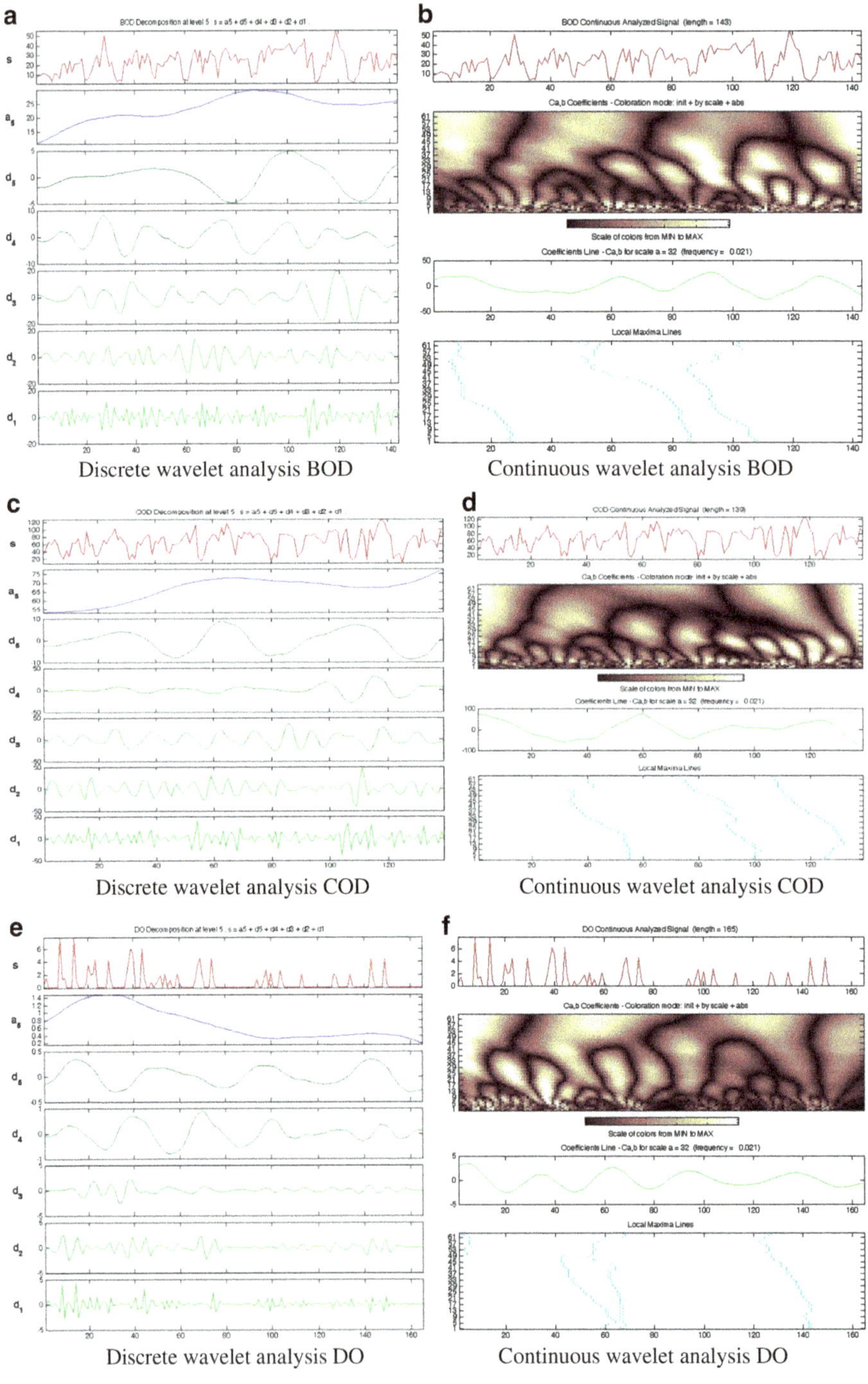

Fig. 3 Discrete and continuous wavelet analysis of water quality parameters

local Lipschitz regularity can be characterized by asymptomatic decay of wavelet transformation at small scales [8, 19, 20]. 1D discrete wavelet is defined as:

$$\psi_{j,k}(t) = 2^{\frac{j}{2}} \psi \left(2^j t - k\right) \tag{10}$$

where j,k are integers.

$$d_{j,k}(t) = \int_{-\infty}^{\infty} f(t)\, \psi_{j,k}(t)\, dt \tag{11}$$

are called the wavelet coefficients of $f(t)$ and $\sum_{j,k} d_{j,k}(t)\, \psi_{j,k}(t)$ is called the wavelet series of f with every orthonormal wavelet.

3 Results and Discussion

3.1 Chemical Oxygen Demand

Stationary R-squared and R-squared values exhibit the similar behavior, thus model is better than the baseline model and RMSE values are high, so dependent series is not closed with its model-predicted level. Using Ljung-Box Q(18) model statistics is 19.035, significance is 0.327, and degree of freedom is 17. ARIMA (1,0,0) fitted and boundary lines at 95 % confidence limits. Predicted, LCL, UCL, and residual value of COD are 64.793, 19.273, 110.314, and 0.2645. 1D discrete wavelet analysis shows that COD has maximum value as 120. The value of amplitude (a_5) and scales at different levels d_5, d_4, d_3, d_2, and d_1 varies between 52 to 78; −10 to +10; −50 to +50; −50 to +50; −50 to +50; −50 to +50, respectively. Using continuous wavelet analysis, it is observed that local maximum exists in May 2004, March 2008, and May 2009 with maximum value of 116, 120, and 103, respectively.

3.2 Biochemical Oxygen Demand

Stationary R-squared and R-squared values exhibit the similar behavior, thus model is better than the baseline model and RMSE values are low, so dependent series is closed with its model-predicted level. Using Ljung-Box Q(18) model statistics is 30.005, significance is 0.026, and degree of freedom is 17. ARIMA (1,0,0) fitted and boundary lines at 95 % confidence limits. Predicted, LCL, UCL, and residual value of BOD are 21.379, 3.534, 39.224, and 0.178. 1D discrete wavelet analysis shows that BOD has maximum value as 50. The value of amplitude (a_5) and scales at different levels d_5, d_4, d_3, d_2, and d_1 varies between 10 to 30; −5 to +5; −10 to +10; −20 to +20; −20 to +20; −20 to +20, respectively. Using continuous wavelet analysis, it is observed that local maximum exists in Feb 2001, Dec 2005, and May 2009 with maximum value of 50, 33.3, and 32.6, respectively.

3.3 *Dissolved Oxygen*

Stationary R-squared and R-squared values exhibit the similar behavior, thus model is better than the baseline model and RMSE values are low, so dependent series is closed with its model-predicted level. Using Ljung-Box Q(18) model statistics is 22.178, significance is 0.138, and degree of freedom is 16. ARIMA (0,0,6) fitted and boundary lines at 95 % confidence limits. Predicted, LCL, UCL, and residual value of DO are 0.6424, −1.9207, 3.2123, and 0.0025. 1D discrete wavelet analysis shows that DO has maximum value as 8. The value of amplitude (a_5) and scales at different levels d_5, d_4, d_3, d_2, and d_1 varies between 0.2 to 1.4; −0.5 to +0.5; −1 to +1; −5 to +5; −5 to +5; −5 to +5, respectively. Using continuous wavelet analysis, it is observed that local maximum exists in Aug 1999, Sep 2011, and Aug 2008 with maximum value of 7.8, 6.1, and 4.4, respectively.

4 Conclusion

1D Daubechies wavelet, time series, and correlation analysis of water quality parameters monitored at the Nizamuddin bridge-mid Stream of Yamuna River in Delhi (India) have been studied. It is observed that the parameters COD–BOD are highly positive correlated and DO–COD; DO–BOD are highly negative correlated.

For all ARIMA model (p,d,q) value of "d," i.e. middle value is zero, thus process is stationary and has constant mean and variance. It is also observed that RMSE values are comparatively very low which show that dependent series is closed with the model-predicted level, thus predictive model is useful at 95 % confidence limits. MAPE, MaxAPE, MAE, MaxAE, normalized BIC are calculated for all parameters and it is observed that all water quality parameters have low value. It concludes that the predicted series is close to the original series, thus it is a perfect fit. Predicted, LCL–UCL values using time series are given as for COD 64.793, 19.273–110.314; for BOD 21.379, 3.534–39.224; for DO 0.6424, −1.9207–3.2123, respectively.

1-D Discrete Daubechies Wavelet analysis explains that the parameters pH, COD, BOD, DO have the maximum value 120, 50, 8 and amplitude (a_5) varies between 52–78, 10–30, 0.2–1.4, respectively. The scale values of Db5, i.e. d_5, d_4, d_3, d_2, and d_1 range between −20 to +20 for all parameters. Using continuous wavelet analysis, it is observed that local maxima exists for COD, BOD, DO in March 2008; Feb 2001; Aug 1999, respectively. It is observed that during summer session the values of COD and DO are at rise due to high rate of evaporation, thus water has become concentrated and contaminants increase.

Therefore, using time series, wavelet and correlation analysis, it is concluded that all parameters cross the prescribed limits of WHO/EPA and water is not fit for drinking, agriculture, and industrial use. River is a natural resource of water, thus the increase of pollution is an alarming situation and preventive measure has to be taken to control the same.

Acknowledgements Authors are thankful to University Grant Commission (UGC), Government of India for financial support (F. 41-803/2012 (SR)); Central Pollution Control Board (CPCB), Government of India for providing the research data; Guru Gobind Singh Indraprastha University, Delhi (India) for providing research facilities. First author is thankful to Sant Baba Bhag Singh Institute of Engineering and Technology for providing study leave to pursue research degree.

References

1. Banu, J.R., Kaliappan, S., Yeom, I.T.: Treatment of domestic wastewater using upflow anaerobic sludge blanket reactor. Int. J. Environ. Sci. Technol. **4**, 363–370 (2007)
2. Bhardwaj, R., Parmar, K.S.: Water quality index and fractal dimension analysis of water parameters. Int. J. Environ. Sci. Technol. (2012). doi:10.1007/s13762-012-0086-y
3. Bhatnagar, A., Vilar, V.J.P., Botelho, C.M.S., Boaventura, R.A.R.: A review of the use of red mud as adsorbent for the removal of toxic pollutants from water and wastewater. Environ. Technol. **32**, 231–249 (2011)
4. Box, G.E.P., Jenkins, G.M., Reinsel, G.C.: Time Series Analysis: Forecasting and Control, 4th edn. Wiley, London (2008)
5. Can, Z., Aslan, Z., Oguz, O., Siddiqi, A.H.: Wavelet transform of meteorological parameter and gravity waves. Ann. Geophys. **23**, 659–663 (2005)
6. Chenini, I., Khemiri, S.: Evaluation of ground water quality using multiple linear regression and structural equation modeling. Int. J. Environ. Sci. Technol. **6**, 509–519 (2009)
7. CPCB, Water Quality Status of Yamuna River (1999–2005): Central Pollution Control Board, Ministry of Environment & Forests, Assessment and Development of River Basin Series: ADSORBS/41/2006–07 (2006)
8. Daubechies, I.: Ten Lectures on Wavelets. SIAM, Philadelphia (1992)
9. DeLurgio, S.A.: Forecasting Principles and Applications, 1st edn. Irwin McGraw-Hill, New York (1998)
10. Grapes, A.: An introduction to wavelets. IEEE Comput. Sci. Eng. Signal Image Process. **2**, 50–61 (1995)
11. Hur, J., Lee, T.H., Lee, B.M.: Estimating the removal efficiency of refractory dissolved organic matter in wastewater treatment plants using a fluorescence technique. Environ. Technol. **32**, 1843–1850 (2011)
12. Imo, T.S., Oomori, T., Toshihiko, M., Tamaki, F.: The comparative study of trihalomethanes in drinking waters. Int. J. Environ. Sci. Technol. **4**, 421–426 (2007)
13. Jain, P., Sharma, J.D., Sohu, D., Sharma, P.: Chemical analysis of drinking water of villages of Sanganer Tehsil, Jaipur District. Int. J. Environ. Sci. Technol. **2**, 373–379 (2005)
14. Ji, M.K., Ahn, Y.T., Khan, M.A., Shanab, R.A.I.A., Cho, Y., Choi, J.Y., Kim, Y.J., Song, H., Jeon, B.H.: Removal of nitrate and ammonium ions from livestock wastewater by hybrid systems composed of zero-valent iron and adsorbents. Environ. Technol. **32**, 1851–1857 (2011)
15. Juang, D.F., Tsai, W.P., Liu, W.K., Lin, J.H.: Treatment of polluted river water by a gravel contact oxidation system constructed under riverbed. Int. J. Environ. Sci. Technol. **5**, 305–314 (2008)
16. Kahya, E., Kalayci, S.: Trend analysis of streamflow in Turkey. J. Hydrol. **289**, 128–144 (2004)

17. Koh, Y.K.K., Chiu, T.Y., Boobis, A., Cartmell, E., Scrimshaw, M.D., Lester, J.N.: Treatment and removal strategies for estrogens from wastewater. Environ. Technol. **29**, 245–267 (2008)
18. Korashey, R.: Using regression analysis to estimate water quality constituents in Bahr El Baqar Drain. J. Appl. Sci. Res. **5**, 1067–1076 (2009)
19. Mallat, S.: A Wavelet Tour of Signal Processing, 2nd edn. Academic, San Diego (2001)
20. Martin, I., Pidou, M., Soares, A., Judd, S., Jefferson, B.: Modelling the energy demands of aerobic and anaerobic membrane bioreactors for wastewater treatment. Environ. Technol. **32**, 921–932 (2011)
21. McCleary, R., Hay, R.A.: Applied Time Series Analysis for the Social Sciences. Sage, Beverly Hills (1980)
22. Mousavi, M., Kiani, S., Lotfi, S., Naeemi, N., Honarmand, M.: Transient and spatial modeling and simulation of polybrominated diphenyl ethers reaction and transport in air, water and soil. Int. J. Environ. Sci. Technol. **5**, 323–330 (2008)
23. Parmar, K.S., Chugh, P., Minhas, P., Sahota, H.S.: Alarming pollution levels in rivers of Punjab. Indian J. Environ. Prot. **29**, 953–959 (2009)
24. Prasad, B.G., Narayana, T.S.: Subsurface water quality of different sampling stations with some selected parameters at Machilipatnam Town. Nat. Environ. Pollut. Technol. **3**, 47–50 (2004)
25. Psargaonkar, A., Gupta, A., Devotta, S.: Multivariate analysis of ground water resources in Ganga- Yamuna Basin (India). J. Environ. Sci. Eng. **50**, 215–222 (2008)
26. Rangarajan, G.: A climate predictability index and its applications. Geophys. Res. Lett. **24**, 1239–1242 (1997)
27. Rangarajan, G., Ding, M.: Integrated approach to the assessment of long range correlation in time series data. Phys. Rev. E **61**, 4991–5001 (2000)
28. Rangarajan, G., Sant, D.A.: Fractal dimensional analysis of Indian climatic dynamics. Chaos Solitons Fractals **19**, 285–291 (2004)
29. Vassilis, Z., Antonopoulos, M., Mitsiou, A.K.: Statistical and trend analysis of water quality and quantity data for the Strymon River in Greece. Hydrol. Earth Syst. Sci. **5**, 679–691 (2001)
30. WHO: International Standards for Drinking Water. World Health Organization, Geneva (1971)
31. Yeon, I.S., Jun, K.W., Lee, H.J.: The improvement of total organic carbon forecasting using neural networks discharge model. Environ. Technol. **30**, 45–51 (2009)

An Efficient Wavelet Based Approximation Method to Film-Pore Diffusion Model Arising in Chemical Engineering

Pandy Pirabaharan, R. David Chandrakumar, and G. Hariharan

Abstract In this paper, we have established an efficient wavelet based approximation method to solve film-pore diffusion model (FPDM) arising in engineering. Film pore diffusion model is widely used to determine study the kinetics of adsorption systems. The use of wavelet based approximation method is found to be accurate, simple, fast, flexible, convenient, and computationally attractive. The present paper focus that FPDM satisfactorily describes the kinetics of methylene blue adsorption onto the three low-cost adsorbents, guava, teak, and gulmohar plant leaf powders used in this study.

Keywords Methylene blue • Adsorption kinetics • Film-pore diffusion model • Low-cost adsorbents • Legendre wavelet method

---

P. Pirabaharan (✉)
Department of Mathematics, Anna University, University College of Engineering-Dindigul, Dindigul 624 622, Tamil Nadu, India
e-mail: ppirabharan@gmail.com

R.D. Chandrakumar
Department of Mathematics, Vikram College of Engineering, Enathi, Sivaganga District, Tamil Nadu, India
e-mail: dckumarmaths@gmail.com

G. Hariharan
Department of Mathematics, School of Humanities and Sciences, SASTRA University, Thanjavur 613 401, Tamil Nadu, India
e-mail: hariharan@maths.sastra.edu

C. Bandt et al. (eds.), *Fractals, Wavelets, and their Applications*, Springer Proceedings in Mathematics & Statistics 92, DOI 10.1007/978-3-319-08105-2_34,

1 Introduction

In recent years, adsorption mechanism has been established to be one of the highly efficient methods for removal of colors, odors, and organic and inorganic pollutants emanating from various industrial processes. Large amounts of dyes are used by textile industry and a significant portion of these dyes is not consumed in the process and therefore let out with the effluent. As the cost of commercial adsorbents is too high, interest for using low-cost adsorbents for removal of dyes from textile effluents is continuously growing. A recent survey indicates that, in India, on an average freshwater consumed and effluent generated per kg of finished textile are 175 L and 125 L, respectively [16]. The presence of dyes in aqueous effluents is highly objectionable as this affects the photosynthetic activity in receiving water body by reducing/preventing light penetration. As the dyes are recalcitrant in nature it is difficult to treat them in conventional biological treatment plant [14, 15]. Therefore, identification of low-cost adsorbents is given more attention by the researchers recently as commercial adsorbents like activated carbon are too costly. Few recent studies investigating application of low-cost adsorbents are: jackfruit peel [2], pineapple stem [5], phoenix tree leaves [6], pomelo peel [4], shells of bittim [1], orange peel [12], broad been peels [3], etc.

In our previous reports we have established the feasibility and adsorption of MB onto three plant leaf powders, namely guava leaf powder (GLP), teak leaf powder (TLP), and gulmohar leaf powder (GUL) [16]. Film-pore diffusion model (FPDM) was employed successfully to describe the kinetics of methylene blue adsorption onto GLP, TLP, and GUL. Diffusion based kinetic models are too complex and require rigorous solution methods. For many of the diffusion models pure analytical solution is not possible. In our previous paper we had employed method of lines to solve FPDM and had shown that film-pore model could describe the kinetics of adsorption of MB onto GLP, TLP, and GUL [16]. In this work, we have proposed a wavelet based approximation method to FPDM.

There is a growing interest in using various wavelets to study problems of greater computational complexity. Among the wavelet transform families the Haar and Legendre wavelets deserve much attention. The basic idea of Legendre wavelet method (LWM) is to convert the PDEs to a system of algebraic equations by the operational matrices of integral or derivative. The main goal is to show how wavelets and multi-resolution analysis can be applied for improving the method in terms of easy implementability and achieving the rapidity of its convergence.

Hariharan et al. [7–10] had introduced the diffusion equation, convection–diffusion equation, Reaction–diffusion equation, nonlinear parabolic equations, fractional Klein–Gordon equations, Sine-Gordon equations, and Fisher's equation by the Haar wavelet method. Mohammadi and Hosseini [13] had showed a new Legendre wavelet operational matrix of derivative in solving singular ordinary differential equations.

In this work, we have applied a LWM for the numerical solution of the FPDM equation.

2 Materials and Methods

Detailed development of FPDM is described earlier by McKay and co-workers [11, 17]. Solution of FPDM by method of lines is described in paper [16]. In the present paper development of LWM is described in detail and the results are compared with previous solution.

2.1 Legendre Wavelets Preliminaries

2.1.1 Wavelets

Wavelets are the family of functions which are derived from the family of scaling function $\{\phi_{j,k}, k \in Z\}$ where:

$$\phi(x) = \sum_k a_k \phi(2x - k) \tag{1}$$

For the continuous wavelets, the following equation can be represented:

$$\Psi_{a,b}(x) = |a|^{\frac{-1}{2}} \Psi(\frac{x-b}{a}) \; a, b \in R, a \neq 0. \tag{2}$$

where a and b are dilation and translation parameters, respectively, such that $\Psi(x)$ is a single wavelet function.

The discrete values are put for a and b in the initial form of the continuous wavelets, i.e:

$$a = a_0^{-j}, \; a_0 > 1, b_0 > 1, \tag{3}$$

$$b = kb_0 a_0^{-j}, \; j, k \in Z \tag{4}$$

Then, a family of discrete wavelets can be constructed as follows:

$$\Psi_{j,k} = |a_0|^{\frac{1}{2}} \Psi(2^j x - k) \tag{5}$$

So, $\Psi_{j,k}(x)$ constitutes an orthonormal basis in $L^2(R)$, where $\Psi(x)$ is a single function.

2.1.2 Legendre Wavelets

The Legendre wavelets are defined by

$$\Psi_{n,m}(t) = \begin{cases} \sqrt{m + \frac{1}{2}} 2^{\frac{k}{2}} p_m(2^k t - \hat{n}), & for \; \frac{\hat{n}-1}{2^k} \leq t \leq \frac{\hat{n}}{2^k} \\ 0 & for \; others \end{cases} \tag{6}$$

where $m = 0, 1, 2, \cdots, M-1$, and $n = 1, 2, \cdots, 2^{k-1}$. The coefficient $\sqrt{m+\frac{1}{2}}$ is for orthonormality, then, the wavelets $\Psi_{k,m}(x)$ form an orthonormal basis for $L^2[0,1]$.

In the above formulation of Legendre wavelets, the Legendre polynomials are in the following way:

$$\begin{aligned} p_0 &= 1 \\ p_1 &= x \\ p_{m+1}(x) &= \frac{2m+1}{m+1} x p_m(x) - \frac{m}{m+1} p_{m-1}(x) \end{aligned} \tag{7}$$

and $\{p_{m+1}(x)\}$ are the orthogonal functions of order m, which is named the well-known shifted Legendre polynomials on the interval $[0, 1]$. Note that, in the general form of Legendre wavelets, the dilation parameter is $a = 2^{-k}$ and the translation parameter is $b = n2^k$.

2.1.3 Block Pulse Functions

The block pulse functions (BPFs) form a complete set of orthogonal functions which defined on the interval $[0, b)$ by

$$b_i(t) = \begin{cases} 1 & \frac{i-1}{m} b \le t \le \frac{i}{m} b \\ 0 & elsewhere \end{cases} \tag{8}$$

for $i = 1, 2, \cdots, m$. It is also known that for any absolutely integrable function $f(t)$ on $[0, b)$ can be expanded in BPFs:

$$f(t) \cong \xi^T B_m(t) \tag{9}$$

$$\xi^T = [f_1, f_2 . \cdots, f_m], B_m(t) = [b_1(t), b_2(t), \cdots, b_m(t)] \tag{10}$$

where f_i are the coefficients of the block pulse function, given by

$$f_i = \frac{m}{b} \int_0^b f(t) b_i(t) dt \tag{11}$$

Remark 1. Let A and B are two matrices of $m \times m$, then $A \otimes B = (a_{ij} \times b_{ij})_{mm}$.

Lemma 1. *Assuming $f(t)$ and $g(t)$ are two absolutely integrable functions, which can be expanded in BPF as $f(t) = FB(t)$ and $g(t) = GB(t)$, respectively, then we have*

$$f(t)g(t) = FB(t)B^T(t)G^T = HB(t) \tag{12}$$

where $H = F \otimes G$.

2.1.4 Approximating the Nonlinear Term

The Legendre wavelets can be expanded into m-set of block pulse functions as

$$\Psi(t) = \phi_{m\times m} B_m(t) \tag{13}$$

Taking the collocation points as following

$$t_i = \frac{i - \frac{1}{2}}{2^{k-1}M}, \quad i = 1, 2, \cdots, 2^{k-1}M \tag{14}$$

The m-square Legendre matrix $\phi_{m\times m}$ is defined as

$$\phi_{m\times m} \cong [\Psi(t_1)\Psi(t_2)\cdots\Psi(t_{2^{k-1}}M)] \tag{15}$$

The operational matrix of product of Legendre wavelets can be obtained by using the properties of BPFs, let $f(x,t)$ and $g(x,t)$ are two absolutely integrable functions, which can be expanded by Legendre wavelets as $f(x,t) = \Psi^T(x)F\Psi(t)$ and $g(x,t) = \Psi^T(x)G\Psi(t)$, respectively. Then

$$f(x,t) = \Psi^T(x)F\Psi(t) = B^T(x)\phi_{mm}^T F\phi_{mm}B(t) \tag{16}$$

$$g(x,t) = \Psi^T(x)G\Psi(t) = B^T(x)\phi_{mm}^T G\phi_{mm}B(t) \tag{17}$$

and $F_b = \phi_{mm}^T F\phi_{mm}, G_b = \phi_{mm}^T G\phi_{mm}, H_b = F_b \otimes G_b$.

Then

$$\begin{aligned} f(x,t)g(x,t) &= B^T H_b B(t), \\ &= B^T(x)\phi_{mm}^T\, inv(\phi_{mm}^T)H_b\, inv(inv(\phi_{mm}^T)H_b\, inv(\phi_{mm}))\phi_{mm}B(t) \\ &= \Psi^T(x)H\Psi(t) \end{aligned} \tag{18}$$

where $H = inv(\phi_{mm}^T)H_b\, inv((\phi_{mm}))$

2.2 *Function Approximation*

A given function $f(x)$ with the domain [0, 1] can be approximated by:

$$f(x) = \sum_{k=1}^{\infty}\sum_{m=0}^{\infty} c_{k,m}\Psi_{k,m}(x) = C^T\Psi(x) \tag{19}$$

Here C and Ψ are the matrices of size $(2^{j-1}M \times 1)$.

$$C = [c_{1,0}, c_{1,1}, \cdots, c_{1,M-1}, c_{2,0}, c_{2,1}, \cdots, c_{2,M-1}, \cdots, c_{2^{j-1},1}, \cdots, c_{2^{j-1},M-1}]^T \tag{20}$$

$$\Psi(x) = [\Psi_{1,0}, \Psi_{1,1}, \Psi_{2,0}, \Psi_{2,1}, \cdots, \Psi_{2,M-1}, \cdots, \Psi_{2^{j-1},M-1}]^T \tag{21}$$

3 Method of Solution

First we give a brief introduction of film-pore model. FPDM assumes that both external film and internal pore diffusion resistances are significant and play a role in controlling the mass transfer. Thus, the governing equations are:

1. Assuming linear driving force the rate of external mass transfer is given by

$$\frac{dC_t}{dt} = -k_j \frac{A_s}{V}(C_t - C_s) \tag{22}$$

2. Within the pore diffusion of solutes follows Flick's law of diffusion. Following equations is obtained by making a mass balance of dye in a spherical

$$\varepsilon \frac{\partial C_i}{\partial t} + \rho_p \frac{\partial q_i}{\partial t} = D_{eff} \left[\frac{\partial^2 C_i}{\partial r^2} + \frac{1}{r} \frac{\partial C_i}{\partial r} \right] \tag{23}$$

corresponding initial condition and boundary conditions are

$$I.C : Att = 0, C_i = 0 \; for \; 0 \leq r \leq R \tag{24}$$

$$B.C.1 : \frac{\partial C_i}{\partial r} = 0 \; at \; r = 0 \tag{25}$$

$$B.C.2 : k_f(C_t - C_s) = D_{eff} \frac{\partial C_i}{\partial r} \; at \; r = R \tag{26}$$

3. Solid phase concentration at any radial location may be expressed as function of aqueous phase concentration at that location as follows:

$$q_i = f(C_i) \tag{27}$$

Assuming equilibrium within the pore Eq. (27) is described by the relevant isotherm expression of the system. Substituting Eq. (27) in Eq. (23) we get

$$\varepsilon \frac{\partial C_i}{\partial t} + \rho_p \frac{\partial f(C_i)}{\partial t} = D_{eff} \left[\frac{\partial^2 C_i}{\partial r^2} + \frac{1}{r} \frac{\partial C_i}{\partial r} \right] \tag{28}$$

Since the system follows Langmuir isotherm [16]

$$q_i = f(C_i) = \frac{q_c K_L C_i}{1 + K_L C_i} \tag{29}$$

Following dimensionless variables were defined to convert the equations into dimensionless form

$$Z = \frac{r}{g};\ \overline{C_i} = \frac{C_i}{C_0};\ \overline{C_t} = \frac{C_t}{C_0};\ B_i = \frac{k_f R}{D_{eff}} \tag{30}$$

After substituting the dimension variables in Eq. (28) can be rewritten as follows:

$$\frac{\partial \overline{C_i}}{\partial \tau} = A(\overline{C_i})\left[\frac{\partial^2 C_i}{\partial Z^2} + \frac{1}{Z}\frac{\partial C_i}{\partial Z}\right] \tag{31}$$

$$A(\overline{C_i}) = \frac{1}{\left(\varepsilon_p + \left(\frac{q_h l_p}{C_0}\right)\left(\frac{-q+C_0}{(1+bC_0\overline{C_i})^2}\right)\right)} \tag{32}$$

Consider the equation

$$\dot{\overline{C_i}}(Z,\tau) = A(\overline{C_i})\left[\overline{C_i}'' + (\frac{1}{Z})\overline{C_i}'\right] \tag{33}$$

$$\begin{aligned} C_i(z,0) &= e^{-z} \\ C_i(z,1) &= e^{-z-0.09} \\ C_i(0,\tau) &= e^{-0.09\tau} \\ C_i(1,\tau) &= e^{-1-0.09\tau} \end{aligned} \tag{34}$$

We solve Eq. (33) by applying the LWM

$$\overline{C_i}''(z,\tau) = C^T P_\tau \Psi(z,\tau) + \overline{C_i}''(z,0) \tag{35}$$

$$\overline{C_i}'(z,\tau) = C^T P_\tau P_z[\Psi(z,\tau) - P_z(1,\tau)] + g_1(z,\tau) \tag{36}$$

$$\dot{\overline{C_i}}(z,\tau) = C^T P_z^2[\Psi(z,\tau) - z\Psi(1,\tau)] + g_2(z,\tau) \tag{37}$$

$$\begin{aligned} \overline{C_i}(z,\tau) = {} & C^T P_\tau P_z^2[\Psi(z,\tau) - z\Psi(1,\tau)] + \overline{C_i}(z,0) - \overline{C_i}(0,0) + \\ & z[\overline{C_i}(1,\tau) - \overline{C_i}(1,0) + \overline{C_i}(0,0) - \overline{C_i}(0,\tau)] + \overline{C_i}(0,\tau) \end{aligned} \tag{38}$$

Table 1 Comparison between Legendre wavelet method (LWM) and method of lines (MOL) by obtaining the mass transfer coefficients using film-pore diffusion model adsorption of MB onto GLp and $k = 2$ and $M = 3, t = 10\,\mathrm{s}$

Temperature	C_0	k_f (m s^{-1})		D_{eff} (m^2 s^{-1})		Error	
(K)	(mg)	MOL(M)	LWM(L)	MOL(M)	LWM(L)	E_M	E_L
303	50	1.00×10^{-6}	4.23×10^{-6}	1.74×10^{-13}	1.24×10^{-14}	1.197	0.038
	100					0.140	0.029
	150					0.935	0.283
	200					1.610	1.541
313	50	1.71×10^{-6}	6.35×10^{-6}	6.46×10^{-13}	7.32×10^{-14}	1.462	1.312
	100					1,120	0.653
	150					1.267	0.120
	200					7.570	3.626
323	50	4.27×10^{-6}	4.01×10^{-6}	3.11×10^{-13}	5.31×10^{-13}	0.856	0.192
	100					0.160	0.001
	150					2.168	1.127
	200					3.164	1.002

E_M error by Method of lines, E_L error by Legendre wavelet method

in which

$$g_1(z,\tau) = \dot{\overline{C_i}}(z,0) - \overline{C_i}(1,0) - \overline{C_i}(0,\tau) + C_i(1,\tau) + 1 \; and$$

$$g_2(z,\tau) = z[\overline{C_i}(1,\tau) - C_i'(0,\tau)] + C_i'(0,\tau)$$

Substitute Eqs. (35)–(38) into Eq. (33), we get

$$C^T P_z^2[\Psi(z,\tau) - z\Psi(1,\tau)] + g_2(z,\tau) =$$
$$A(\overline{C_i})[(C^T P_\tau \Psi(z,\tau) + \overline{C_i}''(z,0)) + \tfrac{1}{z}(C^T P_\tau P_z[\Psi(z,\tau) - P_z\Psi(1,\tau)]$$
$$+ g_1(z,\tau))]$$

From formula (28) the wavelet coefficients C^T can be calculated successfully. Here $A(\overline{C_i})$ are constants (linear) and $\epsilon = 0.5, \rho = 500$.

Table 1 gives a comparison of Legendre wavelet (LW) solutions and method of lines. It is evident that Legendre wavelet solutions are better than that of the method of lines. Value of absolute error decreased when k was increased. The results show that combining with wavelet matrix, the method in this paper can be effectively used in numerical calculus for constant coefficient differential equations, and that the method is feasible. We can see that the numerical solutions are in good agreement with exact solution. The power of the manageable method is thus confirmed. All the numerical experiments presented in this section were computed in double precision with some MATLAB codes on a personal computer system with Processor Inter(R) Core^{TM} 2 Duo CPU T5470 @ 1.60GHz(2CPUs) and 1GB RAM.

4 Conclusion

In the present paper FPDM model equations had been solved by the LWM. It was found that the model could predict the concentration decay curve for all adsorption of methylene blue onto TLP, GUL, and GLP excellently with a small deviation during initial period. In comparison with existing numerical schemes used to solve the nonlinear parabolic equations, the scheme in this paper is an improvement over other methods in terms of accuracy. It is worth mentioning that Legendre wavelet solution provides excellent results even for small values of k. For larger values of k, we can obtain the results closer to the real values.

References

1. Aydin, H., Baysal, G.: Adsorption of acid dyes in aqueous solutions by shells of bittim (Pistacia khinjuk Stocks). Desalination **196**(1–3), 248–259 (2006)
2. Hameed, B.H.: Removal of cationic dye from aqueous solution using jackfruit peel as non-conventional low-cost adsorbent. J. Hazard. Mater. **162**(1), 344–350 (2009)
3. Hameed, B.H., El-Khaiary, M.I.: Sorption kinetics and isotherm studies of a cationic dye using agricultural waste: broad bean peels. J. Hazard. Mater. **154**(1–3), 639–648 (2008)
4. Hameed, B.H., Mahmoud, D.K., Ahmad, A.L.: Sorption of basic dye from aqueous solution by pomelo (Citrus grandis) peel in a batch system. Colloids Surf. A Physicochem. Eng. Asp. **316**(1–3), 78–84 (2008)
5. Hameed, B.H., Krishni, R.R., Sata, S.A.: A novel agricultural waste adsorbent for the removal of cationic dye from aqueous solutions. J. Hazard. Mater. **162**(1), 305–311 (2009)
6. Han, R., Wang, Y., Zhao, X., Wang, Y., Xie, F., Cheng, J., Tang, M.: Adsorption of methylene blue by phoenix tree leaf powder in a fixed-bed column: experiments and prediction of breakthrough curves. Desalination **245**(1-3), 284–297 (2009). http://www.sciencedirect.com/science/journal/00119164/245/1
7. Hariharan, G.: Haar wavelet method for solving sine-Gordon and Klein-Gordon equations. Int. J. Nonlinear Sci. **9**(2), 1–10 (2010)
8. Hariharan, G., Kannan, K.: Haar wavelet method for solving FitzHugh-Nagumo equation. Int. J. Math. Stat. Sci. **2**, 2 (2010)
9. Hariharan, G., Kannan, K.: Haar wavelet method for solving some nonlinear parabolic equations. J. Math. Chem. **48**(4), 1044–1061 (2010)
10. Hariharan, G., Kannan, K.: A comparative study of a Haar wavelet method and a restrictive Taylor's series method for solving convection-diffusion equations. Int. J. Comput. Methods Eng. Sci. Mech. **11**(4), 173–184 (2010)
11. Ho, Y.S., McKay, G.: A comparison of chemisorption kinetic models applied to pollutant removal on various sorbents. Process Saf. Environ. Prot. **76**(B4), 332–340 (1998)
12. Khaled, A., El Nemr, A., El-Sikaily, A., Abdelwaha, O.: Removal of Direct N Blue-106 from artificial textile dye effluent using activated carbon from orange peel: adsorption isotherm and kinetic studies. J. Hazard. Mater. **165**(1–3), 100–110 (2009)
13. Mohammadi, F., Hosseini, M.M.: A new Legendre wavelet operational matrix of derivative and its applications in solving the singular ordinary differential equations. J. Franklin Inst. **348**, 1787–1796 (2011)

14. Ponnusami, V., Krithika, V., Madhuram, R., Srivastava, S.N.: Biosorption of reactive dye using acid-treated rice husk: factorial design analysis. J. Hazard. Mater. **142**, 397–403 (2007)
15. Ponnusami, V., Gunasekar, V., Srivastava, S.N.: Kinetics of methylene blue removal from aqueous solution using gulmohar (Delonix regia) plant leaf powder: multivariate regression analysis. J. Hazard. Mater. **169**, 119–127 (2009)
16. Ponnusami, V., Rajan, K.S., Srivastava, S.N.: Application of film-pore diffusion model, for methylene blue adsorption onto plant leaf powders. Chem. Eng. J. **163**(3), 236–242 (2010)
17. Porter, D.C.K., McKay, G.: Film-pore diffusion model for the fixed-bed sorption of copper and cadmium ions onto bone char. Water Res. **35**, 3876–3886 (2001)

A New Wavelet-Based Hybrid Method for Fisher Type Equation

R. Rajaram and G. Hariharan

Abstract In this paper, we have introduced a new wavelet-based hybrid method for solving the Fisher's type equations. To the best of our knowledge, until now there is no rigorous wavelet solution has been addressed for the Fisher's equations. With the help of wavelets operational matrices, the Fisher's equations are converted into a system of algebraic equations. Some numerical examples are presented to demonstrate the validity and applicability of the method.

Keywords Fisher's equation • Operational matrices • Legendre wavelets • Homotopy analysis method • Haar wavelets

1 Introduction

Wavelet theory possesses many useful properties, such as compact support, orthogonality, dyadic, orthonormality, and multi-resolution analysis (MRA). Fractional Partial Differential Equations (FPDEs) are generalizations of classical partial differential equations of integer order. Mathematical modelling of complex process is a major challenge for contemporary scientist. Analytical methods enable researchers to study the effect of differential variables or parameters on the function under study easily. Recently, there are several new approaches have been proposed for solving nonlinear PDEs, for example, the Adomian Decomposition Method [20], the variational iteration method [16], Differential Transform Method [2], reduced differential transform method [15], Homotopy Analysis method [24, 26], and exp-function method [29].

R. Rajaram (✉) • G. Hariharan
Department of Mathematics, School of Humanities and Sciences,
SASTRA University, Thanjavur, Tamilnadu 613401, India
e-mail: rraja@maths.sastra.edu; hariharan@maths.sastra.edu

C. Bandt et al. (eds.), *Fractals, Wavelets, and their Applications*, Springer Proceedings in Mathematics & Statistics 92, DOI 10.1007/978-3-319-08105-2_35,

In the numerical analysis, wavelet-based methods and hybrid methods become important tools because of the properties of localization. In wavelet-based methods, there are two important ways of improving the approximation of the solutions: (i) Increasing the order of the wavelet family and (ii) The increasing the resolution level of the wavelets. There is a growing interest in using various wavelets [4–12, 14, 19, 21–23, 25, 28] to study problems, of greater computational complexity. Among the wavelet transform families the Haar and Legendre wavelets deserve much attention. [because of its simplicity in its analytical expression]. The basic idea of Legendre wavelet method is to convert the PDEs to a system of algebraic equations by the operational matrices of integral or derivative [19, 21–23, 28]. The main goal is to show how wavelets and multiresolution analysis can be applied for improving the method, in terms of easy implementation and achieving the rapidity [speed] of convergence [towards the exact solution.] Razzagi and Yousefi [22, 23] introduced the Legendre wavelet method for solving variational wavelet method for solving variational problems and constrained optimal control problems. Hariharan et al. [4–7] had introduced the diffusion equation, convection–diffusion, Reaction–diffusion equation, Non-linear parabolic equations, fractional Klein–Gordan equations, Sine–Gordan equations, and Fisher's equation by the Haar wavelet method. Mohammadi and Hosseini [19] had showed a new Legendre wavelet operational matrix of derivative in solving singular ordinary differential equations. Jafari et al. [11] had solved the fractional differential equations by Legendre wavelet method. Parsian[21] introduced two-dimensional Legendre wavelets and operational matrices of integration. In recent years, many analytical/approximation methods have been proposed for solving Fisher's and fractional Fisher's equations. For example, Adomian decomposition method [15], the variational iteration method [16], the Homotopy perturbation method [13, 17], the differential transform method [2], the homotopy analysis method [24, 26], and other methods [1, 3, 18, 27]. Recently, Hariharan and Rajaraman [9] established a new coupled wavelet-based method applied to the nonlinear reaction–diffusion equation arising in mathematical chemistry. Yin et al. [8] introduced a wavelet-based hybrid method for solving Klein–Gordan equations.

In this work, we have applied a wavelet-based coupled method (LLWM) which combines the Laplace transform method and the Legendre wavelets method for the numerical solution of Fisher's equations.

2 Legendre Wavelets and Its Properties

2.1 *Wavelets*

Wavelets are the family of functions which are derived from the family of scaling function $\{\varphi_{j,k}(x), j, k \in Z\}$ where

$$\varphi(x) = \sum_k a_k \varphi(2x - k). \tag{1}$$

For the continues wavelets, the following equation can be represented.

$$\psi_{a,b}(x) = |a|^{\frac{-1}{2}} \psi(\frac{x-b}{a}) a, b \in R, a \neq 0. \tag{2}$$

where "a" and "b" are dilation and translation parameters, respectively, such that $\varphi(x)$ is a single wavelet function. The discrete values are put for "a" and "b" in the initial form of the continues wavelets, i.e.:

$$a = a_0^{-j}, a_0 > 1, b_0 > 1, \tag{3}$$

$$b = k b_0 a_0^{-j}, j, k \in Z. \tag{4}$$

Then, a family of discrete wavelets can be constructed as follows:

$$\psi_{j,k} = |a_0|^{\frac{1}{2}} \psi(2^j x - k) \tag{5}$$

so, $\psi_{j,k}(x)$ constitutes an orthonormal basis in $L^2(R)$, where $\psi(x)$ is a single function.

2.2 Legendre Wavelets

The Legendre wavelets are defined by

$$\psi_{n,m}(t) = \begin{cases} \sqrt{m + \frac{1}{2}} 2^{\frac{k}{2}} L_m(2^k t - n) for \frac{n-1}{2^k} \leq t \leq \frac{n+1}{2^k} \\ 0, otherwise. \end{cases} \tag{6}$$

where $m = 0, 1, 2, .., M - 1$ and $k = 1, 2, .., 2^j - 1$. The coefficient $\sqrt{m + \frac{1}{2}}$ is for orthonormality, then, the wavelets $\psi_n m(x)$ form an orthonormal basis for $L^2[0, 1]$. In the above formulation of Legendre wavelets, the Legendre polynomials are in the following way:
$p_0 = 1, p_1 = x$

$$p_{m+1}(x) = \frac{2m + 1}{m + 1} x p_m(x) - \frac{m}{m + 1} p_{m-1}(x) \tag{7}$$

and $p_{m+1}(x)$ are the orthogonal functions of order m, which is named the well-known shifted Legendre polynomials on the interval [0, 1]. Note that, in general form of Legendre wavelets, the dilation parameter is $a = 2^{-j}$ and the translation parameter is $b = n2^j$.

4 Illustrative Examples

Example 4.1: We consider the well-known Fisher's equation

$$\frac{\partial U}{\partial t} = \frac{\partial^2 U}{\partial x^2} + \alpha U(1-U) \tag{24}$$

subject to the initial conditions

$$U(x,0) = \frac{1}{(1+\exp(\sqrt{\frac{\alpha}{6}}x))^2} \tag{25}$$

Using Homotopy Analysis Method (HAM), the exact solution in closed form is given by

$$U(x,t) = \frac{1}{(1+\exp(\sqrt{\frac{\alpha}{t}}x) - \frac{5\alpha t}{6})^2} \tag{26}$$

x	t	U_{exact}	U_{LLWM}
0.25	0.5	0.81839	0.81855
	1.0	0.98292	0.98305
	2.0	0.99988	0.99999
	5.0	1.0000	1.0000
0.5	0.5	0.77590	0.77602
	1.0	0.97815	0.97824
	2.0	0.99985	0.99996
	5.0	1.0000	1.0000

Our proposed method (LLMW) can be compared with Wazwaz and Gorguis results. (see [24].)

5 Conclusion

In this work, a new coupled wavelet-based method has been successfully employed to obtain the numerical solution of Fisher type equations. The proposed scheme is the capability to overcome the difficulty arising in calculating the integral values while dealing with nonlinear problems. This method shows higher efficiency than the traditional Legendre wavelet method for solving nonlinear PDEs. Numerical example illustrates the powerful of the proposed scheme LLWM. Also this paper illustrates the validity and excellent potential of the LLWM for nonlinear and fractional PDEs. The numerical solutions obtained using the proposed method show that the solutions are in very good coincidence with the exact solution. In addition the calculations involved in LLWM are simple, straightforward, and low computational cost.

References

1. Al-Khaled, K.: Numerical study of Fisher's reaction-diffusion equation by the sinc-collocation method. J. Comput. Appl. Math. **13**, 245–255 (2001)
2. Carey, G.F., Shen, Y.: Least-squares finite element approximationn of Fisher's reaction-diffusion equation. Numer. Meth. Part. Differ. Equat. 175–186 (1995)
3. Hariharan, G.: The homotopy analysis method applied to the Kolmogorov-Petrovskii-Piskunov (KPP) and fractional KPP equations. J. Math. Chem. **51**, 992–1000 (2013)
4. Hariharan, G., Kannan, K., Sharma, K.: Haar wavelet in estimating the depth profile of soil temperature. Appl. Math. Comput. **210**, 119–225 (2009a)
5. Hariharan, G., Kannan, K.: Haar wavelet method for solving Fisher's equation. Appl. Math. Comput. **211**, 284–292 (2009b)
6. Hariharan, G., Kannan, K.: Haar wavelet method for solving nonlinear parabolic equations. J. Math. Chem. **48**, 1044–1061 (2010a)
7. Hariharan, G., Kannan, K.: A comparative study of a Haar wavelet method and a restrictive Taylor's series method for solving convection-diffusion equations. Int. J. Comput. Meth. Eng. Sci. Mech. **11**(4), 173–184 (2010b)
8. Hariharan, G., Rajaraman, R.: A new coupled wavele-based method applied to the nonlinear reaction-diffusion equation arising in mathematical chemistry. J. Math. Chem. **51**, 2386–2400 (2013)
9. He, J.H., Wu, X.H.: Exp-function method for nonlinear wave equations. Chaos, Solitons Fractals **30**, 700–708 (2006)
10. Heydari, M.H., Hooshmandasl, M.R., Maalek Ghaini, F.M., Mohammadi, F.: Wavelet collocation method for solving multiorder fractional differential equations. J. Appl. Math. **2012**, Article ID 163821 (2012)
11. Jafari, H., Soleymanivaraki, M., Firoozjaee, M.A.: Legendre wavelets for solving fractional differential equations. J. Appl. Math. **4**(27), 65–70 (2011)
12. Khan, N.A., Khan, N.-U., Ara, A., Jamil, M.: Approximate analytical solution of fractional reaction-diffussion equations. J. Kind Saud Univ. Sci. **24**, 111–118 (2012)
13. Liao, S.J.: Beyond Perturbation: Introduction to Homotopy Analysis Method. CRC Press/Chapman and Hall, Boca Raton (2004)
14. Maleknejad, K., Sohrabi, S.: Numerical solution of Fredholm integral equations of the first kind by using Legendre wavelets. Appl. Math. Comput. **186**, 836–843 (2007)
15. Matinfar, M., Ghanbari, M.: Solving the Fisher's equations by means of variational iteration method. Int. J. Contemp. Math. Sci. **4**(7), 343–348 (2009a)
16. Matinfar, M., Ghanbari, M.: Homotopy perturbation method for the Fisher's equation and its generalized. Int. J. Nonlinear Sci. **8**(4), 448–455 (2009b)
17. Matinfar, M., Bahar, S.R., Ghasemi, M.: Solving the generalized Fisher's equation by the differential transform method. J. Appl. Math. Inform. **30**(3–4), 555–560 (2012)
18. Mittal, R.C., Jiwari, R.: Numerical study of Fisher's equation by using differential quadrature method. Int. J. Inform. Syst. Sci. **5**(1), 143–160 (2008)
19. Mohammadi, F., Hosseini, M.M.: A new Legendre wavelet operational matrix of derivative and its applications in solving singular ordinary differential equations. J. Franklin Inst. **348**, 1787–1796 (2011)
20. Olmos, D., Shizgal, B.: A spectral method of solution of Fisher's equation. J. Comput. Appl. Math. **193**, 219–242 (2006)

21. Parsian, H.: Two dimension Legendre wavelets and operational matrices of integration. Acta Math. Academiae Paedagogicae Nyireghaziens **21**, 101–106 (2005)
22. Razzaghi, M., Yousefi, S.: The Legendre wavelets direct method for variational problems. Math. Comput. Simulat. **53**, 185–192 (2000)
23. Razzaghi, M., Yousefi, S.: The Legendre wavelets operational matrix of integration. Int. J. Syst. Sci. **32**, 495–502 (2001)
24. Wazwaz, A.M., Gorguis, A.: An analytical study of Fisher's equation by using Adomian decomposition method. Appl. Math. Comput. **154**, 609–620 (2004)
25. Yang, Y.: Solving a nonlinear multi-order fractional differential equation using legendre psedu-spectral method. Appl. Math. **4**, 113–118 (2013)
26. Yildirem, K., Ibis, B., Bayram, M.: New solutions of the nonlinear Fisher type equations by the reduced differential transform. Nonlinear Sci. Lett. A. **3**(1), 29–36 (2012)
27. Yin, F., Song, J., Lu, F.: A coupled method of Laplace transform and Legendre wavelets for nonlinear Klein-Gordan equations. Math. Meth. Appl. Sci. 2013 (Press)
28. Yousefi, S.A.: Legendre wavelets method for solving differential equations of Lane-Emden type. App. Math. Comput. **181**, 1417–1442 (2006)
29. Zhou, X.W.: Exp-function method for solving Fisher's equation. J. Phys. Conf. Ser. **96** (2008)

CPSIA information can be obtained
at www.ICGtesting.com
Printed in the USA
LVHW02*2314080418
572701LV00001B/65/P

9 783319 081045